T0252269

Isuzu
Rodeo & Amigo
Honda
Passport
Automotive
Repair
Manual

by Robert Maddox, Mike Stubblefield and John H Haynes

Member of the Guild of Motoring Writers

Models covered:
Isuzu Rodeo 1991 through 2002
Isuzu Amigo 1989 through 1994 and 1998 through 2002
Honda Passport 1995 through 2002

(47017-8U10)

J H Haynes & Co. Ltd.
Haynes North America, Inc.
www.haynes.com

Acknowledgements

Wiring diagrams produced exclusively for Haynes North America, Inc. by Valley Forge Technical Information Services. Technical writers who contributed to this project include Jeff Kibler and Jay Storer.

A book in the Haynes Automotive Repair Manual Series

ISBN 13: 978-1-56392-481-1
ISBN 10: 1-56392-481-1

Library of Congress Control Number 2002115738

While every attempt is made to ensure that the information in this manual is correct, no liability can be accepted by the authors or publishers for loss, damage or injury caused by any errors in, or omissions from, the information given.

03-416

Contents

Haynes mechanic, author and photographer with 1997 Isuzu Rodeo

About this manual

Its purpose

The purpose of this manual is to help you get the best value from your vehicle. It can do so in several ways. It can help you decide what work must be done, even if you choose to have it done by a dealer service department or a repair shop; it provides information and procedures for routine maintenance and servicing; and it offers diagnostic and repair procedures to follow when trouble occurs.

We hope you use the manual to tackle the work yourself. For many simpler jobs, doing it yourself may be quicker than arranging an appointment to get the vehicle into a shop and making the trips to leave it and pick it up. More importantly, a lot of money can be saved by avoiding the expense the shop must pass on to you to cover its labor and overhead costs. An added benefit is the sense of satisfaction and accomplishment that you feel after doing the job yourself.

Using the manual

The manual is divided into Chapters. Each Chapter is divided into numbered Sections, which are headed in bold type between horizontal lines. Each Section consists of consecutively numbered paragraphs.

At the beginning of each numbered Section you will be referred to any illustrations which apply to the procedures in that Section. The reference numbers used in illustration captions pinpoint the pertinent Section and the Step within that Section. That is, illustration 3.2 means the illustration refers to Section 3 and Step (or paragraph) 2 within that Section.

Procedures, once described in the text, are not normally repeated. When it's necessary to refer to another Chapter, the reference will be given as Chapter and Section number. Cross references given without use of the word "Chapter" apply to Sections and/or paragraphs in the same Chapter. For example, "see Section 8" means in the same Chapter.

References to the left or right side of the vehicle assume you are sitting in the driver's seat, facing forward.

Even though we have prepared this manual with extreme care, neither the publisher nor the author can accept responsibility for any errors in, or omissions from, the information given.

NOTE

A **Note** provides information necessary to properly complete a procedure or information which will make the procedure easier to understand.

CAUTION

A **Caution** provides a special procedure or special steps which must be taken while completing the procedure where the Caution is found. Not heeding a Caution can result in damage to the assembly being worked on.

WARNING

A **Warning** provides a special procedure or special steps which must be taken while completing the procedure where the Warning is found. Not heeding a Warning can result in personal injury.

Introduction to the Isuzu Rodeo, Amigo and the Honda Passport

Isuzu Rodeo and the Honda Passport models are available in four-door station wagon body styles. The Isuzu Amigo is available in two door body style only.

Early four-cylinder engines are equipped with a carburetor. Later four-cylinder engines and 3.1L V6 engines are equipped with fuel injection. All 2.2L four-cylinder and 3.2L V6 engines are equipped with multi-port fuel injection systems.

The engine drives the rear wheels through either a four or five-speed manual or four-speed automatic transmission via a driveshaft and solid rear axle. A transfer case and driveshaft are used to drive the front driveaxles on 4WD models.

The suspension features independent control arms, torsion bars and shock absorbers at the front. A solid axle at the rear is suspended by leaf springs or coil springs and shock absorbers.

The steering box is mounted to the left of the engine and is connected to the steering arms through a series of rods. 1998 and later models are equipped with rack-and-pinion steering. Power assist is optional on some models.

The brakes are disc at the front and drums at the rear, with power assist standard. Some later models are equipped with disc brakes in the rear, instead of drums.

Vehicle identification numbers

Modifications are a continuing and unpublicized process in vehicle manufacturing. Since spare parts manuals and lists are compiled on a numerical basis, the individual vehicle numbers are essential to correctly identify the component required.

The Vehicle Identification Number (VIN) is visible through the driver's side of the windshield

Vehicle Identification Number (VIN)

This very important identification number is stamped on a plate attached to the left side of the dashboard just inside the windshield on the driver's side of the vehicle **(see illustration)**. The VIN also appears on the Vehicle Certificate of Title and Registration. It contains information such as where and when the vehicle was manufactured, the model year and the body style.

Engine identification number

The engine ID number on four-cylinder engines is located on a machined surface on the left rear corner of the block, near the bottom. On V6 engines, the ID number is located on a pad on the left side of the block **(see illustration)**.

The Engine Identification Number is stamped on the left side of the engine block

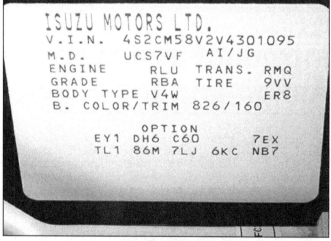

The Service parts and paint code label is affixed to the engine compartment firewall

The Manufacturer's Certification label is affixed to the drivers side door end or post

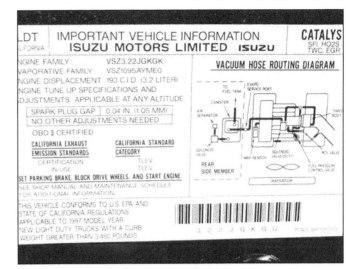

The Vehicle Emissions Control Information (VECI) label is located on the bottom side of the hood

Service parts and paint code identification plate

This plate is located in the engine compartment, usually on the firewall. It tells what options the vehicle came equipped with and the color and type of paint which was originally applied to the vehicle (see illustration).

Manufacturer Certification label

The Manufacturer Certification label is affixed to the left front door pillar. The plate contains the name of the manufacturer, the month and year of production, the Gross Vehicle Weight Rating (GVWR) and the certification statement (see illustration).

Vehicle Emissions Control Information (VECI) label

The emissions control information label is found under the hood. This label contains information on the emissions control equipment installed on the vehicle, as well as tune-up specifications (see illustration).

Buying parts

Replacement parts are available from many sources, which generally fall into one of two categories - authorized dealer parts departments and independent retail auto parts stores. Our advice concerning these parts is as follows:

Retail auto parts stores: Good auto parts stores will stock frequently needed components which wear out relatively fast, such as clutch components, exhaust systems, brake parts, tune-up parts, etc. These stores often supply new or reconditioned parts on an exchange basis, which can save a considerable amount of money. Discount auto parts stores are often very good places to buy materials and parts needed for general vehicle maintenance such as oil, grease, filters, spark plugs, belts, touch-up paint, bulbs, etc. They also usually sell tools and general accessories, have convenient hours, charge lower prices and can often be found not far from home.

Authorized dealer parts department: This is the best source for parts which are unique to the vehicle and not generally available elsewhere (such as major engine parts, transmission parts, trim pieces, etc.).

Warranty information: If the vehicle is still covered under warranty, be sure that any replacement parts purchased - regardless of the source - do not invalidate the warranty!

To be sure of obtaining the correct parts, have engine and chassis numbers available and, if possible, take the old parts along for positive identification.

Maintenance techniques, tools and working facilities

Maintenance techniques

There are a number of techniques involved in maintenance and repair that will be referred to throughout this manual. Application of these techniques will enable the home mechanic to be more efficient, better organized and capable of performing the various tasks properly, which will ensure that the repair job is thorough and complete.

Fasteners

Fasteners are nuts, bolts, studs and screws used to hold two or more parts together. There are a few things to keep in mind when working with fasteners. Almost all of them use a locking device of some type, either a lockwasher, locknut, locking tab or thread adhesive. All threaded fasteners should be clean and straight, with undamaged threads and undamaged corners on the hex head where the wrench fits. Develop the habit of replacing all damaged nuts and bolts with new ones. Special locknuts with nylon or fiber inserts can only be used once. If they are removed, they lose their locking ability and must be replaced with new ones.

Rusted nuts and bolts should be treated with a penetrating fluid to ease removal and prevent breakage. Some mechanics use turpentine in a spout-type oil can, which works quite well. After applying the rust penetrant, let it work for a few minutes before trying to loosen the nut or bolt. Badly rusted fasteners may have to be chiseled or sawed off or removed with a special nut breaker, available at tool stores.

If a bolt or stud breaks off in an assembly, it can be drilled and removed with a special tool commonly available for this purpose.

Most automotive machine shops can perform this task, as well as other repair procedures, such as the repair of threaded holes that have been stripped out.

Flat washers and lockwashers, when removed from an assembly, should always be replaced exactly as removed. Replace any damaged washers with new ones. Never use a lockwasher on any soft metal surface (such as aluminum), thin sheet metal or plastic.

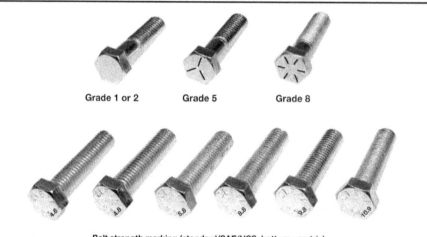

Grade 1 or 2   Grade 5   Grade 8

Bolt strength marking (standard/SAE/USS; bottom - metric)

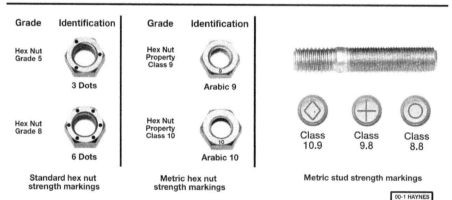

Grade	Identification	Grade	Identification
Hex Nut Grade 5	3 Dots	Hex Nut Property Class 9	Arabic 9
Hex Nut Grade 8	6 Dots	Hex Nut Property Class 10	Arabic 10

Standard hex nut strength markings

Metric hex nut strength markings

Class 10.9   Class 9.8   Class 8.8

Metric stud strength markings

Fastener sizes

For a number of reasons, automobile manufacturers are making wider and wider use of metric fasteners. Therefore, it is important to be able to tell the difference between standard (sometimes called U.S. or SAE) and metric hardware, since they cannot be interchanged.

All bolts, whether standard or metric, are sized according to diameter, thread pitch and length. For example, a standard 1/2 - 13 x 1 bolt is 1/2 inch in diameter, has 13 threads per inch and is 1 inch long. An M12 - 1.75 x 25 metric bolt is 12 mm in diameter, has a thread pitch of 1.75 mm (the distance between threads) and is 25 mm long. The two bolts are nearly identical, and easily confused, but they are not interchangeable.

In addition to the differences in diameter, thread pitch and length, metric and standard bolts can also be distinguished by examining the bolt heads. To begin with, the distance across the flats on a standard bolt head is measured in inches, while the same dimension on a metric bolt is sized in millimeters (the same is true for nuts). As a result, a standard wrench should not be used on a metric bolt and a metric wrench should not be used on a standard bolt. Also, most standard bolts have slashes radiating out from the center of the head to denote the grade or strength of the bolt, which is an indication of the amount of torque that can be applied to it. The greater the number of slashes, the greater the strength of the bolt. Grades 0 through 5 are commonly used on automobiles. Metric bolts have a property class (grade) number, rather than a slash, molded into their heads to indicate bolt strength. In this case, the higher the number, the stronger the bolt. Property class numbers 8.8, 9.8 and 10.9 are commonly used on automobiles.

Strength markings can also be used to distinguish standard hex nuts from metric hex nuts. Many standard nuts have dots stamped into one side, while metric nuts are marked with a number. The greater the number of dots, or the higher the number, the greater the strength of the nut.

Metric studs are also marked on their ends according to property class (grade). Larger studs are numbered (the same as metric bolts), while smaller studs carry a geometric code to denote grade.

It should be noted that many fasteners, especially Grades 0 through 2, have no distinguishing marks on them. When such is the case, the only way to determine whether it is standard or metric is to measure the thread pitch or compare it to a known fastener of the same size.

Standard fasteners are often referred to as SAE, as opposed to metric. However, it should be noted that SAE technically refers to a non-metric fine thread fastener only. Coarse thread non-metric fasteners are referred to as USS sizes.

Since fasteners of the same size (both standard and metric) may have different strength ratings, be sure to reinstall any bolts, studs or nuts removed from your vehicle in their original locations. Also, when replacing a fastener with a new one, make sure that the new one has a strength rating equal to or greater than the original.

Metric thread sizes

	Ft-lbs	Nm
M-6	6 to 9	9 to 12
M-8	14 to 21	19 to 28
M-10	28 to 40	38 to 54
M-12	50 to 71	68 to 96
M-14	80 to 140	109 to 154

Pipe thread sizes

1/8	5 to 8	7 to 10
1/4	12 to 18	17 to 24
3/8	22 to 33	30 to 44
1/2	25 to 35	34 to 47

U.S. thread sizes

1/4 - 20	6 to 9	9 to 12
5/16 - 18	12 to 18	17 to 24
5/16 - 24	14 to 20	19 to 27
3/8 - 16	22 to 32	30 to 43
3/8 - 24	27 to 38	37 to 51
7/16 - 14	40 to 55	55 to 74
7/16 - 20	40 to 60	55 to 81
1/2 - 13	55 to 80	75 to 108

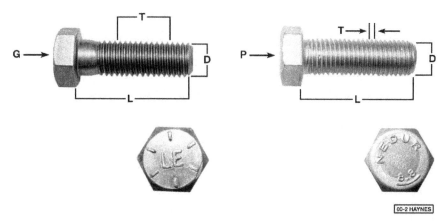

Standard (SAE and USS) bolt dimensions/grade marks

G Grade marks (bolt strength)
L Length (in inches)
T Thread pitch (number of threads per inch)
D Nominal diameter (in inches)

Metric bolt dimensions/grade marks

P Property class (bolt strength)
L Length (in millimeters)
T Thread pitch (distance between threads in millimeters)
D Diameter

Tightening sequences and procedures

Most threaded fasteners should be tightened to a specific torque value (torque is the twisting force applied to a threaded component such as a nut or bolt). Overtightening the fastener can weaken it and cause it to break, while undertightening can cause it to eventually come loose. Bolts, screws and studs, depending on the material they are made of and their thread diameters, have specific torque values, many of which are noted in the Specifications at the beginning of each Chapter. Be sure to follow the torque recommendations closely. For fasteners not assigned a specific torque, a general torque value chart is presented here as a guide. These torque values are for dry (unlubricated) fasteners threaded into steel or cast iron (not aluminum). As was previously mentioned, the size and grade of a fastener determine the amount of torque that can safely be applied to it. The figures listed here are approximate for Grade 2 and Grade 3 fasteners. Higher grades can tolerate higher torque values.

Fasteners laid out in a pattern, such as cylinder head bolts, oil pan bolts, differential cover bolts, etc., must be loosened or tightened in sequence to avoid warping the com-

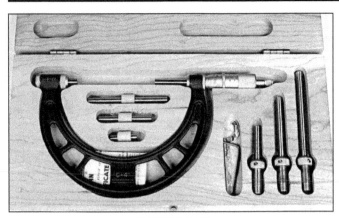

Micrometer set

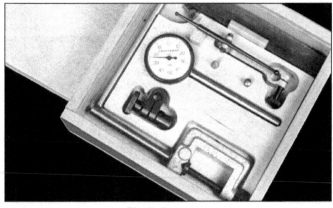

Dial indicator set

ponent. This sequence will normally be shown in the appropriate Chapter. If a specific pattern is not given, the following procedures can be used to prevent warping.

Initially, the bolts or nuts should be assembled finger-tight only. Next, they should be tightened one full turn each, in a criss-cross or diagonal pattern. After each one has been tightened one full turn, return to the first one and tighten them all one-half turn, following the same pattern. Finally, tighten each of them one-quarter turn at a time until each fastener has been tightened to the proper torque. To loosen and remove the fasteners, the procedure would be reversed.

Component disassembly

Component disassembly should be done with care and purpose to help ensure that the parts go back together properly. Always keep track of the sequence in which parts are removed. Make note of special characteristics or marks on parts that can be installed more than one way, such as a grooved thrust washer on a shaft. It is a good idea to lay the disassembled parts out on a clean surface in the order that they were removed. It may also be helpful to make sketches or take instant photos of components before removal.

When removing fasteners from a component, keep track of their locations. Sometimes threading a bolt back in a part, or putting the washers and nut back on a stud, can prevent mix-ups later. If nuts and bolts cannot be returned to their original locations, they should be kept in a compartmented box or a series of small boxes. A cupcake or muffin tin is ideal for this purpose, since each cavity can hold the bolts and nuts from a particular area (i.e. oil pan bolts, valve cover bolts, engine mount bolts, etc.). A pan of this type is especially helpful when working on assemblies with very small parts, such as the carburetor, alternator, valve train or interior dash and trim pieces. The cavities can be marked with paint or tape to identify the contents.

Whenever wiring looms, harnesses or connectors are separated, it is a good idea to identify the two halves with numbered pieces of masking tape so they can be easily reconnected.

Gasket sealing surfaces

Throughout any vehicle, gaskets are used to seal the mating surfaces between two parts and keep lubricants, fluids, vacuum or pressure contained in an assembly.

Many times these gaskets are coated with a liquid or paste-type gasket sealing compound before assembly. Age, heat and pressure can sometimes cause the two parts to stick together so tightly that they are very difficult to separate. Often, the assembly can be loosened by striking it with a soft-face hammer near the mating surfaces. A regular hammer can be used if a block of wood is placed between the hammer and the part. Do not hammer on cast parts or parts that could be easily damaged. With any particularly stubborn part, always recheck to make sure that every fastener has been removed.

Avoid using a screwdriver or bar to pry apart an assembly, as they can easily mar the gasket sealing surfaces of the parts, which must remain smooth. If prying is absolutely necessary, use an old broom handle, but keep in mind that extra clean up will be necessary if the wood splinters.

After the parts are separated, the old gasket must be carefully removed and the gasket surfaces cleaned. If you're working on cast iron or aluminum parts, stubborn gasket material can be soaked with rust penetrant or treated with a special chemical to soften it so it can be easily scraped off. **Caution:** *Never use gasket removal solutions or caustic chemicals on plastic or other composite components.* A scraper can be fashioned from a piece of copper tubing by flattening and sharpening one end. Copper is recommended because it is usually softer than the surfaces to be scraped, which reduces the chance of gouging the part. Some gaskets can be removed with a wire brush, but regardless of the method used, the mating surfaces must be left clean and smooth. If for some reason the gasket surface is gouged, then a gasket sealer thick enough to fill scratches will have to be used during reassembly of the components. For most applications, a non-drying (or semi-drying) gasket sealer should be used.

Hose removal tips

Warning: *If the vehicle is equipped with air conditioning, do not disconnect any of the A/C hoses without first having the system depressurized by a dealer service department or a service station.*

Hose removal precautions closely parallel gasket removal precautions. Avoid scratching or gouging the surface that the hose mates against or the connection may leak. This is especially true for radiator hoses. Because of various chemical reactions, the rubber in hoses can bond itself to the metal spigot that the hose fits over. To remove a hose, first loosen the hose clamps that secure it to the spigot. Then, with slip-joint pliers, grab the hose at the clamp and rotate it around the spigot. Work it back and forth until it is completely free, then pull it off. Silicone or other lubricants will ease removal if they can be applied between the hose and the outside of the spigot. Apply the same lubricant to the inside of the hose and the outside of the spigot to simplify installation.

As a last resort (and if the hose is to be replaced with a new one anyway), the rubber can be slit with a knife and the hose peeled from the spigot. If this must be done, be careful that the metal connection is not damaged.

If a hose clamp is broken or damaged, do not reuse it. Wire-type clamps usually weaken with age, so it is a good idea to replace them with screw-type clamps whenever a hose is removed.

Tools

A selection of good tools is a basic requirement for anyone who plans to maintain and repair his or her own vehicle. For the owner who has few tools, the initial investment might seem high, but when compared to the spiraling costs of professional auto maintenance and repair, it is a wise one.

To help the owner decide which tools are needed to perform the tasks detailed in this manual, the following tool lists are offered: *Maintenance and minor repair, Repair/overhaul* and *Special.*

The newcomer to practical mechanics should start off with the *maintenance and*

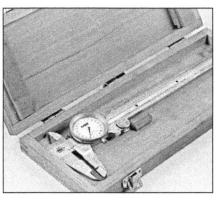

Dial caliper

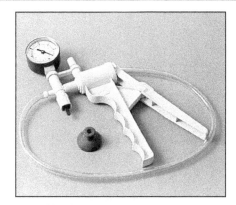

Hand-operated vacuum pump

Timing light

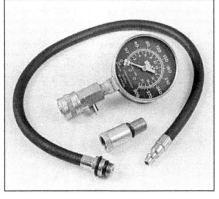

Compression gauge with spark plug hole adapter

Damper/steering wheel puller

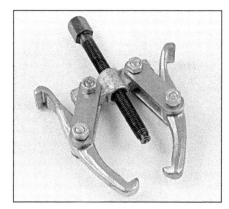

General purpose puller

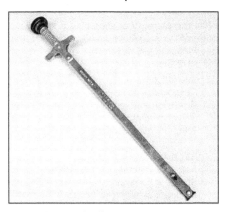

Hydraulic lifter removal tool

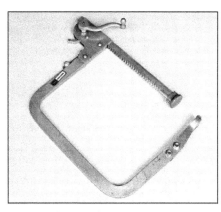

Valve spring compressor

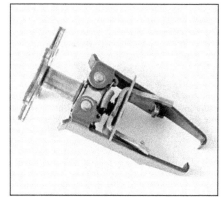

Valve spring compressor

Ridge reamer

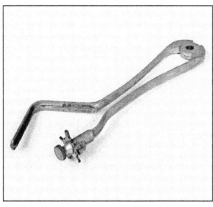

Piston ring groove cleaning tool

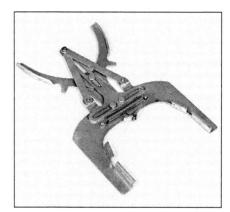

Ring removal/installation tool

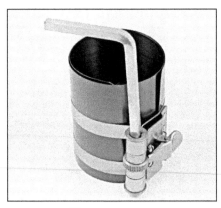

Ring compressor

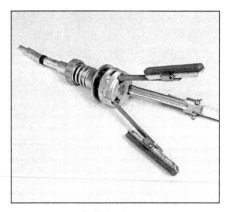

Cylinder hone

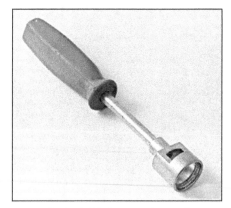

Brake hold-down spring tool

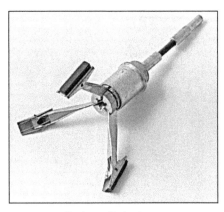

Brake cylinder hone

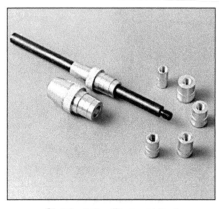

Clutch plate alignment tool

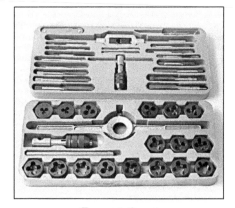

Tap and die set

minor repair tool kit, which is adequate for the simpler jobs performed on a vehicle. Then, as confidence and experience grow, the owner can tackle more difficult tasks, buying additional tools as they are needed. Eventually the basic kit will be expanded into the *repair and overhaul* tool set. Over a period of time, the experienced do-it-yourselfer will assemble a tool set complete enough for most repair and overhaul procedures and will add tools from the special category when it is felt that the expense is justified by the frequency of use.

Maintenance and minor repair tool kit

The tools in this list should be considered the minimum required for performance of routine maintenance, servicing and minor repair work. We recommend the purchase of combination wrenches (box-end and open-end combined in one wrench). While more expensive than open end wrenches, they offer the advantages of both types of wrench.

> *Combination wrench set (1/4-inch to 1 inch or 6 mm to 19 mm)*
> *Adjustable wrench, 8 inch*
> *Spark plug wrench with rubber insert*
> *Spark plug gap adjusting tool*
> *Feeler gauge set*
> *Brake bleeder wrench*
> *Standard screwdriver (5/16-inch x 6 inch)*

> *Phillips screwdriver (No. 2 x 6 inch)*
> *Combination pliers - 6 inch*
> *Hacksaw and assortment of blades*
> *Tire pressure gauge*
> *Grease gun*
> *Oil can*
> *Fine emery cloth*
> *Wire brush*
> *Battery post and cable cleaning tool*
> *Oil filter wrench*
> *Funnel (medium size)*
> *Safety goggles*
> *Jackstands (2)*
> *Drain pan*

Note: *If basic tune-ups are going to be part of routine maintenance, it will be necessary to purchase a good quality stroboscopic timing light and combination tachometer/dwell meter. Although they are included in the list of special tools, it is mentioned here because they are absolutely necessary for tuning most vehicles properly.*

Repair and overhaul tool set

These tools are essential for anyone who plans to perform major repairs and are in addition to those in the maintenance and minor repair tool kit. Included is a comprehensive set of sockets which, though expensive, are invaluable because of their versatility, especially when various extensions and drives are available. We recommend the 1/2-inch drive over the 3/8-inch drive. Although the larger drive is bulky and more expensive,

it has the capacity of accepting a very wide range of large sockets. Ideally, however, the mechanic should have a 3/8-inch drive set and a 1/2-inch drive set.

> *Socket set(s)*
> *Reversible ratchet*
> *Extension - 10 inch*
> *Universal joint*
> *Torque wrench (same size drive as sockets)*
> *Ball peen hammer - 8 ounce*
> *Soft-face hammer (plastic/rubber)*
> *Standard screwdriver (1/4-inch x 6 inch)*
> *Standard screwdriver (stubby - 5/16-inch)*
> *Phillips screwdriver (No. 3 x 8 inch)*
> *Phillips screwdriver (stubby - No. 2)*
> *Pliers - vise grip*
> *Pliers - lineman's*
> *Pliers - needle nose*
> *Pliers - snap-ring (internal and external)*
> *Cold chisel - 1/2-inch*
> *Scribe*
> *Scraper (made from flattened copper tubing)*
> *Centerpunch*
> *Pin punches (1/16, 1/8, 3/16-inch)*
> *Steel rule/straightedge - 12 inch*
> *Allen wrench set (1/8 to 3/8-inch or 4 mm to 10 mm)*
> *A selection of files*
> *Wire brush (large)*
> *Jackstands (second set)*
> *Jack (scissor or hydraulic type)*

Note: *Another tool which is often useful is an electric drill with a chuck capacity of 3/8-inch and a set of good quality drill bits.*

Special tools

The tools in this list include those which are not used regularly, are expensive to buy, or which need to be used in accordance with their manufacturer's instructions. Unless these tools will be used frequently, it is not very economical to purchase many of them. A consideration would be to split the cost and use between yourself and a friend or friends. In addition, most of these tools can be obtained from a tool rental shop on a temporary basis.

This list primarily contains only those tools and instruments widely available to the public, and not those special tools produced by the vehicle manufacturer for distribution to dealer service departments. Occasionally, references to the manufacturer's special tools are included in the text of this manual. Generally, an alternative method of doing the job without the special tool is offered. However, sometimes there is no alternative to their use. Where this is the case, and the tool cannot be purchased or borrowed, the work should be turned over to the dealer service department or an automotive repair shop.

Valve spring compressor
Piston ring groove cleaning tool
Piston ring compressor
Piston ring installation tool
Cylinder compression gauge
Cylinder ridge reamer
Cylinder surfacing hone
Cylinder bore gauge
Micrometers and/or dial calipers
Hydraulic lifter removal tool
Balljoint separator
Universal-type puller
Impact screwdriver
Dial indicator set
Stroboscopic timing light (inductive pick-up)
Hand operated vacuum/pressure pump
Tachometer/dwell meter
Universal electrical multimeter
Cable hoist
Brake spring removal and installation tools
Floor jack

Buying tools

For the do-it-yourselfer who is just starting to get involved in vehicle maintenance and repair, there are a number of options available when purchasing tools. If maintenance and minor repair is the extent of the work to be done, the purchase of individual tools is satisfactory. If, on the other hand, extensive work is planned, it would be a good idea to purchase a modest tool set from one of the large retail chain stores. A set can usually be bought at a substantial savings over the individual tool prices, and they often

come with a tool box. As additional tools are needed, add-on sets, individual tools and a larger tool box can be purchased to expand the tool selection. Building a tool set gradually allows the cost of the tools to be spread over a longer period of time and gives the mechanic the freedom to choose only those tools that will actually be used.

Tool stores will often be the only source of some of the special tools that are needed, but regardless of where tools are bought, try to avoid cheap ones, especially when buying screwdrivers and sockets, because they won't last very long. The expense involved in replacing cheap tools will eventually be greater than the initial cost of quality tools.

Care and maintenance of tools

Good tools are expensive, so it makes sense to treat them with respect. Keep them clean and in usable condition and store them properly when not in use. Always wipe off any dirt, grease or metal chips before putting them away. Never leave tools lying around in the work area. Upon completion of a job, always check closely under the hood for tools that may have been left there so they won't get lost during a test drive.

Some tools, such as screwdrivers, pliers, wrenches and sockets, can be hung on a panel mounted on the garage or workshop wall, while others should be kept in a tool box or tray. Measuring instruments, gauges, meters, etc. must be carefully stored where they cannot be damaged by weather or impact from other tools.

When tools are used with care and stored properly, they will last a very long time. Even with the best of care, though, tools will wear out if used frequently. When a tool is damaged or worn out, replace it. Subsequent jobs will be safer and more enjoyable if you do.

How to repair damaged threads

Sometimes, the internal threads of a nut or bolt hole can become stripped, usually from overtightening. Stripping threads is an all-too-common occurrence, especially when working with aluminum parts, because aluminum is so soft that it easily strips out.

Usually, external or internal threads are only partially stripped. After they've been cleaned up with a tap or die, they'll still work. Sometimes, however, threads are badly damaged. When this happens, you've got three choices:

1) *Drill and tap the hole to the next suitable oversize and install a larger diameter bolt, screw or stud.*
2) *Drill and tap the hole to accept a threaded plug, then drill and tap the plug to the original screw size. You can also buy a plug already threaded to the original size. Then you simply drill a hole to the specified size, then run the threaded*

plug into the hole with a bolt and jam nut. Once the plug is fully seated, remove the jam nut and bolt.
3) *The third method uses a patented thread repair kit like Heli-Coil or Slimsert. These easy-to-use kits are designed to repair damaged threads in straight-through holes and blind holes. Both are available as kits which can handle a variety of sizes and thread patterns. Drill the hole, then tap it with the special included tap. Install the Heli-Coil and the hole is back to its original diameter and thread pitch.*

Regardless of which method you use, be sure to proceed calmly and carefully. A little impatience or carelessness during one of these relatively simple procedures can ruin your whole day's work and cost you a bundle if you wreck an expensive part.

Working facilities

Not to be overlooked when discussing tools is the workshop. If anything more than routine maintenance is to be carried out, some sort of suitable work area is essential.

It is understood, and appreciated, that many home mechanics do not have a good workshop or garage available, and end up removing an engine or doing major repairs outside. It is recommended, however, that the overhaul or repair be completed under the cover of a roof.

A clean, flat workbench or table of comfortable working height is an absolute necessity. The workbench should be equipped with a vise that has a jaw opening of at least four inches.

As mentioned previously, some clean, dry storage space is also required for tools, as well as the lubricants, fluids, cleaning solvents, etc. which soon become necessary.

Sometimes waste oil and fluids, drained from the engine or cooling system during normal maintenance or repairs, present a disposal problem. To avoid pouring them on the ground or into a sewage system, pour the used fluids into large containers, seal them with caps and take them to an authorized disposal site or recycling center. Plastic jugs, such as old antifreeze containers, are ideal for this purpose.

Always keep a supply of old newspapers and clean rags available. Old towels are excellent for mopping up spills. Many mechanics use rolls of paper towels for most work because they are readily available and disposable. To help keep the area under the vehicle clean, a large cardboard box can be cut open and flattened to protect the garage or shop floor.

Whenever working over a painted surface, such as when leaning over a fender to service something under the hood, always cover it with an old blanket or bedspread to protect the finish. Vinyl covered pads, made especially for this purpose, are available at auto parts stores.

Jacking and towing

Jacking

The jack supplied with the vehicle should only be used for raising the vehicle when changing a tire or placing jackstands under the frame. **Warning:** *Never work under the vehicle or start the engine while this jack is being used as the only means of support.*

The vehicle should be on level ground with the hazard flashers on, the wheels blocked, the parking brake applied and the transmission in Park (automatic) or Reverse (manual). If a tire is being changed, loosen the lug nuts one-half turn and leave them in place until the wheel is raised off the ground.

Place the jack under the vehicle suspension in the indicated position **(see illustrations)**. Operate the jack with a slow, smooth motion until the wheel is raised off the ground. Remove the lug nuts, pull off the wheel, install the spare and thread the lug nuts back on with the beveled sides facing in. Tighten them snugly, but wait until the vehicle is lowered to tighten them completely.

Lower the vehicle, remove the jack and tighten the nuts (if loosened or removed) in a criss-cross pattern.

Towing

As a general rule, the vehicle should be towed with professional towing equipment. It should be attached to the main structural members of the vehicle, not the bumpers or brackets. Safety is a major consideration when towing and all applicable state and local laws must be obeyed. A safety chain system must be used at all times.

Before towing a vehicle make sure the drivetrain, axles, steering and suspension systems are in good condition and not damaged. Block the wheels to prevent the vehicle from rolling while attaching the towing equipment. Release the parking brake, place the transmission in neutral and turn the ignition key to the OFF position. On 4WD models, place the transfer case in 2H and unlock the front hubs. Remember that power steering and power brakes will not work with the engine off.

The recommended method of towing is from the rear, with the rear wheels off the ground. When a vehicle is towed with the rear wheels raised, the steering wheel must be clamped in the straight ahead position with a special device designed for use during towing. Remember, the ignition key must be in the OFF position, since the steering lock mechanism isn't strong enough to hold the front wheels straight while towing. On 4WD models, disconnect the front driveshaft from the front differential and safely secure it to the chassis. If conditions require the vehicle to be towed from the front with the rear wheels on the ground, disconnect the rear driveshaft from the rear differential and safely secure it to the chassis.

The vehicle may be towed with either two or four wheels on the ground without disconnecting the driveshaft(s) from the differential(s) as long as the towing speed does not exceed 30 mph and the distance does not exceed 50 miles. For further information on towing requirements, refer to the vehicle owners manual.

Front jacking point - Place the jack on the side of the vehicle under the frame rail

Rear jacking point - Place the jack at the rear of the vehicle under the rear axle

Booster battery (jump) starting

Observe these precautions when using a booster battery to start a vehicle:

a) *Before connecting the booster battery, make sure the ignition switch is in the Off position.*
b) *Turn off the lights, heater and other electrical loads.*
c) *Your eyes should be shielded. Safety goggles are a good idea.*
d) *Make sure the booster battery is the same voltage as the dead one in the vehicle.*
e) *The two vehicles MUST NOT TOUCH each other!*
f) *Make sure the transaxle is in Neutral (manual) or Park (automatic).*
g) *If the booster battery is not a maintenance-free type, remove the vent caps and lay a cloth over the vent holes.*

Connect the red jumper cable to the positive (+) terminals of each battery **(see illustration)**.

Connect one end of the black jumper cable to the negative (-) terminal of the booster battery. The other end of this cable should be connected to a good ground on the vehicle to be started, such as a bolt or bracket on the body.

Start the engine using the booster battery, then, with the engine running at idle speed, disconnect the jumper cables in the reverse order of connection.

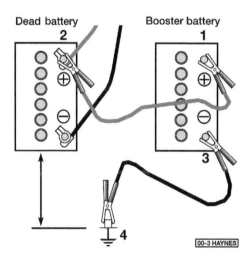

Make the booster battery cable connections in the numerical order shown (note that the negative cable of the booster battery is NOT attached to the negative terminal of the dead battery)

Automotive chemicals and lubricants

A number of automotive chemicals and lubricants are available for use during vehicle maintenance and repair. They include a wide variety of products ranging from cleaning solvents and degreasers to lubricants and protective sprays for rubber, plastic and vinyl.

Cleaners

Carburetor cleaner and choke cleaner is a strong solvent for gum, varnish and carbon. Most carburetor cleaners leave a dry-type lubricant film which will not harden or gum up. Because of this film it is not recommended for use on electrical components.

Brake system cleaner is used to remove brake dust, grease and brake fluid from the brake system, where clean surfaces are absolutely necessary. It leaves no residue and often eliminates brake squeal caused by contaminants.

Electrical cleaner removes oxidation, corrosion and carbon deposits from electrical contacts, restoring full current flow. It can also be used to clean spark plugs, carburetor jets, voltage regulators and other parts where an oil-free surface is desired.

Demoisturants remove water and moisture from electrical components such as alternators, voltage regulators, electrical connectors and fuse blocks. They are non-conductive and non-corrosive.

Degreasers are heavy-duty solvents used to remove grease from the outside of the engine and from chassis components. They can be sprayed or brushed on and, depending on the type, are rinsed off either with water or solvent.

Lubricants

Motor oil is the lubricant formulated for use in engines. It normally contains a wide variety of additives to prevent corrosion and reduce foaming and wear. Motor oil comes in various weights (viscosity ratings) from 5 to 80. The recommended weight of the oil depends on the season, temperature and the demands on the engine. Light oil is used in cold climates and under light load conditions. Heavy oil is used in hot climates and where high loads are encountered. Multi-viscosity oils are designed to have characteristics of both light and heavy oils and are available in a number of weights from 5W-20 to 20W-50.

Gear oil is designed to be used in differentials, manual transmissions and other areas where high-temperature lubrication is required.

Chassis and wheel bearing grease is a heavy grease used where increased loads and friction are encountered, such as for wheel bearings, balljoints, tie-rod ends and universal joints.

High-temperature wheel bearing grease is designed to withstand the extreme temperatures encountered by wheel bearings in disc brake equipped vehicles. It usually contains molybdenum disulfide (moly), which is a dry-type lubricant.

White grease is a heavy grease for metal-to-metal applications where water is a problem. White grease stays soft under both low and high temperatures (usually from -100 to +190-degrees F), and will not wash off or dilute in the presence of water.

Assembly lube is a special extreme pressure lubricant, usually containing moly, used to lubricate high-load parts (such as main and rod bearings and cam lobes) for initial start-up of a new engine. The assembly lube lubricates the parts without being squeezed out or washed away until the engine oiling system begins to function.

Silicone lubricants are used to protect rubber, plastic, vinyl and nylon parts.

Graphite lubricants are used where oils cannot be used due to contamination problems, such as in locks. The dry graphite will lubricate metal parts while remaining uncontaminated by dirt, water, oil or acids. It is electrically conductive and will not foul electrical contacts in locks such as the ignition switch.

Moly penetrants loosen and lubricate frozen, rusted and corroded fasteners and prevent future rusting or freezing.

Heat-sink grease is a special electrically non-conductive grease that is used for mounting electronic ignition modules where it is essential that heat is transferred away from the module.

Sealants

RTV sealant is one of the most widely used gasket compounds. Made from silicone, RTV is air curing, it seals, bonds, waterproofs, fills surface irregularities, remains flexible, doesn't shrink, is relatively easy to remove, and is used as a supplementary sealer with almost all low and medium temperature gaskets.

Anaerobic sealant is much like RTV in that it can be used either to seal gaskets or to form gaskets by itself. It remains flexible, is solvent resistant and fills surface imperfections. The difference between an anaerobic sealant and an RTV-type sealant is in the curing. RTV cures when exposed to air, while an anaerobic sealant cures only in the absence of air. This means that an anaerobic sealant cures only after the assembly of parts, sealing them together.

Thread and pipe sealant is used for sealing hydraulic and pneumatic fittings and vacuum lines. It is usually made from a Teflon compound, and comes in a spray, a paint-on liquid and as a wrap-around tape.

Chemicals

Anti-seize compound prevents seizing, galling, cold welding, rust and corrosion in fasteners. High-temperature ant-seize, usually made with copper and graphite lubricants, is used for exhaust system and exhaust manifold bolts.

Anaerobic locking compounds are used to keep fasteners from vibrating or working loose and cure only after installation, in the absence of air. Medium strength locking compound is used for small nuts, bolts and screws that may be removed later. High-strength locking compound is for large nuts, bolts and studs which aren't removed on a regular basis.

Oil additives range from viscosity index improvers to chemical treatments that claim to reduce internal engine friction. It should be noted that most oil manufacturers caution against using additives with their oils.

Gas additives perform several functions, depending on their chemical makeup. They usually contain solvents that help dissolve gum and varnish that build up on carburetor, fuel injection and intake parts. They also serve to break down carbon deposits that form on the inside surfaces of the combustion chambers. Some additives contain upper cylinder lubricants for valves and piston rings, and others contain chemicals to remove condensation from the gas tank.

Miscellaneous

Brake fluid is specially formulated hydraulic fluid that can withstand the heat and pressure encountered in brake systems. Care must be taken so this fluid does not come in contact with painted surfaces or plastics. An opened container should always be resealed to prevent contamination by water or dirt.

Weatherstrip adhesive is used to bond weatherstripping around doors, windows and trunk lids. It is sometimes used to attach trim pieces.

Undercoating is a petroleum-based, tar-like substance that is designed to protect metal surfaces on the underside of the vehicle from corrosion. It also acts as a sound-deadening agent by insulating the bottom of the vehicle.

Waxes and polishes are used to help protect painted and plated surfaces from the weather. Different types of paint may require the use of different types of wax and polish. Some polishes utilize a chemical or abrasive cleaner to help remove the top layer of oxidized (dull) paint on older vehicles. In recent years many non-wax polishes that contain a wide variety of chemicals such as polymers and silicones have been introduced. These non-wax polishes are usually easier to apply and last longer than conventional waxes and polishes.

Conversion factors

Length (distance)

Inches (in)	X	25.4	= Millimeters (mm)	X 0.0394	= Inches (in)
Feet (ft)	X	0.305	= Meters (m)	X 3.281	= Feet (ft)
Miles	X	1.609	= Kilometers (km)	X 0.621	= Miles

Volume (capacity)

Cubic inches (cu in; in^3)	X	16.387	= Cubic centimeters (cc; cm^3)	X 0.061	= Cubic inches (cu in; in^3)
Imperial pints (Imp pt)	X	0.568	= Liters (l)	X 1.76	= Imperial pints (Imp pt)
Imperial quarts (Imp qt)	X	1.137	= Liters (l)	X 0.88	= Imperial quarts (Imp qt)
Imperial quarts (Imp qt)	X	1.201	= US quarts (US qt)	X 0.833	= Imperial quarts (Imp qt)
US quarts (US qt)	X	0.946	= Liters (l)	X 1.057	= US quarts (US qt)
Imperial gallons (Imp gal)	X	4.546	= Liters (l)	X 0.22	= Imperial gallons (Imp gal)
Imperial gallons (Imp gal)	X	1.201	= US gallons (US gal)	X 0.833	= Imperial gallons (Imp gal)
US gallons (US gal)	X	3.785	= Liters (l)	X 0.264	= US gallons (US gal)

Mass (weight)

Ounces (oz)	X	28.35	= Grams (g)	X 0.035	= Ounces (oz)
Pounds (lb)	X	0.454	= Kilograms (kg)	X 2.205	= Pounds (lb)

Force

Ounces-force (ozf; oz)	X	0.278	= Newtons (N)	X 3.6	= Ounces-force (ozf; oz)
Pounds-force (lbf; lb)	X	4.448	= Newtons (N)	X 0.225	= Pounds-force (lbf; lb)
Newtons (N)	X	0.1	= Kilograms-force (kgf; kg)	X 9.81	= Newtons (N)

Pressure

Pounds-force per square inch (psi; lbf/in^2; lb/in^2)	X	0.070	= Kilograms-force per square centimeter (kgf/cm^2; kg/cm^2)	X 14.223	= Pounds-force per square inch (psi; lbf/in^2; lb/in^2)
Pounds-force per square inch (psi; lbf/in^2; lb/in^2)	X	0.068	= Atmospheres (atm)	X 14.696	= Pounds-force per square inch (psi; lbf/in^2; lb/in^2)
Pounds-force per square inch (psi; lbf/in^2; lb/in^2)	X	0.069	= Bars	X 14.5	= Pounds-force per square inch (psi; lbf/in^2; lb/in^2)
Pounds-force per square inch (psi; lbf/in^2; lb/in^2)	X	6.895	= Kilopascals (kPa)	X 0.145	= Pounds-force per square inch (psi; lbf/in^2; lb/in^2)
Kilopascals (kPa)	X	0.01	= Kilograms-force per square centimeter (kgf/cm^2; kg/cm^2)	X 98.1	= Kilopascals (kPa)

Torque (moment of force)

Pounds-force inches (lbf in; lb in)	X	1.152	= Kilograms-force centimeter (kgf cm; kg cm)	X 0.868	= Pounds-force inches (lbf in; lb in)
Pounds-force inches (lbf in; lb in)	X	0.113	= Newton meters (Nm)	X 8.85	= Pounds-force inches (lbf in; lb in)
Pounds-force inches (lbf in; lb in)	X	0.083	= Pounds-force feet (lbf ft; lb ft)	X 12	= Pounds-force inches (lbf in; lb in)
Pounds-force feet (lbf ft; lb ft)	X	0.138	= Kilograms-force meters (kgf m; kg m)	X 7.233	= Pounds-force feet (lbf ft; lb ft)
Pounds-force feet (lbf ft; lb ft)	X	1.356	= Newton meters (Nm)	X 0.738	= Pounds-force feet (lbf ft; lb ft)
Newton meters (Nm)	X	0.102	= Kilograms-force meters (kgf m; kg m)	X 9.804	= Newton meters (Nm)

Vacuum

Inches mercury (in. Hg)	X	3.377	= Kilopascals (kPa)	X 0.2961	= Inches mercury
Inches mercury (in. Hg)	X	25.4	= Millimeters mercury (mm Hg)	X 0.0394	= Inches mercury

Power

Horsepower (hp)	X	745.7	= Watts (W)	X 0.0013	= Horsepower (hp)

Velocity (speed)

Miles per hour (miles/hr; mph)	X	1.609	= Kilometers per hour (km/hr; kph)	X 0.621	= Miles per hour (miles/hr; mph)

Fuel consumption*

Miles per gallon, Imperial (mpg)	X	0.354	= Kilometers per liter (km/l)	X 2.825	= Miles per gallon, Imperial (mpg)
Miles per gallon, US (mpg)	X	0.425	= Kilometers per liter (km/l)	X 2.352	= Miles per gallon, US (mpg)

Temperature

Degrees Fahrenheit = (°C x 1.8) + 32

Degrees Celsius (Degrees Centigrade; °C) = (°F - 32) x 0.56

It is common practice to convert from miles per gallon (mpg) to liters/100 kilometers (l/100km), where mpg (Imperial) x l/100 km = 282 and mpg (US) x l/100 km = 235

Safety first!

Regardless of how enthusiastic you may be about getting on with the job at hand, take the time to ensure that your safety is not jeopardized. A moment's lack of attention can result in an accident, as can failure to observe certain simple safety precautions. The possibility of an accident will always exist, and the following points should not be considered a comprehensive list of all dangers. Rather, they are intended to make you aware of the risks and to encourage a safety conscious approach to all work you carry out on your vehicle.

Essential DOs and DON'Ts

DON'T rely on a jack when working under the vehicle. Always use approved jackstands to support the weight of the vehicle and place them under the recommended lift or support points.

DON'T attempt to loosen extremely tight fasteners (i.e. wheel lug nuts) while the vehicle is on a jack - it may fall.

DON'T start the engine without first making sure that the transmission is in Neutral (or Park where applicable) and the parking brake is set.

DON'T remove the radiator cap from a hot cooling system - let it cool or cover it with a cloth and release the pressure gradually.

DON'T attempt to drain the engine oil until you are sure it has cooled to the point that it will not burn you.

DON'T touch any part of the engine or exhaust system until it has cooled sufficiently to avoid burns.

DON'T siphon toxic liquids such as gasoline, antifreeze and brake fluid by mouth, or allow them to remain on your skin.

DON'T inhale brake lining dust - it is potentially hazardous (see *Asbestos* below).

DON'T allow spilled oil or grease to remain on the floor - wipe it up before someone slips on it.

DON'T use loose fitting wrenches or other tools which may slip and cause injury.

DON'T push on wrenches when loosening or tightening nuts or bolts. Always try to pull the wrench toward you. If the situation calls for pushing the wrench away, push with an open hand to avoid scraped knuckles if the wrench should slip.

DON'T attempt to lift a heavy component alone - get someone to help you.

DON'T rush or take unsafe shortcuts to finish a job.

DON'T allow children or animals in or around the vehicle while you are working on it.

DO wear eye protection when using power tools such as a drill, sander, bench grinder, etc. and when working under a vehicle.

DO keep loose clothing and long hair well out of the way of moving parts.

DO make sure that any hoist used has a safe working load rating adequate for the job.

DO get someone to check on you periodically when working alone on a vehicle.

DO carry out work in a logical sequence and make sure that everything is correctly assembled and tightened.

DO keep chemicals and fluids tightly capped and out of the reach of children and pets.

DO remember that your vehicle's safety affects that of yourself and others. If in doubt on any point, get professional advice.

Asbestos

Certain friction, insulating, sealing, and other products - such as brake linings, brake bands, clutch linings, torque converters, gaskets, etc. - may contain asbestos. Extreme care must be taken to avoid inhalation of dust from such products, since it is hazardous to health. If in doubt, assume that they do contain asbestos.

Fire

Remember at all times that gasoline is highly flammable. Never smoke or have any kind of open flame around when working on a vehicle. But the risk does not end there. A spark caused by an electrical short circuit, by two metal surfaces contacting each other, or even by static electricity built up in your body under certain conditions, can ignite gasoline vapors, which in a confined space are highly explosive. Do not, under any circumstances, use gasoline for cleaning parts. Use an approved safety solvent.

Always disconnect the battery ground (-) cable at the battery before working on any part of the fuel system or electrical system. Never risk spilling fuel on a hot engine or exhaust component. It is strongly recommended that a fire extinguisher suitable for use on fuel and electrical fires be kept handy in the garage or workshop at all times. Never try to extinguish a fuel or electrical fire with water.

Fumes

Certain fumes are highly toxic and can quickly cause unconsciousness and even death if inhaled to any extent. Gasoline vapor falls into this category, as do the vapors from some cleaning solvents. Any draining or pouring of such volatile fluids should be done in a well ventilated area.

When using cleaning fluids and solvents, read the instructions on the container carefully. Never use materials from unmarked containers.

Never run the engine in an enclosed space, such as a garage. Exhaust fumes contain carbon monoxide, which is extremely poisonous. If you need to run the engine, always do so in the open air, or at least have the rear of the vehicle outside the work area.

If you are fortunate enough to have the use of an inspection pit, never drain or pour gasoline and never run the engine while the vehicle is over the pit. The fumes, being heavier than air, will concentrate in the pit with possibly lethal results.

The battery

Never create a spark or allow a bare light bulb near a battery. They normally give off a certain amount of hydrogen gas, which is highly explosive.

Always disconnect the battery ground (-) cable at the battery before working on the fuel or electrical systems.

If possible, loosen the filler caps or cover when charging the battery from an external source (this does not apply to sealed or maintenance-free batteries). Do not charge at an excessive rate or the battery may burst.

Take care when adding water to a non maintenance-free battery and when carrying a battery. The electrolyte, even when diluted, is very corrosive and should not be allowed to contact clothing or skin.

Always wear eye protection when cleaning the battery to prevent the caustic deposits from entering your eyes.

Household current

When using an electric power tool, inspection light, etc., which operates on household current, always make sure that the tool is correctly connected to its plug and that, where necessary, it is properly grounded. Do not use such items in damp conditions and, again, do not create a spark or apply excessive heat in the vicinity of fuel or fuel vapor.

Secondary ignition system voltage

A severe electric shock can result from touching certain parts of the ignition system (such as the spark plug wires) when the engine is running or being cranked, particularly if components are damp or the insulation is defective. In the case of an electronic ignition system, the secondary system voltage is much higher and could prove fatal.

DECIMALS to MILLIMETERS

Decimal	mm	Decimal	mm
0.001	0.0254	0.500	12.7000
0.002	0.0508	0.510	12.9540
0.003	0.0762	0.520	13.2080
0.004	0.1016	0.530	13.4620
0.005	0.1270	0.540	13.7160
0.006	0.1524	0.550	13.9700
0.007	0.1778	0.560	14.2240
0.008	0.2032	0.570	14.4780
0.009	0.2286	0.580	14.7320
		0.590	14.9860
0.010	0.2540		
0.020	0.5080		
0.030	0.7620		
0.040	1.0160	0.600	15.2400
0.050	1.2700	0.610	15.4940
0.060	1.5240	0.620	15.7480
0.070	1.7780	0.630	16.0020
0.080	2.0320	0.640	16.2560
0.090	2.2860	0.650	16.5100
		0.660	16.7640
0.100	2.5400	0.670	17.0180
0.110	2.7940	0.680	17.2720
0.120	3.0480	0.690	17.5260
0.130	3.3020		
0.140	3.5560		
0.150	3.8100		
0.160	4.0640	0.700	17.7800
0.170	4.3180	0.710	18.0340
0.180	4.5720	0.720	18.2880
0.190	4.8260	0.730	18.5420
		0.740	18.7960
0.200	5.0800	0.750	19.0500
0.210	5.3340	0.760	19.3040
0.220	5.5880	0.770	19.5580
0.230	5.8420	0.780	19.8120
0.240	6.0960	0.790	20.0660
0.250	6.3500		
0.260	6.6040		
0.270	6.8580	0.800	20.3200
0.280	7.1120	0.810	20.5740
0.290	7.3660	0.820	21.8280
		0.830	21.0820
0.300	7.6200	0.840	21.3360
0.310	7.8740	0.850	21.5900
0.320	8.1280	0.860	21.8440
0.330	8.3820	0.870	22.0980
0.340	8.6360	0.880	22.3520
0.350	8.8900	0.890	22.6060
0.360	9.1440		
0.370	9.3980		
0.380	9.6520		
0.390	9.9060		
		0.900	22.8600
0.400	10.1600	0.910	23.1140
0.410	10.4140	0.920	23.3680
0.420	10.6680	0.930	23.6220
0.430	10.9220	0.940	23.8760
0.440	11.1760	0.950	24.1300
0.450	11.4300	0.960	24.3840
0.460	11.6840	0.970	24.6380
0.470	11.9380	0.980	24.8920
0.480	12.1920	0.990	25.1460
0.490	12.4460	1.000	25.4000

FRACTIONS to DECIMALS to MILLIMETERS

Fraction	Decimal	mm	Fraction	Decimal	mm
1/64	0.0156	0.3969	33/64	0.5156	13.0969
1/32	0.0312	0.7938	17/32	0.5312	13.4938
3/64	0.0469	1.1906	35/64	0.5469	13.8906
1/16	0.0625	1.5875	9/16	0.5625	14.2875
5/64	0.0781	1.9844	37/64	0.5781	14.6844
3/32	0.0938	2.3812	19/32	0.5938	15.0812
7/64	0.1094	2.7781	39/64	0.6094	15.4781
1/8	0.1250	3.1750	5/8	0.6250	15.8750
9/64	0.1406	3.5719	41/64	0.6406	16.2719
5/32	0.1562	3.9688	21/32	0.6562	16.6688
11/64	0.1719	4.3656	43/64	0.6719	17.0656
3/16	0.1875	4.7625	11/16	0.6875	17.4625
13/64	0.2031	5.1594	45/64	0.7031	17.8594
7/32	0.2188	5.5562	23/32	0.7188	18.2562
15/64	0.2344	5.9531	47/64	0.7344	18.6531
1/4	0.2500	6.3500	3/4	0.7500	19.0500
17/64	0.2656	6.7469	49/64	0.7656	19.4469
9/32	0.2812	7.1438	25/32	0.7812	19.8438
19/64	0.2969	7.5406	51/64	0.7969	20.2406
5/16	0.3125	7.9375	13/16	0.8125	20.6375
21/64	0.3281	8.3344	53/64	0.8281	21.0344
11/32	0.3438	8.7312	27/32	0.8438	21.4312
23/64	0.3594	9.1281	55/64	0.8594	21.8281
3/8	0.3750	9.5250	7/8	0.8750	22.2250
25/64	0.3906	9.9219	57/64	0.8906	22.6219
13/32	0.4062	10.3188	29/32	0.9062	23.0188
27/64	0.4219	10.7156	59/64	0.9219	23.4156
7/16	0.4375	11.1125	15/16	0.9375	23.8125
29/64	0.4531	11.5094	61/64	0.9531	24.2094
15/32	0.4688	11.9062	31/32	0.9688	24.6062
31/64	0.4844	12.3031	63/64	0.9844	25.0031
1/2	0.5000	12.7000	1	1.0000	25.4000

Troubleshooting

Contents

Engine and performance

1 Engine will not rotate when attempting to start

1 Battery terminal connections loose or corroded. Check the cable terminals at the battery; tighten cable clamp and/or clean off corrosion as necessary (see Chapter 1).
2 Battery discharged or faulty. If the cable ends are clean and tight on the battery posts, turn the key to the On position and switch on the headlights or windshield wipers. If they won't run, the battery is discharged.
3 Automatic transmission not engaged in park (P) or Neutral (N).
4 Broken, loose or disconnected wires in the starting circuit. Inspect all wires and connectors at the battery, starter solenoid and ignition switch (on steering column).
5 Starter motor pinion jammed in flywheel ring gear. If manual transmission, place transmission in gear and rock the vehicle to manually turn the engine. Remove starter (Chapter 5) and inspect pinion and flywheel (Chapter 2) at earliest convenience.
6 Starter solenoid faulty (Chapter 5).
7 Starter motor faulty (Chapter 5).
8 Ignition switch faulty (Chapter 12).
9 Engine seized. Try to turn the crankshaft with a large socket and breaker bar on the pulley bolt.

2 Engine rotates but will not start

1 Fuel tank empty.
2 Battery discharged (engine rotates slowly). Check the operation of electrical components as described in previous Section.
3 Battery terminal connections loose or corroded. See previous Section.
4 Fuel not reaching carburetor or fuel injector. Check for clogged fuel filter or lines and defective fuel pump. Also make sure the tank vent lines aren't clogged (Chapter 4A or 4B).
5 Choke not operating properly (Chapter 1).
6 Faulty distributor components. Check the cap and rotor (Chapter 1).
7 Low cylinder compression. Check as described in Chapter 2.
8 Valve clearances not properly adjusted - Chapter 1 (four-cylinder engine).
9 Water in fuel. Drain tank and fill with new fuel.
10 Defective ignition coil (Chapter 5).
11 Dirty or clogged carburetor jets or fuel injector (Chapter 4B). Carburetor out of adjustment. Check the float level (Chapter 4A).
12 Wet or damaged ignition components (Chapters 1 and 5).
13 Worn, faulty or incorrectly gapped spark plugs (Chapter 1).
14 Broken, loose or disconnected wires in the starting circuit (see previous Section).
15 Loose distributor (changing ignition timing). Turn the distributor body as necessary to start the engine, then adjust the ignition timing as soon as possible (Chapter 1).
16 Broken, loose or disconnected wires at the ignition coil or faulty coil (Chapter 5).
17 Timing chain or belt failure or wear affecting valve timing (Chapter 2).

3 Starter motor operates without turning engine

1 Starter pinion sticking. Remove the starter (Chapter 5) and inspect.
2 Starter pinion or flywheel/driveplate teeth worn or broken. Remove the inspection cover and inspect.

4 Engine hard to start when cold

1 Battery discharged or low. Check as described in Chapter 1.
2 Fuel not reaching the carburetor or fuel injectors. Check the fuel filter, lines and fuel pump (Chapters 1 and 4A, 4B).
3 Choke inoperative (Chapters 1 and 4).
4 Defective spark plugs (Chapter 1).

5 Engine hard to start when hot

1 Air filter dirty (Chapter 1).
2 Fuel not reaching carburetor or fuel injectors (see Section 4). Check for a vapor lock situation, brought about by clogged fuel tank vent lines.
3 Bad engine ground connection.
4 Choke sticking (Chapter 1).
5 Defective pick-up coil in distributor (Chapter 5).
6 Float level too high (Chapter 4A).
7 Faulty coolant temperature sensor (ECT) or intake air temperature sensor (IAT) (Chapter 6).

6 Starter motor noisy or engages roughly

1 Pinion or flywheel/driveplate teeth worn or broken. Remove the inspection cover on the left side of the engine and inspect.
2 Starter motor mounting bolts loose or missing.

7 Engine starts but stops immediately

1 Loose or damaged wire harness connections at distributor, coil or alternator.
2 Intake manifold vacuum leaks. Make sure all mounting bolts/nuts are tight and all vacuum hoses connected to the manifold are attached properly and in good condition (Chapters 2A, 4A, 4B).
3 Insufficient fuel flow to carburetor or fuel injectors (Chapters 4A and 4B).
4 Idle speed incorrect (Chapter 1).

8 Engine 'lopes' while idling or idles erratically

1 Vacuum leaks. Check mounting bolts at the intake manifold for tightness. Make sure that all vacuum hoses are connected and in good condition. Use a stethoscope or a length of fuel hose held against your ear to listen for vacuum leaks while the engine is running. A hissing sound will be heard. A soapy water solution will also detect leaks. Check the intake manifold gasket surfaces.
2 Leaking EGR valve or plugged PCV valve (see Chapters 1 and 6).
3 Air filter clogged (Chapter 1).
4 Fuel pump not delivering sufficient fuel (Chapter 4A, 4B).
5 Leaking head gasket. Perform a cylinder compression check (Chapter 2).
6 Timing belt or chain worn (Chapter 2).
7 Camshaft lobes worn (Chapter 2).
8 Valve clearance out of adjustment - Chapter 1 (four-cylinder engine).
9 Valves burned or otherwise leaking (Chapter 2).
10 Ignition timing out of adjustment (Chapter 1).
11 Ignition system not operating properly (Chapters 1 and 5).
12 Thermostatic air cleaner not operating properly (Chapter 1).
13 Choke not operating properly (Chapters 1 and 4).
14 Dirty or clogged injector(s). Carburetor dirty, clogged or out of adjustment. Check the float level (Chapter 4A, 4B).
15 Idle speed out of adjustment (Chapter 1).

9 Engine misses at idle speed

1 Spark plugs faulty or not gapped properly (Chapter 1).
2 Faulty spark plug wires (Chapter 1).
3 Wet or damaged distributor components (Chapter 1).
4 Short circuits in ignition, coil or spark plug wires.
5 Sticking or faulty emissions systems (see Chapter 6).
6 Clogged fuel filter and/or foreign matter in fuel. Remove the fuel filter (Chapter 1) and inspect.
7 Vacuum leaks at intake manifold or hose connections. Check as described in Section 8.
8 Incorrect idle speed (Chapter 1) or idle mixture (Chapter 4).
9 Incorrect ignition timing (Chapter 1).
10 Low or uneven cylinder compression. Check as described in Chapter 2.

11 Choke not operating properly (Chapter 1).
12 Clogged or dirty fuel injectors (Chapter 4).

10 Excessively high idle speed

1 Sticking throttle linkage (Chapter 4A).
2 Choke opened excessively at idle (Chapter 4A).
3 Idle speed incorrectly adjusted (Chapter 1).
4 Valve clearances incorrectly adjusted - Chapter 1 (four-cylinder engine).

11 Battery will not hold a charge

1 Alternator drivebelt defective or not adjusted properly (Chapter 1).
2 Battery cables loose or corroded (Chapter 1).
3 Alternator not charging properly (Chapter 5).
4 Loose, broken or faulty wires in the charging circuit (Chapter 5).
5 Short circuit causing a continuous drain on the battery.
6 Battery defective internally.

12 Alternator light stays on

1 Fault in alternator or charging circuit (Chapter 5).
2 Alternator drivebelt defective or not properly adjusted (Chapter 1).

13 Alternator light fails to come on when key is turned on

1 Faulty bulb (Chapter 12).
2 Defective alternator (Chapter 5).
3 Fault in the printed circuit, dash wiring or bulb holder (Chapter 12).

14 Engine misses throughout driving speed range

1 Fuel filter clogged and/or impurities in the fuel system. Check fuel filter (Chapter 1) or clean system (Chapter 4).
2 Faulty or incorrectly gapped spark plugs (Chapter 1).
3 Incorrect ignition timing (Chapter 1).
4 Cracked distributor cap, disconnected distributor wires or damaged distributor components (Chapter 1).
5 Defective spark plug wires (Chapter 1).
6 Emissions system components faulty (Chapter 6).
7 Low or uneven cylinder compression pressures. Check as described in Chapter 2.
8 Weak or faulty ignition coil (Chapter 5).
9 Weak or faulty ignition system (Chapter 5).

10 Vacuum leaks at intake manifold or vacuum hoses (see Section 8).
11 Dirty or clogged carburetor or fuel injector (Chapter 4A, 4B).
12 Leaky EGR valve (Chapter 6).
13 Carburetor out of adjustment (Chapter 4A, 4B).
14 Idle speed out of adjustment (Chapter 1).

15 Hesitation or stumble during acceleration

1 Ignition timing incorrect (Chapter 1).
2 Ignition system not operating properly (Chapter 5).
3 Dirty or clogged carburetor or fuel injector (Chapter 4A, 4B).
4 Low fuel pressure. Check for proper operation of the fuel pump and for restrictions in the fuel filter and lines (Chapter 4A, 4B).
5 Carburetor out of adjustment (Chapter 4A).

16 Engine stalls

1 Idle speed incorrect (Chapter 1).
2 Fuel filter clogged and/or water and impurities in the fuel system (Chapter 1).
3 Choke not operating properly (Chapter 1).
4 Damaged or wet distributor cap and wires.
5 Emissions system components faulty (Chapter 6).
6 Faulty or incorrectly gapped spark plugs (Chapter 1). Also check the spark plug wires (Chapter 1).
7 Vacuum leak at the carburetor, intake manifold or vacuum hoses. Check as described in Section 8.
8 Valve clearances incorrect - Chapter 1 (four-cylinder engine).

17 Engine lacks power

1 Incorrect ignition timing (Chapter 1).
2 Excessive play in distributor shaft. At the same time check for faulty distributor cap, wires, etc. (Chapter 1).
3 Faulty or incorrectly gapped spark plugs (Chapter 1).
4 Air filter dirty (Chapter 1).
5 Faulty ignition coil (Chapter 5).
6 Brakes binding (Chapters 1 and 10).
7 Automatic transmission fluid level incorrect, causing slippage (Chapter 1).
8 Clutch slipping (Chapter 8).
9 Fuel filter clogged and/or impurities in the fuel system (Chapters 1 and 4).
10 EGR system not functioning properly (Chapter 6).
11 Use of sub-standard fuel. Fill tank with

proper octane fuel.
12 Low or uneven cylinder compression pressures. Check as described in Chapter 2E.
13 Air leak at carburetor or intake manifold (check as described in Section 8).
14 Dirty or clogged carburetor jets or malfunctioning choke (Chapters 1 and 4).

18 Engine backfires

1 EGR system not functioning properly (Chapter 6).
2 Ignition timing incorrect (Chapter 1).
3 Thermostatic air cleaner system not operating properly (Chapter 6).
4 Vacuum leak (refer to Section 8).
5 Valve clearances incorrect - Chapter 1 (four-cylinder engine).
6 Damaged valve springs or sticking valves (Chapter 2).
7 Intake air leak (see Section 8).
8 Carburetor float level out of adjustment (Chapter 4A).

19 Engine surges while holding accelerator steady

1 Intake air leak (see Section 8).
2 Fuel pump not working properly (Chapter 4A, 4B).

20 Pinging or knocking engine sounds when engine is under load

1 Incorrect grade of fuel. Fill tank with fuel of the proper octane rating.
2 Ignition timing incorrect (Chapter 1).
3 Carbon build-up in combustion chambers. Remove cylinder head(s) and clean combustion chambers (Chapter 2).
4 Incorrect spark plugs (Chapter 1).

21 Engine diesels (continues to run) after being turned off

1 Idle speed too high (Chapter 1).
2 Ignition timing incorrect (Chapter 1).
3 Incorrect spark plug heat range (Chapter 1).
4 Intake air leak (see Section 8).
5 Carbon build-up in combustion chambers. Remove the cylinder head(s) and clean the combustion chambers (Chapter 2).
6 Valves sticking (Chapter 2).
7 Valve clearances incorrect - Chapter 1 (four-cylinder engine).
8 EGR system not operating properly (Chapter 6).
9 Fuel shut-off system not operating properly (Chapter 6).
10 Check for causes of overheating (Section 27).

22 Low oil pressure

1 Improper grade of oil.
2 Oil pump worn or damaged (Chapter 2).
3 Engine overheating (refer to Section 27).
4 Clogged oil filter (Chapter 1).
5 Clogged oil strainer (Chapter 2).
6 Oil pressure gauge not working properly (Chapter 2E).

23 Excessive oil consumption

1 Loose oil drain plug.
2 Loose bolts or damaged oil pan gasket (Chapter 2).
3 Loose bolts or damaged front cover gasket (Chapter 2).
4 Front or rear crankshaft oil seal leaking (Chapter 2).
5 Loose bolts or damaged rocker arm cover gasket (Chapter 2).
6 Loose oil filter (Chapter 1).
7 Loose or damaged oil pressure switch (Chapter 2).
8 Pistons and cylinders excessively worn (Chapter 2).
9 Piston rings not installed correctly on pistons (Chapter 2).
10 Worn or damaged piston rings (Chapter 2).
11 Intake and/or exhaust valve oil seals worn or damaged (Chapter 2).
12 Worn valve stems.
13 Worn or damaged valves/guides (Chapter 2).

24 Excessive fuel consumption

1 Dirty or clogged air filter element (Chapter 1).
2 Incorrect ignition timing (Chapter 1).
3 Incorrect idle speed (Chapter 1).
4 Low tire pressure or incorrect tire size (Chapter 11).
5 Fuel leakage. Check all connections, lines and components in the fuel system (Chapter 4A, 4B).
6 Choke not operating properly (Chapter 1).
7 Dirty or clogged carburetor jets or fuel injectors (Chapter 4A, 4B).

25 Fuel odor

1 Fuel leakage. Check all connections, lines and components in the fuel system (Chapter 4A, 4B).
2 Fuel tank overfilled. Fill only to automatic shut-off.
3 Charcoal canister filter in Evaporative Emissions Control system clogged (Chapter 1).
4 Vapor leaks from Evaporative Emissions Control system lines (Chapter 6).

26 Miscellaneous engine noises

1 A strong dull noise that becomes more rapid as the engine accelerates indicates worn or damaged crankshaft bearings or an unevenly worn crankshaft. To pinpoint the trouble spot, remove the spark plug wire from one plug at a time and crank the engine over. If the noise stops, the cylinder with the removed plug wire indicates the problem area. Replace the bearing and/or service or replace the crankshaft (Chapter 2).
2 A similar (yet slightly higher pitched) noise to the crankshaft knocking described in the previous paragraph, that becomes more rapid as the engine accelerates, indicates worn or damaged connecting rod bearings (Chapter 2). The procedure for locating the problem cylinder is the same as described in Paragraph 1.
3 An overlapping metallic noise that increases in intensity as the engine speed increases, yet diminishes as the engine warms up indicates abnormal piston and cylinder wear (Chapter 2). To locate the problem cylinder, use the procedure described in Paragraph 1.
4 A rapid clicking noise that becomes faster as the engine accelerates indicates a worn piston pin or piston pin hole. This sound will happen each time the piston hits the highest and lowest points in the stroke (Chapter 2). The procedure for locating the problem piston is described in Paragraph 1.
5 A metallic clicking noise coming from the water pump indicates worn or damaged water pump bearings or pump. Replace the water pump with a new one (Chapter 3).
6 A rapid tapping sound or clicking sound that becomes faster as the engine speed increases indicates "valve tapping" or improperly adjusted valve clearances. This can be identified by holding one end of a section of hose to your ear and placing the other end at different spots along the rocker arm cover. The point where the sound is loudest indicates the problem valve. Adjust the valve clearance (Chapter 1 or 2B). If the problem persists, you likely have a collapsed valve lifter or other damaged valve train component. Changing the engine oil and adding a high viscosity oil treatment will sometimes cure a stuck lifter problem. If the problem still persists, the lifters, pushrods and rocker arms must be removed for inspection (see Chapter 2).
7 A steady metallic rattling or rapping sound coming from the area of the timing chain cover indicates a worn, damaged or out-of-adjustment timing chain. Service or replace the chain and related components (Chapter 2).

Cooling system

27 Overheating

1 Insufficient coolant in system (Chapter 1).
2 Drivebelt defective or not adjusted properly (Chapter 1).
3 Radiator core blocked or radiator grille dirty or restricted (Chapter 3).
4 Thermostat faulty (Chapter 3).
5 Fan not functioning properly (Chapter 3).
6 Radiator cap not maintaining proper pressure. Have cap pressure tested by gas station or repair shop.
7 Ignition timing incorrect (Chapter 1).
8 Defective water pump (Chapter 3).
9 Improper grade of engine oil.
10 Inaccurate temperature gauge (Chapter 12).

28 Overcooling

1 Thermostat faulty (Chapter 3).
2 Inaccurate temperature gauge (Chapter 12).

29 External coolant leakage

1 Deteriorated or damaged hoses. Loose clamps at hose connections (Chapter 1).
2 Water pump seals defective. If this is the case, water will drip from the weep hole in the water pump body (Chapter 3).
3 Leakage from radiator core or header tank. This will require the radiator to be professionally repaired (see Chapter 3 for removal procedures).
4 Engine drain plugs or water jacket freeze plugs leaking (see Chapters 1 and 2).
5 Leak from coolant temperature switch (Chapter 3).
6 Leak from damaged gaskets or small cracks (Chapter 2).
7 Damaged head gasket. This can be verified by checking the condition of the engine oil as noted in Section 30.

30 Internal coolant leakage

Note: *Internal coolant leaks can usually be detected by examining the oil. Check the dipstick and inside the rocker arm cover for water deposits and an oil consistency like that of a milkshake.*
1 Leaking cylinder head gasket. Have the system pressure tested or remove the cylinder head (Chapter 2) and inspect.
2 Cracked cylinder bore or cylinder head. Dismantle engine and inspect (Chapter 2).
3 Loose cylinder head bolts (tighten as described in Chapter 2).

31 Abnormal coolant loss

1 Overfilling system (Chapter 1).
2 Coolant boiling away due to overheating (see causes in Section 27).
3 Internal or external leakage (see Sec-

tions 29 and 30).
4   Faulty radiator cap. Have the cap pressure tested.
5   Cooling system being pressurized by engine compression. This could be due to a cracked head or block or leaking head gasket(s).

32   Poor coolant circulation

1   Inoperative water pump. A quick test is to pinch the top radiator hose closed with your hand while the engine is idling, then release it. You should feel a surge of coolant if the pump is working properly (Chapter 3).
2   Restriction in cooling system. Drain, flush and refill the system (Chapter 1). If necessary, remove the radiator (Chapter 3) and have it reverse flushed or professionally cleaned.
3   Loose water pump drivebelt (Chapter 1).
4   Thermostat sticking (Chapter 3).
5   Insufficient coolant (Chapter 1).

33   Corrosion

1   Excessive impurities in the water. Soft, clean water is recommended. Distilled or rainwater is satisfactory.
2   Insufficient antifreeze solution (refer to Chapter 1 for the proper ratio of water to antifreeze).
3   Infrequent flushing and draining of system. Regular flushing of the cooling system should be carried out at the specified intervals as described in (Chapter 1).

Clutch

Note: *All clutch related service information is located in Chapter 8, unless otherwise noted.*

34   Fails to release (pedal pressed to the floor - shift lever does not move freely in and out of Reverse)

1   Freeplay incorrectly adjusted (see Chapter 8).
2   Clutch contaminated with oil. Remove clutch plate and inspect.
3   Clutch plate warped, distorted or otherwise damaged.
4   Diaphragm spring fatigued. Remove clutch cover/pressure plate assembly and inspect.
5   Broken, binding or damaged release cable (cable-operated clutch).
6   Leakage of fluid from clutch hydraulic system. Inspect master cylinder, operating cylinder and connecting lines.
7   Air in clutch hydraulic system. Bleed the system.
8   Insufficient pedal stroke. Check and

adjust as necessary.
9   Piston seal in operating cylinder deformed or damaged.
10   Lack of grease on pilot bearing.

35   Clutch slips (engine speed increases with no increase in vehicle speed)

1   Worn or oil-soaked clutch plate.
2   Clutch plate not broken in. It may take 30 or 40 normal starts for a new clutch to seat.
3   Diaphragm spring weak or damaged. Remove clutch cover/pressure plate assembly and inspect.
4   Flywheel warped (Chapter 2).
5   Debris in master cylinder preventing the piston from returning to its normal position.
6   Clutch hydraulic line damaged.
7   Binding in the release mechanism.

36   Grabbing (chattering) as clutch is engaged

1   Oil on clutch plate. Remove and inspect. Repair any leaks.
2   Worn or loose engine or transmission mounts. They may move slightly when clutch is released. Inspect mounts and bolts.
3   Worn splines on transmission input shaft. Remove clutch components and inspect.
4   Warped pressure plate or flywheel. Remove clutch components and inspect.
5   Diaphragm spring fatigued. Remove clutch cover/pressure plate assembly and inspect.
6   Clutch linings hardened or warped.
7   Clutch lining rivets loose.

37   Squeal or rumble with clutch engaged (pedal released)

1   Improper pedal adjustment. Adjust pedal freeplay (Chapter 8).
2   Release bearing binding on transmission shaft. Remove clutch components and check bearing. Remove any burrs or nicks, clean and relubricate before reinstallation.
3   Pilot bushing worn or damaged.
4   Clutch rivets loose.
5   Clutch plate cracked.
6   Fatigued clutch plate torsion springs. Replace clutch plate.

38   Squeal or rumble with clutch disengaged (pedal depressed)

1   Worn or damaged release bearing.
2   Worn or broken pressure plate diaphragm fingers.

39   Clutch pedal stays on floor when disengaged

Binding linkage or release bearing. Inspect linkage or remove clutch components as necessary.

Manual transmission

Note: *All manual transmission service information is located in Chapter 7, unless otherwise noted.*

40   Noisy in Neutral with engine running

1   Input shaft bearing worn.
2   Damaged main drive gear bearing.
3   Insufficient transmission oil (Chapter 1).
4   Transmission oil in poor condition. Drain and fill with proper grade oil. Check old oil for water and debris (Chapter 1).
5   Noise can be caused by variations in engine torque. Change the idle speed and see if noise disappears.

41   Noisy in all gears

1   Any of the above causes, and/or:
2   Worn or damaged output gear bearings or shaft.

42   Noisy in one particular gear

1   Worn, damaged or chipped gear teeth.
2   Worn or damaged synchronizer.

43   Slips out of gear

1   Transmission loose on clutch housing.
2   Stiff shift lever seal.
3   Shift linkage binding.
4   Broken or loose input gear bearing retainer.
5   Dirt between clutch lever and engine housing.
6   Worn linkage.
7   Damaged or worn check balls, fork rod ball grooves or check springs.
8   Worn mainshaft or countershaft bearings.
9   Loose engine mounts (Chapter 2).
10   Excessive gear end play.
11   Worn synchronizers.

44   Oil leaks

1   Excessive amount of lubricant in transmission (see Chapter 1 for correct checking procedures). Drain lubricant as required.
2   Rear oil seal or speedometer oil seal damaged.

3   To pinpoint a leak, first remove all built-up dirt and grime from the transmission. Degreasing agents and/or steam cleaning will achieve this. With the underside clean, drive the vehicle at low speeds so the air flow will not blow the leak far from its source. Raise the vehicle and determine where the leak is located.

45   Difficulty engaging gears

1   Clutch not releasing completely.
2   Loose or damaged shift linkage. Make a thorough inspection, replacing parts as necessary.
3   Insufficient transmission oil (Chapter 1).
4   Transmission oil in poor condition. Drain and fill with proper grade oil. Check oil for water and debris (Chapter 1).
5   Worn or damaged striking rod.
6   Sticking or jamming gears.

46   Noise occurs while shifting gears

1   Check for proper operation of the clutch (Chapter 8).
2   Faulty synchronizer assemblies.

Automatic transmission

Note: *Due to the complexity of the automatic transmission, it's difficult for the home mechanic to properly diagnose and service. For problems other than the following, the vehicle should be taken to a reputable mechanic.*

47   Fluid leakage

1   Automatic transmission fluid is a deep red color, and fluid leaks should not be confused with engine oil which can easily be blown by air flow to the transmission.
2   To pinpoint a leak, first remove all built-up dirt and grime from the transmission. Degreasing agents and/or steam cleaning will achieve this. With the underside clean, drive the vehicle at low speeds so the air flow will not blow the leak far from its source. Raise the vehicle and determine where the leak is located. Common areas of leakage are:

a) *Fluid pan: tighten mounting bolts and/or replace pan gasket as necessary (Chapter 1).*
b) *Rear extension: tighten bolts and/or replace oil seal as necessary.*
c) *Filler pipe: replace the rubber oil seal where pipe enters transmission case.*
d) *Transmission oil lines: tighten fittings where lines enter transmission case and/or replace lines.*
e) *Vent pipe: transmission overfilled and/or water in fluid (see checking procedures, Chapter 1).*

f) *Speedometer connector: replace the O-ring where speedometer cable enters transmission case.*

48   General shift mechanism problems

Chapter 7 deals with checking and adjusting the shift linkage on automatic transmissions. Common problems which may be caused by out of adjustment linkage are:

a) *Engine starting in gears other than P (park) or N (Neutral).*
b) *Indicator pointing to a gear other than the one actually engaged.*
c) *Vehicle moves with transmission in P (Park) position.*

49   Transmission will not downshift with the accelerator pedal pressed to the floor

Chapter 7 deals with adjusting the TV linkage to enable the transmission to downshift properly.

50   Engine will start in gears other than Park or Neutral

Chapter 7 deals with adjusting the Neutral start switch installed on automatic transmissions.

51   Transmission slips, shifts rough, is noisy or has no drive in forward or Reverse gears

1   There are many probable causes for the above problems, but the home mechanic should concern himself only with one possibility; fluid level.
2   Before taking the vehicle to a shop, check the fluid level and condition as described in Chapter 1. Add fluid, if necessary, or change the fluid and filter if needed. If problems persist, have a professional diagnose the transmission.

Driveshaft

Note: *Refer to Chapter 8, unless otherwise specified, for service information.*

52   Leaks at front of driveshaft

Defective transmission rear seal. See Chapter 7 for replacement procedure. As this is done, check the splined yoke for burrs or roughness that could damage the new seal. Remove burrs with a fine file or whetstone.

53   Knock or clunk when transmission is under initial load (just after transmission is put into gear)

1   Loose or disconnected rear suspension components. Check all mounting bolts and bushings (Chapters 7 and 10).
2   Loose driveshaft bolts. Inspect all bolts and nuts and tighten them securely.
3   Worn or damaged universal joint bearings. Inspect the universal joints (Chapter 8).
4   Worn sleeve yoke and mainshaft spline.

54   Metallic grating sound consistent with vehicle speed

Pronounced wear in the universal joint bearings. Replace U-joints or driveshafts, as necessary.

55   Vibration

Note: *Before blaming the driveshaft, make sure the tires are perfectly balanced and perform the following test.*
1   Install a tachometer inside the vehicle to monitor engine speed as the vehicle is driven. Drive the vehicle and note the engine speed at which the vibration (roughness) is most pronounced. Now shift the transmission to a different gear and bring the engine speed to the same point.
2   If the vibration occurs at the same engine speed (rpm) regardless of which gear the transmission is in, the driveshaft is NOT at fault since the driveshaft speed varies.
3   If the vibration decreases or is eliminated when the transmission is in a different gear at the same engine speed, refer to the following probable causes.
4   Bent or dented driveshaft. Inspect and replace as necessary.
5   Undercoating or built-up dirt, etc. on the driveshaft. Clean the shaft thoroughly.
6   Worn universal joint bearings. Replace the U-joints or driveshaft as necessary.
7   Driveshaft and/or companion flange out of balance. Check for missing weights on the shaft. Remove driveshaft and reinstall 180-degrees from original position, then recheck. Have the driveshaft balanced if problem persists.
8   Loose driveshaft mounting bolts/nuts.
9   Defective center bearing, if so equipped.
10   Worn transmission rear bushing (Chapter 7).

56   Scraping noise

Make sure the dust cover on the sleeve yoke isn't rubbing on the transmission extension housing.

57 Whining or whistling noise

Defective center bearing, if so equipped.

Rear axle and differential

Note: *For differential servicing information, refer to Chapter 8, unless otherwise specified.*

58 Noise - same when in drive as when vehicle is coasting

1   Road noise. No corrective action available.
2   Tire noise. Inspect tires and check tire pressures (Chapter 1).
3   Front wheel bearings loose, worn or damaged (Chapter 1).
4   Insufficient differential oil (Chapter 1).
5   Defective differential.

59 Knocking sound when starting or shifting gears

Defective or incorrectly adjusted differential.

60 Noise when turning

Defective differential.

61 Vibration

See probable causes under Driveshaft. Proceed under the guidelines listed for the driveshaft. If the problem persists, check the rear wheel bearings by raising the rear of the vehicle and spinning the wheels by hand. Listen for evidence of rough (noisy) bearings. Remove and inspect (Chapter 8).

62 Oil leaks

1   Pinion oil seal damaged (Chapter 8).
2   Axleshaft oil seals damaged (Chapter 8).
3   Differential cover leaking. Tighten mounting bolts or replace the gasket as required.
4   Loose filler or drain plug on differential (Chapter 1).
5   Clogged or damaged breather on differential.

Transfer case (4WD models)

Note: *Unless otherwise specified, refer to Chapter 7C for service and repair information.*

63 Gear jumping out of mesh

1   Incorrect control lever freeplay

2   Interference between the control lever and the console.
3   Play or fatigue in the transfer case mounts.
4   Internal wear or incorrect adjustments.

64 Difficult shifting

1   Lack of oil.
2   Internal wear, damage or incorrect adjustment.

65 Noise

1   Lack of oil in transfer case.
2   Noise in 4H and 4L, but not in 2H indicates cause is in the front differential or front axle.
3   Noise in 2H, 4H and 4L indicates cause is in rear differential or rear axle.
4   Noise in 2H and 4H but not in 4L, or in 4L only, indicates internal wear or damage in transfer case.

Brakes

Note: *Before assuming a brake problem exists, make sure the tires are in good condition and inflated properly, the front end alignment is correct and the vehicle is not loaded with weight in an unequal manner. All service procedures for the brakes are included in Chapter 9, unless otherwise noted.*

66 Vehicle pulls to one side during braking

1   Defective, damaged or oil contaminated brake pad on one side. Inspect as described in Chapter 1. Refer to Chapter 10 if replacement is required.
2   Excessive wear of brake pad material or disc on one side. Inspect and repair as necessary.
3   Loose or disconnected front suspension components. Inspect and tighten all bolts securely (Chapters 1 and 11).
4   Defective caliper assembly. Remove caliper and inspect for stuck piston or damage.
5   Brake pad-to-rotor adjustment needed. Inspect automatic adjusting mechanism for proper operation.
6   Scored or out of round rotor.
7   Loose caliper mounting bolts.
8   Incorrect wheel bearing adjustment.

67 Noise (high-pitched squeal)

1   Front brake pads worn out. This noise comes from the wear sensor rubbing against the disc. Replace pads with new ones immediately!
2   Glazed or contaminated pads.
3   Dirty or scored rotor.
4   Bent support plate.

68 Excessive brake pedal travel

1   Partial brake system failure. Inspect entire system (Chapter 1) and correct as required.
2   Insufficient fluid in master cylinder. Check (Chapter 1) and add fluid - bleed system if necessary.
3   Air in system. Bleed system.
4   Excessive lateral rotor play.
5   Brakes out of adjustment. Check the operation of the automatic adjusters.
6   Defective proportioning valve. Replace valve and bleed system.

69 Brake pedal feels spongy when depressed

1   Air in brake lines. Bleed the brake system.
2   Deteriorated rubber brake hoses. Inspect all system hoses and lines. Replace parts as necessary.
3   Master cylinder mounting nuts loose. Inspect master cylinder bolts (nuts) and tighten them securely.
4   Master cylinder faulty.
5   Incorrect shoe or pad clearance.
6   Defective check valve. Replace valve and bleed system.
7   Clogged reservoir cap vent hole.
8   Deformed rubber brake lines.
9   Soft or swollen caliper seals.
10   Poor quality brake fluid. Bleed entire system and fill with new approved fluid.

70 Excessive effort required to stop vehicle

1   Power brake booster not operating properly.
2   Excessively worn linings or pads. Check and replace if necessary.
3   One or more caliper pistons seized or sticking. Inspect and rebuild as required.
4   Brake pads or linings contaminated with oil or grease. Inspect and replace as required.
5   New pads or linings installed and not yet seated. It'll take a while for the new material to seat against the rotor or drum.
6   Worn or damaged master cylinder or caliper assemblies. Check particularly for frozen pistons.
7   Also see causes listed under Section 69.

71 Pedal travels to the floor with little resistance

Little or no fluid in the master cylinder reservoir caused by leaking caliper piston(s) or loose, damaged or disconnected brake lines. Inspect entire system and repair as necessary.

72 Brake pedal pulsates during brake application

1   Wheel bearings damaged, worn or out of adjustment (Chapter 1).
2   Caliper not sliding properly due to improper installation or obstructions. Remove and inspect.
3   Rotor not within specifications. Remove the rotor and check for excessive lateral runout and parallelism. Have the rotors resurfaced or replace them with new ones. Also make sure that all rotors are the same thickness.
4   Out of round rear brake drums. Remove the drums and have them turned or replace them with new ones.

73 Brakes drag (indicated by sluggish engine performance or wheels being very hot after driving)

1   Output rod adjustment incorrect at the brake pedal.
2   Obstructed master cylinder compensator. Disassemble master cylinder and clean.
3   Master cylinder piston seized in bore. Overhaul master cylinder.
4   Caliper assembly in need of overhaul.
5   Brake pads or shoes worn out.
6   Piston cups in master cylinder or caliper assembly deformed. Overhaul master cylinder.
7   Rotor not within specifications (Section 72).
8   Parking brake assembly will not release.
9   Clogged brake lines.
10 Wheel bearings out of adjustment (Chapter 1).
11 Brake pedal height improperly adjusted.
12 Wheel cylinder needs overhaul.
13 Improper shoe to drum clearance. Adjust as necessary.

74 Rear brakes lock up under light brake application

1   Tire pressures too high.
2   Tires excessively worn (Chapter 1).

75 Rear brakes lock up under heavy brake application

1   Tire pressures too high.
2   Tires excessively worn (Chapter 1).
3   Front brake pads contaminated with oil, mud or water. Clean or replace the pads.
4   Front brake pads excessively worn.
5   Defective master cylinder or caliper assembly.

Suspension and steering

Note: *All service procedures for the suspension and steering systems are included in Chapter 10, unless otherwise noted.*

76 Vehicle pulls to one side

1   Tire pressures uneven (Chapter 1).
2   Defective tire (Chapter 1).
3   Excessive wear in suspension or steering components (Chapter 1).
4   Wheel alignment incorrect.
5   Front brakes dragging. Inspect as described in Section 73.
6   Wheel bearings improperly adjusted (Chapter 1 or 8).
7   Wheel lug nuts loose.

77 Shimmy, shake or vibration

1   Tire or wheel out of balance or out of round. Have them balanced on the vehicle.
2   Loose, worn or out of adjustment wheel bearings (Chapter 1 or 8).
3   Shock absorbers and/or suspension components worn or damaged. Check for worn bushings in the upper and lower links.
4   Wheel lug nuts loose.
5   Incorrect tire pressures.
6   Excessively worn or damaged tire.
7   Loosely mounted steering gear housing.
8   Steering gear improperly adjusted.
9   Loose, worn or damaged steering components.
10 Damaged idler arm.
11 Worn balljoint.

78 Excessive pitching and/or rolling around corners or during braking

1   Defective shock absorbers. Replace as a set.
2   Broken or weak leaf springs and/or suspension components.
3   Worn or damaged stabilizer bar or bushings.

79 Wandering or general instability

1   Improper tire pressures.
2   Worn or damaged upper and lower link or tension rod bushings.
3   Incorrect front end alignment.
4   Worn or damaged steering linkage or suspension components.
5   Improperly adjusted steering gear.
6   Out of balance wheels.
7   Loose wheel lug nuts.
8   Worn rear shock absorbers.
9   Fatigued or damaged rear leaf springs.

80 Excessively stiff steering

1   Lack of lubricant in power steering fluid reservoir, where appropriate (Chapter 1).
2   Incorrect tire pressures (Chapter 1).
3   Lack of lubrication at balljoints (Chapter 1).
4   Front end out of alignment.
5   Steering gear out of adjustment or lacking lubrication.
6   Improperly adjusted wheel bearings.
7   Worn or damaged steering gear.
8   Interference of steering column with turn signal switch.
9   Low tire pressures.
10 Worn or damaged balljoints.
11 Worn or damaged steering linkage.
12 See also Section 79.

81 Excessive play in steering

1   Loose wheel bearings (Chapter 1 or 8).
2   Excessive wear in suspension bushings (Chapter 1).
3   Steering gear improperly adjusted.
4   Incorrect wheel alignment.
5   Steering gear mounting bolts loose.
6   Worn steering linkage.

82 Lack of power assistance

1   Steering pump drivebelt faulty or not adjusted properly (Chapter 1).
2   Fluid level low (Chapter 1).
3   Hoses or pipes restricting the flow. Inspect and replace parts as necessary.
4   Air in power steering system. Bleed system.
5   Defective power steering pump.

83 Steering wheel fails to return to straight-ahead position

1   Incorrect front end alignment.
2   Tire pressures low.
3   Steering gears improperly engaged.
4   Steering column out of alignment.
5   Worn or damaged balljoint.
6   Worn or damaged steering linkage.
7   Improperly lubricated idler arm.
8   Insufficient oil in steering gear.
9   Lack of fluid in power steering pump.

84 Steering effort not the same in both directions (power system)

1   Leaks in steering gear.
2   Clogged fluid passage in steering gear.

85 Noisy power steering pump

1   Insufficient oil in pump.
2   Clogged hoses or oil filter in pump.
3   Loose pulley.

4   Improperly adjusted drivebelt (Chapter 1).
5   Defective pump.

86 Miscellaneous noises

1   Improper tire pressures.
2   Insufficiently lubricated balljoint or steering linkage.
3   Loose or worn steering gear, steering linkage or suspension components.
4   Defective shock absorber.
5   Defective wheel bearing.
6   Worn or damaged suspension bushings.
7   Damaged leaf spring.
8   Loose wheel lug nuts.
9   Worn or damaged rear axleshaft spline.
10   Worn or damaged rear shock absorber mounting bushing.
11   Incorrect rear axle end play.

12   See also causes of noises at the rear axle and driveshaft.

87 Excessive tire wear (not specific to one area)

1   Incorrect tire pressures.
2   Tires out of balance. Have them balanced on the vehicle.
3   Wheels damaged. Inspect and replace as necessary.
4   Suspension or steering components worn (Chapter 1).

88 Excessive tire wear on outside edge

1   Incorrect tire pressure.

2   Excessive speed in turns.
3   Front end alignment incorrect (excessive toe-in).

89 Excessive tire wear on inside edge

1   Incorrect tire pressure.
2   Front end alignment incorrect (toe-out).
3   Loose or damaged steering components (Chapter 1).

90 Tire tread worn in one place

1   Tires out of balance. Have them balanced on the vehicle.
2   Damaged or buckled wheel. Inspect and replace if necessary.
3   Defective tire.

Chapter 1
Tune-up and routine maintenance

Contents

Specifications

Recommended lubricants and fluids

Note: *Listed here are manufacturers recommendations at the time this manual was written. Manufacturers occasionally upgrade their fluid and lubricant specifications, so check with your local auto parts store for current recommendations.*

Engine oil	
Type	API "certified for gasoline engines"
Viscosity	See accompanying chart
Automatic transmission fluid	DEXRON II automatic transmission fluid
Manual transmission lubricant	
MUA and MSG models	Engine oil (see accompanying chart)
BW model	DEXRON II or III automatic transmission fluid
Transfer case lubricant	Engine oil (see accompanying chart)
Differential lubricant	SAE 80W-90 GL-5 gear lubricant

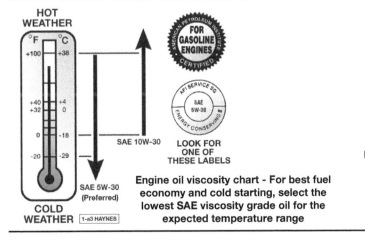

Engine oil viscosity chart - For best fuel economy and cold starting, select the lowest SAE viscosity grade oil for the expected temperature range

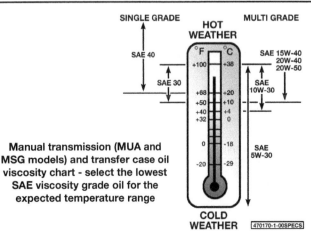

Manual transmission (MUA and MSG models) and transfer case oil viscosity chart - select the lowest SAE viscosity grade oil for the expected temperature range

Recommended lubricants and fluids (continued)

Power steering fluid type... DEXRON II or IIE automatic transmission fluid
Brake fluid type.. DOT 3 brake fluid
Clutch fluid type... DOT 3 brake fluid
Coolant type .... 50/50 mixture of ethylene glycol-based antifreeze and water
Chassis grease .... NLGI no. 2 chassis grease

Capacities*

Engine oil (with new oil filter)
   2.2L engine (1998 and later)... 4.8 qts
   2.3L engine
      1989 through 1991.. 4.2 qts
      1992 through 1993.. 3.7 qts
   2.6L engine
      1989 through 1991.. 5.2 qts
      1992 through 1997.. 4.4 qts
   3.1L V6 engine
      1991 through 1992.. 4.5 qts
   3.2L V6 engine
      1993 through 1997.. 5.7 qts
      1998 and later .... 5.0 qts
Cooling system
   Four-cylinder engines.. 9.5 qts
   3.1L V6 engine.. 11.4 qts
   3.2L V6 engine
      1993 through 1997
         Automatic transmission .... 9.3 qts
         Manual transmission .... 9.7 qts
      1998
         Automatic transmission .... 11.1 qts
         Manual transmission .... 11.2 qts
      1999 and later
         Automatic transmission .... 11.7 qts
         Manual transmission .... 11.6 qts
Automatic transmission
   AW03-72L .... 6.9 qts
   4L30-E .... 9.1 qts
Manual transmission
   1989 through 1997
      BW transmission .... 2.4 qts
      MSG transmission .... 1.6 qts
      MUA transmission.. 3.1 qts
   1998 and 1999
      2WD models .... 2.25 qts
      4WD models .... 3.1 qts
   2000 and later (all models) .... 3.1 qts
Transfer case .... 1.5 qts
Front differential
   1989 through 1997 .... 1.6 qts
   1998 and later .... 1.33 qts
Rear differential
   1989 through 1997
      Banjo model .... 1.6 qts
      Saginaw model .... 2.0 qts
      Dana model.. 1.9 qts
   1998 and later .... 1.87 qts

All capacities approximate. Add as necessary to bring up to appropriate level.

Ignition system

Spark plug type and gap*

1989 through 1997
   Four-cylinder engines
      Isuzu models... Champion RN12YC or equivalent @ 0.044 inch
      Honda Passport .... Champion RN11YC4 or equivalent @ 0.044 inch
   Six-cylinder engines
      1991 and 1992 .... Champion RV15YC4 or equivalent @ 0.044 inch
      1993 through 1995.. Champion RC12YC or equivalent @ 0.044 inch
      1996 and 1997 .... Champion RC10PYP4 or equivalent @ 0.044 inch
1998 and later (all models).. RC10PYP4, PK16PR11 or K16PR-P11 @ 0.040 to 0.043 inch

Ignition timing and idle speed (1989 through 1997)*

2.3L engine (1989 through 1993)	6-degrees BTDC @ 900 rpm**
2.6L engine	
1989 through 1995..	12-degrees BTDC @ 900 rpm***
1996 on	Non adjustable
3.1L engine (1991 and 1992)..	10-degrees BTDC @ 800 rpm****
3.2L engine	Non adjustable

Idle speed (1998 and later) (not adjustable)

2.2L four-cylinder engine	750 rpm
3.2L V6 engine	800 rpm

Note: *All timing adjustments must be made with the canister purge line and the EGR vacuum line disconnected and plugged.*
Use the information printed on the Vehicle Emissions Control Information label, if different than the Specifications listed here.

**Pinch off hot idle air compensator hose*

***Disconnect the vacuum switching valve (VSV)*

****Disconnect the set timing connector at the distributor*

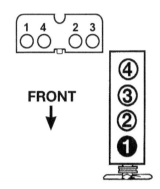

Cylinder location and coil pack terminal identification (2.2L four-cylinder engines)

Firing order and distributor rotation

Four-cylinder engines	
Firing order..	1-3-4-2
Distributor rotation	Counterclockwise
Six-cylinder engines	
Firing order..	1-2-3-4-5-6
Distributor rotation (3.1L V6 only)	Clockwise

Valve clearances (engine cold)

2.2L four-cylinder engine..	No adjustment necessary
2.3L four-cylinder engine	
Intake	0.006 inch
Exhaust	0.010 inch
2.6L four-cylinder engine (1989 through 1992)	
Intake	0.008 inch
Exhaust	0.008 inch
3.2L V6 engine (1998 and later)*	
Intake	0.009 to 0.013 inch
Exhaust	0.010 to 0.014 inch

The valves on pre-1998 3.2L V6 engines have hydraulic valve lash adjusters.

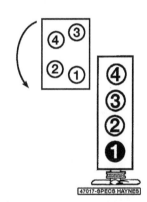

Cylinder location and distributor rotation (2.3L and 2.6L four-cylinder engines with distributor)

Clutch pedal

Amigo	
Cable-operated clutch	
Pedal height..	7-9/32 to 7-11/16 inches
Freeplay...	19/32 to 1 inch
Switch adjustment..	1/32 to 3/64 inch
Hydraulic clutch	
Pedal height..	6-47/64 to 7-1/8 inches
Freeplay...	3/16 to 19/32 inch
Switch adjustment..	1/32 to 3/64 inch

The blackened terminal shown on the distributor cap indicates the Number One spark plug wire position

Cylinder location and distributor rotation - 3.1L V6 engine with distributor

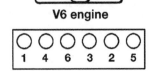

V6 engine

Cylinder location and coil pack terminal locations - 3.2L V6 engine with coil packs (1993 through 1995)

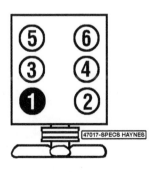

Cylinder locations - 3.2L V6 engine with coils on plugs (1996 and later)

Clutch pedal (continued)

Rodeo and Passport

1991 through 1995

Pedal height

Four-cylinder engine.. 6-47/64 to 7-1/8 inches

Six-cylinder engines .... 7-41/64 to 8-1/32 inches

Freeplay .... 13/64 to 39/64 inch

Switch adjustment .... 1/32 to 1/16 inch

1996 and later

Pedal height .... 7 to 7-13/32 inches

Freeplay .... 13/64 to 39/64 inch

Switch adjustment .... 1/32 to 1/16 inch

Brakes

Brake pedal (1989 through 1997)

Height .... 6-27/32 to 7-1/4 inches

Freeplay... 15/64 to 13/32 inch

Switch adjustment.. 1/32 to 3/64 inch

Brake pedal (1998 and later)

Height .... 6-13/16 to 7-9/32 inches

Freeplay... 15/64 to 25/64 inch

Switch adjustment.. 1/32 to 3/64 inch

Brake lining wear limit

Disc brake pads .... 1/8 inch

Drum brake shoes .... 1/16 inch

Torque specifications

Ft-lbs (unless otherwise indicated)

Automatic transmission pan bolts .... 96 in-lbs

Spark plugs.. 156 in-lbs

Engine oil drain plug 18

Manual transmission check/fill and drain plugs .... 18 to 25

Automatic transmission overfill screw (1996 and later models) 28

Differential check/fill and drain plugs .... 18 to 25

Transfer case check/fill and drain plugs... 14 to 29

Wheel lug nuts

Steel wheels .... 65 to 75

Aluminum wheels .... 80 to 90

Typical engine compartment layout (2.3L and 2.6L four-cylinder models)

1	Engine oil filler cap	5	Coolant reservoir	9	Battery
2	Oil dipstick (not visible)	6	Air cleaner assembly	10	Windshield washer fluid reservoir
3	Brake fluid reservoir	7	Radiator cap	11	Fuse/relay box
4	Clutch fluid reservoir	8	Radiator hose		

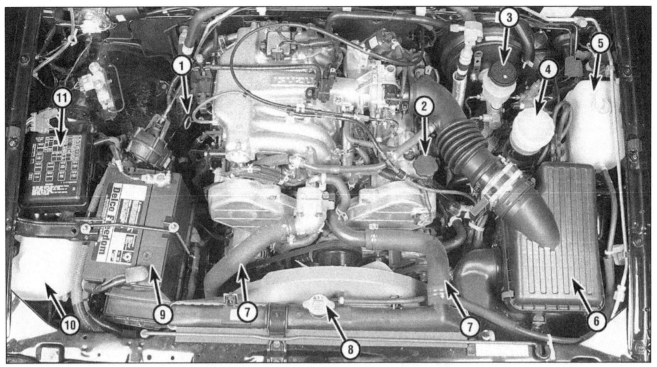

Typical engine compartment layout (1997 and earlier 3.2L V6 models)

1	Engine oil dipstick	5	Coolant reservoir	9	Battery
2	Engine oil filler cap	6	Air cleaner assembly	10	Windshield washer fluid reservoir
3	Brake fluid reservoir	7	Radiator hoses	11	Fuse/relay box
4	Power steering fluid reservoir	8	Radiator cap		

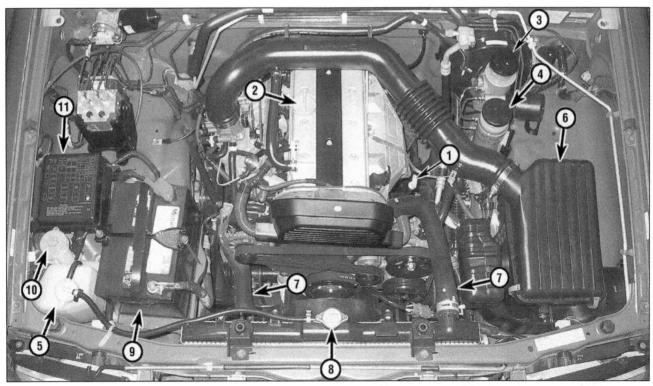

Typical engine compartment layout (2.2L four-cylinder models)

1	Engine oil dipstick	5	Coolant reservoir	9	Battery
2	Engine oil filler cap	6	Air cleaner assembly	10	Windshield washer fluid reservoir
3	Brake fluid reservoir	7	Radiator hoses	11	Fuse/relay box
4	Power steering fluid reservoir	8	Radiator cap		

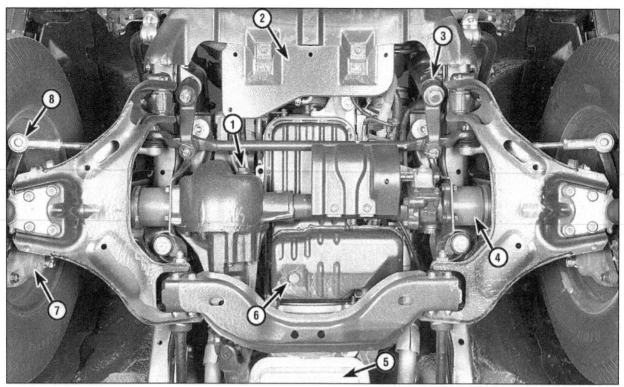

Typical engine compartment underside component locations (1989 through 1997 4WD models; 1998 and later models are equipped with rack-and-pinion steering gear)

1	Front differential check/fill plug	4	Front axle CV joint	7	Front brake caliper
2	Engine oil filter (not visible)	5	Transmission oil pan	8	Steering tie-rod end
3	Steering gear	6	Engine oil drain plug		

Typical engine compartment underside component locations (1989 through 1997 2WD models)

1	Engine oil filter	4	Exhaust pipe	7	Front brake caliper
2	Steering gear	5	Manual transmission drain plug	8	Shock absorber
3	Steering tie-rod	6	Engine oil drain plug		

Typical engine compartment underside component locations (1998 and later 2WD models)

1	Engine oil filter	4	Steering gear tie-rod end	7	Shock absorber
2	Transfer gear	5	Engine oil drain plug	8	Exhaust pipe
3	Rack-and-pinion steering gear	6	Manual transmission drain plug		

Typical rear under side component locations (1989 through 1997 models)

1	Tail pipe	4	Fuel filter	7	Rear brake caliper
2	Rear differential check/fill plug	5	Driveshaft	8	Rear leaf spring
3	Shock absorber	6	Muffler		

Typical rear under side component locations (1998 and later models)

1	Tail pipe	3	Shock absorber	5	Muffler
2	Rear differential check/fill plug	4	Driveshaft	6	Coil spring

1 Maintenance schedule

The following maintenance intervals are based on the assumption that the vehicle owner will be doing the maintenance or service work, as opposed to having a dealer service department do the work. Although the time/mileage intervals are loosely based on factory recommendations, most have been shortened to ensure, for example, that such items as lubricants and fluids are checked/changed at intervals that promote maximum engine/driveline service life. Also, subject to the preference of the individual owner interested in keeping his or her vehicle in peak condition at all times, and with the vehicle's ultimate resale in mind, many of the maintenance procedures may be performed more often than recommended in the following schedule. We encourage such owner initiative.

When the vehicle is new it should be serviced initially by a factory authorized dealer service department to protect the factory warranty. In many cases the initial maintenance check is done at no cost to the owner (check with your dealer service department for more information).

Every 250 miles or weekly, whichever comes first

Check the engine oil level (Section 4)
Check the engine coolant level (Section 4)
Check the windshield washer fluid level (Section 4)
Check the brake and clutch fluid levels (Section 4)
Check the tires and tire pressures (Section 5)

Every 3000 miles or 3 months, whichever comes first

All items listed above plus:
Check the automatic transmission fluid level (Section 6)
Check the power steering fluid level (Section 7)
Change the engine oil and filter (Section 8)*

Every 7500 miles or 6 months, whichever comes first

All items listed above plus:

Check and service the battery (Section 9)
Check the cooling system (Section 10)
Inspect and replace if necessary the windshield wiper blades (Section 11)
Inspect and replace if necessary all underhood hoses (Section 12)
Check and lubricate the accelerator linkage (Section 13)
Rotate the tires (Section 14)
Inspect the suspension and steering components(Section 15)*
Lubricate the chassis components (Section 16)
Inspect the exhaust system (Section 17)*

Every 15,000 miles or 12 months, whichever comes first

All items listed above plus:

Adjust the valves (four-cylinder engines) (Section 18)
Check and adjust if necessary, the clutch pedal and brake pedal freeplay (Section 19)
Check the manual transmission lubricant (Section 20)*
Check the transfer case lubricant level (Section 21)*
Check the differential lubricant level (Section 22)*
Check the brakes (Section 23)*
Inspect the fuel system (Section 24)
Check the engine drivebelts (Section 25)

Every 30,000 miles or 24 months, whichever comes first

Check the operation of the thermostatic air cleaner system (Section 26)
Replace the air filter (Section 27)
Check the operation of the carburetor choke (Section 28)
Inspect the PCV system (Section 29)
Check the Exhaust Gas Recirculation (EGR) system (Section 30)
Replace the fuel filter (Section 31)
Replace the spark plugs (Section 32)
Inspect the spark plug wires, distributor cap and rotor (Section 33)
Check and lubricate the distributor advance mechanism (four-cylinder engines) (Section 34)

Check and adjust if necessary, the idle speed (four-cylinder models only) (Section 35)
Check and adjust, if necessary, the ignition timing (Section 36)
Check, repack and adjust the front wheel bearings (Section 37)
Change the power steering fluid (Section 38)
Service the cooling system (drain, flush and refill) (Section 39)
Change the automatic transmission fluid and filter (Section 40)**
Change the manual transmission lubricant (Section 41)**
Change the transfer case lubricant (4WD models) (Section 42)**
Change the differential lubricant (Section 43)**
Check the steering gear freeplay (Chapter 10)

Every 60,000 miles

Replace the timing belt (Chapter 2)
Adjust the valves (1998 and later V6 engine) (Section 44)

Every 90,000 miles

Reset the emissions indicator (Section 45)
Replace the oxygen sensor (Chapter 6)

* *This item is affected by "severe" operating conditions as described below. If your vehicle is operated under severe conditions, perform all maintenance indicated with an asterisk (*) at 3000 mile/3 month intervals. Severe conditions are indicated if you mainly operate your vehicle under one or more of the following:*

In dusty areas
Towing a trailer
Idling for extended periods and/or low speed operation
When outside temperatures remain below freezing and most trips are less than 4 miles

** *If operated under one or more of the following conditions, change the indicated fluids every 15,000 miles or 12 months:*

In heavy city traffic where the outside temperature regularly reaches 90-degrees F (32-degrees C) or higher
In hilly or mountainous terrain
Frequent trailer pulling

2 Introduction

This Chapter is designed to help the home mechanic maintain the Isuzu Rodeo, Amigo and Honda Passport with the goals of maximum performance, economy, safety and reliability in mind.

Included is a master maintenance schedule followed by procedures dealing specifically with every item on the schedule.

Visual checks, adjustments, component replacement and other helpful items are included. Refer to the **accompanying illustrations** of the engine compartment and the underside of the vehicle for the locations of various components.

Servicing your vehicle in accordance with the planned mileage/time maintenance schedule and the step-by-step procedures should result in maximum reliability and extend the life of your vehicle. Keep in mind

that it's a comprehensive plan - maintaining some items but not others at the specified intervals will not produce the same results.

As you perform routine maintenance procedures, you'll find that many can, and should, be grouped together because of the nature of the procedures or because of the proximity of two otherwise unrelated components or systems.

For example, if the vehicle is raised for chassis lubrication, you should inspect the

exhaust, suspension, steering and fuel systems while you're under the vehicle. When you're rotating the tires, it makes good sense to check the brakes since the wheels are already removed. Finally, let's suppose you have to borrow or rent a torque wrench. Even if you only need it to tighten the spark plugs, you might as well check the torque of as many critical fasteners as time allows.

The first step in this maintenance program is to prepare yourself before the actual work begins. Read through all the procedures you're planning to do, then gather up all the parts and tools needed. If it looks like you might run into problems during a particular job, seek advice from a mechanic or experienced do-it-yourselfer.

3   Tune-up general information

The term tune-up is used in this manual to represent a combination of individual operations rather than one specific procedure.

If, from the time the vehicle is new, the routine maintenance schedule is followed closely and frequent checks are made of fluid levels and high wear items, as suggested throughout this manual, the engine will be kept in relatively good running condition and the need for additional work will be minimized.

More likely than not, however, there will be times when the engine is running poorly due to lack of regular maintenance. This is even more likely if a used vehicle, which has not received regular and frequent maintenance checks, is purchased. In such cases, an engine tune-up will be needed outside of the regular routine maintenance intervals.

The first step in any tune-up or diagnostic procedure to help correct a poor running engine is a cylinder compression check. A compression check (see Chapter 2 Part E) will help determine the condition of internal engine components and should be used as a guide for tune-up and repair procedures. If, for instance, a compression check indicates serious internal engine wear, a conventional tune-up will not improve the performance of the engine and would be a waste of time and money. Because of its importance, the compression check should be done by someone with the right equipment and the knowledge to use it properly.

The following procedures are those most often needed to bring a generally poor running engine back into a proper state of tune.

Minor tune-up

Check all engine related fluids
Clean and check the battery (Section 9)
Check and service the cooling system (Section 10)
Adjust the valve clearances (Section 18)
Check the fuel system (Section 24 and Chapter 4)
Check and adjust the drivebelts (Section 25)

4.2a The oil dipstick (arrow) on four-cylinder engines is located on the drivers side of the engine compartment

Major tune-up

All items listed under Minor tune-up plus . . .
Check the air cleaner thermostatic control valve (Section 26)
Replace the air filter (Section 27)
Check the carburetor choke (Section 28)
Check the PCV system (Section 29)
Check the EGR system (Section 30 and Chapter 6)
Replace the fuel filter (Section 31)
Replace the spark plugs (Section 32)
Replace the spark plug wires, distributor cap and rotor (Section 33)
Lubricate the distributor advance mechanism (four-cylinder models only) (Section 34)
Check and adjust the idle speed (Section 35)
Check and adjust the ignition timing (Section 36)
Check the charging system (Chapter 5)
Check the ignition system (Chapter 5)

4   Fluid level checks (every 250 miles or weekly)

Note: *The following are fluid level checks to be done on a 250 mile or weekly basis. Additional fluid level checks can be found in specific maintenance procedures which follow. Regardless of how often the fluid levels are checked, watch for puddles under the vehicle - if leaks are noted, make repairs immediately.*
1   Fluids are an essential part of the lubri-

4.2b The oil dipstick (arrow) on the 3.2L V6 engine is located on the passenger's side of the engine compartment (on the 3.1L V6 engine it's on the driver's side)

cation, cooling, brake, clutch and windshield washer systems. Because the fluids gradually become depleted and/or contaminated during normal operation of the vehicle, they must be periodically replenished. See Recommended lubricants and fluids at the beginning of this Chapter before adding fluid to any of the following components. **Note:** *The vehicle must be on level ground when fluid levels are checked.*

Engine oil

Refer to illustrations 4.2a, 4.2b, 4.4 and 4.6
2   The engine oil level is checked with a dipstick that extends through a tube and into the oil pan at the bottom of the engine **(see illustrations)**.
3   The oil level should be checked before the vehicle has been driven, or about 15 minutes after the engine has been shut off. If the oil is checked immediately after driving the vehicle, some of the oil will remain in the upper engine components, resulting in an inaccurate reading on the dipstick.
4   Pull the dipstick from the tube and wipe all the oil from the end with a clean rag or paper towel. Insert the clean dipstick all the way back into the tube, then pull it out again. Note the oil at the end of the dipstick. Add oil as necessary to keep the level between the L and F marks or within the hatched area on the dipstick **(see illustration)**.
5   Don't overfill the engine by adding too much oil since this may result in oil-fouled spark plugs, oil leaks or oil seal failures.

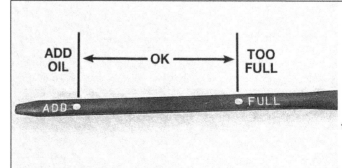

4.4 The oil level must be maintained in the OK range at all times - it takes one quart of oil to raise the level from the ADD mark to the FULL mark

4.6 Unscrew the oil filler cap (arrow) to remove it - make sure the area around the opening is clean before unscrewing the cap; this will prevent dirt from entering the engine

4.8a On 1989 through 1997 models, the coolant reservoir (arrow) is located on the left side of the engine compartment; maintain the coolant level between the upper and lower marks

4.8b On 1998 and later models, the coolant reservoir is located in the right front corner of the engine compartment; maintain the coolant level between the upper and lower marks

6   Oil is added to the engine after removing a threaded cap from the rocker arm cover (see illustration). A funnel will help to reduce spills.

7   Checking the oil level is an important preventive maintenance step. A consistently low oil level indicates oil leakage through damaged seals, defective gaskets or past worn rings or valve guides. If the oil looks milky in color or has water droplets in it, the cylinder head gasket(s) may be blown or the head(s) or block may be cracked. The engine should be checked immediately. The condition of the oil should also be checked. Whenever you check the oil level, slide your thumb and index finger up the dipstick before wiping off the oil. If you see small dirt or metal particles clinging to the dipstick, the oil should be changed (Section 8).

Engine coolant

Refer to illustrations 4.8a and 4.8b
Warning: *Don't allow antifreeze to come in contact with your skin or painted surfaces of the vehicle. Flush contaminated areas immediately with plenty of water. Don't store new coolant or leave old coolant lying around where it's accessible to children or pets - they're attracted by its sweet taste. Ingestion of even a small amount of coolant can be fatal! Wipe up garage floor and drip pan coolant spills immediately. Keep antifreeze containers covered and repair leaks in your cooling system immediately.*

8   All vehicles covered by this manual are equipped with a pressurized coolant recovery system. A white plastic coolant reservoir located in the engine compartment is connected by a hose to the radiator filler neck (see illustration). If the engine overheats, coolant escapes through a valve in the radiator cap and travels through the hose into the reservoir. As the engine cools, the coolant is automatically drawn back into the cooling system to maintain the correct level.

9   The coolant level in the reservoir should be checked regularly. **Warning:** *Do not remove the radiator cap to check the coolant*

level when the engine is warm. The level in the reservoir varies with the temperature of the engine. When the engine is cold, the coolant level should be at or slightly above the lower mark on the reservoir. Once the engine has warmed up, the level should be at or near the upper mark. If it isn't, allow the engine to cool, then remove the cap from the reservoir and add a 50/50 mixture of ethylene glycol-based antifreeze and water.

10   Drive the vehicle and recheck the coolant level. If only a small amount of coolant is required to bring the system up to the proper level, water can be used. However, repeated additions of water will dilute the antifreeze and water solution. In order to maintain the proper ratio of antifreeze and water, always top up the coolant level with the correct mixture. An empty plastic milk jug or bleach bottle makes an excellent container for mixing coolant. Do not use rust inhibitors or additives.

11   If the coolant level drops consistently, there may be a leak in the system. Inspect the radiator, hoses, filler cap, drain plugs and water pump (see Section 10). If no leaks are

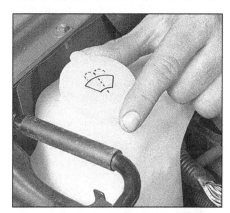

4.14a The reservoir for the windshield washer fluid is located on the right side of the engine compartment and fluid is added after pulling the top up and out - how often you use the washers will dictate how often you need to check the reservoir

noted, have the radiator cap pressure tested by a service station.

12   If you have to remove the radiator cap, wait until the engine has cooled, then wrap a thick cloth around the cap and turn it to the first stop. If coolant or steam escapes, let the engine cool down longer, then remove the cap.

13   Check the condition of the coolant as well. It should be relatively clear. If it's brown or rust colored, the system should be drained, flushed and refilled. Even if the coolant appears to be normal, the corrosion inhibitors wear out, so it must be replaced at the specified intervals.

Windshield and rear washer fluid

Refer to illustrations 4.14a and 4.14b
14   Fluid for the windshield washer system is stored in a plastic reservoir located on the passenger's side of the engine compartment (see illustration). If necessary, refer to the underhood component illustration(s) at the beginning of this Chapter to locate the reservoir. On vehicles with rear window washers, the fluid reservoir is located under a cover on the left side of the cargo area (see illustration).

4.14b The rear washer fluid reservoir (arrow) is located in the rear cargo area, next to the tire jack

15   In milder climates, plain water can be used in the reservoir, but it should be kept no more than 2/3 full to allow for expansion if the water freezes. In colder climates, use windshield washer system antifreeze, available at any auto parts store, to lower the freezing point of the fluid. Mix the antifreeze with water in accordance with the manufacturer's directions on the container. **Caution:** *Don't use cooling system antifreeze - it will damage the vehicle's paint.*

16   To help prevent icing in cold weather, warm the windshield with the defroster before using the washer.

Battery electrolyte

17   Most vehicles covered by this manual are equipped with a battery which is permanently sealed (except for vent holes) and has no filler caps. Water does not have to be added to these batteries at any time; however, if a maintenance-Type battery has been installed on the vehicle since it was new, remove all the cell caps (on top of the battery (usually there are two caps that cover three cells each). If the electrolyte level is low, add distilled water until the level is above the plates. There is usually a split-ring indicator in each cell to help you judge when enough water has been added. Add water until the electrolyte level is just up to the bottom of the split ring indicator. Do not overfill the battery or it will spew out electrolyte when it is charging.

Brake and clutch fluid

Refer to illustrations 4.19a and 4.19b

18   The brake master cylinder is mounted on the front of the power booster unit in the engine compartment. The clutch cylinder used on manual transmissions is mounted adjacent to it on the firewall.

19   The fluid inside is readily visible. The level should be between the upper and lower marks on the reservoirs **(see illustrations)**. If a low level is indicated, be sure to wipe the top of the reservoir cover with a clean rag to prevent contamination of the brake and/or clutch system before removing the cover.

20   When adding fluid, pour it carefully into the reservoir to avoid spilling it onto surrounding painted surfaces. Be sure the specified fluid is used, since mixing different types of brake fluid can cause damage to the system. See *Recommended lubricants and fluids* at the front of this Chapter or your owner's manual. **Warning:** *Brake fluid can harm your eyes and damage painted surfaces, so be very careful when handling or pouring it. Don't use brake fluid that's been standing open or is more than one year old. Brake fluid absorbs moisture from the air. Excess moisture can cause a dangerous loss of brake efficiency.*

21   At this time the fluid and master cylinder can be inspected for contamination. The system should be drained and refilled if deposits, dirt particles or water droplets are seen in the fluid.

4.19a Keep the brake fluid level near the MAX mark on the reservoir - fluid can be added after unscrewing the cap

22   After filling the reservoir to the proper level, make sure the cover is on tight to prevent fluid leakage.

23   The brake fluid level in the master cylinder will drop slightly as the pads and the brake shoes at each wheel wear down during normal operation. If the master cylinder requires repeated additions to keep it at the proper level, it's an indication of leakage in the brake system, which should be corrected immediately. Check all brake lines and connections (see Section 23 for more information).

24   If, upon checking the master cylinder fluid level, you discover one or both reservoirs empty or nearly empty, the brake system should be bled (Chapter 9).

5   Tire and tire pressure checks (every 250 miles or weekly)

Refer to illustrations 5.2, 5.3, 5.4a, 5.4b and 5.8

1   Periodic inspection of the tires may spare you the inconvenience of being stranded with a flat tire. It can also provide you with vital information regarding possible problems in the steering and suspension systems before major damage occurs.

2   The original tires on this vehicle are equipped with wear indicator bars that will appear when tread depth reaches a predetermined limit, usually 1/16-inch. Tread wear can be monitored with a simple, inexpensive device known as a tread depth indicator **(see illustration)**.

3   Note any abnormal tread wear **(see illustration)**. Tread pattern irregularities such as cupping, flat spots and more wear on one side than the other are indications of front end alignment and/or balance problems. If any of these conditions are noted, take the vehicle to a tire shop or service station to correct the problem.

4   Look closely for cuts, punctures and embedded nails or tacks. Sometimes a tire will hold air pressure for a short time or leak

4.19b The clutch fluid level must be kept between the upper and lower marks on the reservoir - unscrew the cap to add fluid

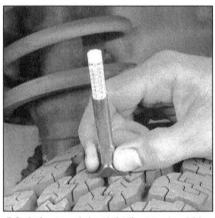

5.2 A tire tread depth indicator should be used to monitor tire wear - they are available at auto parts stores and service stations and cost very little

down very slowly after a nail has embedded itself in the tread. If a slow leak persists, check the valve stem core to make sure it's tight **(see illustration)**. Examine the tread for an object that may have embedded itself in the tire or for a "plug" that may have begun to leak (radial tire punctures are repaired with a plug that's installed in a puncture). If a puncture is suspected, it can be easily verified by spraying a solution of soapy water onto the puncture area **(see illustration)**. The soapy solution will bubble if there's a leak. Unless the puncture is unusually large, a tire shop or service station can usually repair the tire.

5   Carefully inspect the inner sidewall of each tire for evidence of brake fluid leakage. If you see any, inspect the brakes immediately.

6   Correct air pressure adds miles to the lifespan of the tires, improves mileage and enhances overall ride quality. Tire pressure cannot be accurately estimated by looking at a tire, especially if it's a radial. A tire pressure gauge is essential. Keep an accurate gauge in the vehicle. The pressure gauges attached to the nozzles of air hoses at gas stations are often inaccurate.

UNDERINFLATION

CUPPING

Cupping may be caused by:

- Underinflation and/or mechanical irregularities such as out-of-balance condition of wheel and/or tire, and bent or damaged wheel.
- Loose or worn steering tie-rod or steering idler arm.
- Loose, damaged or worn front suspension parts.

OVERINFLATION

INCORRECT TOE-IN OR EXTREME CAMBER

FEATHERING DUE TO MISALIGNMENT

5.3 This chart will help you determine the condition of your tires, the probable cause(s) of abnormal wear and the corrective action necessary

5.4a If a tire loses air on a steady basis, check the valve core first to make sure it's snug (special inexpensive wrenches are commonly available at auto parts stores)

5.4b If the valve core is tight, raise the corner of the vehicle with the low tire and spray a soapy water solution onto the tread as the tire is turned slowly - slow leaks will cause small bubbles to appear

5.8 To extend the life of your tires, check the air pressure at least once a week with an accurate gauge (don't forget the spare!)

7   Always check tire pressure when the tires are cold. Cold, in this case, means the vehicle has not been driven over a mile in the three hours preceding a tire pressure check. A pressure rise of four to eight pounds is not uncommon once the tires are warm.

8   Unscrew the valve cap protruding from the wheel or hubcap and push the gauge firmly onto the valve stem **(see illustration)**. Note the reading on the gauge and compare

the figure to the recommended tire pressure shown on the placard on the glove compartment door. Be sure to reinstall the valve cap to keep dirt and moisture out of the valve stem mechanism. Check all four tires and, if necessary, add enough air to bring them up to the recommended pressure.

9   Don't forget to keep the spare tire inflated to the specified pressure (refer to your owner's manual or the tire sidewall).

6   Automatic transmission fluid level check (every 3000 miles or 3 months)

1   The automatic transmission fluid level should be carefully maintained. Low fluid level can lead to slipping or loss of drive, while overfilling can cause foaming and loss of fluid.

2   With the parking brake set, start the engine, then move the shift lever through all the gear ranges, ending in Park. The fluid level must be checked with the vehicle level and the engine running at idle. **Note:** *Incorrect fluid level readings will result if the vehicle has just been driven at high speeds for an extended period, in hot weather in city traffic, or if it has been pulling a trailer. If any of these conditions apply, wait until the fluid has cooled (about 30 minutes).*

1989 through 1995 models

Refer to illustration 6.6

3   With the transmission at normal operating temperature, remove the dipstick from the filler tube. The dipstick is located at the rear of the engine compartment on the passenger side.

4   Carefully touch the fluid at the end of the dipstick to determine if it's cool (86 to 122-degrees F) or hot (123 to 176-degrees F). Wipe the fluid from the dipstick with a clean rag and push it back into the filler tube until the cap seats.

5   Pull the dipstick out again and note the fluid level.

6   If the fluid felt cool, the level should be within the cold (C) range - between the lower holes **(see illustration)**. If the fluid was hot, the level should be within the hot (H) range.

7   If additional fluid is required, add it directly into the tube using a funnel. It takes about one pint to raise the level from the L mark to the H mark with a hot transmission, so add the fluid a little at a time and keep checking the level until it's correct.

8   The condition of the fluid should also be checked along with the level. If the fluid at the end of the dipstick is a dark reddish-brown color, or if it smells burned, it should be changed. If you are in doubt about the condition of the fluid, purchase some new fluid and compare the two for color and smell.

1996 and later

Refer to illustration 6.10

9   Checking the automatic transmission fluid on 1996 and later models requires the use of a special fluid temperature gauge which may be ordered at your dealer parts department. **Caution:** *Do not attempt to perform this procedure without the use of this tool or damage may occur to the transmission*.

10   Run the engine at idle speed for at least three minutes then open the overfill screw **(see illustration). Caution:** *Do not open the transmission overfill screw with the engine stopped.*

11   Place the fluid gauge in position then measure the fluid temperature until it reaches between 90-degrees F and 135-degrees F.

12   With the fluid at the above temperature, use a syringe or small pump to add fluid through the overfill screw opening until it just starts to run out of the opening.

13   Tighten the overfill screw to the torque listed in this Chapter's Specifications.

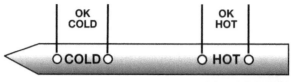

6.10 On 1996 and later models, add fluid through the overfill screw opening (arrow) with the transmission at the specified operating temperature

7   Power steering fluid level check (every 3000 miles or 3 months)

Refer to illustration 7.4

1   Unlike manual steering, the power steering system relies on fluid which may, over a period of time, require replenishing.

2   The fluid reservoir for the power steering pump is located on the pump body at the front of the engine.

3   For the check, the front wheels should be pointed straight ahead and the engine should be off. The engine and power steering fluid should be cold.

4   The fluid level in the translucent plastic reservoir can be checked without removing the cap **(see illustration)**.

5   If additional fluid is required, pour the specified type directly into the reservoir, using a funnel to prevent spills.

6   If the reservoir requires frequent fluid additions, all power steering hoses, hose connections and the power steering pump should be carefully checked for leaks.

8   Engine oil and filter change (every 3000 miles or 3 months)

Refer to illustrations 8.3, 8.9, 8.14 and 8.18

1   Frequent oil changes are the most important preventive maintenance procedures that can be done by the home mechanic. As engine oil ages, it becomes diluted and contaminated, which leads to premature engine wear.

6.6 On 1989 through 1995 models, the automatic transmission fluid level must be between the two upper holes in the dipstick with the fluid at normal operating temperature

7.4 Unscrew the power steering reservoir cap to add fluid - the level must be kept above the lower mark on the housing

2   Although some sources recommend oil filter changes every other oil change, the minimal cost of an oil filter and the fact that it's easy to install dictate that a new filter be used every time the oil is changed.

3   Gather all necessary tools and materials before beginning this procedure **(see illustration)**.

4   You should have plenty of clean rags and newspapers handy to mop up any spills. Access to the underside of the vehicle is greatly improved if the vehicle can be lifted on a hoist, driven onto ramps or supported by jackstands. **Warning:** *Do not work under a vehicle which is supported only by a bumper, hydraulic or scissors-type jack!*

5   If this is your first oil change, get under the vehicle and familiarize yourself with the locations of the oil drain plug and the oil filter. The engine and exhaust components will be warm during the actual work, so note how they're situated to avoid touching them when working under the vehicle.

6   Warm the engine to normal operating temperature. If the new oil or any tools are needed, use the warm-up time to obtain everything necessary for the job. The correct oil for your application can be found in *Recommended lubricants and fluids* at the beginning of this Chapter.

7   With the engine oil warm (warm engine oil will drain better and more built-up sludge will be removed with it), raise and support the vehicle. Make sure it's safely supported!

8   Move all necessary tools, rags and newspapers under the vehicle. Set the drain pan under the drain plug. Keep in mind that the oil will initially flow from the pan with

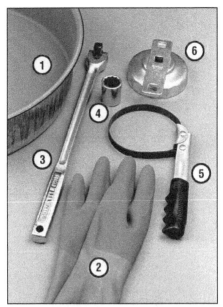

8.3 These tools are required when changing the engine oil and filter

1   **Drain pan** - *It should be fairly shallow in depth, but wide in order to prevent spills*
2   **Rubber gloves** - *When removing the drain plug and filter, it is inevitable that you will get oil on your hands (the gloves will prevent burns)*
3   **Breaker bar** - *Sometimes the oil drain plug is pretty tight and a long breaker bar is needed to loosen it*
4   **Socket** - *To be used with the breaker bar or a ratchet (must be the correct size to fit the drain plug)*
5   **Filter wrench** - *This is a metal band-type wrench, which requires clearance around the filter to be effective*
6   **Filter wrench** - *This type fits on the bottom of the filter and can be turned with a ratchet or breaker bar (different size wrenches are available for different types of filters)*

some force; position the pan accordingly.

9   Being careful not to touch any of the hot exhaust components, use a wrench to remove the drain plug near the bottom of the oil pan **(see illustration)**. Depending on how hot the oil is, you may want to wear gloves while unscrewing the plug the final few turns.

10   Allow the old oil to drain into the pan. It may be necessary to move the pan as the oil flow slows to a trickle.

11   After all the oil has drained, wipe off the drain plug with a clean rag. Small metal particles may cling to the plug and would immediately contaminate the new oil.

12   Clean the area around the drain plug opening and reinstall the plug. Tighten the plug securely with the wrench. If a torque wrench is available, use it to tighten the plug.

13   Move the drain pan into position under the oil filter.

14   Use the filter wrench to loosen the oil filter **(see illustration)**. Chain or metal band filter wrenches may distort the filter canister, but it doesn't matter since the filter will be discarded anyway.

15   Completely unscrew the old filter. Be careful; it's full of oil. Empty the oil inside the filter into the drain pan. Make sure the rubber gasket came off with the filter. If not, remove it from the oil filter adapter.

16   Compare the old filter with the new one to make sure they're the same type.

17   Use a clean rag to remove all oil, dirt and sludge from the area where the oil filter mounts to the engine. Check the old filter to make sure the rubber gasket isn't stuck to the engine. If the gasket is stuck to the engine, remove it.

18   Apply a light coat of clean oil to the rubber gasket on the new oil filter **(see illustration)**.

19   Attach the new filter to the engine, following the tightening directions printed on the filter canister or packing box. Most filter manufacturers recommend against using a filter wrench due to the possibility of over-tightening and damage to the seal.

20   Remove all tools, rags, etc. from under

the vehicle, being careful not to spill the oil in the drain pan, then lower the vehicle.

21   Move to the engine compartment and locate the oil filler cap.

22   Pour the fresh oil into the filler opening. A funnel may be used.

23   Pour three or four quarts of fresh oil into the engine. Wait a few minutes to allow the oil to drain into the pan, then check the level on the oil dipstick (see Section 4 if necessary). If the oil level is above the L mark, start the engine and allow the new oil to circulate.

24   Run the engine for only about a minute and then shut it off. Immediately look under the vehicle and check for leaks at the oil pan drain plug and around the oil filter. If either one is leaking, tighten it a little more.

25   With the new oil circulated and the filter now completely full, recheck the level on the dipstick and add more oil as necessary.

26   During the first few trips after an oil change, make it a point to check frequently for leaks and correct oil level.

27   The old oil drained from the engine cannot be reused in its present state and should be disposed of. Check with your local refuse disposal company, disposal facility or environmental agency to see if they will accept the oil for recycling. Don't pour used oil into drains or onto the ground. After the oil has cooled, it can be drained into a suitable container (capped plastic jugs, topped bottles, milk cartons, etc.) for transport to one of these disposal sites.

9   Battery check, maintenance and charging (every 7500 miles or 6 months)

Refer to illustrations 9.1, 9.6a, 9.6b, 9.7a and 9.7b

Warning: *Certain precautions must be followed when checking and servicing the battery. Hydrogen gas, which is highly flammable, is always present in the battery cells, so keep lighted tobacco and all other*

8.9 The engine oil drain plug is located at the rear of the oil pan (four-cylinder engine shown) - it's usually very tight, so use a box-end wrench or six-point socket to avoid rounding off the hex

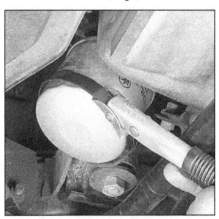

8.14 The oil filter is usually on very tight as well and will require a special wrench for removal - DO NOT use the wrench to tighten the new filter!

8.18 Lubricate the oil filter gasket with clean engine oil before installing the filter on the engine

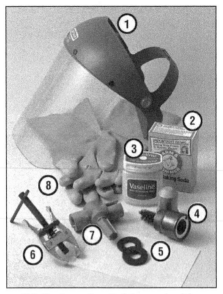

9.1 Tools and materials required for battery maintenance

1   *Face shield/safety goggles - When removing corrosion with a brush, the acidic particles can easily fly up into your eyes*

2   *Baking soda - A solution of baking soda and water can be used to neutralize corrosion*

3   *Petroleum jelly - A layer of this on the battery posts will help prevent corrosion*

4   *Battery post/cable cleaner - This wire brush cleaning tool will remove all traces of corrosion from the battery posts and cable clamps*

5   *Treated felt washers - Placing one of these on each post, directly under the cable clamps, will help prevent corrosion*

6   *Puller - Sometimes the cable clamps are very difficult to pull off the posts, even after the nut/bolt has been completely loosened. This tool pulls the clamp straight up and off the post without damage*

7   *Battery post/cable cleaner - Here is another cleaning tool which is a slightly different version of Number 4 above, but it does the same thing*

8   *Rubber gloves - Another safety item to consider when servicing the battery; remember that's acid inside the battery!*

open flames and sparks away from the battery. The electrolyte inside the battery is actually dilute sulfuric acid, which will cause injury if splashed on your skin or in your eyes. It will also ruin clothes and painted surfaces. When removing the battery cables, always detach the negative cable first and hook it up last!

Maintenance

1   A routine preventive maintenance program for the battery in your vehicle is the only way to ensure quick and reliable starts. But

9.6a Battery terminal corrosion usually appears as light, fluffy powder

9.7a When cleaning the cable clamps, all corrosion must be removed (the inside of the clamp is tapered to match the taper on the post, so don't remove too much material)

before performing any battery maintenance, make sure that you have the proper equipment necessary to work safely around the battery **(see illustration)**.

2   There are also several precautions that should be taken whenever battery maintenance is performed. Before servicing the battery, always turn the engine and all accessories off and disconnect the cable from the negative terminal of the battery.

3   The battery produces hydrogen gas, which is both flammable and explosive. Never create a spark, smoke or light a match around the battery. Always charge the battery in a ventilated area.

4   Electrolyte contains poisonous and corrosive sulfuric acid. Do not allow it to get in your eyes, on your skin or on your clothes. Never ingest it. Wear protective safety glasses when working near the battery. Keep children away from the battery.

5   Note the external condition of the battery. If the positive terminal and cable clamp on your vehicle's battery is equipped with a rubber protector, make sure that it's not torn or damaged. It should completely cover the terminal. Look for any corroded or loose connections, cracks in the case or cover or loose

9.6b Removing the cable from a battery post with a wrench - sometimes special battery pliers are required for this procedure if corrosion has caused deterioration of the nut hex (always remove the ground cable first and hook it up last!)

9.7b Regardless of the type of tool used on the battery posts, a clean, shiny surface should be the end result

hold-down clamps. Also check the entire length of each cable for cracks and frayed conductors.

6   If corrosion, which looks like white, fluffy deposits **(see illustration)** is evident, particularly around the terminals, the battery should be removed for cleaning. Loosen the cable clamp bolts with a wrench, being careful to remove the ground cable first, and slide them off the terminals **(see illustration)**. Then disconnect the hold-down clamp bolt and nut, remove the clamp and lift the battery from the engine compartment.

7   Clean the cable clamps thoroughly with a battery brush or a terminal cleaner and a solution of warm water and baking soda **(see illustration)**. Wash the terminals and the top of the battery case with the same solution but make sure that the solution doesn't get into the battery. When cleaning the cables, terminals and battery top, wear safety goggles and rubber gloves to prevent any solution from coming in contact with your eyes or hands. Wear old clothes too - even diluted, sulfuric acid splashed onto clothes will burn holes in

them. If the terminals have been extensively corroded, clean them up with a terminal cleaner **(see illustration)**. Thoroughly wash all cleaned areas with plain water.

8   Make sure that the battery tray is in good condition and the hold-down clamp bolts are tight. If the battery is removed from the tray, make sure no parts remain in the bottom of the tray when the battery is reinstalled. When reinstalling the hold-down clamp bolts, do not overtighten them.

9   Any metal parts of the vehicle damaged by corrosion should be covered with a zinc-based primer, then painted.

10   Information on removing and installing the battery can be found in Chapter 5. Information on jump starting can be found at the front of this manual. For more detailed battery checking procedures, refer to the *Haynes Automotive Electrical Manual*.

Charging

Warning: *When batteries are being charged, hydrogen gas, which is very explosive and flammable, is produced. Do not smoke or allow open flames near a battery. Wear eye protection when near the battery during charging. Also, make sure the charger is unplugged before connecting or disconnecting the battery from the charger.*

Note: *The manufacturer recommends the battery be removed from the vehicle for charging because the gas that escapes during this procedure can damage the paint. Fast charging with the battery cables connected can result in damage to the electrical system.*

11   Slow-rate charging is the best way to restore a battery that's discharged to the point where it will not start the engine. It's also a good way to maintain the battery charge in a vehicle that's only driven a few miles between starts. Maintaining the battery charge is particularly important in the winter when the battery must work harder to start the engine and electrical accessories that drain the battery are in greater use.

12   It's best to use a one or two-amp battery charger (sometimes called a "trickle" charger). They are the safest and put the least strain on the battery. They are also the least expensive. For a faster charge, you can use a higher amperage charger, but don't use one rated more than 1/10th the amp/hour rating of the battery. Rapid boost charges that claim to restore the power of the battery in one to two hours are hardest on the battery and can damage batteries not in good condition. This type of charging should only be used in emergency situations.

13   The average time necessary to charge a battery should be listed in the instructions that come with the charger. As a general rule, a trickle charger will charge a battery in 12 to 16 hours.

14   Remove all the cell caps (if equipped) and cover the holes with a clean cloth to prevent spattering electrolyte. Disconnect the negative battery cable and hook the battery charger cable clamps up to the battery posts

(positive to positive, negative to negative), then plug in the charger. Make sure it is set at 12-volts if it has a selector switch.

15   If you're using a charger with a rate higher than two amps, check the battery regularly during charging to make sure it doesn't overheat. If you're using a trickle charger, you can safely let the battery charge overnight after you've checked it regularly for the first couple of hours.

16   If the battery has removable cell caps, measure the specific gravity with a hydrometer every hour during the last few hours of the charging cycle. Hydrometers are available inexpensively from auto parts stores - follow the instructions that come with the hydrometer. Consider the battery charged when there's no change in the specific gravity reading for two hours and the electrolyte in the cells is gassing (bubbling) freely. The specific gravity reading from each cell should be very close to the others. If not, the battery probably has a bad cell(s).

17   Some batteries with sealed tops have built-in hydrometers on the top that indicate the state of charge by the color displayed in the hydrometer window. Normally, a bright-colored hydrometer indicates a full charge and a dark hydrometer indicates the battery still needs charging.

18   If the battery has a sealed top and no built-in hydrometer, you can hook up a voltmeter across the battery terminals to check the charge. A fully charged battery should read 12.6 volts or higher after the surface charge has been removed.

19   Further information on the battery and jump starting can be found in Chapter 5 and at the front of this manual.

---

10   Cooling system check (every 7500 miles or 6 months)

Refer to illustration 10.4

1   Many major engine failures can be attributed to a faulty cooling system. If the vehicle is equipped with an automatic transmission, the cooling system also cools the transmission fluid and thus plays an important role in prolonging transmission life.

2   The cooling system should be checked with the engine cold. Do this before the vehicle is driven for the day or after it has been shut off for at least three hours.

3   Remove the radiator cap by turning it to the left until it reaches a stop. If you hear a hissing sound (indicating there is still pressure in the system), wait until this stops. Now press down on the cap with the palm of your hand and continue turning to the left until the cap can be removed. Thoroughly clean the cap, inside and out, with clean water. Also clean the filler neck on the radiator. All traces of corrosion should be removed. The coolant inside the radiator should be relatively transparent. If it is rust colored, the system should be drained and refilled (see Section 39). If the coolant level is not up to the top, add additional antifreeze/coolant mixture (see Section 4).

Check for a chafed area that could fail prematurely.

Check for a soft area indicating the hose has deteriorated inside.

Overtightening the clamp on a hardened hose will damage the hose and cause a leak.

Check each hose for swelling and oil-soaked ends. Cracks and breaks can be located by squeezing the hose.

10.4   Hoses, like drivebelts, have a habit of failing at the worst possible time - to prevent the inconvenience of a blown radiator or heater hose, inspect them carefully as shown here

---

4   Carefully check the large upper and lower radiator hoses along with the smaller diameter heater hoses which run from the engine to the firewall. Inspect each hose along its entire length, replacing any hose which is cracked, swollen or shows signs of deterioration. Cracks may become more apparent if the hose is squeezed **(see illustration)**. Regardless of condition, it's a good idea to replace hoses with new ones every two years.

5   Make sure all hose connections are tight. A leak in the cooling system will usually show up as white or rust colored deposits on the areas adjoining the leak. If wire-type clamps are used at the ends of the hoses, it may be a good idea to replace them with more secure screw-type clamps.

6   Use compressed air or a soft brush to remove bugs, leaves, etc. from the front of the radiator or air conditioning condenser. Be careful not to damage the delicate cooling

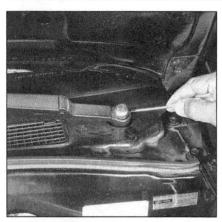

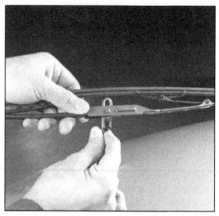

11.3 Gently pry off the trim cap and check the tightness of the wiper arm retaining nut

11.5a On earlier models, depress the tab (arrow) and pull the blade assembly off the wiper arm

11.5b On later models, press on the release tab and push the blade assembly down out of the hook in the arm

fins or cut yourself on them.

7   Every other inspection, or at the first indication of cooling system problems, have the cap and system pressure tested. If you don't have a pressure tester, most gas stations and repair shops will do this for a minimal charge.

11   Wiper blade inspection and replacement (every 7500 miles or 6 months)

Refer to illustrations 11.3, 11.5a, 11.5b and 11.6

1   The windshield wiper and blade assembly should be inspected periodically for damage, loose components and cracked or worn blade elements.
2   Road film can build up on the wiper blades and affect their efficiency, so they should be washed regularly with a mild detergent solution.
3   The action of the wiping mechanism can loosen bolts, nuts and fasteners, so they should be checked and tightened, as necessary **(see illustration)**, at the same time the wiper blades are checked.
4   If the wiper blade elements are cracked, worn or warped, or no longer clean adequately, they should be replaced with new ones.

11.6 Use needle-nose pliers to compress the rubber element, then slide the element out - slide the new element in and lock the blade assembly fingers into the notches of the wiper element

5   Lift the arm assembly away from the glass for clearance, press on the release lever, then slide the wiper blade assembly out of the arm **(see illustrations)**.
6   Use needle-nose pliers to compress the blade element, then slide the element out of the frame and discard it **(see illustration)**.
7   Installation is the reverse of removal.

12   Underhood hose check and replacement (every 7500 miles or 6 months)

General

1   **Warning:** *Replacement of air conditioning hoses must be left to a dealer service department or air conditioning shop that has the equipment to depressurize the system safely. Never remove air conditioning components or hoses until the system has been depressurized.*
2   High temperatures in the engine compartment can cause the deterioration of the rubber and plastic hoses used for engine, accessory and emission systems operation. Periodic inspection should be made for cracks, loose clamps, material hardening and leaks. Information specific to the cooling system hoses can be found in Section 10.

3   Some, but not all, hoses are secured to the fittings with clamps. Where clamps are used, check to be sure they haven't lost their tension, allowing the hose to leak. If clamps aren't used, make sure the hose has not expanded and/or hardened where it slips over the fitting, allowing it to leak.

Vacuum hoses

4   It's quite common for vacuum hoses, especially those in the emissions system, to be color coded or identified by colored stripes molded into them. Various systems require hoses with different wall thickness, collapse resistance and temperature resistance. When replacing hoses, be sure the new ones are made of the same material.
5   Often the only effective way to check a hose is to remove it completely from the vehicle. If more than one hose is removed, be sure to label the hoses and fittings to ensure correct installation.
6   When checking vacuum hoses, be sure to include any plastic T-fittings in the check. Inspect the fittings for cracks and the hose where it fits over the fitting for distortion, which could cause leakage.
7   A small piece of vacuum hose (1/4-inch inside diameter) can be used as a stethoscope to detect vacuum leaks. Hold one end of the hose to your ear and probe around vacuum hoses and fittings, listening for the "hissing" sound characteristic of a vacuum leak. **Warning:** *When probing with the vacuum hose stethoscope, be very careful not to come into contact with moving engine components such as the drivebelt, cooling fan, etc.*

Fuel hose

Warning: *There are certain precautions which must be taken when inspecting or servicing fuel system components. Work in a well ventilated area and do not allow open flames (cigarettes, appliance pilot lights, etc.) or bare light bulbs near the work area. Mop up any spills immediately and do not store fuel soaked rags where they could ignite.*

13.1 Check the accelerator cable and linkage for free movement

8   Check all rubber fuel lines for deterioration and chafing. Check especially for cracks in areas where the hose bends and just before fittings, such as where a hose attaches to the fuel filter.
9   Only high quality fuel line should be used for fuel line replacement. Never, under any circumstances, use unreinforced vacuum line, clear plastic tubing or water hose for fuel lines.
10   Spring-type clamps are commonly used on fuel lines. These clamps often lose their tension over a period of time, and can be "sprung" during removal. Replace all spring-type clamps with screw clamps whenever a hose is replaced.

Metal lines

11   Sections of metal line are often used in the fuel system. Check carefully to be sure the line has not been bent or crimped and that cracks have not started in the line.
12   If a section of metal fuel line must be replaced, only seamless steel tubing should be used, since copper and aluminum tubing don't have the strength necessary to withstand normal engine vibration.
13   Check the metal brake lines where they

enter the master cylinder and brake proportioning unit (if used) for cracks in the lines or loose fittings. Any sign of brake fluid leakage calls for an immediate thorough inspection of the brake system.

13   Accelerator linkage check and lubrication (every 7500 miles or 6 months)

Refer to illustration 13.1

1   At the specified intervals, inspect the accelerator linkage and check it for free movement to make sure it is not binding **(see illustration)**.
2   Lubricate the linkage with a few drops of oil.

14   Tire rotation (every 7500 miles or 6 months)

Refer to illustration 14.2

1   The tires should be rotated at the specified intervals and whenever uneven wear is noticed.
2   Refer to the **accompanying illustration** for the preferred tire rotation pattern.
3   Refer to the information in *Jacking and towing* at the front of this manual for the proper procedures to follow when raising the

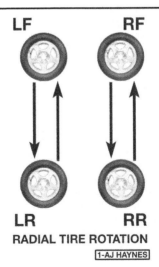

LF   RF

LR   RR
RADIAL TIRE ROTATION
1-AJ HAYNES

14.2 Tire rotation diagram for radial tires

vehicle and changing a tire. If the brakes are to be checked, don't apply the parking brake as stated. Make sure the tires are blocked to prevent the vehicle from rolling as it's raised.
4   Preferably, the entire vehicle should be raised at the same time. This can be done on a hoist or by jacking up each corner and then lowering the vehicle onto jackstands placed under the frame rails. Always use four jackstands and make sure the vehicle is safely supported.
5   After rotation, check and adjust the tire pressures as necessary and be sure to tighten the lug nuts to the torque listed in this Chapter's Specifications.

15   Suspension and steering check (every 7500 miles or 6 months)

Refer to illustrations 15.6a, 15.6b, 15.9 and 15.11

1   Whenever the front of the vehicle is raised for any reason, it's a good idea to visually check the suspension and steering components for wear.
2   Indications of steering or suspension problems include excessive play in the steering wheel before the front wheels react, excessive swaying around corners or body movement over rough roads and binding at some point as the steering wheel is turned.
3   Before the vehicle is raised for inspection, test the shock absorbers by pushing down aggressively at each corner. If the vehicle doesn't come back to a level position within one or two bounces, the shocks are worn and should be replaced. As this is done listen for squeaks and other noises from the suspension components. Information on shock absorber and suspension components can be found in Chapter 10.
4   Raise the front end of the vehicle and support it on jackstands. Make sure it's safely supported!
5   Crawl under the vehicle and check for loose bolts, broken or disconnected parts and deteriorated rubber bushings on all suspension and steering components. Look for grease or fluid leaking from around the steering gear assembly and shock absorbers. If equipped, check the power steering hoses and connections for leaks.
6   The balljoint boots should be checked at this time **(see illustration)**. This includes not only the upper and lower suspension balljoints, but those connecting the steering linkage parts as well **(see illustration)**. After cleaning around the balljoints, inspect the seals for cracks and damage.
7   Grip the top and bottom of each wheel and try to move it in and out. It won't take a lot of effort to be able to feel any play in the wheel bearings. If the play is noticeable it would be a good idea to adjust it right away or it could confuse further inspections. Grip each side of the wheel and try rocking it laterally. Steady pressure will, of course, turn the steering, but back-and-forth pressure will reveal a loose steering joint. If some play is

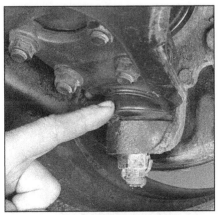

15.6a Push on the lower balljoint boots to check for damage and leaks

15.6b The grease oozing from this tie-rod end boot indicates that it needs replacing

15.9 On 1998 and later models with rack-and-pinion steering, inspect the dust boots that protect the inner end of each tie-rod; look for cracks and tears between the pleats of each dust boot

15.11 Push on the driveaxle boots to check for cracks

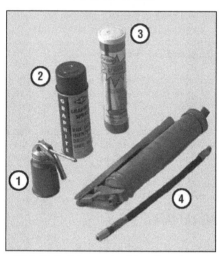

16.1 Materials required for chassis and body lubrication

*1   **Engine oil** - Light engine oil in a can like this can be used for door and hood hinges*
*2   **Graphite spray** - Used to lubricate lock cylinders*
*3   **Grease** - Grease, in a variety of types and weights, is available for use in a grease gun. Check the Specifications for your requirements*
*4   **Grease gun** - A common grease gun, shown here with a detachable hose and nozzle, is needed for chassis lubrication. After use, clean it thoroughly*

felt it would be easier to get assistance from someone so while one person rocks the wheel from side to side, the other can look at the joints, bushings and connections in the steering linkage. Generally speaking, there are eight places where the play may occur. The two outer balljoints on the tie-rod ends are the most likely on all vehicles. On 1989 through 1997 models, also inspect the two inner joints of the tie-rods (where they join the center rod). Any play in them means replacement of the tie-rod end. Also, on 1989 through 1997 models, check the steering gear arm balljoint and the one on the idler arm, which supports the steering center link on the side opposite the steering box. This unit is bolted to the side of the frame member and any play calls for replacement.

8   To check the steering box on 1989 through 1997 models, first make sure the bolts holding the steering box to the frame are tight. Then get another person to help examine the mechanism. One should look at, or hold onto, the arm at the bottom of the steering box while the other turns the steering wheel a little from side to side. The amount of lost motion between the steering wheel and the gear arm indicates the degree of wear in the steering box mechanism. This check should be carried out with the wheels first in the straight ahead position and then at nearly full lock on each side. If the play only occurs noticeably in the straight ahead position then the wear is most likely in the worm and/or nut. If it occurs at all positions, then the wear is probably in the sector shaft bearing. Oil leaks from the unit are another indication of such wear. In either case the steering box will need removal for closer examination and repair.

9   On 1998 and later vehicles, which have rack-and-pinion steering, there are less high-wear areas. Inspect the steering gear assembly; make sure that it's fastened securely to the lower crossmember. Inspect the pinch clamp that connects the transfer gear output shaft to the steering gear input shaft; make sure that both bolts are tight. Also make sure that the power steering line connections at the steering gear are tight and are not leaking

power steering fluid. And inspect the dust boot covering the inner end of each tie-rod **(see illustration)**. Make sure that there are no cracks or tears between the pleats.

10   Moving to the vehicle interior, check the play in the steering wheel by turning it slowly in both directions until the wheels can just be felt turning. The steering wheel freeplay should be less than 1-3/8 inch (35 mm). Excessive play is another indication of wear in the steering gear or linkage. On 1989 through 1997 models, the steering box can be adjusted for wear (see Chapter 10).

11   On 4WD models, inspect the front driveaxle CV joint boots for tears and leakage of grease **(see illustration)**

12   Following the inspection of the front, a similar inspection should be made of the rear suspension components, again checking for loose bolts, damaged or disconnected parts and deteriorated rubber bushings.

16   Chassis lubrication (every 7500 miles or 6 months)

Refer to illustrations 16.1, 16.3a, 16.3b and 16.3c

1   A grease gun and cartridge filled with the recommended grease are the only items required for chassis lubrication other than some clean rags and equipment needed to raise and support the vehicle safely **(see illustration)**.

2   There are several points on the vehicle's suspension, steering and drivetrain components that must be periodically lubricated with lithium based multi-purpose grease, depending on model and year. Included are the upper and lower suspension balljoints, the swivel joints on the steering linkage and, on 4WD models, the front and rear driveshafts.

3   The grease point for each upper suspension balljoint (if equipped) is on top of the balljoint and is accessible by removing the front wheel and tire **(see illustration)**. On

some models the inner and outer tie-rod ends are designed to be lubricated as well as the driveshaft slip yoke and universal joints **(see illustrations)**.

4   For easier access under the vehicle, raise it with a jack and place jackstands under the frame. Make sure the vehicle is safely supported on the stands!

5   If grease fittings aren't already installed, the plugs will have to be removed and fittings screwed into place.

6   Force a little of the grease out of the gun nozzle to remove any dirt, then wipe it clean with a rag.

16.3a Look under the vehicle for grease fittings (arrow) located on the balljoints, tie-rod ends and driveline components

16.3b Front driveshaft grease fitting (4WD models)

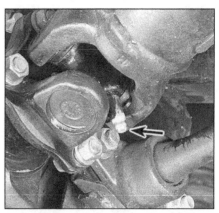

16.3c The rear driveshaft grease fitting (arrow) may be difficult to reach - rotating the driveshaft may improve access

7   Wipe the grease fitting and push the nozzle firmly over it. Squeeze the trigger on the grease gun to force grease into the component. Both the balljoints and swivel joints should be lubricated until the rubber reservoir is firm to the touch. Don't pump too much grease into the fittings or it could rupture the reservoir. If the grease seeps out around the grease gun nozzle, the fitting is clogged or the nozzle isn't seated all the way. Resecure the gun nozzle to the fitting and try again. If necessary, replace the fitting.

8   Wipe excess grease from the components and the grease fittings.

9   While you're under the vehicle, clean and lubricate the parking brake cable along with the cable guides and levers. This can be done by smearing some of the chassis grease onto the cable and its related parts with your fingers.

10   Lower the vehicle to the ground for the remaining body lubrication process.

11   Open the hood and rear tailgate and smear a little chassis grease on the latch mechanisms. Have an assistant pull the release knob from inside the vehicle as you lubricate the cable at the latch.

12   Lubricate all the hinges (door, hood, hatch) with a few drops of light engine oil to keep them in proper working order.

13   The key lock cylinders can be lubricated with spray-on graphite, which is available at auto parts stores.

17   Exhaust system check (every 7500 miles or 6 months)

Refer to illustrations 17.2 and 17.4

1   With the engine cold (at least three hours after the vehicle has been driven), check the complete exhaust system from its starting point at the engine to the end of the tailpipe. This should be done on a hoist where unrestricted access is available.

2   Check the pipes and connections for evidence of leaks **(see illustration)**, severe corrosion or damage. Make sure that all brackets and hangers are in good condition and tight.

3   At the same time, inspect the underside of the body for holes, corrosion, open seams, etc. which may allow exhaust gases to enter the passenger compartment. Seal all body openings with silicone or body putty.

4   Rattles and other noises can often be traced to the exhaust system, especially the mounts and hangers **(see illustration)**, Try to move the pipes, muffler and catalytic converter. If the components can come in con-

tact with the body or suspension parts, secure the exhaust system with new mounts.

5   Check the running condition of the engine by inspecting inside the end of the tailpipe. The exhaust deposits here are an indication of engine state-of-tune. If the pipe is black and sooty or coated with white deposits, the engine is in need of a tune-up, including a thorough fuel system inspection.

18   Valve clearance adjustment (four-cylinder engines) (every 15,000 miles or 12 months)

Refer to illustrations 18.4 and 18.9

Note: *This procedure does not apply to 2.2L four-cylinder engines, which are equipped with hydraulic valve lifters that require no adjustment.*

1   The valve clearances are checked and adjusted with the engine cold.

2   Remove the air cleaner assembly (see Chapter 4).

3   Remove the valve cover (see Chapter 2).

4   Before adjusting valve clearance, check that the rocker arm pedestal retaining nuts **(see illustration)** are tightened to the specified torque listed in Chapter 2A.

17.2 Check the flange connections (arrow) for exhaust leaks - also check that the retaining nuts are securely tightened

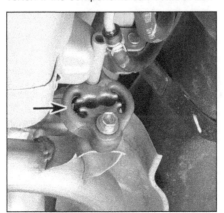

17.4 Check the exhaust system hangers (arrow) for damage and cracks

18.4 Make sure the rocker arm pedestal retaining nuts are tight before adjusting the valves

18.9 To adjust the valve clearance, loosen the adjuster lock nut with a box end wrench and back off the adjuster screw with a screwdriver, Carefully tighten the adjuster screw until you feel a slight drag when withdrawing the feeler gauge, then tighten the adjuster locknut while still holding the adjuster screw with a screwdriver

5   Rotate the crankshaft to bring the number one piston to the Top Dead Center (TDC) position on the compression stroke (see Chapter 2, part A for details). The number one cylinder rocker arms (closest to the front of the engine) should be loose (able to move up and down slightly) and the camshaft lobes should be facing away from the rocker arms.

6   With the crankshaft in this position, the valves on the following cylinders can be checked and adjusted:

No. 1 cylinder	Intake and exhaust
No. 2 cylinder	Intake
No. 3 cylinder	Exhaust

7   First, check/adjust the intake valve clearance. Insert the appropriate size feeler gauge (see the Specifications) between the intake valve stem and the adjusting screw. If the clearance is correct, there should be a slight drag on the feeler gauge as it is inserted between the valve stem and the

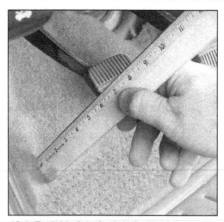

19.1 Pedal height is the distance between the pedal pad and the floor

adjusting screw.

8   If adjustment is required, loosen the adjusting screw locknut. Carefully tighten the adjusting screw until you can feel a slight drag on the feeler gauge as you withdraw it from between the stem and adjusting screw.

9   Hold the adjusting screw with a screwdriver (to keep it from turning) and tighten the locknut **(see illustration)**. Recheck the clearance to make sure it hasn't changed.

10   Repeat the procedure on the exhaust valve on the number 1 cylinder, the intake valve on number 2 cylinder and the exhaust valve on the number 3 cylinder.

11   Rotate the crankshaft one full turn clockwise until the number 4 piston is at TDC on the compression stroke. The number 4 cylinder rocker arms (closest to the rear of the engine) should be loose, with the camshaft lobes facing away from the rocker arms.

12   With the crankshaft in this position, the valves on the following cylinders can be checked and adjusted:

No. 2 cylinder	Exhaust
No. 3 cylinder	Intake
No. 4 cylinder	Intake and exhaust

13   Install the valve cover and the air cleaner assembly.

19   Clutch/brake pedal height and freeplay adjustment (every 15,000 miles or 12 months)

Pedal height

Refer to illustrations 19.1 and 19.2

1   The height of the clutch and brake pedal is the distance the pedal sits off the floor **(see illustration)**. If the pedal height is not within the specified range, it must be adjusted.

2   To adjust the clutch pedal loosen the locknut and back the stopper bolt or switch out for clearance, then loosen the locknut on the clutch pushrod. Turn the pushrod to adjust the pedal height in the middle of the specified range, then retighten the locknut **(see illustration)**. **Note:** *When adjusting a cable actuated clutch pedal simply loosen the locknut and back the stopper bolt or switch in or out to achieve the proper height.*

3   Before measuring the brake pedal height make sure the pedal is in the fully returned position, then start the engine and depress the accelerator several times to activate the brake power booster. Measure the pedal height and adjust if necessary. To adjust the pedal height loosen the locknut and back the stopper bolt or switch out for clearance, then loosen the locknut on the clutch pushrod. Turn the pushrod to adjust the pedal height in the middle of the specified range, then retighten the locknut.

4   Adjust the pedal switch by turning it clockwise until the switch body just contacts the pedal arm, then rotate it counterclockwise to gain the specified clearance listed at the beginning of this Chapter and tighten the switch locknut.

Pedal freeplay

Refer to illustration 19.5

5   The freeplay is the pedal slack, or the distance the pedal can be depressed before it begins to have any effect on the clutch or brake system **(see illustration)**. If the pedal freeplay is not within the specified range, it

19.2 Back-out the stopper bolt or switch (A) for clearance, then loosen the locknut on the pushrod (B). Turn the pushrod to adjust the pedal height

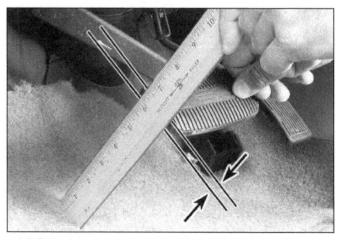

19.5 Pedal freeplay is the distance from the natural resting point of the pedal to the point at which resistance is felt

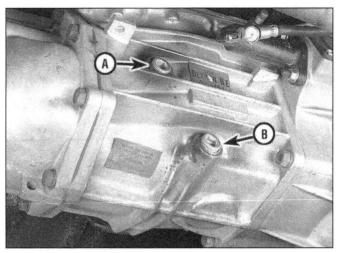

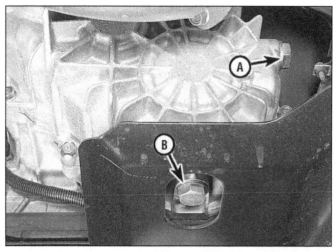

**20.1 Manual transmission check/fill plug (A) and drain plug (B)
(2WD model shown)**

**21.1 Transfer case check/fill plug (A) and drain plug (B)
(1997 model shown)**

must be adjusted.

6   To adjust the pedal freeplay on vehicles equipped with a hydraulic clutch loosen the locknut on the clutch pushrod. Then back off the pushrod to adjust the pedal freeplay to the specified range and retighten the locknut.

7   On vehicles equipped with a cable actuated clutch, pedal freeplay is achieved by adjusting the clutch cable. Refer to Chapter 8 for cable adjustment procedures.

8   Before measuring the brake pedal freeplay turn off the engine and depress the brake pedal a half dozen times. Measure the pedal freeplay and adjust if necessary. loosen the locknut on the brake pushrod. Then back off the pushrod to adjust the pedal freeplay to the specified range and retighten the locknut.

20   Manual transmission lubricant level check (every 15,000 miles or 12 months)

Refer to illustration 20.1

1   Manual transmissions don't have a dipstick. The oil level is checked by removing a plug from the side of the transmission case **(see illustration)**. Locate the plug and use a rag to clean the plug and the area around. **Note:** *On 4WD vehicles with MUA type transmissions the fill plugs are located on the passenger side while drain plugs are located on the driver side.*

it. If the vehicle is raised to gain access to the plug, be sure to support it safely on jackstands - DO NOT crawl under the vehicle when it's supported only by a jack!

2   With the engine and transmission cold, remove the plug. If lubricant immediately starts leaking out, thread the plug back into the transmission - the level is correct. If it doesn't, reach inside the hole with your little finger. The level should be even with the bottom of the plug hole.

3   If the transmission needs more lubricant, use a syringe or small pump to add it through the plug hole.

4   Thread the plug back into the transmission and tighten it securely. Drive the vehicle, then check for leaks around the plug.

21   Transfer case lubricant level check (every 15,000 miles or 12 months)

Refer to illustration 21.1

1   Remove the transfer case rock guard (if equipped). The lubricant level is checked by removing a plug from the side of the case **(see illustration)**. If the vehicle is raised to gain access to the plug, be sure to support it safely on jackstands - DO NOT crawl under the vehicle when it's supported only by a jack!

2   With the engine and transfer case cold, remove the plug. If lubricant immediately starts leaking out, thread the plug back into the case - the level is correct. If it doesn't, completely remove the plug and reach inside the hole with your little finger. The level should be even with the bottom of the plug hole.

3   If more lubricant is needed, use a syringe or small pump to add it through the opening.

4   Thread the plug back into the case and tighten it securely. Drive the vehicle, then check for leaks around the plug. Install the rock guard.

22   Differential lubricant level check (every 15,000 miles or 12 months)

Refer to illustration 22.2

Note: *Models equipped with 4WD have two differentials, be sure to check the lubricant level in both differentials.*

1   The differential has a check/fill plug which must be removed to check the lubricant level. If the vehicle is raised to gain access to the plug, be sure to support it safely

**22.2 The differential check/fill plug
(arrow) is located on the back cover**

on jackstands - DO NOT crawl under the vehicle when it's supported only by a jack.

2   Remove the check/fill plug from the differential **(see illustration)**.

3   The lubricant level should be at the bottom of the plug opening. If not, use a syringe to add the recommended lubricant until it just starts to run out of the opening.

4   Install the plug and tighten it securely.

23   Brake check (every 15,000 miles or 12 months)

Warning: *The dust created by the brake system is harmful to your health. Never blow it out with compressed air and don't inhale any of it. An approved filtering mask should be worn when working on the brakes. Do not, under any circumstances, use petroleum-based solvents to clean brake parts. Use brake system cleaner only! Try to use non-asbestos replacement parts whenever possible.*

Note: *For detailed photographs of the brake system, refer to Chapter 9.*

1   In addition to the specified intervals, the brakes should be inspected every time the

wheels are removed or whenever a defect is suspected. Any of the following symptoms could indicate a potential brake system defect: The vehicle pulls to one side when the brake pedal is depressed; the brakes make squealing or dragging noises when applied; brake pedal travel is excessive; the pedal pulsates; brake fluid leaks, usually onto the inside of the tire or wheel.

2   The disc brake pads have built-in wear indicators which should make a high pitched squealing or scraping noise when they are worn to the replacement point. When you hear this noise, replace the pads immediately or expensive damage to the discs can result.

3   Loosen the wheel lug nuts.

4   Raise the vehicle and place it securely on jackstands.

5   Remove the wheels.

Disc brakes

Refer to illustrations 23.6 and 23.11

6   There are two pads (an outer and an inner) in each caliper. The pads are visible through inspection holes in each caliper **(see illustration)**.

7   Check the pad thickness by looking at each end of the caliper and through the inspection hole in the caliper body. If the lining material is less than the thickness listed in this Chapter's Specifications, replace the pads. **Note:** *Keep in mind that the lining material is riveted or bonded to a metal backing plate and the metal portion is not included in this measurement.*

8   If it is difficult to determine the exact thickness of the remaining pad material by the above method, or if you are at all concerned about the condition of the pads, remove the caliper(s), then remove the pads from the calipers for further inspection (refer to Chapter 9).

9   Once the pads are removed from the calipers, clean them with brake cleaner and re-measure them with a ruler or a vernier caliper.

10   Measure the disc thickness with a micrometer to make sure that it still has ser-

23.6 You'll find an inspection hole like this in each caliper - placing a ruler across the hole should enable you to determine the thickness of the remaining lining material for both inner and outer pads - the lining can also be inspected by looking at the end of each caliper

vice life remaining. If any disc is thinner than the specified minimum thickness, replace it (refer to Chapter 9). Even if the disc has service life remaining, check its condition. Look for scoring, gouging and burned spots. If these conditions exist, remove the disc and have it resurfaced (see Chapter 9).

11   Before installing the wheels, check all brake lines and hoses for damage, wear, deformation, cracks, corrosion, leakage, bends and twists, particularly in the vicinity of the rubber hoses at the calipers **(see illustration)**. Check the clamps for tightness and the connections for leakage. Make sure that all hoses and lines are clear of sharp edges, moving parts and the exhaust system. If any of the above conditions are noted, repair, reroute or replace the lines and/or fittings as necessary (see Chapter 9).

Drum brakes

Refer to illustrations 23.15 and 23.17

12   On rear drum brakes, make sure the

23.11 Check along the brake hoses and at each fitting (arrow) for deterioration and cracks

parking brake is off then proceed to tap on the outside of the drum with a rubber mallet to loosen it.

13   Remove the brake drums.

14   With the drums removed, carefully clean the brake assembly with brake system cleaner. **Warning:** *Don't blow the dust out with compressed air and don't inhale any of it (it may contain asbestos, which is harmful to your health).*

15   Note the thickness of the lining material on both front and rear brake shoes. If the material has worn away to within 1/16-inch of the recessed rivets or metal backing, the shoes should be replaced **(see illustration)**. The shoes should also be replaced if they're cracked, glazed (shiny areas), or covered with brake fluid.

16   Make sure all the brake assembly springs are connected and in good condition.

17   Check the brake components for signs of fluid leakage. With your finger or a small screwdriver, carefully pry back the rubber cups on the wheel cylinder located at the top of the brake shoes **(see illustration)**. Any leakage here is an indication that the wheel cylinders should be overhauled immediately

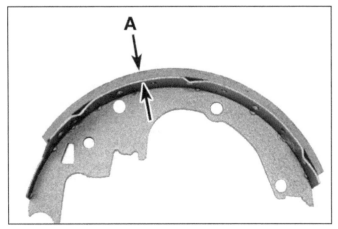

23.15 If the lining is bonded to the brake shoe, measure the lining thickness from the outer surface to the metal shoe, as shown here; if the lining is riveted to the shoe, measure from the lining outer surface to the rivet heads

23.17 Check the wheel cylinder boots (arrows) for leaking fluid indicating that the cylinder must be replaced or rebuilt

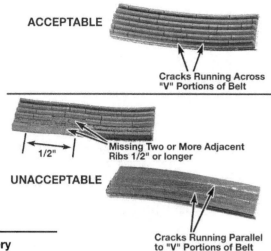

ACCEPTABLE

Cracks Running Across "V" Portions of Belt

1/2"

Missing Two or More Adjacent Ribs 1/2" or longer

UNACCEPTABLE

Cracks Running Parallel to "V" Portions of Belt

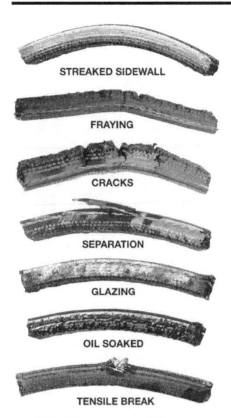

STREAKED SIDEWALL

FRAYING

CRACKS

SEPARATION

GLAZING

OIL SOAKED

TENSILE BREAK

25.3a Here are some of the more common problems associated with V-belts (check the belts very carefully to prevent an untimely breakdown)

25.3b Small cracks in the underside of a serpentine (or "V-ribbed") belt are acceptable - lengthwise cracks, or missing pieces are cause for replacement

24 Fuel system check (every 15,000 miles or 12 months)

Warning: *Gasoline is extremely flammable, so take extra precautions when you work on any part of the fuel system. Don't smoke or allow open flames or bare light bulbs near the work area, and don't work in a garage where a gas-type appliance (such as a water heater or clothes dryer) is present. Since gasoline is carcinogenic, wear latex gloves when there's a possibility of being exposed to fuel, and, if you spill any fuel on your skin, rinse it off immediately with soap and water. Mop up any spills immediately and do not store fuel-soaked rags where they could ignite. The fuel system on fuel-injected models is under constant pressure, so, if any fuel lines are to be disconnected, the fuel pressure in the system must be relieved first (see Chapter 4 for more information). When you perform any kind of work on the fuel system, wear safety glasses and have a Class B type fire extinguisher on hand.*

1   If you smell gasoline while driving or after the vehicle has been sitting in the sun, inspect the fuel system immediately.

2   Remove the gas filler cap and inspect it for damage and corrosion. The gasket should have an unbroken sealing imprint. If the gasket is damaged or corroded, remove it and install a new one.

3   Inspect the fuel feed and return lines for cracks. Make sure the threaded flare nut type connectors which secure the metal fuel lines to fuel injection system and the clamps which secure the hoses to the in-line fuel filter are tight.

4   Since some components of the fuel system - the fuel tank and part of the fuel feed and return lines, for example - are underneath the vehicle, they can be inspected more easily with the vehicle raised on a hoist. If that's not possible, raise the vehicle and support it securely on jackstands.

5   With the vehicle raised and safely supported, inspect the gas tank and filler neck for punctures, cracks and other damage. The connection between the filler neck and the tank is particularly critical. Sometimes a rubber filler neck will leak because of loose

clamps or deteriorated rubber. These are problems a home mechanic can usually rectify. **Warning:** *Do not, under any circumstances, try to repair a fuel tank (except rubber components). A welding torch or any open flame can easily cause fuel vapors inside the tank to explode.*

6   Carefully check all rubber hoses and metal lines leading away from the fuel tank. Check for loose connections, deteriorated hoses, crimped lines and other damage. Carefully inspect the lines from the tank to the fuel injection system. Repair or replace damaged sections as necessary (see Chapter 4B).

25 Drivebelt check, adjustment and replacement (every 15,000 miles or 12 months)

Check

Refer to illustrations 25.3a, 25.3b and 25.4

1   The drivebelts, or V-belts as they are sometimes called, are located at the front of the engine and play an important role in the overall operation of the vehicle and its components. Due to their function and material make-up, the belts are prone to failure after a period of time and should be inspected and adjusted periodically to prevent major engine damage.

2   The number of belts used on a particular vehicle depends on the accessories installed. Drivebelts are used to turn the alternator, power steering pump, water pump and air conditioning compressor. Depending on the pulley arrangement, more than one of these components may be driven by a single belt.

3   With the engine off, open the hood and locate the various belts at the front of the engine. Using your fingers (and a flashlight, if necessary), move along the belts checking for cracks and separation of the belt plies. Also check for fraying and glazing, which gives the belt a shiny appearance **(see illustrations)**. Both sides of each belt should be inspected, which means you will have to twist the belt to check the underside.

(see Chapter 9). Also, check all hoses and connections for signs of leakage.

18   Wipe the inside of the drum with a clean rag and denatured alcohol or brake cleaner. Again, be careful not to breathe the dangerous asbestos dust.

19   Check the inside of the drum for cracks, score marks, deep scratches and "hard spots" which will appear as small discolored areas. If imperfections cannot be removed with fine emery cloth, the drum must be taken to an automotive machine shop for resurfacing.

20   Repeat the procedure for the remaining wheel. If the inspection reveals that all parts are in good condition, reinstall the brake drums, install the wheels and lower the vehicle to the ground.

Parking brake

21   The parking brake is operated by the parking brake lever which is mounted next to the steering column or by the floor console and locks the rear brake system. The easiest, and perhaps most obvious method of periodically checking the operation of the parking brake assembly is to park the vehicle on a steep hill with the parking brake set and the transmission in Neutral. If the parking brake cannot prevent the vehicle from rolling within the specified number of clicks listed in this chapter's specifications, it's in need of adjustment (see Chapter 9).

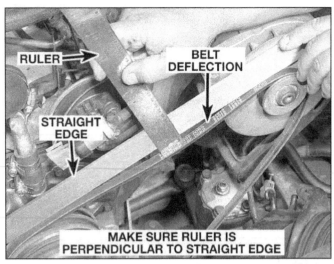

25.6a Some components will have an adjusting mechanism - loosen the lock bolt (A) and the pivot bolt (B) then turn the adjusting bolt (C) counterclockwise to loosen or clockwise to tighten the belt (3.2L V6 shown)

25.4 Measuring drivebelt deflection with a straightedge and ruler

4   The tension of each belt is checked by pushing on the belt at a distance halfway between the pulleys. Push firmly with your thumb and see how much the belt moves (deflects) **(see illustration)**. As rule of thumb, if the distance from pulley center-to-pulley center is between 7 and 11 inches, the belt should deflect 1/4-inch. If the belt travels between pulleys spaced 12 to 16 inches apart, the belt should deflect 1/2-inch for a V-belt or 1/4-inch for a serpentine belt.

Adjustment

Refer to illustrations 25.6a, 25.6b, 25.6c and 25.6d

5   If it is necessary to adjust the belt tension, either to make the belt tighter or looser, it is done by moving the belt-driven accessory on the bracket.

6   Some components have an adjusting bolt and a pivot bolt which tensions the belt. Both bolts must be loosened slightly to enable you to move the component. Other components are rigidly mounted and require an adjustable idler pulley which is mounted between the components to tension the belt **(see illustrations)**.

7   After the bolts have been loosened, move the component away from the engine to tighten the belt or toward the engine to loosen the belt. Measure the belt tension in accordance with one of the above methods. Repeat this step until the drivebelt is adjusted.

Replacement

8   To replace a belt, follow the above procedures for drivebelt adjustment but slip the belt off the pulleys and remove it. Since belts tend to wear out more or less at the same time, it's a good idea to replace all of them at the same time. Mark each belt and the corresponding pulley grooves so the replacement belts can be installed properly.

9   Take the old belts with you when purchasing new ones in order to make a direct comparison for length, width and design.

10   Adjust the belts as described earlier in this Section.

26   Thermostatic air cleaner check (every 30,000 miles or 24 months)

1   Carbureted models and 3.1L V6 models are equipped with a thermostatically controlled air cleaner, which draws air to the carburetor or throttle body from different locations depending on engine temperature.

2   This is a simple visual check. However, if access is tight, a small mirror may have to be used.

3   Open the hood and find the air control valve on the air cleaner assembly. It's located inside the long snorkel portion of the metal air cleaner housing.

4   If there's a flexible air duct attached to the end of the snorkel, disconnect it so you can look through the end of the snorkel and see the air control valve inside. A mirror may

25.6b Other components require an adjustable idler pulley to adjust belt tension - loosen the idler pulley lock nut (A) then turn the adjusting bolt (B) to loosen or tighten the belt (3.2L V6 shown)

25.6c Some power steering pumps require loosening the adjusting bolt (A) and inserting a 1/2-inch drive breaker bar into the square hole (B) and pry upward to tension the drivebelt

25.6d To relieve tension on the serpentine belt used on 2.2L four-cylinder engines and 1998 and later 3.2L V6 engines, simply turn the pulley bolt in a clockwise direction; to restore tension, simply allow the spring-loaded tensioner to return to its normal position

27.8 Detach the air cleaner housing cover clips (arrows) (fuel-injected models)

27.9 With the hoses still attached, pull the cover up, then lift the element out of the housing

be needed if you can't safely look directly into the end of the snorkel.

5   The check should be done when the engine and outside air are cold. Start the engine and watch the air control valve, which should move up and close off the snorkel air passage. With the valve closed, air can't enter through the end of the snorkel, but instead enters the air cleaner through the hot air duct attached to the exhaust manifold.

6   As the engine warms up to operating temperature, the valve should open to allow air through the snorkel end. Depending on outside air temperature, this may take 10 to 15 minutes. To speed up the check you can reconnect the snorkel air duct, drive the vehicle and then check the position of the valve.

7   If the thermostatic air cleaner isn't operating properly, see Chapter 6 for more information.

---

27   Air filter replacement (every 30,000 miles or 24 months)

---

1   At the specified intervals, the air filter should be replaced with a new one. A thorough program of preventive maintenance would also call for the filter to be inspected periodically between changes, especially if the vehicle is often driven in dusty conditions.

2   The air filter is located inside the air cleaner housing, which is mounted on top of the carburetor or at the front of the engine compartment next to the battery on fuel-injected models.

2.3L four-cylinder and 3.1L V6 models

3   Remove the wing nut(s) that hold the top plate to the air cleaner body, release the clips (if equipped) and lift it off.

4   Lift the air filter out of the housing. If it's covered with dirt, it should be replaced.

5   Wipe the inside of the air cleaner housing with a rag.

6   Place the old filter (if in good condition) or the new filter (if replacement is necessary) into the air cleaner housing.

7   Reinstall the top plate on the air cleaner

and tighten the wing nut(s), then snap the clips into place.

2.6L four-cylinder models

Refer to illustrations 27.8 and 27.9

8   Detach the three filter cover retaining clips **(see illustration)**.

9   Lift the cover up for access to the filter element **(see illustration)**.

10   Lower the new filter into the housing, seat the cover and snap the retaining clips into place.

2.2L four-cylinder and 3.2L V6 models

Refer to illustrations 27.11 and 27.12

11   Detach the three filter cover retaining clips **(see illustration)**.

12   Lift up the cover for access to the filter element **(see illustration)**.

13   Lower the new filter into the housing, seat the cover and snap the retaining clips into place.

27.11 To detach the air cleaner housing cover on 2.2L four-cylinder and 3.2L V6 engines, remove these three clips (arrows) (2.2L four-cylinder model shown, 3.2L V6 models similar)

27.12 Remove the old filter element from the lower half of the air cleaner housing (2.2L four-cylinder model shown, 3.2L V6 models similar)

28   Carburetor choke check (every 30,000 miles or 24 months)

1   The choke operates only when the engine is cold, so this check should be performed before the engine has been started for the day.

2   Open the hood and remove the top plate of the air cleaner assembly. It's held in place by one or two wing nuts at the center and several spring clips around the edge. If any vacuum hoses must be disconnected, tag them to ensure reinstallation in their original positions.

3   Look at the center of the air cleaner housing. You'll notice a flat plate at the carburetor opening.

4   Have an assistant press the throttle pedal to the floor. The plate should close completely. Start the engine while you watch the plate at the carburetor. Don't position your face near the carburetor, as the engine could backfire, causing serious burns! When the engine starts, the choke plate should open slightly.

5   Allow the engine to continue running at an idle speed. As the engine warms up to operating temperature, the plate should slowly open, allowing more air to enter through the top of the carburetor.

6   After a few minutes, the choke plate should be completely open to the vertical position. Blip the throttle to make sure the fast idle cam disengages.

7   You'll notice that engine speed corresponds to the plate opening. With the plate closed, the engine should run at a fast idle speed. As the plate opens and the throttle is moved to disengage the fast idle cam, the engine speed will decrease.

8   With the engine off and the throttle held half-way open, open and close the choke several times. Check the linkage to see if it's hooked up correctly and make sure it doesn't bind.

9   If the choke or linkage binds, sticks or works sluggishly, clean it with choke cleaner (an aerosol spray available at auto parts stores). If the condition persists after cleaning, replace the troublesome parts.

10   Visually inspect all vacuum hoses to be sure they're securely connected and look for cracks and deterioration. Replace as necessary.

11   If the choke fails to operate normally, but no mechanical causes can be found, check the choke electrical circuits.

29   Positive Crankcase Ventilation (PCV) valve check and replacement (every 30,000 miles or 24 months)

Refer to illustrations 29.2a, 29.2b, 29.2c, 29.4a and 29.4b

1   The Positive Crankcase Ventilation (PCV) system directs blowby gases from the crankcase through the PCV valve and hose

29.2a   On some four-cylinder models, the PCV valve screws into the valve cover, so a wrench is needed for removal

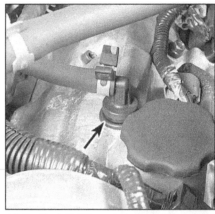

29.2b   On 3.2L V6 models the PCV valve is located at the front of the driver's side valve cover

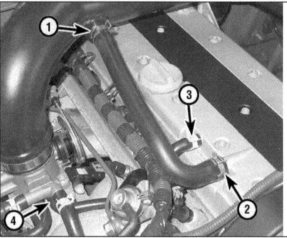

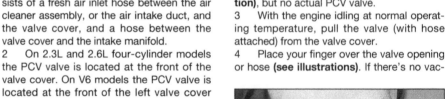

29.2c   Typical PCV system on a 2.2L four-cylinder engine

1   *Fresh air is diverted from the air intake duct into the PCV fresh air inlet hose*

2   *Fresh air is drawn from the fresh air inlet hose into the crankcase*

3   *Crankcase vapors are drawn from the crankcase into the PCV hose*

4   *Crankcase vapors are drawn from the PCV hose into the intake manifold*

back into the intake manifold so they can be burned in the engine. The PCV system consists of a fresh air inlet hose between the air cleaner assembly, or the air intake duct, and the valve cover, and a hose between the valve cover and the intake manifold.

2   On 2.3L and 2.6L four-cylinder models the PCV valve is located at the front of the valve cover. On V6 models the PCV valve is located at the front of the left valve cover

(see illustrations). On 2.2L four-cylinder models, there is a PCV system (see illustration), but no actual PCV valve.

3   With the engine idling at normal operating temperature, pull the valve (with hose attached) from the valve cover.

4   Place your finger over the valve opening or hose (see illustrations). If there's no vac-

29.4a   With the engine running, put your finger over the end of the PCV valve; you should feel vacuum

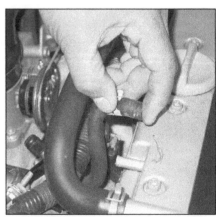

29.4b   Even on PCV systems without a PCV valve, such as the system on 2.2L four-cylinder models, the principle is the same: when you put your finger over the PCV hose, you should feel vacuum

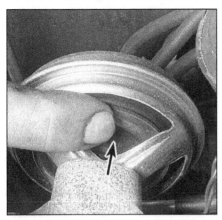

30.2 Push on the EGR valve diaphragm to make sure it moves easily with no binding

31.5 Detach the hose clamps and filter retaining bolt - always install the filter with the directional arrow pointed towards the engine

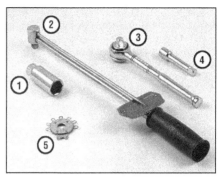

32.2 Tools required for changing spark plugs

1 **Spark plug socket** - This will have special padding inside to protect the spark plug's porcelain insulator
2 **Torque wrench** - Although not mandatory, using this tool is the best way to ensure the plugs are tightened properly
3 **Ratchet** - Standard hand tool to fit the spark plug socket
4 **Extension** - Depending on model and accessories, you may need special extensions and universal joints to reach one or more of the plugs
5 **Spark plug gap gauge** - This gauge for checking the gap comes in a variety of styles. Make sure the gap for your engine is included

uum, check for a plugged hose, manifold port, or the valve itself. Replace any plugged or deteriorated hoses.

5   Turn off the engine and shake the PCV valve, listening for a rattle. If the valve doesn't rattle, replace it with a new one.

6   To replace the valve, pull it from the end of the hose, noting its installed position.

7   When purchasing a replacement PCV valve, make sure it's for your particular vehicle and engine size. Compare the old valve with the new one to make sure they're the same.

8   Push the valve into the end of the hose until it's seated.

9   Inspect all rubber hoses and grommets for damage and hardening. Replace them, if necessary.

10   Press the PCV valve and hose securely into position. For further information on the PCV system refer to Chapter 6.

30   Exhaust Gas Recirculation (EGR) system check (1996 and earlier)(every 30,000 miles or 24 months)

Refer to illustration 30.2

1   The EGR valve is located on the intake manifold. Most of the time, when a problem develops in the emissions system, it is due to a stuck or corroded EGR valve.

2   With the engine cold to prevent burns, push on the valve's diaphragm **(see illustration)**. Using moderate pressure, you should be able to move the diaphragm up-and-down within the housing. **Note:** *This check does not apply to electronically-controlled EGR valves used on 1997 and later models, i.e. 3.2L V6 and 2.2L four-cylinder models.*

3   If the diaphragm does not move or moves only with much effort, replace the EGR valve with a new one. If in doubt about the quality of the valve, compare the free movement of your EGR valve with a new valve.

4   Refer to Chapter 6 for more information on the EGR system.

31   Fuel filter replacement (every 30,000 miles or 24 months)

Refer to illustration 31.5

Warning: *Gasoline is extremely flammable, so take extra precautions when you work on any part of the fuel system. Don't smoke or allow open flames or bare light bulbs near the work area, and don't work in a garage where a gas-type appliance (such as a water heater or clothes dryer) is present. Since gasoline is carcinogenic, wear latex gloves when there's a possibility of being exposed to fuel, and, if you spill any fuel on your skin, rinse it off immediately with soap and water. Mop up any spills immediately and do not store fuel-soaked rags where they could ignite. The fuel system on fuel-injected models is under constant pressure, so, if any fuel lines are to be disconnected, the fuel pressure in the system must be relieved first (see Chapter 4 for more information). When you perform any kind of work on the fuel system, wear safety glasses and have a Class B type fire extinguisher on hand.*

1   Fuel filters often become clogged due to excessive dirt or foreign material in the fuel system and must be replaced according to the maintenance interval suggestions.

2   Considering the volatile nature of gasoline, precautions should be exercised when a fuel filter is replaced. The job should be done with the engine cold or after sitting at least three hours. Be prepared to immediately clean up any spilled gasoline. In addition, you need to obtain a replacement filter for your specific vehicle.

3   Remove the fuel filler cap to relieve any built-up pressure in the tank. On fuel-injected models, depressurize the fuel system (see Chapter 4B). **Warning:** *Before attempting to remove the fuel filter, disconnect the negative cable from the battery and position it out of the way so it can't accidentally contact the battery post.*

4   Raise the vehicle and support it securely on jackstands. The fuel filter is located under the vehicle on the right side frame rail, adja-

cent to the fuel tank. 5

To replace the filter, loosen the clamps and slide them down the hoses, past the fittings on the filter **(see illustration)**.

6   Carefully twist and pull on the hoses to separate them from the filter. If the hoses are in bad shape, now would be a good time to replace them with new ones.

7   Connect the filter to the hoses and tighten the clamps securely. If spring-type clamps were originally installed, it would be a good idea to replace them with screw-type clamps. Make sure that the arrow on the filter housing faces toward the engine. Start the engine and check carefully for leaks at the filter hose connections.

32   Spark plug replacement (every 30,000 miles or 24 months)

Refer to illustrations 32.2, 32.5a, 32.5b, 32.6a, 32.6b, 32.6c, 32.8, 32.9 and 32.10

1   The spark plugs are located on both sides of the engine on V6 models and on the driver's side on four-cylinder models. Label each spark plug wire before proceeding any further.

2   The only tools necessary for spark plug replacement are: a spark plug socket (spark plug sockets are padded inside to prevent damage to the porcelain insulators on the new plugs), a ratchet, an extension of suitable length and a gap gauge to check and adjust the gaps on the new plugs **(see illustration)**. A special plug wire removal tool is

available for separating the wire boots from the spark plugs, but it isn't absolutely necessary. A torque wrench should be used to tighten the new plugs. On 2.2L engines, you'll also need a 5mm Allen wrench to unscrew the spark plug cover bolts.

3   The best approach when replacing the spark plugs is to purchase the new ones in advance, adjust them to the proper gap and replace them one at a time. When buying the new spark plugs, be sure to obtain the correct plug type for your particular engine. This information can be found on the *Emission Control Information* label located under the hood, in the factory owner's manual and in this Chapter's Specifications. If differences exist between the plug specified on the emissions label and in the owner's manual, assume that the emissions label is correct.

4   Allow the engine to cool completely before attempting to remove any of the plugs. While you're waiting for the engine to cool, check the new plugs for defects and adjust the gaps.

5   The gap is checked by inserting the proper thickness gauge between the electrodes at the tip of the plug **(see illustration)**. The gap between the electrodes should be the same as the one specified on the *Emissions Control Information* label or in this Chapter's Specifications. The wire should just slide between the electrodes with a slight amount of drag. If the gap is incorrect, use the adjuster on the gauge body to bend the curved side electrode slightly until the proper gap is obtained **(see illustration)**. If the side electrode is not exactly over the center electrode, bend it with the adjuster until it is. Check for cracks in the porcelain insulator (if any are found, the plug should not be used).

6   With the engine cool, remove the spark plug wire from one spark plug. Pull only on the boot at the end of the wire - do not pull on the wire. A plug wire removal tool should be used if available **(see illustration)**. Later model 3.2L V6 engines are equipped with individual coil packs which must be removed first to access the spark plugs **(see illustrations)**.

7   If compressed air is available, use it to

32.5a Spark plug manufacturers recommend using a wire type gauge when checking the gap - if the wire does not slide between the electrodes with a slight drag, adjustment is required

32.5b To change the gap, bend the *side* electrode only, as indicated by the arrows, and be very careful not to crack or chip the porcelain insulator surrounding the center electrode

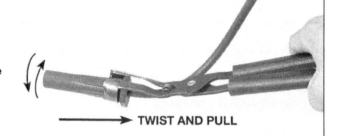

32.6a Using a spark plug boot puller tool like this one will make the job of removing the spark plug boots much easier

TWIST AND PULL

blow any dirt or foreign material away from the spark plug hole. The idea here is to eliminate the possibility of debris falling into the cylinder as the spark plug is removed. **Warning:** *Wear eye protection when using compressed air.*

8   Place the spark plug socket over the plug and remove it from the engine by turning it in a counterclockwise direction **(see illustration)**.

9   Compare the spark plug with the chart on the inside back cover of this manual to get an indication of the general running condition of the engine. Before installing the new plugs,

it is a good idea to apply a thin coat of anti-seize compound to the threads **(see illustration)**.

10   Thread one of the new plugs into the hole until you can no longer turn it with your fingers, then tighten it with a torque wrench (if available) or the ratchet. It might be a good idea to slip a length of rubber hose over the end of the plug to use as a tool to thread it into place **(see illustration)**. The hose will grip the plug well enough to turn it, but will start to slip if the plug begins to cross-thread in the hole - this will prevent damaged threads and the accompanying repair costs.

32.6b On later model 3.2L V6 engines the coil packs must be removed to access the spark plugs - disconnect the electrical connector (A) and remove the coil pack retaining screws (B)

32.6c Pull straight up on the coil pack to remove it

32.8 Use a special spark plug socket with a long extension to unscrew the spark plugs

32.9 Apply a thin coat of anti-seize compound to the spark plug threads

32.10 A length of snug-fitting rubber hose will save time and prevent damaged threads when installing the spark plugs

11   Before pushing the spark plug wire onto the end of the plug, inspect it following the procedures outlined in Section 33.

12   Attach the plug wire to the new spark plug, again using a twisting motion on the boot until it's seated on the spark plug.

13   Repeat the procedure for the remaining spark plugs, replacing them one at a time to prevent mixing up the spark plug wires.

33   Spark plug wire, distributor cap and rotor check and replacement (every 30,000 miles or 24 months)

Refer to illustrations 33.11 and 33.12

1   The spark plug wires should be checked whenever new spark plugs are installed.

2   Begin this procedure by making a visual check of the spark plug wires while the engine is running. In a darkened garage (make sure there is ventilation) start the engine and observe each plug wire. Be careful not to come into contact with any moving engine parts. If there is a break in the wire, you will see arcing or a small spark at the damaged area. If arcing is noticed, make a note to obtain new wires, then allow the engine to cool and check the distributor cap and rotor.

3   The spark plug wires should be inspected one at a time to prevent mixing up the order, which is essential for proper engine operation. Each original plug wire should be numbered to help identify its location. If the number is illegible, a piece of tape can be marked with the correct number and wrapped around the plug wire.

4   Disconnect the plug wire from the spark plug. A removal tool can be used for this purpose or you can grasp the rubber boot, twist the boot half a turn and pull the boot free. Do not pull on the wire itself **(see illustration 32.6a)**.

5   Check inside the boot for corrosion, which will look like a white crusty powder.

6   Push the wire and boot back onto the end of the spark plug. It should fit tightly onto the end of the plug. If it doesn't, remove the wire and use pliers to carefully crimp the metal connector inside the wire boot until the fit is snug.

7   Using a clean rag, wipe the entire length of the wire to remove built-up dirt and grease. Once the wire is clean, check for burns, cracks and other damage. Do not bend the wire sharply, because the conductor within the wire might break.

8   Disconnect the wire from the distributor. Again, pull only on the rubber boot. Check for corrosion and a tight fit. Press the wire back into the distributor.

9   Inspect the remaining spark plug wires, making sure that each one is securely fastened at the distributor and spark plug when the check is complete.

10   If new spark plug wires are required, purchase a set for your specific engine model. Pre-cut wire sets with the boots already installed are available. Remove and replace the wires one at a time to avoid mix-ups in the firing order. If a mix-up does occur refer to the firing order specifications at the beginning of this Chapter.

11   On vehicles equipped with a distributor, detach the distributor cap by prying off the two cap retaining clips or removing the screws. Look inside it for cracks, carbon tracks and worn, burned or loose contacts **(see illustration)**.

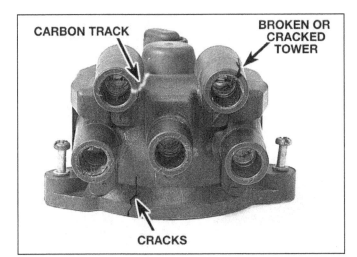

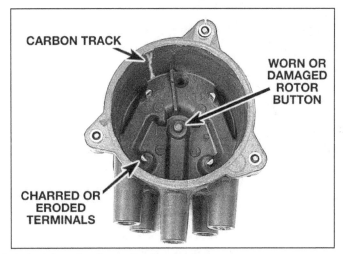

33.11 Shown here are some of the common defects to look for when inspecting the distributor cap (if in doubt about its condition, install a new one

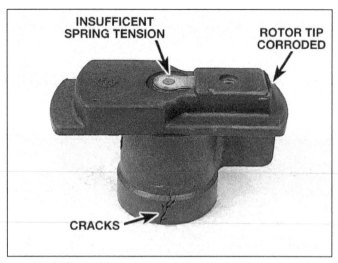

33.12 The ignition rotor should be checked for wear and corrosion as indicated here (if in doubt about its condition, buy a new one)

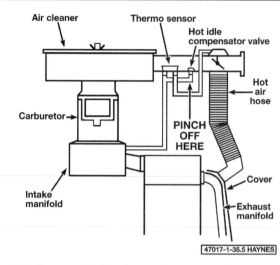

35.5 Use a small clamping device to pinch off the hot idle compensator valve vacuum hose (carbureted models only)

12   Pull the rotor off the distributor shaft and examine it for cracks and carbon tracks **(see illustration)**. On some models it will be necessary to remove a retaining screw before the rotor can be pulled off. Replace the cap and rotor if any damage or defects are noted.

13   It is common practice to install a new cap and rotor whenever new spark plug wires are installed, but if you wish to continue using the old cap, clean the terminals first.

14   When installing a new cap, remove the wires from the old cap one at a time and attach them to the new cap in the exact same location - do not simultaneously remove all the wires from the old cap or firing order mix-ups may occur.

34 Distributor advance mechanism check and lubrication (every 30,000 miles or 24 months)

1   Remove the distributor cap and rotor.

2   Use a small screwdriver to check that the distributor advance mechanism moves smoothly and easily and that the springs aren't damaged or stretched.

3   Lubricate the mechanism with a couple of drops of light oil and install the rotor and cap.

35 Idle speed check and adjustment (early four-cylinder engines only) (every 30,000 miles or 24 months)

Refer to illustrations 35.5, 35.7a and 35.7b
Note: *This procedure does not apply to 3.1L and 3.2L V6 models, or to 2.2L four-cylinder models. On these models, the minimum idle speed is preset at the factory and does not require periodic adjustment.*

1   Engine idle speed is the speed at which the engine operates when no throttle pedal pressure is applied. The idle speed is critical

to the performance of the engine itself, as well as many engine sub-systems.

2   A hand-held tachometer must be used when adjusting idle speed to get an accurate reading. The exact hook-up for these meters varies with the manufacturer, so follow the particular directions included.

3   Set the parking brake and block the wheels. Be sure the transmission is in Neutral (manual transmission) or Park (automatic transmission).

4   Turn off the headlights and any other accessories during this procedure.

5   On carbureted models, the distributor vacuum, evaporative canister purge and EGR vacuum hoses must be disconnected and plugged (see Chapter 6). The hot idle compensator vacuum line must be closed by pinching the rubber hose **(see illustration)**.

6   On air-conditioned carbureted models, turn the air conditioning On to Max cold and High blower, then open the throttle about 1/3 to allow the speed up solenoid to reach full

35.7a Disconnect the vacuum switching valve electrical connector (arrow) located next to the battery (fuel-injected models only)

35.7b The idle speed adjusting screw on fuel-injected models is located on the back side of the throttle body

travel. The idle speed adjusting screw is located at the base of the carburetor by the accelerator linkage (see Chapter 4A).

7   On fuel-injected models, the idle speed is to be checked with the throttle valve closed and the evaporative canister purge and EGR vacuum hoses disconnected and plugged (see Chapter 6). Unplug the connector from the pressure regulator Vacuum Switching Valve (VSV). The idle speed adjusting screw is located on the throttle body **(see illustrations)**.

8   Start the engine and allow it to reach normal operating temperature.

9   Check the engine idle speed with the tachometer and compare it to the specifications. If the idle speed is incorrect, turn the idle speed adjusting screw until the idle speed is correct.

10   Most models have a tune-up decal or Vehicle Emission Control Information (VECI) label located in the engine compartment with the specifications for setting idle speed. If no VECI label is found refer to the Specifications Section at the beginning of this Chapter.

36   Ignition timing check and adjustment (every 30,000 miles or 24 months)

Refer to illustrations 36.1, 36.4 and 36.9
Note: *This Section contains information for adjusting the ignition timing on 2.3L and 2.6L four-cylinder engines and 3.1L V6 engines only. It is imperative that the procedures included on the tune-up or Vehicle Emissions Control Information (VECI) label be followed when adjusting the ignition timing. The label will include all information concerning preliminary steps to be performed before adjusting the timing, as well as the timing specifications. If no VECI label is found refer to the specifications chart at the beginning of this Chapter. The ignition timing on 3.2L V6 engines and on 2.2L four-cylinder engines is controlled by the PCM and is non-adjustable, therefore it will not be covered in this Section.*

1   At the specified intervals, or when the distributor has been removed, the ignition timing should be checked and adjusted. Before attempting to check the timing some special tools will be needed for this procedure **(see illustration)**.

2   Various procedures such as idle speed adjustment must be performed before attempting to check the timing. Locate the Tune-up or VECI label under the hood and read through and perform all preliminary instructions concerning ignition timing. If no VECI label is found refer to the Specifications Section at the beginning of this Chapter.

3   Connect a timing light in accordance with the manufacturer's instructions. Generally, the light will be connected to power and ground sources and to the number one spark plug wire (refer to the cylinder location diagram in this Chapter's Specifications).

4   Locate the numbered timing tag on the

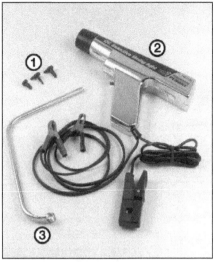

36.1 Tools needed to check and adjust the ignition timing

1   *Vacuum plugs - Vacuum hoses will, in most cases, have to be disconnected and plugged. Molded plugs in various shapes and sizes are available for this*

2   *Inductive pick-up timing light - Flashes a bright, concentrated beam of light when the number one spark plug fires. Connect the leads according to the instructions supplied with the light*

3   *Distributor wrench - On some models, the hold-down bolt for the distributor is difficult to reach and turn with conventional wrenches or sockets. A special wrench like this must be used*

front cover of the engine **(see illustration)**. It's located just behind the lower crankshaft pulley. Clean it off with solvent if necessary so you can see the numbers and small grooves.

5   Use chalk or paint to mark the groove in the crankshaft pulley.

6   Mark the timing tab in accordance with the number of degrees called for on the VECI label or the tune-up label in the engine compartment.

7   Make sure the timing light is clear of all moving engine components, then start the engine and warm it up to normal operating temperature.

8   Aim the flashing timing light at the timing mark by the crankshaft pulley, again being careful not to come in contact with moving parts. The marks should appear to be stationary. If the marks are in alignment, the timing is correct.

9   If the notch on the crankshaft pulley is not aligned with the correct mark on the timing tab, loosen the distributor hold-down bolt and rotate the distributor until the notch is aligned with the correct timing mark **(see illustration)**.

10   Retighten the hold-down bolt and recheck the timing.

36.4 The ignition timing indicator is located at the front of the engine by the crankshaft pulley

36.9 Loosen the hold-down nut (arrow) and turn the distributor to adjust the ignition timing

11   Turn off the engine and disconnect the timing light. Reconnect the vacuum hoses, and any other components which were disconnected.

37   Front wheel bearing check, repack and adjustment (every 30,000 miles or 24 months)

Note: *This Section contains information for 2WD models only. For information on 4WD models, see Chapter 8.*

Check and repack

Refer to illustrations 37.1, 37.6, 37.7, 37.8, 37.9, 37.11 and 37.15

1   In most cases the front wheel bearings will not need servicing until the brake shoes or pads are changed. However, the bearings should be checked whenever the front of the vehicle is raised for any reason. Several items, including a torque wrench and special grease, are required for this procedure **(see illustration)**. **Note:** *This procedure requires a*

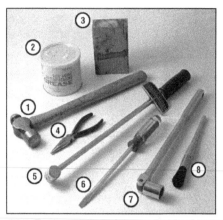

37.1 Tools and materials needed for front wheel bearing maintenance

1   **Hammer** - A common hammer will do just fine
2   **Grease** - High-temperature grease that is formulated specially for front wheel bearings should be used
3   **Wood block** - If you have a scrap piece of 2x4, it can be used to drive the new seal into the hub
4   **Needle-nose pliers** - Used to straighten and remove the cotter pin in the spindle
5   **Torque wrench** - This is used to ensure all play is taken out of the bearing by applying a slight preload before final adjustment
6   **Screwdriver** - Used to remove the seal from the hub (a long screwdriver is preferred)
7   **Socket/breaker bar** - Needed to loosen the nut on the spindle if it's extremely tight
8   **Brush** - Together with some clean solvent, this will be used to remove old grease from the hub and spindle

37.6 Unscrew the retaining bolts (arrows) and remove the dust cap

37.7 Unscrew the retaining screws (arrows) and remove the locking ring

37.8 Remove the wheel bearing adjusting nut with a special removal socket (available at automotive parts stores), or a hammer and punch

37.9 Pull the brake disc out slightly, then push it back in to disengage the outer wheel bearing

37.11 Use a screwdriver or seal removal tool to pry out the grease seal

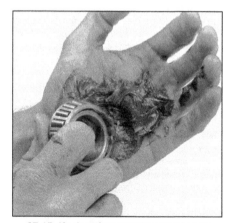

37.15 If a bearing packing tool is not available, work grease into the bearing rollers by pressing it against the palm of your hand

special socket (available at most automotive parts stores) to remove and retighten the wheel bearing adjusting nut.
2   With the vehicle securely supported on jackstands, spin each wheel and check for noise, rolling resistance and freeplay.
3   Grasp the top of each tire with one hand and the bottom with the other. Move the wheel in and out on the spindle. If there's any noticeable movement, the bearings should be checked and then repacked with grease, or replaced if necessary.
4   Remove the wheel.
5   Remove the brake caliper and anchor bracket (see Chapter 9). Hang the caliper out of the way on a piece of wire.
6   Unscrew the retaining screws and remove the dust cap **(see illustration)**.
7   Unscrew the lock ring retaining screws and remove the lock ring **(see illustration)**,
8   Using a special removal socket or a hammer and punch to remove the outer adjusting nut, **(see illustration)**.
9   Pull the brake disc out slightly, then push it back in. This should force the outer

bearing outward on the spindle so it can be removed **(see illustration)**.
10   Remove the disc assembly from the spindle.
11   Use a screwdriver or a seal puller tool to pry the seal out of the rear of the hub **(see illustration)**. Note how the seal is installed.
12   Remove the inner wheel bearing from the hub.

13   Use solvent to remove all traces of the old grease from the bearings, hub and spindle. A small brush may prove helpful; however make sure no bristles from the brush embed themselves inside the bearing rollers. Allow the parts to air dry.
14   Carefully inspect the bearings for cracks, heat discoloration, worn rollers, etc.

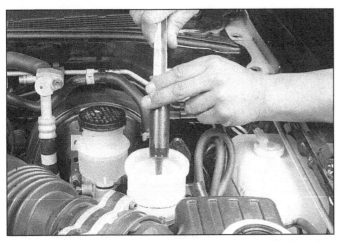

38.4 Use a hand suction pump or similar device to withdraw the fluid from the power steering pump reservoir

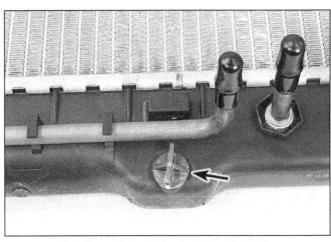

39.4 The radiator drain plug is located at the bottom of the radiator

Check the bearing races inside the hub for wear and damage. If the bearing races are defective, drive the bearing race out of the hub using a brass drift. Drive the new race into the hub using the appropriate size bearing driver (inexpensive bearing driver sets are available at most automotive parts stores). Note that the bearings and races are replaced as matched sets; used bearings should never be installed on new races.

15   Use only high-temperature front wheel bearing grease to pack the bearings. Inexpensive bearing packing tools are available at automotive parts stores, but not entirely necessary. If one is not available, pack the grease by hand completely into the bearings, forcing it between the rollers, cone and cage from the back side **(see illustration)**.

16   Apply a thin coat of grease to the spindle at the outer bearing seat, inner bearing seat, shoulder and seal seat.

17   Place a small quantity of grease inboard of each bearing race inside the hub. Using your finger, form a dam at these points to provide extra grease availability and to keep thinned grease from flowing out of the bearing.

18   Place the grease-packed inner bearing into the rear of the hub and put a little more grease outboard of the bearing.

19   Place a new seal over the inner bearing and tap the seal evenly into place with a hammer and block of wood until it's flush with the hub.

Adjustment

20   Carefully place the brake disc/hub assembly onto the spindle and push the grease-packed outer bearing into position.

21   Install the adjusting nut. Then tighten the adjusting nut to 22 ft-lbs while rotating the disc/hub assembly to seat the bearings.

22   Back off the adjuster nut one full turn, then retighten the adjuster nut hand tight. Make sure the disc rotates freely with no excessive freeplay.

23   Install the locking ring, if the bolt holes in the locking ring don't line up with the holes in

the adjuster nut turn the locking ring over. If the locking ring still does not line up turn the adjuster nut just enough to align the holes. Reinstall the locking ring retaining screws and recheck the freeplay.

24   The remainder of the installation is the reverse of removal.

38 Power steering fluid replacement (every 30,000 miles or 24 months)

Refer to illustration 38.4

1   At the specified time intervals, the power steering fluid should be drained and replaced. Since the power steering fluid may drip or splash when pouring it, place plenty of rags around the reservoir to protect any surrounding painted surfaces.

2   Before beginning work, purchase the specified power steering fluid (see *Recommended lubricants and fluids* at the beginning of this Chapter).

3   Remove the cap from the power steering fluid reservoir.

4   Using a hand suction pump or similar device, withdraw the fluid from the reservoir **(see illustration)**.

5   Add new fluid to the reservoir until it rises to the MAX line on the reservoir then install the reservoir cap.

6   Bleed the power steering system as described in Chapter 10.

39 Cooling system servicing (draining, flushing and refilling) (every 30,000 miles or 24 months)

Refer to illustration 39.4

Warning: *Do not allow antifreeze to come in contact with your skin or painted surfaces of the vehicle. Rinse off spills immediately with plenty of water. Antifreeze is highly toxic if ingested. Never leave antifreeze lying around in an open container or in puddles on the floor; children and pets are attracted by it's*

sweet smell and may drink it. Check with local authorities about disposing of used antifreeze. Many communities have collection centers which will see that antifreeze is disposed of safely. Never dump used antifreeze on the ground or into drains.
Note: *Non-toxic antifreeze is now manufactured and available at local auto parts stores, but even these types should be disposed of properly.*

1   Periodically, the cooling system should be drained, flushed and refilled to replenish the antifreeze mixture and prevent formation of rust and corrosion, which can impair the performance of the cooling system and cause engine damage. When the cooling system is serviced, all hoses and the radiator cap should be checked and replaced if necessary.

2   Apply the parking brake and block the wheels. If the vehicle has just been driven, wait several hours to allow the engine to cool down before beginning this procedure.

3   Once the engine is completely cool, remove the radiator cap. Place the heater temperature control in the maximum heat position.

4   Move a large container under the radiator drain to catch the coolant, then unscrew the drain plug (a pair of pliers may be required to turn it) **(see illustration)**.

5   After the coolant stops flowing out of the radiator, move the container under the engine block drain plug(s) (on four-cylinder engines there is one plug on the side of the block; on V6 engines there are two plugs, located on both sides of the block). Remove the plug(s) and allow the coolant in the block to drain.

6   While the coolant is draining, check the condition of the radiator hoses, heater hoses and clamps (refer to Section 10 if necessary).

7   Replace any damaged clamps or hoses.

8   Once the system is completely drained, flush the radiator with fresh water from a garden hose until it runs clear at the drain. The flushing action of the water will remove sediments from the radiator but will not remove rust and scale from the engine and cooling

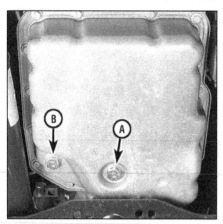

40.7 The transmission drain plug (A) is located on the bottom of the fluid pan - The overfill screw (B) is used on 1996 and later models only

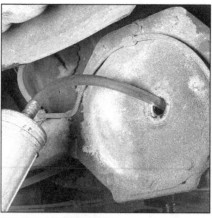

43.6 This is the easiest way to remove the lubricant - work the end of the hose to the bottom of the differential housing and draw out the old lubricant with a suction pump

tube surfaces.

9   These deposits can be removed with a chemical cleaner. Follow the procedure outlined in the manufacturer's instructions. If the radiator is severely corroded, damaged or leaking, it should be removed (see Chapter 3) and taken to a radiator repair shop.

10   Remove the overflow hose from the coolant recovery reservoir. Drain the reservoir and flush it with clean water, then reconnect the hose.

11   Reinstall and tighten the radiator drain plug. Install and tighten the block drain plug(s).

12   Slowly add new coolant (a 50/50 mixture of water and antifreeze) to the radiator until it's full. Add coolant to the reservoir up to the lower mark.

13   Leave the radiator cap off and run the engine in a well-ventilated area until the thermostat opens (coolant will begin flowing through the radiator and the upper radiator hose will become hot).

14   Turn the engine off and let it cool. Add more coolant mixture to bring the level back up to the lip on the radiator filler neck.

15   Squeeze the upper radiator hose to expel air, then add more coolant mixture if necessary. Replace the radiator cap.

16   Start the engine, allow it to reach normal operating temperature and check for leaks.

40   Automatic transmission fluid change (every 30,000 miles or 24 months)

Refer to illustration 40.7

Caution: *Refilling the automatic transmission fluid on 1996 and later models requires the use of a special fluid temperature gauge (available from your dealer parts department). Do not attempt to perform this procedure without this tool or damage to the transmission may occur.*

1   At the specified time intervals, the automatic transmission fluid should be drained and replaced.

2   Before beginning work, purchase the specified transmission fluid (see *Recommended fluids and lubricants*, and *Capacities* at the beginning of this Chapter).

3   Other tools necessary for this job include jackstands to support the vehicle in a raised position, a wrench, a drain pan capable of holding at least four quarts, newspapers and clean rags.

4   The fluid should be drained after the vehicle has been driven and brought to operating temperature. Hot fluid is more effective than cold fluid at removing built up sediment. **Warning:** *Fluid temperature can exceed 350-degrees F in a hot transmission. Wear protective gloves.*

5   Raise the vehicle and place it on jackstands.

6   Move the necessary equipment under the vehicle, being careful not to touch any of the hot exhaust components.

7   Place the drain pan under the drain plug in the transmission housing or fluid pan and remove the drain plug **(see illustration)**. Be sure the drain pan is in position, as fluid will come out with some force. Once the fluid is drained, reinstall the drain plug securely.

8   Lower the vehicle.

9   With the engine off, add new fluid to the transmission through the dipstick tube (1995 and earlier models). Use a funnel to prevent spills. It is best to add a little fluid at a time, continually checking the level with the dipstick (see Section 6). Allow the fluid time to drain into the pan.

10   Start the engine and shift the selector into all positions from Park through Low then shift into Park and apply the parking brake.

11   With the engine idling, check the fluid level. Add fluid up to the proper level on the dipstick (see Section 6).

12   On 1996 and later vehicles add fluid through the overfill screw opening **(see illustration 40.7)** until it just starts to run out of the opening. Reinstall the over fill screw plug and refer to Section 6 for the remainder of the check and refill procedure.

41   Manual transmission lubricant change (every 30,000 miles or 24 months)

1   Raise the vehicle and support it securely on jackstands.

2   Move a drain pan, rags, newspapers and wrenches under the transmission.

3   Remove the transmission drain plug at the bottom of the case and allow the lubricant to drain into the pan **(see illustration 20.1)**.

4   After the lubricant has drained completely, reinstall the plug and tighten it securely.

5   Remove the fill plug from the side of the transmission case. Using a hand pump, syringe or funnel, fill the transmission with the specified lubricant until it is level with the lower edge of the filler hole. Reinstall the fill plug and tighten it securely.

6   Lower the vehicle.

7   Drive the vehicle for a short distance, then check the drain and fill plugs for leakage.

42   Transfer case lubricant change (4WD models only) (every 30,000 miles or 24 months)

1   Drive the vehicle for at least 15 minutes to warm the lubricant in the case. Perform this warm-up procedure with 4WD engaged, if possible. Use all gears, including Reverse, to ensure the lubricant is sufficiently warm to drain completely.

2   Raise the vehicle and support it securely on jackstands.

3   Remove the drain plug from the lower part of the case and allow the old lubricant to drain completely **(see illustration 21.1)**.

4   After the lubricant has drained completely, reinstall the plug and tighten it securely.

5   Remove the filler plug from the case.

6   Fill the case with the specified lubricant until it is level with the lower edge of the filler hole.

7   Install the filler plug and tighten it securely.

8   Drive the vehicle for a short distance and recheck the lubricant level. In some instances a small amount of additional lubricant will have to be added.

43   Differential lubricant change (every 30,000 miles or 24 months)

Drain

1   This procedure should be performed after the vehicle has been driven so the lubricant will be warm and therefore flow out of the differential more easily.

2   Raise the vehicle and support it securely on jackstands.

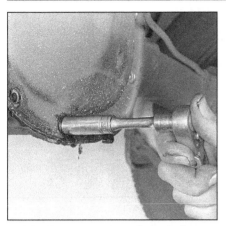

43.8a Remove the bolts from the lower edge of the cover . . .

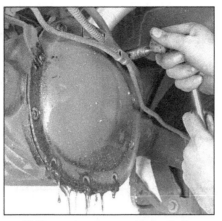

43.8b . . . then loosen the top bolts, allowing the lubricant drain

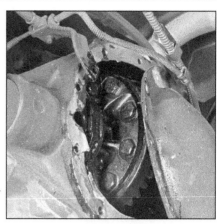

43.8c After the lubricant has drained, remove the bolts and the cover

3   The easiest way to drain the differential(s) is to remove the lubricant through the filler plug hole with a suction pump. If the rear differential cover gasket is leaking, it will be necessary to remove the cover to drain the lubricant (which will also allow you to inspect the differential.

Changing the lubricant with a suction pump

Refer to illustration 43.6
4   Remove the filler plug from the differential (see Section 22).
5   Insert the flexible hose.
6   Work the hose down to the bottom of the differential housing and pump the lubricant out **(see illustration)**.

Changing the lubricant by removing the cover

Refer to illustrations 43.8a, 43.8b, 43.8c and 43.10
7   Move a drain pan, rags, newspapers and wrenches under the vehicle.
8   Remove the bolts on the lower half of the cover. Loosen the bolts on the upper half and use them to loosely retain the cover. Allow the oil to drain into the pan, then completely remove the cover **(see illustrations)**.
9   Using a lint-free rag, clean the inside of the cover and the accessible areas of the differential housing. As this is done, check for chipped gears and metal particles in the lubricant, indicating that the differential should be more thoroughly inspected and/or repaired.
10   Thoroughly clean the gasket mating surfaces of the differential housing and the cover plate. Use a gasket scraper or putty knife to remove all traces of the old gasket **(see illustration)**.
11   Apply a thin layer of RTV sealant to the cover flange, then press a new gasket into position on the cover. Make sure the bolt holes align properly.

Refill

12   Use a hand pump, syringe or funnel to fill the differential housing with the specified

43.10 Carefully scrape the old gasket material off to ensure a leak-free seal

lubricant until it's level with the bottom of the filler plug hole.
13   Install the fill plug and tighten it securely.

---

44   Valve clearance adjustment (1998 and later 3.2L V6 engine) (every 60,000 miles)

Check

Refer to illustration 44.7
1   The valve clearances are checked and adjusted with the engine cold.
2   Position the number 1 piston at Top Dead Center on the compression stroke (see Chapter 2A).
3   Disconnect the cable from the negative terminal of the battery (see Chapter 5).
4   Remove the spark plugs (this will make the engine easier to rotate) (see Section 32).
5   Remove the valve covers (see Chapter 2D).
6   Before checking the valve clearances, check that the camshaft bearing cap bolts are tightened to the torque listed in the Chapter 2D Specifications.
7   Measure the clearances of the number

44.7 Check the clearance of each valve with a feeler gauge of the specified thickness - if the clearance is correct, you will feel a slight drag on the gauge as you pull it out

one cylinder valves with feeler gauges **(see illustration)**. Record the measurements which are out of specification. They will be used later to determine the required replacement shims.
8   Turn the crankshaft 1/3 revolution (120-degrees) and check the valve clearances for cylinder number 2. Turn the crankshaft 1/3 revolution again and check the valve clearances for cylinder number 3. After checking the clearances for that cylinder's valves, turn the crankshaft 1/3 turn and check the valve clearances for the next cylinder in the firing order, and so on, until all of the clearances have been checked. Make notes of the cylinder number and the valve type (intake or exhaust) that need to be adjusted.

Adjustment

Refer to illustrations 44.10a, 44.10b and 44.11
9   After all the valves have been measured, turn the crankshaft pulley until the camshaft lobe above the first valve which you intend to adjust is pointing upward, away from the shim.
10   Position the notch in the valve lifter toward the spark plug. Then depress the

44.10a Install the valve lifter tool as shown and squeeze the handles together to depress the valve lifter, hold the lifter down with the smaller tool . . .

44.10b . . . then remove the shim with a small screwdriver

valve lifter with the special valve lifter tools **(see illustrations)**. Place the special valve lifter tool in position as shown, with the longer jaw of the tool gripping the lower edge of the cast lifter boss and the upper, shorter jaw gripping the upper edge of the lifter itself. Depress the valve lifter by squeezing the handles of the valve lifter tool together, then hold the lifter down with the smaller tool and remove the larger one. Remove the adjusting shim with a small screwdriver or a pair of tweezers **(see illustrations)**. Note that the wire hook on the end of some valve lifter tool handles can be used to clamp both handles together to keep the lifter depressed while the shim is removed.

11   Measure the thickness of the shim with a micrometer **(see illustration)**. To calculate the correct thickness of a replacement shim that will place the valve clearance within the specified value, use the following formula:

   $N = T + (A - V)$
   T = *thickness of the old shim*
   A = *valve clearance measured*
   N = *thickness of the new shim*
   V = *desired valve clearance* (see this Chapter's Specifications)

12   Select a shim with a thickness as close as possible to the valve clearance calculated. Shims, which are available in 41 sizes in increments of 0.0008-inch (0.020 mm), range in size from 0.0945-inch (2.40 mm) to 0.1260-inch (3.20 mm). **Note:** *Through careful analysis of the shim sizes needed to bring the out-of-specification valve clearance within specification, it is often possible to simply move a shim that has to come out anyway to another valve lifter requiring a shim of that particular size, thereby reducing the number of new shims that must be purchased.*

13   Place the special valve lifter tool in position as shown in **illustration 44.10a**, with the longer jaw of the tool gripping the lower edge of the cast lifter boss and the upper, shorter jaw gripping the upper edge of the lifter itself, press down the valve lifter by squeezing the handles of the valve lifter tool together and install the new adjusting shim (note that the

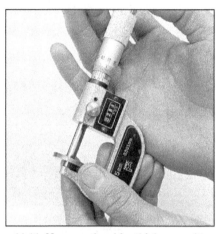

44.11 Measure the shim thickness with a micrometer

wire hook on the end of one valve lifter tool handle can be used to clamp the handles together to keep the lifter depressed while the shim is inserted. Measure the clearance with a feeler gauge to make sure that your calculations are correct.

14   Repeat this procedure until all the valves which are out of clearance have been corrected.

15   Reinstall the valve cover and spark plugs and any other components which were removed.

45   Emissions maintenance indicator - resetting (every 90,000 miles)

Refer to illustration 45.3
Note: *This procedure is required for 1989 through 1995 models equipped with four-cylinder engines and 1993 through 1995 models equipped with V6 engines.*
1   When the specified mileage has elapsed, an O2 indicator light on the dash will come on as a reminder that the oxygen sensor must be replaced (see Chapter 6). After the O2 sensor has been replaced it is necessary to reset the O2 indicator light to remind the operator to replace the sensor at the next 90,000 mile interval.
2   Remove the instrument cluster (see Chapter 12).
3   Locate the two holes marked KSW on the reverse side of the instrument cluster **(see illustration)**.
4   Remove the masking tape from hole KSW "B" (if equipped).
5   Remove the screw from the KSW hole "A" and insert it into the KSW hole "B". **Note:** *depending on the vehicle mileage and the service record of the vehicle it may be necessary to reverse this procedure.*
6   Install new masking tape to the open hole.
7   Reinstall the instrument cluster.

45.3 Remove the screw from KSW hole "A" and insert it into KSW hole "B" - hole location may vary depending on the year, make and model of the vehicle

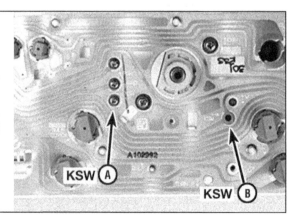

Chapter 2 Part A
2.3L and 2.6L four-cylinder engines

Contents

Specifications

General

Cylinder numbers (front-to-rear)	1-2-3-4
Firing order	1-3-4-2
Displacement	
4ZD1	2.3L (138 cu in)
4ZE1	2.6L (156.2 cu in)
Valve clearances	See Chapter 1

Camshaft

Journal diameter	1.3397 to 1.3324 inches
Endplay	0.002 to 0.0059 inch
Cam lobe height	
4ZD1 2.3L	1.4520 to 1.4320 inches
4ZE1 2.6L	1.4560 to 1.4320 inches
Cam lobe wear limit	0.002 inch or less
Camshaft runout	0.002 inch
Bearing inside diameter	Not available
Bearing oil clearance	0.0026 to 0.0043 inch
Valve spring free length	1.8951 to 1.8321 inches

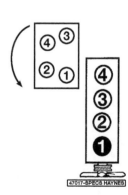

Cylinder location and distributor rotation

Torque specifications

	Ft-lbs (unless otherwise indicated)
Camshaft sprocket bolt	43
Timing belt tensioner lock bolt	168 in-lbs
Crankshaft drivebelt pulley bolt	87
Cylinder head bolts	
Step 1	58
Step 2	72
Exhaust manifold bolts/nuts	16
Flywheel/driveplate-to-crankshaft bolts	40
Valve cover bolts	87 in-lbs
Timing belt cover-to-block bolts	48 in-lbs
Intake manifolds	
Lower intake manifold bolts/nut	16
Air intake plenum (upper intake manifold)	20
Oil pan mounting bolts and nuts	10 to 15
Oil pump mounting bolts	156 in-lbs
Oil pump sprocket nut	55
Rocker arm assembly mounting nuts	16
Engine mount nut to engine bracket	62
Engine mount to chassis bolts	37

1 General information

This Part of Chapter 2 is devoted to in-vehicle repair procedures for the 2.3L (4ZD1) and 2.6L (4ZE1) four-cylinder engines. All information concerning engine removal and installation and engine block and cylinder head overhaul can be found in Part E of this Chapter.

The following repair procedures are based on the assumption the engine is installed in the vehicle. If the engine has been removed from the vehicle and mounted on a stand, many of the steps outlined in this Part of Chapter 2 will not apply.

The Specifications included in this Part of Chapter 2 apply only to the procedures contained in this Part. Part E of Chapter 2 contains the Specifications necessary for cylinder head and engine block rebuilding.

2 Repair operations possible with the engine in the vehicle

Many major repair operations can be accomplished without removing the engine from the vehicle.

Clean the engine compartment and the exterior of the engine with some type of pressure washer before any work is done. It'll make the job easier and help keep dirt out of the internal areas of the engine.

Depending on the components involved, it may be necessary to remove the hood to improve access to the engine as repairs are performed (refer to Chapter 11 if necessary).

If vacuum, exhaust, oil or coolant leaks develop, indicating a need for gasket or seal replacement, the repairs can generally be made with the engine in the vehicle. The oil pan gasket, cylinder head gasket, intake and exhaust manifold gaskets, engine front cover gaskets and the crankshaft oil seals are accessible with the engine in place.

Exterior engine components, such as

the water pump, the starter motor, the alternator, the distributor and the fuel system components, as well as the intake and exhaust manifolds, can be removed for repair with the engine in place.

Since the cylinder head can be removed without pulling the engine, camshaft and valve component servicing can also be accomplished with the engine in the vehicle.

Replacement of, repairs to or inspection of the timing belt and sprockets and the oil pump are all possible with the engine in place.

In extreme cases caused by a lack of necessary equipment, repair or replacement of piston rings, pistons, connecting rods and rod bearings is possible with the engine in the vehicle. However, this practice is not recommended because of the cleaning and preparation work that must be done to the components involved.

3 Top Dead Center (TDC) for number one piston - locating

Refer to illustration 3.6
Note: *The following procedure is based on the assumption that the distributor is correctly installed. If you are trying to locate TDC to install the distributor correctly, piston position must be determined by feeling for compression at the number one spark plug hole (or using a compression gauge) then aligning the ignition timing marks as described in Step 6.*

1   Top Dead Center (TDC) is the highest point in the cylinder that each piston reaches as it travels up-and-down when the crankshaft turns. Each piston reaches TDC on the compression stroke and again on the exhaust stroke, but TDC generally refers to piston position on the compression stroke.

2   Positioning the piston(s) at TDC on the compression stroke is an essential part of many procedures such as valve train component removal and distributor removal.

3   Before beginning this procedure, be

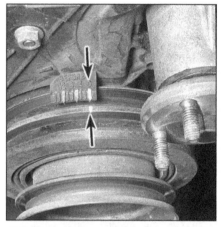

3.6 The number one piston is at TDC on the compression stroke when the notch in the crankshaft pulley is aligned with the 0 on the timing plate and the rotor is pointing at the number one spark plug terminal in the distributor cap

sure to place the transaxle in Neutral and apply the parking brake or block the rear wheels. Also, disconnect the cable from the negative terminal of the battery.

4   When looking at the drivebelt end of the engine, normal crankshaft rotation is clockwise. In order to bring any piston to TDC, the crankshaft must be turned with a socket and ratchet attached to the bolt threaded into the center of the lower drivebelt pulley (vibration damper) on the crankshaft.

5   Install a compression tester into the cylinder number one spark plug hole. The type that threads into the spark plug hole is preferred over the type that requires hand pressure to hold it in.

6   When the piston approaches TDC on the compression stroke, pressure will register on the gauge. Have your assistant stop turning the crankshaft when the timing marks at the crankshaft pulley are aligned **(see illustration)**.

7   If the timing marks are bypassed, turn

4.4 Unbolt the throttle linkage bracket (left arrow) and the clamp (right arrow) to the air injection pipe, then remove the nut from the valve cover and pivot the clamp out of the way

4.5 Valve cover mounting bolts

the crankshaft two complete revolutions clockwise until the timing marks are properly aligned again.

8   After the number one piston has been positioned at TDC on the compression stroke, TDC for any of the remaining pistons can be located by turning the crankshaft one-half turn (180-degrees) on four-cylinder engines or one-third turn (120-degrees) on V6 engines to get to TDC for the next cylinder in the firing order.

4   Valve cover - removal and installation

Refer to illustrations 4.4 and 4.5

1   Remove the air cleaner (Chapter 4).
2   Remove the spark plug wires from all spark plugs and remove the wire supports from the rocker arm cover. Don't disconnect the wires from the supports.
3   Disconnect the brake booster vacuum hose and position it out of the way.
4   Remove the throttle linkage bracket from the valve cover, and the loosen the clamp from the injection pipe on the left **(see illustration)**. Remove the nut that fastens the clamp to the valve cover and pivot it out of the way.
5   Remove the valve cover mounting bolts **(see illustration)**.
6   Detach the valve cover. **Caution:** *If the cover is stuck to the head, bump the end with a block of wood and a hammer to jar it loose. If that doesn't work, try to slip a flexible putty knife between the head and cover to break the gasket seal. Don't pry at the cover-to-head joint or damage to the sealing surfaces may occur, leading to oil leaks in the future.*
7   The mating surfaces of the head and valve cover must be perfectly clean when the cover is installed. Use a gasket scraper to remove all traces of sealant and old gasket material, then clean the mating surfaces with lacquer thinner or acetone. If there's sealant or oil on the mating surfaces when the cover is installed, oil leaks may develop.

8   Apply a continuous bead of sealant to the cover-to-head mating surface of the cover. Be sure to apply it to the inside of the mounting bolt holes.
9   Place the new gasket in position on top of the cylinder head, then place the valve cover on the gasket. While the sealant is still wet, install the mounting bolts and tighten them to the torque listed in this Chapter's Specifications.
10   Complete the installation by reversing the removal procedure.

5   Intake manifold - removal and installation

Note: *To prevent air leaks and possible damage to the valves, the gasket must be replaced whenever either manifold is removed. The engine must be completely cool when this procedure is done.*

Removal

1   Disconnect the negative battery cable from the battery.
2   Drain the cooling system.
3   Remove the air cleaner (see Chapter 4).
4   Mark all lines and hoses before removal to ensure correct installation. Disconnect the fuel and vacuum lines from the carburetor or fuel pipe and throttle valve assembly on fuel injected models. Plug the fuel lines immediately to prevent fuel leakage and to keep dirt from entering the lines.
5   Disconnect all hoses from the intake manifold.
6   Unplug all wiring connectors leading to the carburetor or fuel injection system.
7   Disconnect the brake booster hose from the intake manifold.

Carbureted engines

8   Remove the carburetor (see Chapter 4).
9   Remove the EGR pipe clamp bolt at the rear of the cylinder head.
10   Raise the vehicle and support it securely on jackstands.

11   Remove the EGR pipe from the intake and exhaust manifolds.
12   Working under the vehicle, remove the EGR valve and bracket assembly from the lower side of the intake manifold.
13   Remove the jackstands and lower the vehicle.
14   Disconnect the upper radiator hose from the engine. Disconnect the heater hose at the upper side of the manifold. **Note:** *The hoses should be plugged after disconnection.*
15   Loosen them in 1/4-turn increments, then remove all intake manifold mounting nuts.

Fuel-injected engines

Air intake plenum (upper intake manifold)

Refer to illustrations 5.17 and 5.18

16   Remove the throttle body from the air intake plenum.
17   Disconnect the EGR line from the underside of the plenum **(see illustration)**.

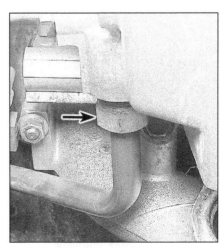

5.17 Before removing the air intake plenum (upper intake manifold), unscrew the EGR fitting from the underside of the plenum, but don't try to pull the EGR line out of the way or you will kink the line

5.18 To detach the air intake plenum (upper intake manifold) from the lower intake manifold, remove the plenum retaining nuts (some nuts, on the underside of the plenum, are not visible in this photo)

5.22 Unscrew the EGR pipe from the exhaust manifold

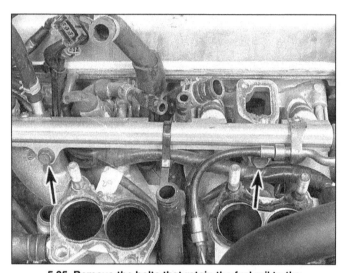

5.25 Remove the bolts that retain the fuel rail to the intake manifold

5.26 Remove the intake manifold upper nuts

18   Remove the plenum retaining nuts **(see illustration)**. (Some of the nuts are on the underside of the plenum.)

19   Lift the intake plenum off the intake manifold.

20   Stuff shop towels into the holes in the intake manifold. Using a gasket scraper or putty knife, remove all traces of old gasket material from the gasket mating surfaces on the intake manifold and common chamber. Be careful not to damage the delicate aluminum surfaces.

Lower intake manifold

Refer to illustrations 5.22, 5.25 and 5.26

21   Remove the upper intake manifold (see Steps 16 through 20).

22   Disconnect the EGR pipe from the exhaust manifold **(see illustration)**.

23   Carefully mark and remove any vacuum lines and hoses.

24   Detach the upper radiator hose from the engine.

25   Remove the fuel rail bolts **(see illustra-**

tion) and separate the fuel rail from the injectors.

26   Loosen them in 1/4-turn increments, then remove the intake manifold upper nuts **(see illustration)**. Carefully reach under the intake manifold and remove the lower retaining nuts.

27   Lift the manifold off the engine.

Installation

28   Before installing the manifold(s), stuff clean, lint-free rags into the engine ports and remove all traces of old gasket and sealant with a scraper. Clean the mating surfaces with lacquer thinner or acetone.

29   Place the (lower) intake manifold in position against the cylinder head and install the bolts, tightening them only enough to support the manifold.

30   Once the manifold is installed, tighten the mounting bolts to the torque listed in this Chapter's Specifications. Work from the center out toward the ends and work up to the final torque in three or four steps.

31   When installing the air intake plenum (upper intake manifold) on fuel-injected engines, install a new gasket and tighten the plenum retaining nuts to the torque listed in this Chapter's Specifications.

32   The remainder of installation is the reverse of the removal procedure.

33   Fill the radiator with coolant, start the engine and check for leaks. Also check and adjust the idle speed and throttle linkage as described in Chapter 1.

6   Exhaust manifold - removal and installation

Refer to illustrations 6.3 and 6.8

1   Disconnect the negative battery cable from the battery.

2   Remove the air cleaner assembly and the hot air hose.

3   Remove the three bolts attaching the exhaust manifold heat shield and detach the

6.3 Remove the exhaust manifold heat shield bolts and detach the head shield

6.8 Remove the exhaust manifold mounting nuts and detach the manifold from the cylinder head

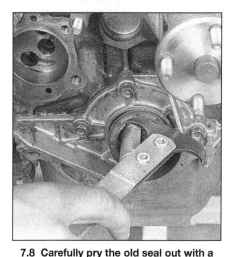

7.6 Use a bolt-type puller that applies force to the hub as shown here - if a jaw-typed puller that applies force to the outer edge is used, the vibration damper will be damaged

7.7 The sprocket should slide off the crankshaft

7.8 Carefully pry the old seal out with a seal puller or screwdriver - don't damage the crankshaft

heat shield **(see illustration)**.

4   On carbureted engines, remove the EGR pipe clip from the upper portion of the transmission and disconnect the EGR pipe from the exhaust manifold.

5   Disconnect the oxygen sensor wire at the connector (see Chapter 1).

6   Remove the nuts and detach the exhaust pipe from the manifold.

7   Remove the spark plug wires from the exhaust side spark plugs for access to the exhaust manifold bolts/nuts.

8   Remove the bolts/nuts that retain the exhaust manifold to the cylinder head **(see illustration)**.

9   Detach the exhaust manifold from the head.

10   Before installing the manifold, remove all traces of the old gasket with a scraper. Clean the mating surfaces with lacquer thinner or acetone.

11   Place a new exhaust manifold gasket in position on the cylinder head, then hold the

manifold in place and install the mounting bolts/nuts finger-tight.

12   Tighten the mounting bolts/nuts, in three or four steps, to the specified torque. Work from the center out toward the ends to prevent distortion of the manifold.

13   Install the remaining components. The remainder of installation is the reverse of the removal procedure.

14   Start the engine and check for exhaust leaks between the manifold and cylinder head and between the manifold and exhaust pipe.

---

7   Crankshaft front oil seal - replacement

Refer to illustrations 7.6, 7.7, 7.8 and 7.9

1   Disconnect the negative battery cable from the battery.

2   Remove the radiator and shroud (see Chapter 3).

3   Unbolt the lower front cover or skidplate (if equipped).

4   Remove the drivebelts (Chapter 1).

5   On manual transmission equipped models, put the transmission in High gear and apply the parking brake. On automatics, remove the starter and wedge a large screwdriver in the ring gear teeth. Remove the pulley mounting bolt. They're usually very tight, so use a six-point socket and a 1/2-inch drive breaker bar.

6   Use a puller to remove the pulley from the crankshaft **(see illustration)**. Do not use a gear puller that applies force to the outer edge of the pulley!

7   Remove the timing belt (Section 8) and detach the sprocket from the nose of the crankshaft **(see illustration)**.

8   Carefully pry the oil seal out of the seal housing with a seal removal tool or screwdriver **(see illustration)**. Don't scratch the seal bore or damage the crankshaft in the process (if the crankshaft is damaged the new seal will end up leaking).

9   Clean the bore in the cover or housing and coat the outer edge of the new seal with engine oil or multi-purpose grease. Using a socket with an outside diameter slightly smaller than the outside diameter of the seal, carefully drive the new seal into place with a hammer **(see illustration)**. If a socket isn't available, a short section of large diameter pipe will work. Check the seal after installation to be sure the spring didn't pop out.
10   Reinstall the sprocket.
11   Installation is the reverse of removal. Be sure to apply multi-purpose grease or clean engine oil to the seal contact surface on the pulley hub before installing the pulley on the crankshaft.
12   Tighten the pulley bolt to the torque listed in this Chapter's Specifications.
13   Run the engine and check for leaks.

8   Timing belt and sprockets - removal, inspection installation and adjustment

Removal

> **** CAUTION ****
>
> The timing system is complex. Severe engine damage will occur if you make any mistakes. Do not attempt this procedure unless you are highly experienced with this type of repair. If you are at all unsure of your abilities, consult an expert. Double-check all your work and be sure everything is correct before you attempt to start the engine.

Refer to illustrations 8.11, 8.12a, 8.12b, 8.13, 8.14 and 8.16
1   Disconnect the negative cable from the battery.
2   Remove the cooling fan/shroud as described in Chapter 3.
3   Refer to Chapter 3 for removal of the water pump pulley.
4   Position the No. 4 piston at Top Dead Center (Section 3).
5   Remove the drivebelts and spark plugs (see Chapter 1). **Note:** *Some early models have an air pump. If so, remove the air pump drivebelt.*
6   Refer to Section 7 and remove the crankshaft pulley.
7   If equipped, disconnect the air conditioning drivebelt idler pulley.
8   If equipped with power steering, remove the pump from the bracket and position it out of the way (see Chapter 10). The hydraulic hoses should not be disconnected from the pump.
9   Remove the crankshaft pulley (see Section 7).
10   If so equipped, remove the air injection tube that runs in front of the timing belt cover.
11   Unbolt and remove the upper and lower timing belt covers **(see illustration)**. Don't lose the rubber seals.
12   Be sure the camshaft timing mark lines up with the lug on the front housing and the

7.9 Use a large socket or piece of pipe and a hammer to install the new seal

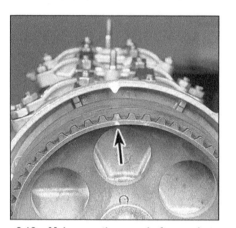

8.12a Make sure the camshaft sprocket timing mark is aligned with the upper back cover

crankshaft sprocket notch lines up with the pointer on the front oil seal housing **(see illustrations)**.
13   If you plan on reinstalling the timing belt, make an arrow on it to indicate the direction of rotation **(see illustration)**.
14   Loosen the tensioner bolt and use a large prybar to push the belt tensioner back

8.13 Mark the direction of rotation on the timing belt if the same belt will be used again

8.11 Remove the timing belt cover retaining bolts

8.12b Crankshaft sprocket timing marks - align the crank mark (lower arrow) with the cast-in mark on the front seal retainer (upper arrow)

against the spring, until there's enough room to slip the timing belt off **(see illustration)**.
15   Slip the belt off the sprockets and remove it from the engine.
16   Inspect the belt for wear, peeling, cracks, hardness, crimping and signs of oil or other fluid contamination **(see illustration)**. The belt should be replaced if any of these

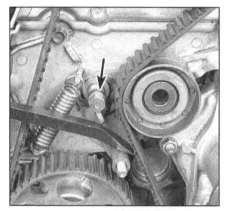

8.14 Wedge the tensioner back to allow room for the timing belt to be removed - arrow indicates the tensioner bolt

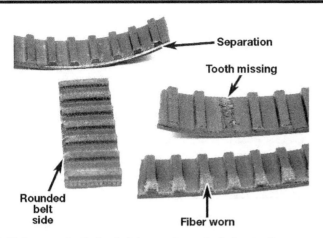

8.16 Inspect the timing belt for cracks and missing teeth; wear on one side of the belt indicates sprocket misalignment problems

8.23 You should be able to twist the timing belt no further than 90-degrees if it's properly adjusted

conditions exist or if the specified mileage has elapsed (see Chapter 1). **Caution:** *Unless the engine has very low mileage, it's common practice to replace the timing belt with a new one every time it's removed. Don't reinstall the original belt unless it's in like-new condition. A belt that breaks during engine operation can cause serious valve/piston damage.*

17   Inspect the sprockets for signs of wear or damage. If sprockets need to be replaced, they should be replaced as a set. Refer to Section 7 for removal of the crankshaft sprocket, to Section 9 for removal of the camshaft sprocket, and Section 13 for removal of the oil pump sprocket.

Installation

** CAUTION **

Before starting the engine, carefully rotate the crankshaft by hand through at least two full revolutions (use a socket and breaker bar on the crankshaft pulley center bolt). If you feel any resistance, STOP! There is something wrong - most likely, valves are contacting the pistons. You must find the problem before proceeding. Check your work and see if any updated repair information is available.

18   Make sure the crankshaft and camshaft sprocket timing marks are still aligned **(see illustrations 8.12a and 8.12b)**. **Note:** *When the mark on the camshaft is facing up, the engine is at TDC, on the compression stroke, for the number four cylinder.*

19   If the old belt is being reinstalled, make sure the arrow is pointed in the proper direction.

20   Slip the belt on, first onto the crankshaft sprocket, then the oil pump sprocket, the camshaft sprocket, and finally the tensioner. Wedge a screwdriver into the pivot area of the tensioner to temporarily release the tension **(see illustration 8.14)** to slip the belt on.

Adjustment

Refer to illustration 8.23

21   Loosen the tensioner lock bolt so only

the spring is applying pressure, then tighten the bolt.

22   Install the crankshaft sprocket bolt and use it to turn the crankshaft in the normal direction of rotation (clockwise) through two complete revolutions (720-degrees) so equal tension is applied to both sides of the timing belt, then repeat Step 21.

23   Recheck the timing mark alignment and check the tension of the belt midway between the oil pump and camshaft sprocket **(see illustration)**. If the tension is incorrect, repeat the adjustment operation.

24   Reinstall the remaining parts in the reverse order of removal.

---

9   Rocker arm assembly and camshaft - removal and installation

Rocker arm assembly removal

Refer to illustration 9.3

1   Remove the valve cover as described in Section 4.

2   Remove the timing belt, with the engine

set to TDC for the No. 4 cylinder (see Section 8).

3   The rocker arm assembly is attached to the cylinder head by several mounting nuts, plus two bolts at the front of the number 1 rocker shaft pedestal. The rocker arm assembly nuts/bolts should be loosened 1/4-turn at a time, working toward the center from both ends **(see illustration)**, to avoid distortion of the shafts by valve spring pressure. Loosen the bolts until they're free, but don't pull them out.

4   Lift the rocker arm assembly off the cylinder head.

Camshaft removal

Refer to illustrations 9.6 and 9.7

5   If you intend to remove the camshaft only, without disassembling the cylinder head, DO NOT alter the timing sprocket position in relationship to the belt! Mark both the timing belt and the camshaft sprocket to preserve the original relationship between the two (see Section 8). The belt must be reinstalled on the camshaft sprocket in the exact same relationship. If a major overhaul is being done, then the sprocket/belt relationship doesn't have to be maintained.

9.3 Loosen the rocker arm shaft-to-head nuts in 1/4-turn increments to avoid distorting the shaft

9.6 Position a screwdriver against a bolt head to prevent the camshaft sprocket from turning

9.7 The camshaft collar slides off the camshaft - arrow indicates the camshaft's Woodruff key; remove the key before replacing the camshaft seal

9.12 Lubricate the lobes and journals of the camshaft with camshaft installation lube before installation

6   Remove the camshaft sprocket bolt. A screwdriver can be inserted through one of the sprocket holes and wedged against a bolt head to keep the sprocket from turning while the bolt is loosened **(see illustration)**.

7   Remove the camshaft sprocket and collar **(see illustration)**.

8   Remove the rocker arm assembly (see Steps 1 through 3).

9   The camshaft sprocket's back cover (aluminum housing bolted to the cylinder head) retains the camshaft oil seal. If the seal is to be reused, pry the Woodruff key from the camshaft **(see illustration 9.7)**, to allow the camshaft to come out of the cover from the rear. Lift the camshaft out of the head.

10   If the camshaft is being removed for replacement due to high-mileage wear, the camshaft front oil seal should also be replaced. Pry the old seal from the camshaft sprocket's back cover with a screwdriver (while the camshaft is out). Install the new seal after the new camshaft is installed by tapping the seal into the cover with an-appropriate-size socket. After the seal is installed, replace the Woodruff key in the camshaft.

Camshaft installation

Refer to illustrations 9.12 and 9.15

11   The camshaft, rocker arms, and rocker shafts should be cleaned with solvent, dried, and inspected for wear or damage. Refer to Part E of this Chapter for inspection and measurement procedures. **Note:** *If the distributor drive gear on the camshaft is worn or damaged, replace the driven gear on the distributor also.*

12   Prior to installing the camshaft, apply a coat of moly-base grease or engine assembly lube to the bearing surfaces and lobes **(see illustration)**.

13   Carefully install the camshaft in the head.

14   Reinstall the rocker arm assembly (see Steps 18 through 21).

15   Make sure the camshaft timing mark is facing up **(see illustration)** and the crankshaft pulley timing mark is aligned with the TDC mark on the front cover (see Section 8). **Note:** *When the mark on the camshaft is facing up, the engine is at TDC, on the compression stroke, for the number four cylinder.*

16   Attach the sprocket to the camshaft.

17   Install the sprocket bolt and washer and tighten the bolt to the torque listed in this Chapter's Specifications.

Rocker arm assembly installation

Refer to illustration 5.20

18   Lubricate the rocker arms and shafts with plenty of clean engine oil.

19   Install the rocker arm assemblies with the front marks on the shafts facing up. **Note:** *The longer shaft (15.3937 inches) must be positioned on the exhaust side and the short shaft (15.2756 inches) on the intake manifold side.*

20   When installing the nuts/bolts, first tighten them until they're finger-tight, starting at the center and working towards the ends. Continue to tighten them in 1/4-turn increments until they're all at the torque listed in this Chapter's Specifications. Use an adjustable wrench to steady the rocker arm springs when tightening the nuts **(see illustration)**. The remaining installation steps are the reverse of removal.

9.15 Turn the camshaft until the mark is facing up

9.20 Hold the rocker arm springs with a wrench while tightening the nuts to prevent damage to the springs

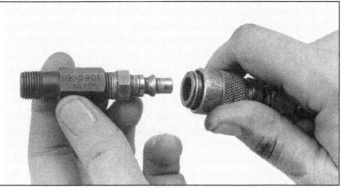

10.4 Compressed air can be used to hold the valves shut as the springs are compressed - this adapter is available at most auto parts stores

10.6 Compress the spring and use needle-nose pliers or a small magnet to remove the keepers

21   Adjust the valve clearances to the cold specifications as described in Chapter 1.
22   Reinstall the valve cover.

10   Valve springs, retainers and seals - replacement

Refer to illustrations 10.4, 10.6 and 10.14
Note: *Broken valve springs and defective valve stem seals can be replaced without removing the cylinder head. Two special tools and a compressed air source are normally required to perform this operation, so read through this Section carefully and rent or buy the tools before beginning the job.*
1   Refer to Section 4 and remove the valve cover. Disconnect the timing belt (see Section 8) and remove the rocker arm assembly and camshaft (see Section 9).
2   Remove the spark plug from the cylinder with the defective component. If all of the valve stem seals are being replaced, all of the spark plugs should be removed.
3   Turn the crankshaft until the piston in the affected cylinder is at top dead center on the compression stroke (refer to Section 3). Since the rocker arm assembly and camshaft have been removed, there is no danger of valves hitting pistons. If you're replacing all of the valve stem seals, begin with cylinder number one and work on the valves for one cylinder at a time. Move from cylinder-to-cylinder following the firing order sequence (1-3-4-2).
4   Thread an adapter into the spark plug hole and connect an air hose from a compressed air source to it **(see illustration)**. Most auto parts stores can supply the air hose adapter. **Note:** *Many cylinder compression gauges utilize a screw-in fitting that may work with your air hose quick-disconnect fitting.*
5   Apply compressed air to the cylinder. The valves should be held in place by the air pressure. If the valve faces or seats are in poor condition, leaks may prevent the air pressure from retaining the valves - refer to the alternative procedure below.
6   Stuff shop rags into the cylinder head holes to prevent parts and tools from falling into the engine, then use a valve spring compressor to compress the spring. Remove the keepers with small needle-nose pliers or a

magnet **(see illustration)**.
7   Remove the valve, spring retainer and valve springs, then remove the umbrella type guide seal. **Note:** *If air pressure fails to hold the valve in the closed position during this operation, the valve face or seat is probably damaged. If so, the cylinder head will have to be removed for additional repair operations.*
8   Wrap a rubber band or tape around the top of the valve stem so the valve won't fall into the combustion chamber, then release the air pressure.
9   Inspect the valve stem for damage. Rotate the valve in the guide and check the end for eccentric movement, which would indicate that the valve is bent.
10   Move the valve up-and-down in the guide and make sure it doesn't bind. If the valve stem binds, either the valve is bent or the guide is damaged. In either case, the head will have to be removed for repair.
11   Reapply air pressure to the cylinder to retain the valve in the closed position, then remove the tape or rubber band from the valve stem.
12   Lubricate the valve stem with engine oil and install a new valve guide oil seal.
13   Measure the valve spring free height and check with the measurements listed in this Chapter's specifications. Replace if the valve spring height exceeds the limit. Install the springs in position over the valve. Make sure the narrow pitch end of the outer spring is against the cylinder head. **Note:** *When the springs are marked with a factory paint mark, the marked end goes toward the cylinder head.*
14   Install the valve spring retainer. Compress the valve spring and position the keepers in the groove. Apply a small dab of grease to the inside of each keeper to hold it in place if necessary **(see illustration)**. Remove the pressure from the spring tool and make sure the keepers are seated.
15   Disconnect the air hose and remove the adapter from the spark plug hole.
16   Install the rocker arm assembly.
17   Install the spark plug(s) and hook up the wire(s).
18   Refer to Section 4 and install the valve cover.
19   Start and run the engine, then check for oil leaks and unusual sounds coming from the valve cover area.

10.14 Apply a small dab of grease to each keeper as shown here - it will keep them in place on the valve stem as the spring is released

11   Cylinder head - removal and installation

Removal
Refer to illustration 11.13
Note: *Depending on the reason for the disassembly, the cylinder head can be removed with the intake and exhaust manifolds still attached, or with the manifolds removed.*
1   Remove the intake and exhaust manifolds as described in Sections 5 and 6.
2   Drain the cooling system (see Chapter 1).
3   Remove the upper radiator hose.
4   Remove the thermostat housing (see Chapter 3).
5   Disconnect the heater hoses from the fitting on the rear of the cylinder head and the intake manifold.
6   Disconnect the spark plug wires from the spark plugs and position them out of the way, then remove the spark plugs.
7   Disconnect the accelerator linkage, fuel line, and all wires and vacuum hoses. Be sure to mark each one carefully to ensure proper installation.

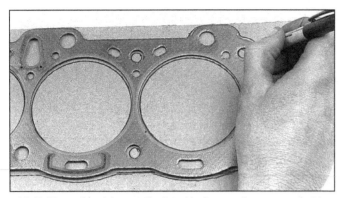

11.13 To avoid mixing up the head bolts, use the new gasket to transfer the hole pattern to a piece of cardboard - punch holes to accept the bolts and push each bolt through the matching hole in the cardboard

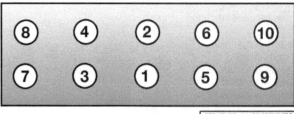

11.23 Cylinder head bolt TIGHTENING sequence

8   On early models with a carburetor, remove the fuel pump.

9   Raise the vehicle and support it on jackstands. Disconnect the exhaust pipe at the exhaust manifold (see Chapter 4). Lower the vehicle.

10   Remove the valve cover (see Section 4).

11   Bring the No. 4 piston to TDC (see Section 3) and remove the camshaft sprocket - be sure to follow the procedure in Sections 8 and 9.

12   On engines with an air injection system, disconnect the AIR hose and the check valve on the exhaust manifold.

13   Using a new head gasket, outline the cylinders and bolt pattern on a piece of cardboard **(see illustration)**. Be sure to indicate the front of the engine for reference. Punch holes at the bolt locations. Loosen the ten cylinder head bolts in 1/4-turn increments until they can be removed by hand (start at the ends and work toward the center of the head). Store the bolts in the cardboard holder as they're removed - this will ensure that they're reinstalled in their original locations.

14   Lift the head off the engine. If it's stuck, DO NOT pry between the head and block - damage to the mating surfaces will result! To dislodge the head, position a block of wood against it and strike the wood block with a hammer.

15   Remove the cylinder head gasket. Place the head on a block of wood on the bench to prevent damage.

16   Refer to Part E for cylinder head inspection and overhaul procedures.

Installation

Refer to illustration 11.23

17   The mating surfaces of the cylinder head and block must be perfectly clean when the head is installed. Use a gasket scraper to remove all traces of carbon and old gasket material, then clean the mating surfaces with lacquer thinner or acetone. If there's oil on the mating surfaces when the head is installed, the gasket may not seal correctly and leaks could develop. When working on the block, stuff the cylinders with clean shop rags to keep out debris. Use a vacuum

cleaner to remove any debris that falls into the cylinders.

18   Check the block and head mating surfaces for nicks, deep scratches and other damage. If damage is slight, it can be removed with a file; if it's excessive, machining may be the only alternative.

19   Use a tap of the correct size to chase the threads in the head bolt holes. Mount each bolt in a vise and run a die down the threads to remove corrosion and restore the threads. Dirt, corrosion, sealant and damaged threads will affect torque readings.

20   Install the gasket over the engine block dowel pins.

21   Make sure the No. 4 piston is still at TDC, then carefully lower the cylinder head onto the engine, over the dowel pins and the gasket. Be careful not to move the gasket.

22   Install the cylinder head mounting bolts. Make sure they're installed in the correct holes. **Note:** *Lightly apply engine oil to the threads and under the bolt heads before installation.*

23   Tighten the bolts following the recommended sequence **(see illustration)**, until the Step 1 specified torque is reached. Repeat the procedure, tightening all bolts to the Step 2 specified torque.

24   Install the camshaft sprocket as

12.7a   Remove the oil pan mounting bolts

described in Section 9.

25   The remainder of the installation procedure is the reverse of the removal procedure.

26   Fill the radiator with coolant, change the engine oil and filter (see Chapter 1), then start the engine and check for leaks. Be sure to recheck the coolant level once the engine has warmed up to operating temperature. Also check and adjust the idle speed as described in Chapter 1.

27   Readjust the valve clearances to the cold specifications as described in Chapter 1.

12   Oil pan - removal and installation

Refer to illustrations 12.7a, 12.7b and 12.10
Note: *Amigos with 2.3L engines are equipped with a one-piece oil pan while 2.6L engines have a two-piece oil pan.*

1   Drain the engine oil.

2   Raise the front of the vehicle and support it on jackstands placed under the frame. Apply the parking brake and block the rear wheels to keep the vehicle from rolling.

3   Remove the splash shield, if equipped.

4   On two-wheel-drive models, disconnect the steering idler arm and Pitman arm (see Chapter 10) and drop the center link and steering damper.

5   On 4WD models, where necessary for oil pan clearance, support the vehicle on jackstands, remove the front wheel and lower the front axle with a jack.

6   Remove the flywheel dust cover (refer to Chapter 7).

7   Remove the bolts and detach the oil pan **(see illustration)**. Don't pry between the block and the oil pan or damage to the sealing surface may result and oil leaks could develop. Use a block of wood and a hammer to dislodge the pan if it's stuck. **Note:** *Several of the oil pan bolts are engine-to-bellhousing bolts* **(see illustration)**.

8   Use a scraper to remove all traces of old gasket material and sealant from the block and oil pan. Clean the gasket sealing surfaces with lacquer thinner or acetone and make sure the bolt holes in the block are clean.

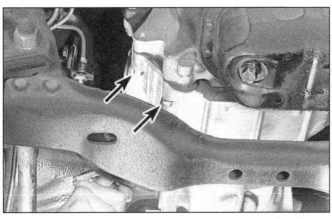

12.7b Several of the oil pan bolts are bellhousing bolts indicate two on the right side)

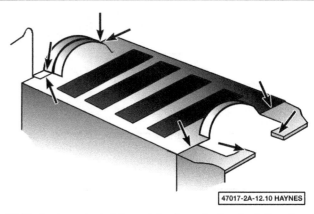

47017-2A-12.10 HAYNES

12.10 Apply sealant at the points shown before installing the oil pan - 2.3L engines

9   Check the oil pan flange for distortion, particularly around the bolt holes. If necessary, place the pan on a block of wood and use a hammer to flatten and restore the gasket surface.

10   On 2.3L engines, there is a one-piece oil pan gasket, while 2.6L engines use RTV sealant. Before installing the oil pan gasket on 2.3L engines, apply a thin coat of RTV sealant to the flange and to the corners of the engine block **(see illustration)**. Attach the new gasket to the pan (make sure the bolt holes are aligned).

11   Position the oil pan against the engine block. On 2.3L engines install the two stiffener plates, the mounting bolts, nuts and lockwashers. On 2.6L models, install the mounting bolts. Tighten them to the specified torque in a criss-cross pattern.

12   Wait at least 30 minutes before filling the engine with oil, then start the engine and check for leaks.

13   Oil pump - removal and installation

Removal

Refer to illustrations 13.3a and 13.3b

1   Disconnect negative battery cable.

2   Refer to Section 8 for timing belt removal.

3   Loosen the oil pump sprocket nut **(see illustration)**, then remove the pump-to-block bolts **(see illustration)** and withdraw the pump from the block cavity. If necessary, the nut can be removed and the sprocket can be detached from the pump shaft.

4   If the oil pump must be inspected or overhauled, refer to Part E of this Chapter.

5   Prior to installation, make sure the number one piston is still at TDC.

Installation

Refer to illustration 13.7

6   Lubricate the outer rotor with clean engine oil, then install it in the block cavity, chamfered side OUT.

13.3a Use a screwdriver to lock the oil pump sprocket when loosening the nut

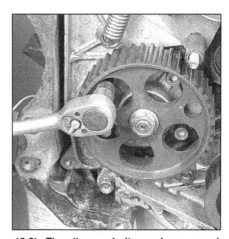

13.3b The oil pump bolts can be removed through the sprocket holes

7   Apply engine oil to the O-ring and insert it into the groove in the oil pump cover **(see illustration)**.

8   Lubricate the inner rotor with clean engine oil, then install the pump cover, engaging the inner rotor with the outer rotor in the engine block cavity.

9   Tighten the mounting bolts to the torque listed in this Chapter's Specifications. Tighten them gradually, in 1/4-turn increments, to avoid distorting the cover.

10   Make sure the oil pump shaft turns smoothly. If it doesn't, refer to Part E of this Chapter.

11   Install the sprocket, apply Loctite 262 to the first shaft threads, then install the nut. Tighten it to the torque listed in this Chapter's specifications.

12   The remainder of the assembly is the reverse of the removal procedure.

13   Start the engine and check for proper oil pressure.

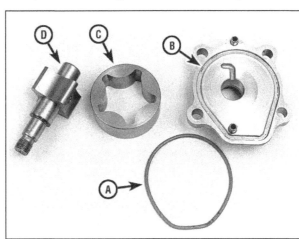

13.7 When reinstalling the oil pump, press the new O-ring (A) into the groove (B) in the oil pump cover, insert the outer rotor (C) into the block cavity, and put the inner rotor (D) into the pump cover

14 Flywheel/driveplate - removal and installation

Refer to illustrations 14.2 and 14.4

1   If the engine is in the vehicle, remove the clutch cover and clutch disc as described in Chapter 8, or the automatic transmission as described in Chapter 7.

2   Flatten the lockplate tabs (if used) and remove the bolts that secure the flywheel/driveplate to the crankshaft rear flange. Be careful, the flywheel is very heavy and shouldn't be dropped. If the crankshaft turns as the bolts are loosened, wedge a screwdriver in the starter ring gear teeth **(see illustration)**.

3   Detach the flywheel/driveplate from the crankshaft flange.

4   Unbolt and remove the engine rear plate **(see illustration)**. Now is a good time to check the engine block rear core plug for leakage.

5   If the teeth on the flywheel/driveplate starter ring gear are badly worn, or if some are missing, install a new flywheel or driveplate.

6   Refer to Chapter 8 for the flywheel inspection procedure.

7   Before installing the flywheel/driveplate, clean the mating surfaces.

8   If removed, reinstall the engine rear plate.

9   Position the flywheel/driveplate against the crankshaft, using a new spacer, if equipped, and insert the mounting bolts. Use Loctite 262 on the bolts. **Note:** *The crankshaft/flywheel bolts will leak oil if the threads are not properly sealed.*

10   Tighten the bolts in a criss-cross pattern to the torque listed in this Chapter's specifications.

11   The remainder of installation is the reverse of removal.

15 Rear main oil seal - replacement

Refer to illustration 15.5

1   The rear crankshaft oil seal can be replaced without removing the oil pan or crankshaft.

2   Remove the transmission (Chapter 7).

3   If equipped with a manual transmission, remove the pressure plate and clutch disc (Chapter 8).

4   Remove the flywheel or driveplate (Section 14).

5   Using a seal removal tool or a screwdriver, carefully pry the seal out of the block **(see illustration)**. Don't scratch or nick the crankshaft in the process!

6   Clean the bore in the block and the seal contact surface on the crankshaft. Check the crankshaft surface for scratches and nicks that could damage the new seal lip and cause oil leaks. If the crankshaft is damaged, the only alternative is a new or different crankshaft.

14.2 Wedge a screwdriver in the flywheel teeth to prevent the flywheel from turning as the bolts are loosened

7   Apply a light coat of engine oil or multi-purpose grease to the outer edge of the new seal. Lubricate the seal lip with moly-base grease.

8   The seal lip must face toward the front of the engine. Carefully work the seal lip over the end of the crankshaft and tap the seal in with a hammer and punch until it's seated squarely in the bore, to the same depth as the original seal.

9   Install the flywheel or driveplate.

10   If equipped with a manual transmission, reinstall the clutch disc and pressure plate.

11   Reinstall the transmission as described in Chapter 7.

16 Engine mounts - replacement

Refer to illustration 16.3

Warning: *Don't position any part of your body under the engine when the engine mounts are unbolted!*

1   Engine mounts are non-adjustable and seldom require service. Periodically they should be inspected for hardness and cracks in the rubber and separation of the rubber

15.5 Use a screwdriver to carefully pry the rear main oil seal out

14.4 Remove the engine rear plate bolts

from the metal backing.

2   To replace the engine mounts with the engine in the vehicle, use the following procedure.

3   Loosen the nuts and bolts that retain the front mounting insulator to the engine mount bracket and frame. Do this on both sides **(see illustration)**.

4   Next, the weight of the engine must be taken off the engine mounts. This can be done from beneath using a jack and wooden block positioned under the oil pan, or from above by removing the air cleaner and using an engine hoist attached to the two engine brackets. The engine should be raised slowly and carefully, while keeping a constant check on clearances around the engine to prevent anything from binding or breaking. Pay particular attention to areas such as the fan, ignition coil wires, vacuum lines leading to the engine and rubber hoses and ducts.

5   Raise the engine just enough to provide adequate room to remove the mounting insulator.

6   Remove the nuts and bolts retaining the insulator, then lift it out, noting how it's installed.

7   Installation is the reverse of removal, but be sure the insulator is installed in the same position it was in before removal.

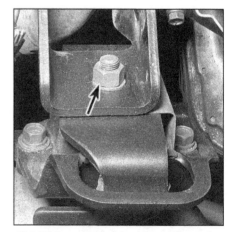

16.3 Remove the nut to release the engine mount bracket from the frame mount

Chapter 2 Part B
2.2L four-cylinder engine

Contents

Specifications

General

Cylinder numbers (front-to-rear)	1-2-3-4
Firing order	1-3-4-2
Displacement	2.2L (134.1 cu in)
Valve clearances	Hydraulic adjusters - no adjustment necessary

Camshaft

Journal diameter	Not available
Endplay	Not available
Cam lobe height	1.7602 inches
Cam lobe wear limit	1.7582 inches
Camshaft runout	Not available
Bearing inside diameter	Not available
Bearing oil clearance	Not available

Valves

Valve spring free length	Not available

Oil pump

Inner gear-to-outer gear teeth clearance	0.004 to 0.008 inch
Gear endplay	0.002 to 0.004 inch

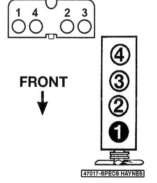

FRONT

Cylinder location and coil pack
terminal identification

Torque specifications

	Ft-lbs (unless otherwise indicated)
Camshaft bearing cap bolts	71 in-lbs
Camshaft sprocket bolts	
Step 1	36
Step 2	60 degrees
Step 3	15 degrees
Crankshaft pulley bolts	168 in-lbs
Crankshaft timing belt sprocket bolt	
Step 1	94
Step 2	45 degrees
Cylinder head bolts	
Step 1	18
Step 2	90 degrees
Step 3	90 degrees
Step 4	90 degrees
Exhaust manifold heat shield bolts..	71 in-lbs
Exhaust manifold nuts	168 in-lbs
Flywheel/driveplate-to-crankshaft bolts	
Step 1	48
Step 2	30 degrees
Step 3	15 degrees
Intake manifold bolt/nuts	16
Oil pan bolts	
Step 1	70 in-lbs
Step 2	30 degrees
Oil pan support bolts	168 in-lbs
Oil pressure switch	18
Oil pump	
Oil pump cover screws	48 in-lbs
Oil pump mounting bolts..	53 in-lbs
Oil pump pickup tube bolts	70 in-lbs
Oil pump pressure relief valve plug	22
Rear timing belt cover bolts..	53 in-lbs
Timing belt tensioner lock bolt	18
Valve cover bolts	83 in-lbs

1   General information

This Part of Chapter 2 is devoted to in-vehicle repair procedures for the 2.2L (X22SE) four-cylinder engine. All information concerning engine removal and installation and engine block and cylinder head overhaul can be found in Part E of this Chapter.

The following repair procedures are based on the assumption the engine is installed in the vehicle. If the engine has been removed from the vehicle and mounted on a stand, many of the steps outlined in this Part of Chapter 2 will not apply.

The Specifications included in this Part of Chapter 2 apply only to the procedures contained in this Part. Part E of Chapter 2 contains the Specifications necessary for cylinder head and engine block rebuilding.

2   Repair operations possible with the engine in the vehicle

Many major repair operations can be accomplished without removing the engine from the vehicle.

Clean the engine compartment and the exterior of the engine with some type of pressure washer before any work is done. It'll make the job easier and help keep dirt out of the internal areas of the engine.

Depending on the components involved, it may be necessary to remove the hood to improve access to the engine as repairs are performed (refer to Chapter 11 if necessary).

If vacuum, exhaust, oil or coolant leaks develop, indicating a need for gasket or seal replacement, the repairs can generally be made with the engine in the vehicle. The oil pan gasket, cylinder head gasket, intake and exhaust manifold gaskets, engine front cover gaskets and the crankshaft oil seals are accessible with the engine in place.

Exterior engine components, such as the water pump, the starter motor, the alternator and the fuel system components, as well as the intake and exhaust manifolds, can be removed for repair with the engine in place.

Since the cylinder head can be removed without pulling the engine, camshaft and valve component servicing can also be accomplished with the engine in the vehicle.

Replacement of, repairs to or inspection of the timing belt and sprockets and the oil pump are all possible with the engine in place.

In extreme cases caused by a lack of necessary equipment, repair or replacement of piston rings, pistons, connecting rods and rod bearings is possible with the engine in the vehicle. However, this practice is not recommended because of the cleaning and preparation work that must be done to the components involved.

3   Top Dead Center (TDC) for number one piston - locating

1   Top Dead Center (TDC) is the highest point in the cylinder that each piston reaches as it travels up-and-down when the crankshaft turns. Each piston reaches TDC on the compression stroke and again on the exhaust stroke, but TDC generally refers to piston position on the compression stroke.

2   Positioning the piston(s) at TDC on the compression stroke is an essential part of many procedures such as timing belt replacement, camshaft removal and installation and cylinder head removal and installation.

3   Before beginning this procedure, be sure to disable the ignition system by disconnecting the electrical connector from the ignition coil/ignition control module (see Chapter 5).

4   Remove the spark plugs (see Chapter 1) and then install a compression gauge in the number one cylinder (the cylinder that's closest to the timing belt end of the engine).

5   In order to bring any piston to TDC, the crankshaft must be turned using one of the following methods. When looking at the front of the engine, normal crankshaft rotation is clockwise. **Warning:** *Before beginning this procedure, be sure to place the transmission in Park.*

a) *The preferred method is to turn the crankshaft with a large socket and breaker bar attached to the pulley bolt threaded into the front of the crankshaft. Apply pressure onto the bolt in a clockwise direction only. Never turn the bolt counterclockwise.*

b) *A remote starter switch, which may save some time, can also be used. Attach the switch leads to the small ignition switch terminal and the positive (red) battery cable terminal on the starter solenoid (mounted near the battery). Once the piston is close to TDC, use a socket and breaker bar as described above.*

c) *If an assistant is available to turn the ignition switch to the START position in short bursts, you can get the piston close to TDC without a remote starter switch. Use a socket and breaker bar as described in Paragraph a) to complete the procedure.*

6   Turn the crankshaft clockwise until you see compression building up on the gauge - you are on the compression stroke for that cylinder. If you did not see compression build up, continue with one more complete revolution to achieve TDC for the number one cylinder.

7   Remove the compression gauge. Through the number one cylinder spark plug hole insert a section of wooden dowel or plastic rod and slowly push it down until it reaches the top surface of the piston crown. **Caution:** *Don't insert a metal or sharp object into the spark plug hole as the piston crown may be damaged.*

8   With the dowel or rod in place on top of the piston crown, slowly rotate the crankshaft clockwise until the dowel or rod is pushed upward, stops, and then starts to move back down. At this point, rotate the crankshaft slightly counterclockwise until the dowel or rod has reached it upper most travel. At this point the number one piston is at the TDC position.

9   After the number one piston has been positioned at TDC on the compression stroke, TDC for the next cylinder can be located by turning the crankshaft another 180 degrees (refer to the firing order in this Chapter's Specifications).

10   Another way to bring the number one piston to TDC is to align the timing marks on the camshaft and crankshaft timing belt sprockets with their corresponding stationary marks on the rear timing cover. But you can't

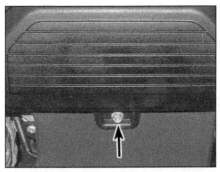

4.3 To remove the cover for the engine wiring harness from the top of the timing belt cover, remove this retaining nut

see the marks on the camshaft sprockets until the timing belt cover has been removed. If you're going to replace the timing belt or if you're going to remove the belt in order to replace a camshaft or crankshaft front oil seal, refer to Section 4. On the 2.2L engine, determining TDC is an integral part of timing belt removal and installation.

---

4   Timing belt and sprockets - removal, inspection and installation

---

Removal

** CAUTION **

The timing system is complex. Severe engine damage will occur if you make any mistakes. Do not attempt this procedure unless you are highly experienced with this type of repair. If you are at all unsure of your abilities, consult an expert. Double-check all your work and be sure everything is correct before you attempt to start the engine.

Refer to illustrations 4.3, 4.4, 4.5, 4.6, 4.7a, 4.7b and 4.9

1   Disconnect the negative battery cable.

2   Remove the serpentine drivebelt (see Section 25 in Chapter 1).

4.5 Using the chain wrench again, remove the four crankshaft pulley bolts and remove the pulley

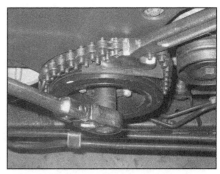

4.4 If you're going to remove the crankshaft timing belt sprocket, put a chain wrench on the crank pulley now and then *loosen* (but don't remove) the crankshaft sprocket bolt

3   Remove the engine harness cover from the top of the timing belt cover **(see illustration)**.

4   If you're going to remove the crankshaft timing belt sprocket, put a chain wrench on the crank pulley now **(see illustration)** and then loosen the crankshaft sprocket bolt. **Note:** *It is not necessary to remove the timing belt sprocket if you're just replacing the timing belt.* **Caution:** *Be sure to place a strip of rubber between the chain and the pulley to protect the pulley from damage.*

5   Using the chain wrench again, remove the four crankshaft pulley bolts **(see illustration)** and remove the pulley. **Caution:** *Be sure to place a strip of rubber between the chain and the pulley to protect the pulley from damage.*

6   Remove the two timing belt front cover bolts **(see illustration)**, and then remove the front cover.

7   Bring the number one piston to Top Dead Center (TDC) on its compression stroke as follows:

a) *Rotate the engine in the direction of normal operation (clockwise, as seen from the front of the vehicle) until the notches on the camshaft pulleys are aligned with the stationary marks on the cylinder head cover* **(see illustration)**.

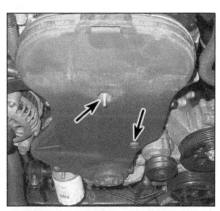

4.6 Remove the timing belt front cover nut and bolt and then remove the front cover

4.7a Rotate the engine in the direction of normal operation (clockwise, as seen from the front of the vehicle) until the notches on the camshaft pulleys are aligned with the stationary marks on the cylinder head cover

4.7b Also, make sure that the timing mark on the crankshaft sprocket is aligned with the mark on the timing cover

b) *Also, make sure that the timing mark on the crankshaft sprocket is aligned with the mark on the timing cover* **(see illustration)**.

8   If you plan to re-use the timing belt, put directional marks on the belt and the timing belt rear cover **(see illustration 8.13 in Chapter 2A)** to indicate the direction of belt travel.

9   Loosen the timing belt tensioner bolt **(see illustration)** and then remove tension from the tensioner by rotating the tensioner arm clockwise. When all tension has been removed from the tensioner, tighten the tensioner bolt to lock the tensioner in this position until the belt is installed again.

10   Remove the timing belt. **Caution:** *Do not bend, twist or turn the timing belt inside out. Do not allow it to come into contact with oil, coolant or fuel. Do not turn the crankshaft or camshaft while the timing belt is removed.*

Inspection

Timing belt

11   Inspect the timing belt for evidence of coolant or lubricant contamination. If the belt is contaminated, locate the source before proceeding. Inspect the timing belt for signs of wear or damage, especially around the drive side (the side pushed by the sprocket teeth) of each belt tooth **(see illustration 8.16 in Chapter 2A)**. Always replace the belt if its condition is in doubt; the cost of belt replacement is negligible compared with the potential cost of engine repairs if the belt should fail while the engine is operating. If you know that the belt has been in service more than 50,000 miles, it's a good idea (but not necessarily required) to replace it even if it looks okay (Isuzu specifies a 100,000 interval for belt replacement). **Caution:** *If the belt appears to be in good enough condition for further service, make sure that you install it with the drive sides of the teeth facing the same way as before. If the belt is installed the*

4.9 Loosen the timing belt tensioner bolt (1), and then rotate tensioner arm clockwise to remove tension from the tensioner; when all tension is removed, lock the tensioner in place by tightening the tensioner bolt

other way around, the belt will fail prematurely.

Pulleys

12   The tensioner pulley and the idler pulleys use sealed bearings, so they should be relatively trouble-free. Spin each pulley and listen carefully to the sound it makes as it spins. The pulleys should operate quietly and smoothly and they should have no runout. If any of the pulleys is noisy, rough or has excessive runout, replace it.

Tensioner pulley

Refer to illustration 4.13

13   To remove the tensioner pulley, remove the pulley bolt and pull off the tensioner. Note the position of the tang on the back of the tensioner in relation to the stopper that's cast into the oil pump cover. The tensioner is spring-loaded, so when installing it, the tang on the tensioner must be positioned on the left (driver's) side of this stop **(see illustra-**

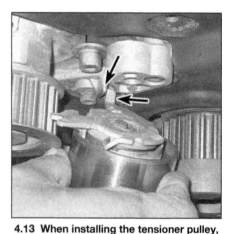

4.13 When installing the tensioner pulley, make sure that the tang on the back of the tensioner is positioned to the left (driver's) side of the stopper cast into the oil pump cover

tion), or the tensioner will not function correctly.

Idler pulleys

Refer to illustration 4.14

14   The idler pulleys use sealed bearings, so they are relatively trouble-free. To remove either idler pulley, simply remove the idler pulley retaining bolt **(see illustration)**.

Timing belt sprockets

15   Also inspect the timing belt sprockets for any obvious damage. On timing-belt engines, the camshaft and crankshaft sprockets are virtually trouble-free, but it's still a good idea to inspect them anytime the belt is removed. Even though the sprockets are likely to be in good condition, there are some procedures which necessitate the removal of the camshaft sprockets (camshaft seal or cam replacement) or the crankshaft sprocket (front main seal replacement or crank removal and installation).Replace all worn parts as necessary.

4.14 To remove either timing belt pulley, simply remove the pulley retaining bolt

4.16 Using a back-up wrench on the hexagonal part of the camshaft (between the timing belt sprocket and the cylinder head), remove the sprocket retaining bolt

4.19 When installing a camshaft timing belt sprocket, make sure that the slot on the backside of the sprocket is aligned with the guide pin on the end of the cam; the intake and exhaust sprockets are the same, but the intake slot or exhaust slot must be aligned with the pin, depending on which cam sprocket is being installed (exhaust cam sprocket shown)

Camshaft sprockets

Refer to illustrations 4.16 and 4.19

Note: *This procedure applies to either camshaft sprocket.*

16   Using a back-up wrench on the hexagonal part of the camshaft (between the timing belt sprocket and the cylinder head), remove the sprocket retaining bolt **(see illustration)**.

17   Remove the timing belt sprocket. If the sprocket is stuck to the end of the cam, use a small puller to remove it.

18   Inspect and, if necessary, replace the camshaft seals (see Section 8).

19   To install the sprocket, align the timing mark on the sprocket with the lug on the camshaft bearing cap. Also, make sure that the correct guide slot on the backside of the sprocket is aligned with the guide pin on the end of the cam **(see illustration)**.

20   Installation is otherwise the reverse of removal. Install the washer, and then tighten the sprocket bolt to the torque listed in this Chapter's Specifications.

Crankshaft sprocket

Refer to illustration 4.21

21   If you loosened the crankshaft timing belt sprocket bolt in Step 4 as instructed, remove the bolt, washer and sprocket **(see illustration)**. Don't lose the key.

22   Inspect and, if necessary, replace the crankshaft front oil seal (see Section 11).

23   Installation is the reverse of removal. Install the key, the sprocket, the washer and bolt, and then tighten the sprocket bolt to the torque listed in this Chapter's Specifications.

Installation

> ## ** CAUTION **
>
> Before starting the engine, carefully rotate the crankshaft by hand through at least two full revolutions (use a socket and breaker bar on the crankshaft pulley center bolt). If you feel any resistance, STOP! There is something wrong - most likely, valves are contacting the pistons. You must find the problem before proceeding. Check your work and see if any updated repair information is available.

Refer to illustrations 4.24, 4.25, 4.27a and 4.27b

24   Verify that the timing marks on the camshaft sprockets and on the crankshaft sprocket are aligned with their stationary marks on the rear timing cover **(see illustrations 4.7a and 4.7b)**. Also verify that the lug on the water pump is aligned with the lug on the engine block **(see illustration)**. If the lugs are not aligned, the pump has turned and must be re-positioned with the lugs aligned as shown (see Chapter 3).

25   While holding the sprockets stationary, install the timing belt. When installing the belt, make sure that the "drive" side of the belt (the left, or driver's, side) is taut **(see**

4.21 Remove the bolt, washer and crankshaft timing belt sprocket; don't lose the key

4.24 Before installing the timing belt, make sure that the lugs on the water pump and on the block are aligned; if they're not, the pump has turned and must be realigned (see Chapter 3)

4.25 When you install the timing belt, make sure that the "drive" side (left, or driver's, side) of the belt is taut

4.27a If you're installing a *used* belt (over 60 minutes of operation), the tensioner pointer should be about 0.16 inch (5/32-inch) to the *left* of the center of the V-notch (viewed from the front of the vehicle)

illustration). After the belt is installed and all timing marks are aligned, rotate the timing belt tensioner arm counterclockwise until the tensioner pointer is adjacent to the V-notch. (The exact alignment between the pointer and the V-notch isn't critical now; you just want to put enough tension on the belt to allow the engine to be rotated several times in order to verify that the timing marks are all still aligned.)

26   Rotate the crankshaft through at least four complete revolutions and then verify that the timing marks are still aligned. If they are, you may proceed. If not, re-install the belt. **Caution:** *If you feel any resistance while rotating the crankshaft, stop immediately. The valves may be hitting the pistons.*

27   Loosen the tensioner bolt and rotate the tensioner arm until the pointer is aligned correctly. If you're installing a used belt (over 60 minutes of operation), the tensioner pointer should be about 0.16 inch (5-32-inch) to the

left of the center of the V-notch, as viewed from the front of the vehicle **(see illustration)**. If you're installing a new belt, the pointer must be at the *center* of the V-notch, as viewed from the front of the vehicle **(see illustration)**.

28   Tighten the timing belt tensioner bolt to the torque listed in this Chapter's Specifications.

29   Install the timing belt cover and tighten the bolt and nut securely.

30   Install the engine wiring harness cover on the front of the engine and tighten the nut securely.

31   Install the crankshaft pulley and tighten the pulley bolts to the torque listed in this Chapter's Specifications.

32   Rotate the drivebelt tensioner pulley clockwise to remove tension, and then install the serpentine drivebelt. Release the tensioner.

33   Reconnect the negative battery cable.

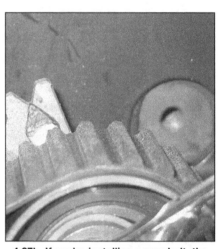

4.27b If you're installing a *new* belt, the pointer must be at the center of the V-notch

5.5 To detach the intake duct bracket from the cylinder head, remove these two bolts (don't forget to reattach the ground wire to the forward bolt during reassembly)

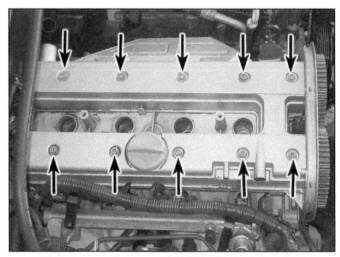

5.8 To detach the valve cover from the cylinder head, remove these 10 bolts

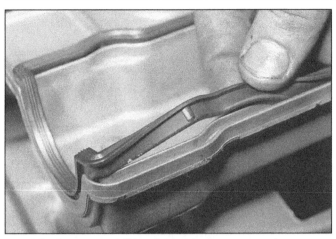

5.9a Remove and discard the old valve cover gasket . . .

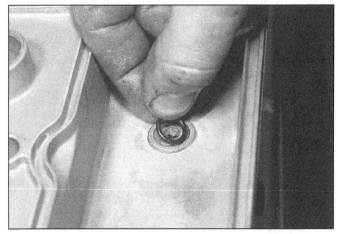

5.9b . . . and remove and discard the old sealing rings for the bolt holes

5 Valve cover - removal and installation

Refer to illustrations 5.5, 5.8, 5.9a and 5.9b

1   Disconnect the negative battery cable.
2   Remove the air intake duct (see Section 9 in Chapter 4B).
3   The wire routed across the forward end of the valve cover is the Crankshaft Position (CKP) sensor wire. Disengage it from the small guides on the valve cover.
4   Disconnect the two PCV hoses from the valve cover (see Chapter 6).
5   Detach the intake duct bracket from the cylinder head **(see illustration)**. Note the ground wire attached to the forward bracket bolt. Don't forget to reattach this ground wire when installing the bracket.
6   Remove the spark plug wire cover, disconnect the spark plug wires from the plugs (see Chapter 5) and set the plug wires aside.
7   Disconnect the electrical connector from the Camshaft Position (CMP) sensor (see Chapter 6) and set the CMP sensor electrical lead aside.
8   Remove the 10 valve cover bolts **(see illustration)** and remove the valve cover.

9   Remove the old valve cover gasket and the sealing rings around each bolt hole **(see illustrations)**.
10   Installation is the reverse of removal. Be sure to use a new valve cover gasket and new sealing rings for the bolt holes and tighten the valve cover bolts to the torque listed in this Chapter's Specifications.

6 Intake manifold - removal and installation

Note: *To prevent air leaks and possible damage to the valves, the gasket must be replaced whenever the exhaust manifold is removed. The engine must be completely cool when this procedure is done.*

Removal

Refer to illustration 6.3, 6.14, 6.15 and 6.16

1   Disconnect the negative battery cable.
2   Drain the cooling system (see Chapter 1).
3   Disconnect the power brake booster vacuum hose **(see illustration)** and then remove the air intake duct (see Section 9 in Chapter 4B).

4   Detach the coolant hoses from the throttle body **(see illustration 6.3)**.
5   Unplug the electrical connectors from the Throttle Position (TP) sensor and the Idle Air Control (IAC) valve **(see illustration 6.3)**.
6   Disconnect the vacuum line from the fuel pressure regulator **(see illustration 6.3)**.
7   Remove the bolts from the two fuel injector harness brackets **(see illustration 6.3)**, disconnect the electrical connectors from the fuel injectors (see Section 11 in Chapter 4B) and set the fuel injector harness aside.
8   Remove the fuel supply and return line banjo bolts from the fuel rail **(see illustration 6.3)**, remove the sealing washers and disconnect the fuel supply and return lines from the fuel rail (see Section 11 in Chapter 4B).
9   Remove the fuel rail assembly (see Section 11 in Chapter 4).
10   Disconnect the accelerator cable from the throttle body (see Section 9 in Chapter 4B).
11   Disconnect the EVAP canister purge valve hose from the throttle body **(see illustration 6.3)**.
12   Disconnect the EVAP hose from the throttle body **(see illustration 6.3)**.

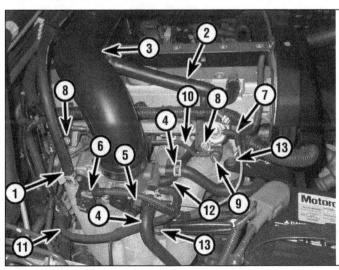

6.3 Intake manifold assembly installation details

1   *Power brake booster vacuum hose*
2   *Positive Crankcase Ventilation (PCV) fresh air intake hose to crankcase*
3   *Air intake duct*
4   *Coolant hoses*
5   *Throttle Position (TP) sensor electrical connector*
6   *Idle Air Control (IAC) valve electrical connector*
7   *Fuel pressure regulator vacuum line*
8   *Fuel injector harness bracket bolts*
9   *Fuel return line banjo bolt (fuel supply line banjo bolt shown, supply line bolt not visible in this photo)*
10   *Fuel return line (fuel supply line not visible in this photo)*
11   *EVAP canister purge valve hose*
12   *PCV inlet hose to intake manifold*
13   *Alternator mounting bracket bolts*

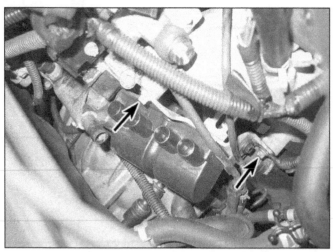

6.14 To detach the ignition coil bracket from the intake manifold, remove these two bolts (spark plug wires removed for clarity)

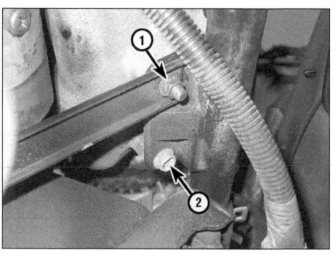

6.15 Working from underneath, remove the intake manifold support bracket nut (1) and the coolant pipe bracket bolt (2) and detach the manifold bracket and the coolant pipe from the manifold

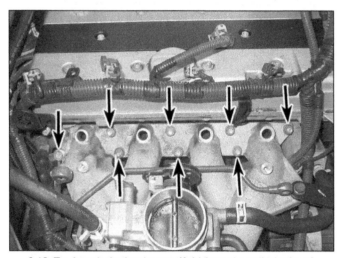

6.16 To detach the intake manifold from the cylinder head, remove this single bolt and these seven nuts

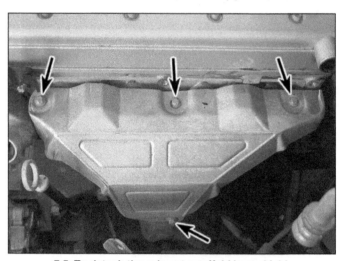

7.5 To detach the exhaust manifold heat shield, remove these four bolts

13   Remove the alternator mounting brackets **(see illustration 6.3)**.

14   Remove the two bolts that attach the ignition coil bracket to the intake manifold **(see illustration)**.

15   Working from underneath the vehicle, remove the intake manifold support bracket nut and the coolant pipe bracket bolt **(see illustration)** and separate the manifold bracket and the coolant pipe from the manifold.

16   Remove the intake manifold bolt and seven nuts **(see illustration)** and remove the intake manifold.

Installation

17   Before installing the manifold, stuff clean, lint-free rags into the intake ports and remove all traces of old gasket and sealant with a scraper. Clean the mating surfaces with lacquer thinner or acetone.

18   Place the intake manifold in position against the cylinder head and install the bolt and nuts, tightening them only enough to support the manifold.

19   Once the manifold is attached to the head, tighten the mounting bolt and nuts to the torque listed in this Chapter's Specifications. Work from the center out toward the ends and work up to the final torque in three or four steps.

20   The remainder of installation is the reverse of the removal procedure.

21   Fill the radiator with coolant, start the engine and check for leaks. Also check and adjust the idle speed and throttle linkage as described in Chapter 1.

7   Exhaust manifold - removal and installation

Refer to illustrations 7.5, 7.6 and 7.7

1   Disconnect the negative battery cable from the battery.

2   Disconnect the PCV fresh air inlet hose from the air intake duct (see Chapter 6).

3   Remove the nut from the air intake duct bracket, loosen the intake duct hose clamps

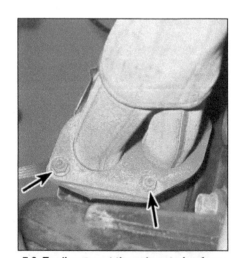

7.6 To disconnect the exhaust pipe from the exhaust manifold, remove the four front exhaust pipe-to-exhaust manifold flange nuts and bolts (two upper fasteners not visible in this photo)

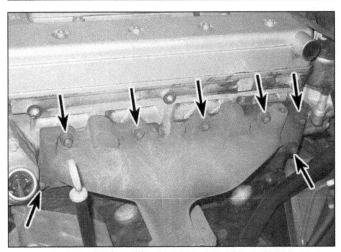

7.7 To separate the exhaust manifold from the cylinder head, remove these exhaust manifold retaining nuts (lower fasteners not visible in this photo)

8.4 If you don't have a seal removal tool, insert a couple of self-tapping screws into the old seal and pull out the seal with pliers

at the air cleaner and the throttle body, and then remove the air intake duct and the air cleaner cover (see Section 9 in Chapter 4B).

4   Remove the air intake duct bracket and the ground wire attached to it **(see illustration 5.5)**.

5   Remove the four exhaust manifold heat shield bolts **(see illustration)**.

6   Remove the four front exhaust pipe-to-exhaust manifold flange nuts and bolts **(see illustration)**.

7   Remove the exhaust manifold retaining nuts **(see illustration)** and then remove the exhaust manifold.

Installation

8   Install the exhaust manifold and tighten the manifold retaining nuts to the torque listed in this Chapter's Specifications.

9   Reattach the front exhaust pipe to the exhaust manifold and tighten the bolts and nuts to the torque listed in this Chapter's Specifications.

10   Install the exhaust manifold heat shield and tighten the bolts in two stages to the torque listed in this Chapter's Specifications. Tighten from the center, working outward in a spiral pattern.

11   Install the intake duct bracket, reattach the ground cable and install - but don't tighten - the bracket retaining nut.

12   Install the intake duct assembly between the air cleaner and the throttle body, tighten the hose clamps at the air cleaner and throttle body, and then tighten the intake duct bracket nut securely

13   Reattach the PCV hose to the air intake duct (see Chapter 6).

14   Reconnect the negative battery cable.

8   Camshaft seals - replacement

Refer to illustrations 8.4 and 8.5
Note: *This procedure applies to either camshaft seal.*

1   Disconnect the negative battery cable from the battery.

2   Remove the timing belt and the camshaft timing belt sprocket (see Section 4).

3   Using a back-up wrench on the hexagonal part of the camshaft (between the timing belt sprocket and the cylinder head), remove the sprocket retaining bolt **(see illustration 4.16)**. Remove the timing belt sprocket. If the sprocket is stuck to the end of the cam, use a small puller to remove it.

4   Using a seal removal tool, pry out the old seal. If you don't have a seal removal tool, carefully punch or drill two small holes 180 degrees apart in the seal, then screw self-tapping screws into the holes and use pliers to extract the seal **(see illustration)**.

5   Using a large deep socket or a short section of pipe or PCV tubing, install the new seal **(see illustration)**.

6   Install the camshaft timing belt sprocket and the timing belt (see Section 4).

7   Reconnect the negative battery cable.

9   Camshafts and valve lifters - removal, inspection and installation

Check

1   If the valvetrain is making unfamiliar noises (usually a pronounced clattering sound that's louder than normal), the hydraulic valve lifters might be the cause. But before removing the lifters for inspection, make sure that the engine oil level is correct. And be sure to allow enough time after a cold start for the lifters to quiet down. If the noise is still excessive, check the lifters - but make sure that the engine is warmed up (and be careful not to burn yourself when working on a hot engine).

2   Remove the valve cover (see Section 5).

3   Turn the crankshaft until all cam lobes

for the number one cylinder are pointing away from the lifters, and then try to push the lifters down with a wood or plastic tool. If a lifter can be depressed by hand pressure alone, replace it. The lifters are neither adjustable nor repairable. Repeat this check for all lifters (and make sure that the cam lobes are pointing up before you check each pair of lifters).

Removal

Refer to illustrations 9.6a, 9.6b and 9.9

4   Remove the timing belt and the camshaft timing belt sprockets (see Section 4).

5   Each camshaft rides on five bearings. Before unbolting any of the five bearing caps, inspect the caps for identification marks and for orientation. Be sure to store the bearing caps in the order in which they're removed, and make sure that they're facing the same direction they face when installed. An egg carton makes a handy storage box for this task. If the bearing caps are numbered, this will help you keep them in order. Make sure that the numbers all face the same way that they face when the caps are installed.

8.5 Using a large deep socket or a short section of pipe or PCV tubing, install the new camshaft seal

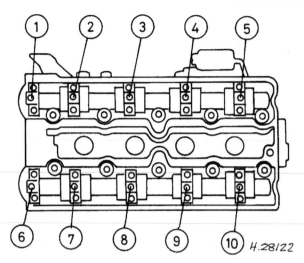

9.6a Camshaft bearing cap numbering sequence

9.6b The identification numbers should be marked on both the bearing caps and on the cylinder head; if the caps aren't numbered, be sure to make your own marks with a permanent marker before removing them

6   The camshaft bearing caps should already be numbered **(see illustrations)**. If the caps aren't numbered, be sure to number them with a permanent marker pen before removing them. Put the marks on the same ends of all the caps to prevent incorrect orientation of the caps during installation.

7   Starting with the middle bearing caps, and working in a criss-cross pattern, *slowly and gradually* loosen all 20 camshaft bearing cap bolts to relieve valve spring tension evenly until the caps are loose. If any of the caps stick, gently tap them with a soft-face hammer. **Caution:** *Failure to follow this procedure exactly as described could tilt the camshafts in the bearings, which could damage the bearing caps or bend the camshafts.* Store the camshaft bearing caps in an egg carton.

8   Before removing either camshaft, look for identification marks, such as "I" for intake and "E" for exhaust. If there are no identification marks, mark the camshafts with your permanent marker to indicate which cam is

the intake cam and which cam is the exhaust cam. Lift out the camshafts, remove the old camshaft seals, wipe the cams off with a clean shop towel and set them aside.

9   Number each valve lifter with a permanent marker (number one intake; number one exhaust; number two intake; etc.). Remove each lifter **(see illustration)** and set it aside in labeled plastic bags, or in the egg carton with the bearing caps, to keep them from getting mixed up. To prevent the hydraulic lifters from "bleeding down" (draining), place them with the camshaft contact surface facing DOWN. If the lifters are not bagged, make sure that they're covered with a clean shop towel to protect them from dirt.

10   To inspect the lifters, refer to Section 10 in Chapter 2E; to inspect the camshafts, refer to Section 16 in Chapter 2E.

Installation

11   Thoroughly clean the camshafts, the bearing surfaces in the head, the camshaft bearing caps, and the valve lifters. Remove all sludge and dirt. Wipe off all components with a clean, lint-free cloth.

12   Lightly lubricate the valve lifter bores with assembly lube or moly-base grease. Install the lifters, referring to the identification marks you made on the lifters prior to disassembly. Lubricate the upper side of the lifters with assembly lube or moly-base grease.

13   Lubricate the camshaft bearing surfaces in the head and the bearing caps, and lubricate the bearing journals and lobes on the camshaft with assembly lube or moly-base grease. Slide a new oil seal onto the front of each camshaft, then carefully lower the camshafts into position with the lobes for the number one cylinder pointing away from the valve lifters. **Caution:** *Failure to adequately lubricate the camshaft and related components can cause damage to bearing and friction surfaces during the first few seconds after engine start-up, when the oil pressure is low for the first few seconds of operation.*

14   Apply a thin coat of assembly lube or

moly-base grease to the bearing surfaces of the camshaft bearing caps and install the caps in their original locations.

15   Install the bearing cap bolts. Working in a criss-cross pattern from the center toward the ends of the cylinder head, slowly and evenly tighten the bearing cap bolts until they're all snug.

16   Working in a criss-cross pattern from the center toward the ends of the head, gradually and evenly tighten all 20 bearing cap bolts to the torque listed in this Chapter's Specifications.

17   Install the valve cover (see Section 5) and the camshaft sprockets and the timing belt (see Section 4).

18   Remove the spark plugs and rotate the crankshaft by hand to make sure the valve timing is correct. After two revolutions, the timing marks on the sprockets should still be aligned. If they're not, re-index the timing belt to the sprockets (see Section 4). **Caution:** *If you feel resistance while rotating the crankshaft, stop immediately and check the valve timing by referring to Section 4.*

19   Install the timing belt cover (see Section 4).

20   Reconnect the negative battery cable.

9.9 Number each valve lifter with a permanent marker (number one intake; number one exhaust; number two intake; etc.) and then remove each lifter assembly and set it aside in a labeled plastic bag, or in the egg carton with the bearing caps (use a magnet or suction cup to remove the lifters)

10.7 To remove the idler pulley bracket, remove this bolt

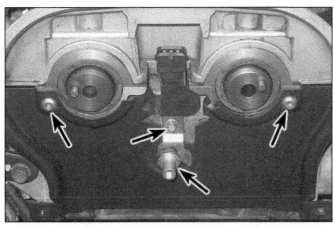

10.8 To remove the Camshaft Position (CMP) sensor, remove the sensor retaining bolt and loosen the stud bolt (note how the CMP sensor bracket slips between the stud bolt and the upper rear timing belt cover); to remove the upper rear timing belt cover, remove the two Torx bolts and the stud bolt;

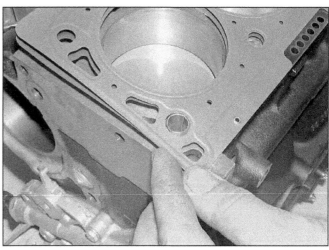

10.24a When installing the cylinder head gasket, make sure that it fits over the dowel pins . . .

10 Cylinder head - removal and installation

Removal

Refer to illustration 10.7 and 10.8

Note: *If you're planning to overhaul the cylinder head, you must remove the intake and exhaust manifolds, the timing belt cover and timing belt, the camshaft timing belt sprockets, the camshafts, the valve lifters and the valve spring assembly. If, on the other hand, you're simply removing the head in order to replace the cylinder head gasket, the head can be removed with the manifolds still installed (though you might need assistance lifting the head and manifolds out of the engine compartment as a single assembly). If that's the case, simply disregard any steps in the following procedure that pertain to stripping the head.*

1   Drain the cooling system (see Chapter 1). Remove the upper radiator hose (see Chapter 3).
2   Remove the air intake duct (see Section 9 in Chapter 4B).
3   Remove the alternator (see Chapter 5) and the alternator mounting brackets **(see illustration 6.3)**.
4   Remove the fuel rail, fuel pressure regulator and fuel injectors (see Chapter 4).
5   Remove the intake manifold (see Section 6).
6   Remove the timing belt and camshaft sprockets (see Section 4).
7   Remove the two idler pulleys **(see illustration 4.14)** and remove the left idler pulley bracket **(see illustration)**.
8   Remove the camshaft position sensor **(see illustration)**.
9   Remove the two bolts and the stud bolt and remove the upper rear timing belt cover **(see illustration 10.8)**.
10   Remove the exhaust manifold (see Section 7).
11   Disconnect the Crankshaft Position

(CKP) sensor and the knock sensor electrical connectors (see Chapter 6).
12   Remove the valve cover (see Section 5).
13   Remove the camshafts and valve lifters (see Section 9).
14   Disconnect the electrical connector and the hoses from the EVAP canister purge valve solenoid (see Chapter 6).
15   Remove the ignition coil/ignition control module and the coil/module mounting bracket, along with the EVAP canister purge valve, as a single assembly (see Chapter 5).
16   Disconnect the electrical connectors from the oil pressure switch (see Chapter 2E), the power steering pressure switch (see Chapter 6) and any other switches or sensors with electrical leads that are part of the main engine harness and then set the harness aside.
17   Using a new head gasket, outline the cylinders and bolt pattern on a piece of cardboard **(see illustration 11.13 in Chapter 2A)**. Be sure to indicate the front of the engine for reference. Punch holes at the bolt locations. Loosen the ten cylinder head bolts slowly and evenly, in 1/4-turn increments, in the reverse order of the tightening sequence **(see illustration 10.27a)** until they can be removed by

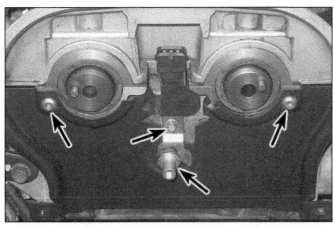

10.24b . . . and that the words OBEN/TOP are facing up

hand. Store the bolts in the cardboard holder as they're removed - this will ensure that they're reinstalled in their original locations.
18   Lift the head off the engine. If it's stuck, DO NOT pry between the head and block - damage to the mating surfaces will result! To dislodge the head, position a block of wood against it and strike the wood block with a hammer.
19   Remove the cylinder head gasket. Place the head on a block of wood on the bench to prevent damage.
20   Refer to Part E for cylinder head inspection and overhaul procedures.

Installation

Refer to illustrations 10.24a, 10.24b, 10.27a and 10.27b

21   The mating surfaces of the cylinder head and block must be perfectly clean when the head is installed. Use a gasket scraper to remove all traces of carbon and old gasket material, then clean the mating surfaces with lacquer thinner or acetone. If there's oil on the mating surfaces when the head is installed, the gasket may not seal correctly and leaks could develop. When working on the block, stuff the cylinders with clean shop rags to keep out debris. Use a vacuum cleaner to remove any debris that falls into the cylinders.
22   Check the block and head mating surfaces for nicks, deep scratches and other damage. If damage is slight, it can be removed with a file; if it's excessive, machining may be the only alternative.
23   Use a tap of the correct size to chase the threads in the head bolt holes. Mount each bolt in a vise and run a die down the threads to remove corrosion and restore the threads. Dirt, corrosion, sealant and damaged threads will affect torque readings.
24   Install the gasket over the engine block dowel pins **(see illustration)**. Make sure that the words "OBEN/TOP" are facing up **(see illustration)**.

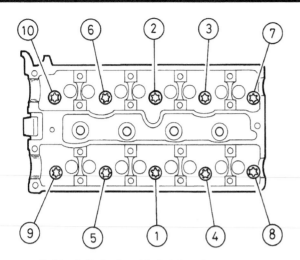

10.27a Cylinder head bolt tightening sequence

10.27b You'll need an angle gauge to tighten the cylinder head bolts

25   Make sure that the number one piston is still at TDC, then carefully lower the cylinder head onto the engine, over the dowel pins and the gasket. Be careful not to move the gasket.

26   Install the cylinder head mounting bolts and tighten them by hand. **Note:** *Lightly apply engine oil to the threads and under the bolt heads before installation.*

27   Tighten the bolts in the recommended sequence **(see illustrations)** to the initial torque listed in this Chapter's Specifications. Repeat this procedure three more times, tightening the bolts each time to the angle torque listed in this Chapter's Specifications.

28   The remainder of the installation procedure is the reverse of the removal procedure.

29   Fill the radiator with coolant, change the engine oil and filter (see Chapter 1), then start the engine and check for leaks. Be sure to recheck the coolant level once the engine has warmed up to operating temperature. Also check and adjust the idle speed as described in Chapter 1.

11   Crankshaft front oil seal - replacement

Refer to illustrations 11.4 and 11.5

1   Disconnect the negative battery cable.
2   Drain the engine oil (see Chapter 1).
3   Remove the timing belt and the crankshaft timing belt sprocket (see Section 4).
4   Carefully pry the crankshaft front oil seal out of the oil pump housing with a seal removal tool or screwdriver. Or use a couple of self-tapping screws to remove the seal **(see illustration)**. Don't scratch the seal bore or damage the crankshaft in the process (if the crankshaft is damaged the new seal will end up leaking).
5   Clean the bore in the oil pump housing and coat the outer edge of the new seal with engine oil or multi-purpose grease. Using a socket with an outside diameter slightly smaller than the outside diameter of the seal,

carefully drive the new seal into place with a hammer **(see illustration)**. If a socket isn't available, a short section of large diameter pipe will work. Check the seal after installation to be sure the spring didn't pop out.

6   Reinstall the crankshaft timing belt sprocket and the timing belt (see Section 4).
7   Refill the engine with oil (see Chapter 1).
8   Reconnect the negative battery cable.
9   Run the engine and check for leaks.

12   Oil pan - removal and installation

Refer to illustrations 12.4, 12.6a, 12.6b and 12.7

1   Disconnect the negative battery cable.
2   Drain the engine oil (see Chapter 1). Remove the dipstick from the dipstick tube.
3   Apply the parking brake and block the rear wheels to keep the vehicle from rolling. Raise the front of the vehicle and support it on jackstands placed under the frame.

11.4 If you don't have a seal removal tool or screwdriver, drill a couple of small holes in the seal, screw in a couple of self-tapping screws and then use a pair of pliers to pry out the seal by levering the pliers against the nose of the crankshaft

11.5 Using a socket with an outside diameter slightly smaller than the outside diameter of the seal, carefully drive the new seal into place with a hammer (if a socket isn't available, a short section of large diameter pipe will work)

12.4 To detach the splash shield, remove these fasteners

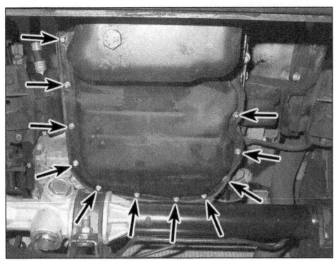

12.6a Oil pan retaining bolts

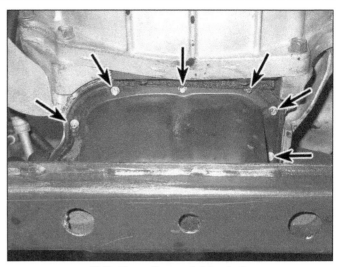

12.6b Rear oil pan retaining bolts

4   Remove the splash shield, if equipped **(see illustration)**.
5   Remove the flywheel dust cover (see Chapter 7).
6   Remove the oil pan retaining bolts **(see illustrations)** and detach the oil pan. Don't pry between the oil pan support (the upper part of the oil pan) and the oil pan or damage to the sealing surface may result and oil leaks could develop. Use a block of wood and a hammer to dislodge the pan if it's stuck. Remove and discard the old oil pan gasket.
7   Remove the oil pan support bolts **(see illustration)** and detach the oil pan support. **Note:** *Several of the oil pan support bolts are screwed into the bellhousing.* Remove and discard the old oil pan support gasket.
8   Use a scraper to remove all traces of oil, old gasket material and sealant from the gasket sealing surfaces of the block, the oil pan support and the oil pan. Clean the gasket sealing surfaces with lacquer thinner or acetone and make sure that the bolt holes in the block and the oil pan support are clean.

9   Check the oil pan flange for distortion, particularly around the bolt holes. If necessary, place the pan on a block of wood and use a hammer to flatten and restore the gasket surface.
10   Before installing the oil pan support gasket, apply a thin coat of RTV sealant to the oil pan support flange and to the gasket sealing surface of the engine block. Attach the new gasket to the pan support; make sure the bolt holes are aligned.
11   Position the oil pan support on the block. Install the mounting bolts. Working in a criss-cross pattern, gradually and evenly tighten them to the torque listed in this Chapter's Specifications.
12   Before installing the oil pan gasket, apply a thin coat of RTV sealant to the oil pan flange and to the gasket sealing surface of the oil pan support. Attach the new gasket to the pan; make sure the bolt holes are aligned.
13   Position the oil pan on the oil pan support. Install the mounting bolts. Working in a criss-cross pattern, gradually and evenly

tighten them to the torque listed in this Chapter's Specifications.
14   Reconnect the negative battery cable.
15   Wait at least 30 minutes before filling the engine with oil, then start the engine and check for leaks.

13   Oil pump - removal and installation

Removal

Refer to illustrations 13.2, 13.3, 13.5 and 13.6

1   Drain the engine oil (see Chapter 1).
2   Remove the oil pressure switch from the oil pump **(see illustration)**.
3   Remove the timing belt and the crankshaft timing belt sprocket (see Section 4). Remove the lower rear timing belt cover **(see illustration)**.
4   Remove the oil pan and the oil pan support (see Section 12).

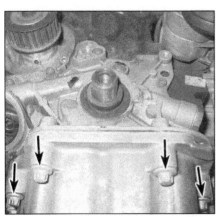

12.7 Front oil pan support bolts (the rest of the oil pan support bolts not visible in this photo)

13.2 Before removing the oil pump, be sure to disconnect the electrical connector from the oil pressure switch and unscrew the switch from the pump

13.3 To detach the lower rear timing belt cover from the engine, remove this bolt

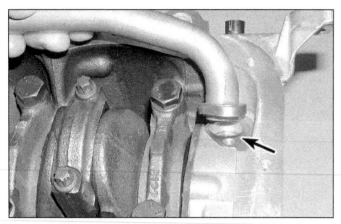

13.5 Remove the pickup tube and discard the old gasket

13.6 To detach the oil pump, remove the pump retaining bolts

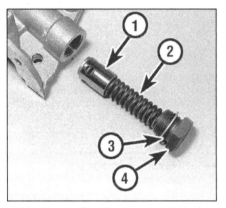

13.8 Remove the oil pressure relief valve from the oil pump

1	Plunger	4	Pressure
2	Spring		relief valve
3	Sealing		plug
	washer		

5   Remove the oil pump pickup tube bolts and remove the pickup tube **(see illustration)**. Remove and discard the old pickup tube gasket.

6   Remove the oil pump retaining bolts **(see illustration)** and carefully remove the pump. Try not to damage the crankshaft front oil seal.

7   Remove the old pump gasket.

Inspection

Refer to illustrations 13.8, 13.11, 13.12, 13.15a, 13.15b, 13.17 and 13.19

8   Remove the oil pressure relief valve **(see illustration)** from the oil pump. Thoroughly clean the relief valve parts and then store them in a plastic bag.

9   Remove the oil pump dowel pins and put them in the plastic bag.

10   Inspect and, if necessary, replace the crankshaft front oil seal. If the crank seal is damaged or excessively worn, pry it out with a seal removal tool or a big screwdriver, or remove it with a pair of self-tapping screws (see Section 11).

11   Remove the oil pump cover **(see illustration)**.

13.11 To detach the cover from the backside of the oil pump, remove these screws

12   Look for punch marks on the inner and outer pump gears **(see illustration)**. If there are none, make your own. The gears must be installed facing the same way that they were facing before disassembly.

13   Remove the inner and outer pump gears.

14   Thoroughly clean all parts in fresh solvent and then inspect the gears and pump body for any signs of scoring or wear. If either of the gears or the gear cavity in the pump

13.15a Measure the clearance between the inner and outer gears of the oil pump with a feeler gauge

13.12 Before removing the inner or outer pump gear (shown), make sure that there are punch marks on both gears; if not, make your own to ensure that the gears are reassembled facing in the correct direction

body is excessively worn or damaged, replace the pump assembly (the parts for each pump are matched at the factory; none of them is available separately).

15   If the parts are still in good condition, measure the clearance between the inner and outer gears with feeler blades and measure the gear endplay **(see illustrations)**. Also

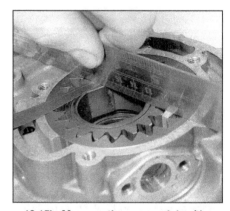

13.15b Measure the gear endplay (the freeplay between the gears and the cover) with a straightedge and a feeler gauge

13.17 Using a socket slightly smaller in diameter than the crankshaft front oil seal, carefully tap the new seal into its recess in the front of the oil pump housing

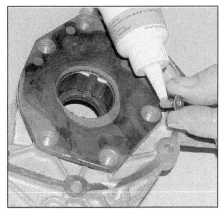

13.19 Make sure that the oil pump cover screws don't back out by applying Loctite, to the threads

13.23 When installing the oil pump onto the engine, make sure that the lip of the new crank front oil seal isn't damaged by dragging it over the step behind the nose (1) of the crank and also make sure that the flats (2) on the crank are correctly engaged with the flats inside the inner gear hub

measure the cover for flatness. Compare your measurements to the pump clearances listed in this Chapter's Specifications.

16   If the clearance between the gear teeth, or between the face of the gears and the cover, is excessive, replace the pump. If the clearances are still within specification, reassemble the pump.

17   Install a new crankshaft front oil seal, if necessary **(see illustration)**.

18   Install the pump inner and outer gears. Make sure that both gears, and the gear cavity in the pump housing, are clean.

19   Install the pump cover, apply Loctite, to the threads of the screws **(see illustration)** and tighten the screws to the torque listed in this Chapter's Specifications.

20   Install the oil pressure relief valve and tighten the relief valve plug to the torque listed in this Chapter's Specifications.

Installation

Refer to illustration 13.23

21   Install the oil pump dowel pins in the block.

22   Using the dowel pins as a guide, place the new gasket in position.

23   Very carefully, install the oil pump on the engine **(see illustration)**. Make sure that the new crankshaft front oil seal lip isn't damaged by the step on the nose of the crank, and make sure that the flats behind the step are correctly engaged with flats inside the hub of the inner gear.

24   Install the oil pump mounting bolts and tighten them to the torque listed in this Chapter's Specifications.

25   Thoroughly wash the oil pump pickup tube in clean solvent and blow it out with compressed air. Using a new gasket, install the pickup tube and tighten the pickup tube retaining bolts to the torque listed in this Chapter's Specifications.

26   Install the oil pan support and the oil pan (see Section 12.).

27   Install the crankshaft timing belt sprocket and the timing belt (see Section 4).

28   Screw the oil pressure switch into the oil

pump and tighten it to the torque listed in this Chapter's Specifications.

29   Refill the engine with oil (see Chapter 1).

30   Start the engine and check for proper oil pressure.

14   Flywheel/driveplate - removal and installation

Refer to Section 14 in Chapter 2A.

15   Rear main oil seal - replacement

Refer to Section 15 in Chapter 2A.

16   Balancer unit - removal, installation and adjustment

Removal and installation

Refer to illustrations 16.1, 16.8 and 16.10

1   The 2.2L engine is equipped with a balancer unit, which is bolted to the crankshaft main bearing caps. The balance unit **(see illustration)** consists of a pair of contra-rotating balance shafts, both of which are driven by the crankshaft. A shim between the balancer unit and the bottom of the block maintains the correct gear tooth backlash.

2   Remove the oil pan and oil pan support (see Section 12).

3   Remove the oil pump pick-up tube (see Section 13).

4   Remove the balance unit mounting bolts and remove the balance unit and shim.

5   Installation is the reverse of removal. However, if you have just replaced the crankshaft, the main bearing caps and/or the balance unit, the gear tooth backlash *must* be measured and, if necessary, adjusted.

Adjustment

6   You'll need to purchase or borrow a special tool (J-43038, or a suitable equiva-

lent) which consists of a pair of adjustment knobs that can be screwed into the ends of the balance shafts. You'll also need a dial gauge with a magnetic base to measure gear tooth backlash, but even with the special tool and the dial gauge, the adjustment procedure is tricky. We recommend that you take the balance unit and (assembled) lower end of the engine to a dealer to have the backlash measured and, if necessary, adjusted. However, if you want to tackle this job yourself, here's how it's done:

7   Make sure that the crankshaft is at TDC on the No. 1 cylinder (see Section 3).

8   Position the two balance shafts so that, when viewed from the front of the engine, the

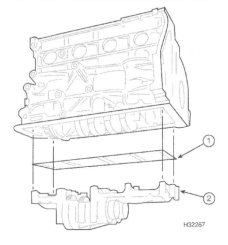

16.1 The gear tooth backlash between the drive gear on the crankshaft and the driven gear on the balance unit (2) is adjusted by installing a thicker or thinner shim (1) (a thicker shim provides more clearance; a thinner shim provides less)

two flat machined surfaces are exactly horizontal **(see illustration)**.

9   Install the old shim (the one already in use) and the balancer unit. Install the balancer unit mounting bolts and tighten them to the torque listed in this Chapter's Specifications.

10   Screw the longer of the two special bolts into the end of the balancer shaft that's on the same side as the intake manifold. Position the measuring arm so that it points to the 9 o'clock position, as viewed from the front of the engine **(see illustration)**. Hand-tighten the bolt.

11   Screw the shorted of the two special bolts into the balancer shaft that's on the same side as the exhaust manifold. Hand-tighten the bolt.

12   Mount the dial gauge on the balancer unit or engine block, with the probe touching the measuring arm between the notches in the flat machined surface of the arm **(see illustration 16.10)**.

13   Determine the "beginning" and the "end" of the backlash by turning the bolt screwed into the exhaust-side balancer shaft backward and forward. When you have a good feel for the beginning of the backlash, turn the exhaust-side balancer shaft to that spot and then zero the dial gauge.

14   Rotate the exhaust-side balancer shaft to the "end" of the backlash and note the indicated measurement on the dial gauge.

15   Measure the backlash at four different balancer shaft positions. Using the crankshaft sprocket bolt, turn the crankshaft clockwise until the measuring arm on the intake-side balancer shaft points to the 6 o'clock position. Loosen the long bolt and reposition the arm back at 9 o'clock. Repeat the backlash measurement procedure in this position, making two additional readings repositioning the balancer shaft each time.

16   If any of the four readings are outside the acceptable range of backlash listed in this Chapter's Specifications, adjust the backlash. Each shim has a code number that represents its thickness. This code is also stamped onto the shim.

Code	Thickness of shim (in mm)
55	0.535 to 0.565
58	0.565 to 0.595
61	0.595 to 0.625
64	0.625 to 0.655
67	0.655 to 0.685
70	0.685 to 0.715
73	0.715 to 0.745
76	0.745 to 0.775
79	0.775 to 0.805
82	0.805 to 0.835
85	0.835 to 0.865

17   Once you've established the backlash, and the thickness of the current shim (from the code number), you can easily determine the thickness (and code number) needed to bring the backlash within the acceptable range. Each shim changes the backlash by 0.02 mm (0.008 inch). For example, if the measured backlash is 0.08 mm (0.032 inch) with a code 70 shim, replace this shim with a code 67 shim to reduce the backlash by 0.06

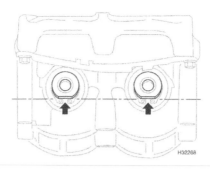

16.8 When installing the balancer unit, make sure that the two flat machined surfaces are horizontal

mm (0.024 inch). **Caution:** *Only one shim can be installed at a time. Do NOT try to adjust backlash by using more than one shim.*

18 After selecting the correct shim and installing the balancer unit and shim as described previously, repeat your measurement to verify that the new shim brings the backlash within an acceptable range.

19 Remove the special tools and the dial indicator.

20   Installation is otherwise the reverse of removal.

17   Engine mounts - replacement

Refer to illustration 17.3a and 17.3b

Warning: *Don't position any part of your body under the engine when the engine mounts are unbolted!*

1   Engine mounts are non-adjustable and seldom require service. Periodically they should be inspected for hardness and cracks in the rubber and separation of the rubber from the metal backing.

2   To replace the engine mounts with the engine in the vehicle, use the following procedure.

3   Loosen the nuts and bolts that retain the left and right mounting insulators to the engine mount brackets and to the frame. Do this on both sides **(see illustrations)**.

4   Next, the weight of the engine must be

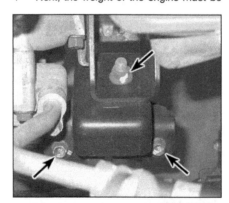

17.3a Loosen the nuts and bolts that attach the left front mounting insulator to the engine mount bracket and frame

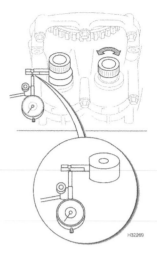

16.10 Screw the longer of the two special bolts into the end of the balancer shaft that's on the same side as the intake manifold, position the measuring arm so that it's at 9 o'clock and then screw the shorter bolt into the exhaust-side balancer shaft; when setting up the dial indicator, position the probe so that it's touching the arm between the notches

taken off the engine mounts. This can be done from beneath using a jack and wooden block positioned under the oil pan, or from above by removing the air cleaner and using an engine hoist attached to the two engine brackets. The engine should be raised slowly and carefully, while keeping a constant check on clearances around the engine to prevent anything from binding or breaking. Pay particular attention to areas such as the fan, ignition coil wires, vacuum lines leading to the engine and rubber hoses and ducts.

5   Raise the engine just enough to provide adequate room to remove the mounting insulator.

6   Remove the nuts and bolts retaining the insulator, then lift it out, noting how it's installed.

7   Installation is the reverse of removal, but be sure the insulator is installed in the same position it was in before removal.

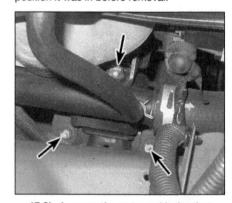

17.3b Loosen the nuts and bolts that attach the right front mounting insulator to the engine mount bracket and frame

Chapter 2 Part C
3.1L V6 engine

Contents

Specifications

General

Displacement	189 cu in (3.1 liters)
Cylinder numbers (front-to-rear)	
Left bank (driver's side)	2-4-6
Right bank	1-3-5
Firing order	1-2-3-4-5-6

Camshaft

Lobe lift	
Intake	0.2306 inch
Exhaust	0.2619 inch
Journal diameter	1.868 to 1.882 inches
Journal-to-bearing (oil) clearance	0.001 to 0.0039 inch

Torque specifications

	Ft-lbs (unless otherwise indicated)
Camshaft sprocket bolt	17
Camshaft cover (rear) bolts	84 to 108 in-lbs
Crankshaft pulley-to-vibration damper bolts	20 to 30

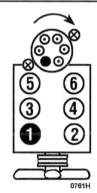

Cylinder location and distributor rotation

The blackened terminal shown on the distributor cap indicates the Number One spark plug wire position

Torque specifications (continued)

Ft-lbs (unless otherwise indicated)

Cylinder head bolts
    Step 1 41
    Step 2 .... Turn an additional 90-degrees (1/4-turn)
Exhaust manifold bolts 25
Flywheel/driveplate bolts 52
Intake manifold bolts/nuts 19
Oil pan mounting bolts/nuts
    Bolts 18
    Nuts .... 84 in-lbs
Oil pump 30
Rocker arm nuts 18
Tensioner bracket-to-engine block bolts .... 168 in-lbs
Timing chain cover bolts 20
Valve cover nuts 72-in-lbs
Vibration damper center bolt 70
Engine mount nut to engine bracket 62
Engine mount to chassis bolts 37

1   General information

This Part of Chapter 2 is devoted to in-vehicle repair procedures for the 3.1L V6 engine in 1991 and 1992 Rodeo models and 1993 and earlier Amigo models. All information concerning engine removal and installation and engine block and cylinder head overhaul can be found in Part E of this Chapter.

The following repair procedures are based on the assumption the engine is installed in the vehicle. If the engine has been removed from the vehicle and mounted on a stand, many of the steps outlined in this Part of Chapter 2 will not apply.

The specifications included in this Part of Chapter 2 apply only to the procedures contained in this Part. Part E of this Chapter contains the specifications necessary for cylinder head and engine block rebuilding.

2   Repair operations possible with the engine in the vehicle

Many major repair operations can be accomplished without removing the engine from the vehicle.

Clean the engine compartment and the exterior of the engine with some type of degreaser before any work is done. It will make the job easier and help keep dirt out of the internal areas of the engine.

Depending on the components involved, it may be helpful to remove the hood to improve access to the engine as repairs are performed (refer to Chapter 11 if necessary). Cover the fenders to prevent damage to the paint. Special pads are available, but an old bedspread or blanket will also work.

If vacuum, exhaust, oil or coolant leaks develop, indicating a need for gasket or seal replacement, the repairs can generally be made with the engine in the vehicle. The intake and exhaust manifold gaskets, timing cover gasket, oil pan gasket, crankshaft oil seals and cylinder head gasket are all accessible with the engine in place.

Exterior engine components, such as the intake and exhaust manifolds, the oil pan (and the oil pump), the water pump, the starter motor, the alternator, the distributor and the fuel system components can be removed for repair with the engine in place.

Since the cylinder heads can be removed without pulling the engine, valve component servicing can also be accomplished with the engine in the vehicle. Replacement of the timing chain, sprockets and camshaft is also possible with the engine in the vehicle.

In extreme cases caused by a lack of necessary equipment, repair or replacement of piston rings, pistons, connecting rods and rod bearings is possible with the engine in the vehicle. However, this practice is not recommended because of the cleaning and preparation work that must be done to the components involved.

3   Top Dead Center (TDC) for number one piston - locating

See Chapter 2, Part A for this procedure. The timing plate is attached to the timing chain cover. Be sure to use the firing order in the Specifications in this Part of Chapter 2 for the V6 engine.

4   Valve covers - removal and installation

1   Disconnect the negative cable from the battery.
2   Remove the air cleaner assembly (see Chapter 4).
3   Label and disconnect the wires and hoses which would interfere with removal of the valve cover(s).
4   Label and detach the spark plug wires and unclip the wire retainers from the studs.
5   Disconnect the throttle, cruise control and TVS cables and bracket at the TBI unit. If necessary for clearance, remove the air conditioning compressor bracket.
6   Disconnect the PCV valve.
7   Disconnect the vacuum pipe at the manifold. If you're removing a left cover, it may be necessary to disconnect the fuel lines from the throttle body (see Chapter 4).
8   If necessary for clearance, remove the alternator (see Chapter 5). If removing the right cover, remove the ignition coil and bracket (see Chapter 5).
9   Remove the six valve cover mounting bolts/nuts and detach the cover. Note that some of the bolts have studs attached to the ends - you'll have to use a deep socket to remove them. If the cover is stuck, use a soft-face hammer or a block of wood and a hammer to dislodge it. If the cover still won't come loose, pry on it carefully with a putty knife, but don't distort the sealing flange surface.
10   Remove all traces of old gasket and sealant with a scraper, then clean the mating surfaces with lacquer thinner or acetone.
11   On models that use RTV sealant in place of a gasket, apply a 1/8-inch bead of sealant to the cover flange. Be sure to apply it around the inside of the bolt holes or oil will leak past the bolt threads. **Caution:** *When applying RTV sealant, keep it out of the bolt holes.*
12   On models that use gaskets, place a small amount of RTV sealant on the seam area where the cylinder head and intake manifold meet before installing the gasket.
13   Install the cover(s) while the RTV is still wet. Tighten the bolts in 1/4-turn increments until the torque listed in this Chapter's Specifications is reached.
14   Reinstall the remaining parts in the reverse order of removal.
15   Run the engine and check for oil leaks.

5.2 To remove the pushrods, simply loosen the rocker arm nuts, rotate the rocker arms out of the way and pull out the pushrods

5.4 When removing the pushrods, be sure to store them separately to ensure reinstallation in their original locations

5 Rocker arms and pushrods - removal, inspection and installation

Removal

Refer to illustrations 5.2 and 5.4

1   Refer to Section 4 and detach the valve cover(s) from the cylinder head(s).

2   Beginning at the front of one cylinder head, remove the rocker arm stud nuts. Store them separately in marked containers to ensure they'll be reinstalled in their original locations. **Note:** *If the pushrods are the only items being removed, loosen each rocker arm nut just enough to allow the rocker arms to be rotated to the side so the pushrods can be lifted out* (see illustration).

3   Lift off the rocker arms and pivot balls and store them in the marked containers with the nuts (they must be reinstalled in their original locations).

4   Remove the pushrods and store them separately to make sure they don't get mixed up during installation (see illustration).

Inspection

5   Check each rocker arm for wear, cracks and other damage, especially where the pushrods and valve stems contact the rocker arm faces.

6   Make sure the hole at the pushrod end of each rocker arm is open.

7   Check each rocker arm pivot area for wear, cracks and galling. If the rocker arms are worn or damaged, replace them with new ones and use new pivot balls as well.

8   Inspect the pushrods for cracks and excessive wear at the ends. Roll each pushrod across a piece of plate glass to see if it's bent (if it wobbles, it's bent).

Installation

Refer to illustrations 5.10 and 5.11

9   Lubricate the lower end of each pushrod with clean engine oil or moly-base grease and install them in their original locations. Make sure each pushrod seats completely in

5.10 The ends of the pushrods and the valve stems should be lubricated with moly-base grease prior to installation of the rocker arm

the lifter.

10   Apply moly-base grease to the ends of the valve stems and the upper ends of the pushrods before positioning the rocker arms over the studs (see illustration).

11   Set the rocker arms in place, then install the pivot balls and nuts. Apply moly-base grease to the pivot balls to prevent damage to the mating surfaces before engine oil pressure builds up (see illustration). Be sure to install each nut with the flat side against the pivot ball.

12   Adjust the valve lash.

Valve adjustment

Refer to illustration 5.15

13   If the valve train components have been serviced just prior to this procedure, make sure they're completely reassembled.

14   Rotate the crankshaft until the number one piston is at top dead center (TDC) on the compression stroke (see Section 3). To make sure you don't mix up the TDC positions of the number one and four pistons, place your fingers on the number one rocker arms as the timing marks line up at the crankshaft pulley. If the rocker arms aren't moving, the number

5.11 Moly-base grease applied to the pivot balls will ensure adequate lubrication until oil pressure builds up when the engine is started

one piston is at TDC. If they move as the timing marks line up, the number four piston is at TDC.

15   Back off the rocker arm nut until play is felt at the pushrod, then turn it back in until all play is removed. This can be determined by

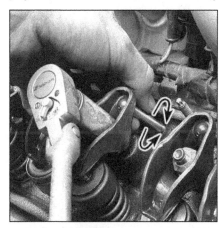

5.15 Rotate the pushrod back-and-forth and tighten the nut until you feel resistance to movement

7.4a The power brake hose fitting is located at the rear of the intake manifold

7.4b Use a screwdriver to remove a stubborn heater hose from the intake manifold

7.12 Use a large screwdriver or prybar to break the manifold gasket seal - DO NOT pry between the mating surfaces

rotating the pushrod while tightening the nut. At the point where the pushrod exhibits drag, all lash has been removed **(see illustration)**. Now tighten the nut an additional 3/4-turn.

16   Adjust the number one, five and six cylinder intake valves and the number one, two and three cylinder exhaust valves with the crankshaft in this position, using the method just described.

17   Rotate the crankshaft until the number four piston is at TDC on the compression stroke and adjust the number two, three and four cylinder intake valves and the number four, five and six cylinder exhaust valves.

18   Install the valve covers.

6   Valve springs, retainers and seals - replacement

Refer to Chapter 2, Part A for this procedure. Remove the valve covers, rocker arms and pushrods; adjust the valve lash following the procedure in Section 5 of this Part.

7.15 Apply a bead of RTV sealant to the block ridges between the heads (arrows) at the front and rear

7   Intake manifold - removal and installation

Refer to illustrations 7.4a, 7.4b, 7.12, 7.15 and 7.20

1   Disconnect the negative cable from the battery.

2   Drain the coolant from the radiator (see Chapter 1).

3   Remove the wires, hoses, cables and fuel lines at the TBI unit (see Chapter 4).

4   Label and disconnect all wires and hoses connected to the intake manifold **(see illustrations)**.

5   If equipped, unbolt the air conditioning compressor and set it aside without disconnecting the hoses (see Chapter 3).

6   Label and disconnect the spark plug wires from the plugs, then remove the distributor cap (see Chapter 1).

7   Remove the distributor (see Chapter 5).

8   Detach the EGR vacuum line (see Chapter 6).

9   Detach the evaporative emission hoses.

10   Remove the valve covers (see Section 4).

11   Disconnect the upper radiator and

heater hoses.

12   Unbolt the intake manifold and lift it off the engine. If it's stuck, pry carefully against a protrusion on the manifold **(see illustration)**. Do not pry between the gasket surfaces.

13   Remove all traces of old gasket material and sealant from the manifold and cylinder head mating surfaces, then clean them with lacquer thinner or acetone. Clean the intake manifold bolt holes in the cylinder head by chasing them with a tap. Compressed air can be used to remove the debris from the holes. **Warning:** *Wear eye protection when using compressed air.*

14   Check the underside of the rear of the intake manifold to see if a machined groove is present. This groove was added to later production models to improve oil sealing.

15   Apply a 3/16-inch bead of RTV sealant to each of the ridges between the heads **(see illustration)**.

16   The intake manifold gaskets will have to be cut where indicated to position the tops behind the pushrods. Cut only those areas necessary to clear the pushrods.

17   Install the new intake manifold gaskets, noting that they're marked Right and Left. Be sure to install them as indicated on the gas-

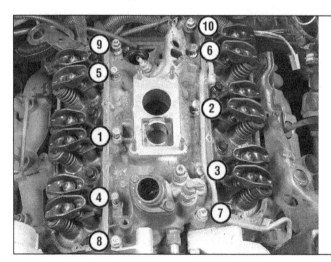

7.20 Intake manifold bolt tightening sequence

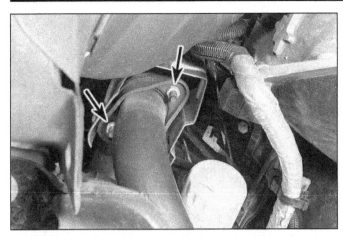

8.3 From underneath, remove the two nuts at the exhaust pipe-to-manifold flange

8.8 Locations of the three upper exhaust manifold bolts (arrows) - right side shown, left side similar

ket manufacturer's instructions. **Note:** *This is only necessary when the gaskets are installed with pushrods in place.*

18   Hold the gaskets in place by extending the RTV sealant bead up 1/4-inch onto the gasket ends.

19   Install the intake manifold on the engine and hand tighten the bolts. Make sure the areas between the block ridges and intake manifold are completely sealed.

20   Following the sequence shown **(see illustration)**, tighten the manifold bolts, in several steps, to the torque listed in this Chapter's specifications.

21   Reinstall the remaining parts in the reverse order of removal.

22   Run the engine, adjust the ignition timing and check for oil and vacuum leaks.

8   Exhaust manifolds - removal and installation

Warning: *Allow the engine to cool completely before performing this procedure.*

Both manifolds

Refer to illustration 8.3

1   Detach the cable from the negative battery terminal.

2   Raise the front of the vehicle and support it securely on jackstands. Apply the parking brake and block the rear wheels to keep the vehicle from rolling.

3   Remove the bolts attaching the exhaust pipe to the exhaust manifold, then separate the pipe from the manifold **(see illustration)**. **Note:** *It's a good idea to apply penetrating oil to these nuts before attempting to loosen them.*

4   Remove the four rear manifold bolts.

5   Remove the jackstands and lower the vehicle.

Right manifold

Refer to illustration 8.8

6   Disconnect the air injection pump and alternator brackets from the manifold.

7   Disconnect the spark plug wires from the spark plugs, labeling them as they're disconnected to simplify installation.

8   Remove the exhaust manifold mounting bolts **(see illustration)** and separate the manifold from the engine.

Left manifold

9   Remove the heat stove tube.

10   Remove the air cleaner assembly (see Chapter 4), labeling all hoses.

11   Disconnect the hoses leading to the air injection valve.

12   Disconnect and label all wires that will interfere with removal of the manifold.

13   Remove the power steering pump bracket from the cylinder head. Loosen the adjusting bracket bolt and detach the drivebelt from the pulley first. After removing the bracket from the cylinder head, place the pump assembly aside, out of the way. Do not disconnect any hoses and be sure to keep the top of the pump up so no fluid spills.

14   Remove the remaining manifold bolts and separate the manifold and heat shield from the engine.

Both manifolds

15   Installation is the reverse of removal. Be sure to clean the cylinder head and manifold surfaces thoroughly before installing the manifold. Tighten the bolts to the specified torque.

9   Cylinder heads - removal and installation

Refer to illustrations 9.14, 9.16, 9.19, 9.21 and 9.22

Warning: *Allow the engine to cool completely before performing this procedure.*

1   Remove the intake manifold (refer to Section 7).

2   Raise the vehicle and support it securely on jackstands.

3   Locate the engine block drain plugs to the rear of the engine mounts (the plug on the

left side is just above the oil filter). Remove the plugs and drain the block.

4   Disconnect the exhaust pipe from the exhaust manifold.

5   If you're working on the left cylinder head, unbolt and remove the oil dipstick tube assembly from the left side of the engine. If you're working on the right cylinder head, remove the drivebelt, alternator and AIR pump with the mounting bracket from the head. Remove the lifting "eye" from the rear of the head (necessary only if the head is going to be replaced with a new one).

6   Remove the jackstands and lower the vehicle.

7   **Note:** *Steps 8 through 11 should be followed if the head is going to be replaced with a new one. The Steps can be performed either before or after the head has been removed. In the accompanying illustrations, the procedures were performed before the head was removed.*

8   Remove the exhaust manifold (refer to Section 8).

9   Remove the ground strap at the rear of the head and the sensor connector at the front of the head.

10   Detach the power steering pump and bracket from the cylinder head.

11   Remove the air-conditioner compressor bracket from the front of the cylinder head, if equipped.

12   Loosen the rocker arm nuts enough to allow removal of the pushrods, then remove the pushrods (see Section 5).

13   Using a new head gasket, outline the cylinders and bolt pattern on a piece of cardboard (see Chapter 2, Part A). Be sure to indicate the front of the engine for reference. Punch holes at the bolt locations. Loosen the ten cylinder head bolts in 1/4-turn increments until they can be removed by hand (start at the ends and work toward the center of the head). Store the bolts in the cardboard holder as they're removed - this will ensure that they're reinstalled in their original locations.

14   Lift the head off the engine. If it's stuck, DO NOT pry between the head and block - damage to the mating surfaces will result! To

9.14 Be careful not to damage the gasket sealing surface when breaking the head loose with a prybar or large screwdriver

9.16 Remove the old gasket (use a putty knife if necessary) and carefully scrape off all the old gasket material and sealant

9.19 Position the new gasket over the dowel pins (arrows)

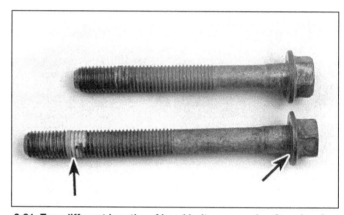

9.21 Two different lengths of head bolts are used - after cleaning the threads, apply sealant to the threads and underside of the bolt heads

dislodge the head, use a long screwdriver or prybar under the cast "ears" to pry up on it **(see illustration)**. Don't damage the cylinder head sealing surface.

15   If a new cylinder head is being installed, attach the components previously removed from the old head.

16   The mating surfaces of the cylinder head and block must be perfectly clean when the head is installed. Use a gasket scraper to remove all traces of carbon and old gasket material, then clean the mating surfaces with lacquer thinner or acetone **(see illustration)**. If there's oil on the mating surfaces when the head is installed, the gasket may not seal correctly and leaks could develop. When working on the block, stuff the cylinders with clean shop rags to keep out debris. Use a vacuum cleaner to remove any debris that falls into the cylinders.

17   Check the block and head mating surfaces for nicks, deep scratches and other damage. If damage is slight, it can be removed with a file; if it's excessive, machining may be the only alternative.

18   Use a tap of the correct size to chase the threads in the head bolt holes. Mount each bolt in a vise and run a die down the threads to remove corrosion and restore the

threads. Dirt, corrosion, sealant and damaged threads will affect torque readings.

19   Install the gasket over the engine block dowel pins with the mark THIS SIDE UP visible **(see illustration)**.

20   Position the cylinder head over the gasket.

21   Coat the cylinder head bolts with a non-hardening sealant and install them **(see illustration)**.

22   Tighten the bolts in the recommended sequence **(see illustration)** to the torque

listed in this Chapter's specifications. Work up to the final torque in three steps.

23   Install the pushrods and rocker arms as described in Section 5.

24   The remaining installation steps are the reverse of removal. Before installing the valve covers, adjust the valve lash (refer to Section 5).

25   Refill the cooling system and change the engine oil and filter (see Chapter 1). Start the engine and set the ignition timing as described in Chapter 1.

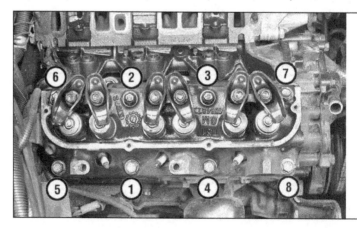

9.22 Cylinder head bolt tightening sequence

10.7 Use a scribe or magnet to remove the lifters - they should be stored separately to ensure reinstallation in their original locations

10.9a If the bottom of any lifter is worn concave, pitted, scratched or galled, replace the entire set with new lifters

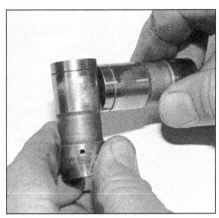

10.9b The foot of each lifter should be slightly convex - the side of another lifter can be used as a straightedge to check it; if it appears flat or concave, it's worn and must not be reused

10 Valve lifters - removal, inspection and installation

1 A noisy valve lifter can be isolated when the engine is idling. Hold a mechanic's stethoscope or a length of hose near the location of each valve while listening at the other end. Another method is to remove the valve cover and, with the engine idling, place a finger on each of the valve spring retainers, one at a time. If a valve lifter is defective, it will be evident from the shock felt at the retainer as the valve seats.

2 The most likely cause of noisy valve lifters is dirt trapped between the plunger and the lifter body or lack of oil flow, viscosity or pressure. Before condemning the lifters, we recommend checking the oil for fuel contamination, correct level, cleanliness and correct viscosity.

Removal
Refer to illustration 10.7

3 Remove the valve cover(s) as described in Section 4.

4 Remove the intake manifold as described in Section 7.

5 Remove the rocker arms and pushrods (Section 5).

6 There are several ways to extract the lifters from the bores. A special tool designed to grip and remove lifters is manufactured by many tool companies and is widely available, but it may not be required in every case. On newer engines without a lot of varnish buildup, the lifters can often be removed with a small magnet or even with your fingers. A machinist's scribe with a bent end can be used to pull the lifters out by positioning the point under the retainer ring in the top of each lifter. **Caution:** *Don't use pliers to remove the lifters unless you intend to replace them with new ones (along with the camshaft). The pliers may damage the precision machined and hardened lifters, rendering them useless.*

7 Before removing the lifters, arrange to store them in a clearly labeled box to ensure they can be reinstalled in their original locations. **Note:** *Some engines may have both standard and 0.25 mm (0.010-inch) oversize lifters installed at the factory. If so, they are marked accordingly. The lifter boss will be marked with a dab of white paint and will have 0.25 (mm) OS stamped on it.* Remove the lifters and store them where they won't get dirty **(see illustration)**.

Inspection and installation
Refer to illustrations 10.9a, 10.9b and 10.9c

8 Clean the lifters with solvent and dry them thoroughly without mixing them up.

9 Check each lifter wall, pushrod seat and foot for pitting, scuffing, score marks and uneven wear **(see illustration)**. Each lifter foot (the surface that rides on the cam lobe) must be slightly convex, although this can be difficult to determine by eye. If the base of the lifter is concave **(see illustration)**, the lifters and camshaft must be replaced. If the lifter walls are damaged or worn (which isn't very likely), inspect the lifter bores in the engine block as well. If the pushrod seats **(see illustration)** are worn, check the pushrod ends.

10 If new lifters are being installed, a new camshaft must also be installed. If the camshaft is replaced, then install new lifters as well (see Section 15). Never install used lifters unless the original camshaft is used and the lifters can be installed in their original locations! When installing lifters, make sure they're coated with moly-base grease or engine assembly lube.

11 Soak new lifters in oil to remove trapped air.

12 The remaining installation steps are the reverse of removal.

13 Run the engine and check for oil leaks.

11 Vibration damper - removal and installation

Refer to illustrations 11.4 and 11.5

1 Remove the bolts and separate the radiator shroud from the radiator.

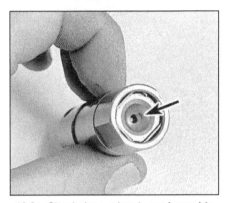

10.9c Check the pushrod seat (arrow) in the top of each lifter for wear

2 Remove the mounting bolts and detach the cooling fan and the radiator shroud. On pick-up models, unbolt the suspension crossmember, if necessary, for clearance.

3 Loosen the adjusting bolts as necessary, then remove the drivebelts, tagging each one as it's removed to simplify installation.

4 Remove the bolts from the crankshaft pulley (a screwdriver can be used to lock the starter ring gear on the flywheel so the

11.4 Remove the four bolts holding the drivebelt pulley to the crankshaft damper

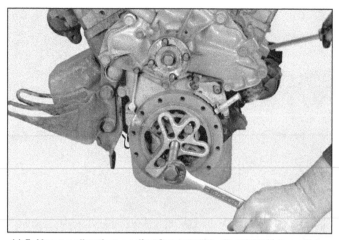

11.5 Use a puller that applies force to the vibration damper hub - be sure the large puller bolt doesn't damage the crankshaft threads

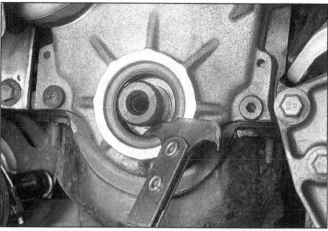

12.1 A seal puller can be used to pry the old seal out - be careful not to nick the crankshaft seal surface

12.3 Drive the new seal in squarely with a large socket, to the same depth as the original seal

12.7 Driving the seal out of the front cover

12.9 Installing the new oil seal with a block of wood and a hammer

crankshaft won't rotate), then detach the pulley hub **(see illustration)**. Remove the vibration damper-to-crankshaft bolt. **Note:** *Remove the flywheel inspection cover or starter motor and have an assistant hold a large screwdriver in the ring gear teeth to hold the engine from turning while you loosen the damper bolt with a breaker bar and socket.*
5   Attach a puller to the damper and draw it off the crankshaft. Be careful not to drop it as it comes free. A common gear puller should not be used - it may separate the outer portion of the damper from the hub. Use only a puller that bolts to the hub **(see illustration)**.
6   Before installation, coat the oil seal journal on the damper with moly-base grease.
7   Place the damper in position over the key on the crankshaft. Make sure the damper keyway lines up with the key.
8   Using a damper installation tool (no. J-29113 or equivalent), push the damper onto the crankshaft. The special tool distributes the pressure evenly around the hub. The bolt that threads into the crankshaft and a large washer can be used if the tool isn't available.

9   Remove the installation tool and install the damper bolt. Tighten the bolt to the specified torque.
10   To install the remaining components, reverse the removal procedure.
11   Adjust the drivebelts (refer to Chapter 1).

12 Crankshaft front oil seal - replacement

With timing chain cover installed on engine

Refer to illustrations 12.1 and 12.3
1   With the vibration damper removed (see Section 11), pry the old seal out of the cover with a large screwdriver or seal puller hub **(see illustration)**. Be very careful not to damage the seal journal on the crankshaft.
2   Place the new seal in position with the open end facing toward the inside of the cover.
3   Drive the seal into the cover until it's seated. Tool J-35468 is recommended for this purpose. The tool is designed to exert even pressure around the entire circumfer-

ence of the seal as it's hammered into place. A section of large-diameter pipe or a large socket can also be used. Be careful not to distort the front cover hub **(see illustration)**.
4   Reinstall the remaining parts in the reverse order of removal.

With timing chain cover removed from engine

Refer to illustrations 12.7 and 12.9
5   This method is preferred, as the cover can be supported while the old seal is removed and the new one is installed.
6   Remove the timing chain cover (refer to Section 13).
7   Support the cover and drive the seal out from the rear **(see illustration)**. Be careful not to damage the cover.
8   With the front of the cover facing up, place the new seal in position with the open end facing toward the inside of the cover.
9   Using a block of wood and hammer, drive the new seal into the cover until it's completely seated **(see illustration)**.
10   Install the timing chain cover by reversing the removal procedure in Section 13.

13.6a Location of the front cover bolts (after the water pump has been removed) - the top bolt isn't visible here

13.6b Some of the timing cover bolts pass through the oil pan

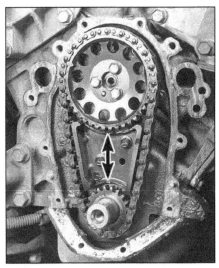

14.8 The timing marks (arrows) on the sprockets should be aligned as shown - a straight line should pass through the center of the camshaft, the camshaft sprocket timing mark, the crankshaft sprocket timing mark and the center of the crankshaft

13 Timing chain cover - removal and installation

Refer to illustrations 13.6a and 13.6b

1   Remove the water pump as described in Chapter 3.
2   If equipped with air conditioning, remove the compressor from the mounting bracket and secure it out of the way. **Warning:** *Do not disconnect any of the air conditioning system hoses without having the system depressurized by a dealer service department or service station.*
3   Remove the compressor mounting bracket.
4   Remove the vibration damper as described in Section 11.
5   Disconnect the lower radiator hose at the timing chain cover.
6   Remove the timing chain cover mounting bolts and separate the cover from the engine **(see illustrations)**.
7   Clean all oil, dirt and old gasket material off the sealing surfaces of the cover and engine block. Replace the oil seal as described in Section 12.
8   Apply a continuous 3/32-inch bead of

anaerobic sealant (Loctite 515 or equivalent) to both mating surfaces of the cover (except the mating surface where the cover engages the oil pan lip). Apply RTV-type sealant to the cover-to-oil pan area. Also apply anaerobic sealant to the areas surrounding the coolant passages.
9   Place the timing chain cover in position on the engine block and install the mounting bolts. Tighten the bolts to the specified torque in a criss-cross pattern (to avoid distorting the cover).
10   The remaining installation steps are the reverse of removal.

14 Timing chain and sprockets - inspection, removal and installation

Refer to illustrations 14.8, 14.9, 14.10 and 14.12

1   Disconnect the cable from the negative battery terminal.
2   Remove the vibration damper (see Section 11).
3   Remove the timing chain cover (see Section 13).

14.9 A screwdriver will keep the camshaft sprocket from turning while loosening the mounting bolts

4   Before removing the chain and sprockets, visually inspect the teeth on the sprockets for signs of wear and check the chain for looseness.
5   If either or both sprockets show any signs of wear (edges on the teeth of the camshaft sprocket rounded, bright or blue areas on the teeth of either sprocket, chipping, pitting, etc.), they should be replaced with new ones. Wear in these areas is very common. Failure to replace a worn timing chain and sprockets may result in erratic engine performance, loss of power and lowered gas mileage.
6   If any one component (timing chain or either sprocket) requires replacement, the other two components should be replaced as well.
7   If it's determined the components require replacement, proceed as follows.
8   Reinstall the vibration damper mounting bolt and use it to turn the crankshaft clockwise until the marks on the camshaft and crankshaft are in exact alignment **(see illustration)**. At this point the number one and four pistons will be at top dead center with the number four piston in the firing position (verify by checking the position of the rotor in the distributor, which should point to the number four spark plug wire terminal). **Note:** *Do not attempt to remove either sprocket or the timing chain until this is done and do not turn the crankshaft or camshaft after the sprockets/chain are removed.*
9   Remove the three camshaft sprocket mounting bolts **(see illustration)** and detach the camshaft sprocket and timing chain from the front of the engine. You may have to tap the sprocket with a soft-face hammer to dislodge it.

14.10 A gear puller will be needed to remove the crankshaft sprocket

14.12 Lubricate the thrust (rear) surface of the camshaft sprocket before installing it

10   If you have to remove the crankshaft sprocket, it can be withdrawn with a gear puller **(see illustration)**.

11   Push the crankshaft sprocket onto the crankshaft with the vibration damper bolt and a large washer or washers.

12   Lubricate the thrust (rear) surface of the camshaft sprocket with moly-base grease or engine assembly lube **(see illustration)**. Install the timing chain over the camshaft sprocket with slack in the chain hanging down.

13   With the timing marks aligned, slip the chain over the crankshaft sprocket and then draw the camshaft sprocket into place with the three bolts. Do not hammer or attempt to drive the camshaft sprocket into place, as it could dislodge the welch plug at the rear of the engine.

14   With the chain and both sprockets in place, check again to make sure the timing marks on the two sprockets are properly aligned. If not, remove the camshaft sprocket and move the chain until the marks align.

15   Lubricate the chain with engine oil and install the remaining components in the reverse order of removal.

15   Camshaft and bearings - removal, inspection and installation

Camshaft lobe lift check

Refer to illustration 15.3

1   To determine the extent of cam lobe wear, the lobe lift should be checked prior to camshaft removal. Refer to Section 4 and remove the valve covers.

2   Position the number one piston at TDC on the compression stroke (see Section 3).

3   Beginning with the number one cylinder valves, mount a dial indicator on the engine and position the plunger against the top surface of the first rocker arm. The plunger should be directly above and in-line with the pushrod **(see illustration)**.

4   Zero the dial indicator, then very slowly turn the crankshaft in the normal direction of rotation until the indicator needle stops and begins to move in the opposite direction. The point at which it stops indicates maximum cam lobe lift.

5   Record this figure for future reference, then reposition the piston at TDC on the compression stroke.

6   Move the dial indicator to the remaining number one cylinder rocker arm and repeat the check. Be sure to record the results for each valve.

7   Repeat the check for the remaining valves. Since each piston must be at TDC on the compression stroke for this procedure, work from cylinder-to-cylinder following the firing order sequence.

8   After the check is complete, compare the results to the Specifications. If camshaft lobe lift is less than specified, cam lobe wear has occurred and a new camshaft should be installed.

Removal

Refer to illustrations 15.16 and 15.17

9   Remove the cable from the negative battery terminal.

10   Drain the oil (Chapter 1).

11   Remove the radiator (Chapter 3).

12   If equipped with air conditioning, remove the condenser (Chapter 3).

13   Remove the valve lifters (Section 10).

15.3 When checking the camshaft lobe lift, the dial indicator plunger must be positioned directly above the pushrod

15.16 Long bolts can be threaded into the camshaft bolt holes to provide a handle for removal and installation - support the cam near the block as it's withdrawn

15.17 A length of wire with a hook on it can be used to support the camshaft as you guide it out of the block

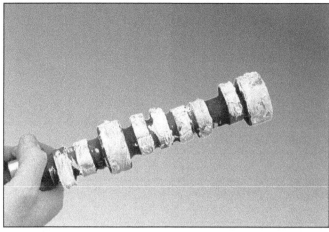

15.19 Be sure to apply camshaft installation lube to the cam lobes and bearing journals before installing the camshaft

14   Remove the timing chain cover (Section 13).
15   Remove the timing chain and camshaft sprocket (Section 14).
16   Install two long bolts in the end of the camshaft to use as a handle to pull on and support the camshaft (see illustration).
17   Carefully draw the camshaft out of the engine block. Do this very slowly to avoid damage to the camshaft bearings as the lobes pass over them. Support the camshaft with one hand near the engine block and the other with a wire hook at the other end (see illustration).

Inspection and installation

Refer to illustration 15.19

18   Refer to Chapter 2, Part E, for camshaft and bearing inspection procedures.
19   Prior to installing the camshaft, coat each of the lobes and journals with camshaft installation lube (see illustration).
20   Slide the camshaft into the engine block, again taking extra care so you don't damage the bearings.
21   Install the camshaft sprocket and timing chain as described in Section 14.
22   Install the remaining components in the reverse order of removal by referring to the appropriate Chapter or Section.
23   Adjust the valve lash (Section 5).
24   Have the air conditioning system (if equipped) evacuated and recharged.

16   Oil pan - removal and installation

Note: *The following procedure is based on the assumption the engine is in the vehicle. If it's been removed, simply unbolt the oil pan and detach it from the block.*

Removal

1   Disconnect the negative battery cable from the battery, then refer to Chapter 1 and drain the engine oil.
2   Refer to Chapter 5 and remove the starter motor.
3   If necessary for clearance, separate the exhaust pipes from the manifolds.
4   Remove the front skidplate (4WD only).
5   Detach the front crossmember.
6   On Trooper models, remove the catalytic converter bolts.
7   Detach the front driveshaft from the differential pinion flange (4WD only).
8   Remove the braces surrounding the flywheel cover plate.
9   Use paint to mark the Pitman arm and shaft for reassembly.
10   Remove the Pitman arm from the shaft (4WD only).
11   Detach the idler arm from the shaft (4WD only).
12   Remove the rubber tube from the front axle vent (4WD only).
13   Use jackstands to support the front axle assembly.
14   Remove the bolts from the right and left axle housing isolators and lower the front axle housing (4WD only).
15   Remove the oil pan mounting nuts/bolts.
16   Carefully separate the pan from the block. Don't pry between the block and pan or damage to the sealing surfaces may result and oil leaks could develop. You may have to turn the crankshaft slightly to maneuver the front of the pan past the crankshaft counterweights.

Installation

17   Clean the gasket sealing surfaces with lacquer thinner or acetone. Make sure the bolt holes in the block are clean.
18   Check the oil pan flanges for distortion, particularly around the bolt holes and corners. If necessary, place the pan on a block of wood and use a hammer to flatten and restore the gasket surface.
19   Check the corners of the oil pan for distortion with a straightedge across the cradle opening. If distortion is severe, install a new oil pan.
20   Apply RTV sealant to the rear cradle corners and use oil pan gasket kit number 8983500853 (or equivalent).
21   Carefully position the pan against the block and install the bolts finger-tight. Tighten the nuts/bolts in three steps to the torque listed in this Chapter's specifications. Start at the center of the pan and work out toward the ends in a spiral pattern.
22   The remaining steps are the reverse of removal. **Caution:** *Don't forget to refill the engine with oil before starting it (see Chapter 1).*
23   Start the engine and check carefully for oil leaks at the oil pan.

17   Oil pump - removal and installation

Refer to illustration 17.2

1   Remove the oil pan (refer to Section 16).
2   Remove the pump-to-rear main bearing cap bolt (see illustration) and separate the pump and extension shaft from the engine.
3   To install the pump, hold it in position and align the top end of the hexagonal extension shaft with the hexagonal socket in the

17.2 Remove the oil pump-to-rear main bearing cap bolt and detach the pump and extension shaft

lower end of the distributor drive gear. The distributor drives the oil pump, so it's essential that the alignment is correct.

4   Install the oil pump-to-rear main bearing cap bolt and tighten it to the torque listed in this Chapter's specifications.

5   Reinstall the oil pan.

18   Flywheel/driveplate - removal and installation

Refer to Chapter 2, Part A for this procedure. Be sure to use the torque specifications in this Part of Chapter 2 for the V6 engine.

19   Crankshaft rear oil seal - replacement

Refer to illustration 19.3

Note: *Special tools, as noted in the Steps which follow, are required for this procedure. They are available from your dealer or may, in some cases, be rented from an auto parts store or tool rental shop.*

1   Remove the transmission (see Chapter 7A or 7B).

2   Remove the flywheel or driveplate (see Section 18).

19.3 Pry the rear main seal out of the bore - be very careful not to scratch the crankshaft seal journal or the edge of the seal bore

3   A one-piece seal is utilized, which allows the oil pan to remain in place during this procedure. Pry out the old seal, taking care not to damage the crankshaft seal journal **(see illustration)**. Check the crankshaft for scratches, burrs and nicks on the seal journal.

4   A special seal installation tool (no.

J-34686) is recommended to properly seat the seal in the bore without damaging it. Lubricate the seal bore, seal lip and seal journal on the crankshaft with engine oil. Slide the seal over the mandrel on the tool until the dust lip on the seal bottoms squarely against the collar on the tool. **Note:** *A large-diameter section of pipe can also be used, if it is the appropriate size.*

5   Position the dowel pin on the tool in the alignment hole in the crankshaft and secure the tool to the crankshaft.

6   Turn the T-handle of the tool until the collar pushes the seal into the bore. Make sure the seal is installed squarely and seated completely.

7   To complete the operation, install the flywheel (or driveplate) and transmission, then start the engine and check for leaks.

20   Engine mounts - check and replacement

Refer to Chapter 2, Part A for this procedure, but note that the V6 engine mounts are slightly different in ways that don't affect the check and replacement procedures. When tightening the fasteners, refer to the Specifications at the beginning of this Chapter for 3.1L V6 torque figures.

Chapter 2 Part D
3.2L V6 engine

Contents

Specifications

General

Displacement	193 cubic inches (3.2 liters)
Firing order	1-2-3-4-5-6
Cylinder numbers (drivebelt end-to-transmission end)	
Right	1-3-5
Left	2-4-6

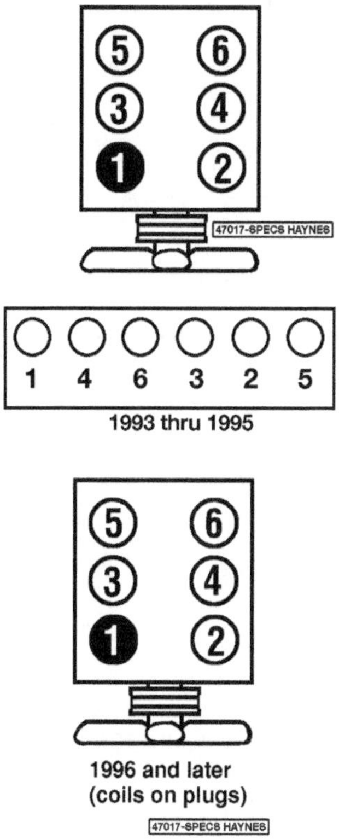

Camshaft - SOHC engine

Lobe height	
Intake	1.3480 inches
Exhaust	1.4638 inches
Lobe wear limit	0.002 inch
Journal oil clearance	
Standard	0.0016 to 0.0043 inch
Limit	0.0197 inch
Runout (maximum)	0.0039 inch

Camshaft - DOHC engine

Lobe height (intake and exhaust)	1.7602 inch
Lobe wear limit	0.002 inch
Journal diameter	
Standard	1.0225 to 1.0233 inch
Limit	1.0157 inch
Taper limit	0.002 inch
Journal oil clearance	
Standard	0.0011 inch
Limit	0.0043 inch
Runout (maximum)	0.0039 inch

Oil pump

Case-to-outer rotor clearance	0.0039 to 0.0071 inch
Clearance over rotors	0.0012 to 0.0035 inch
Inner-to-outer rotor clearance	0.0043 to 0.0094 inch

Torque specifications

Ft-lbs (unless otherwise indicated)

Rocker arm shaft bolts 9SOHC engine)	156 in-lbs
Intake manifolds nuts/bolts	17
Engine mount bracket to engine	30
Engine mount insulator to chassis	30
Engine mount insulator to engine bracket nuts	37

⑤ ⑥
③ ④
① ②

47017-SPECS HAYNES

○ ○ ○ ○ ○ ○
1 4 6 3 2 5

1993 thru 1995

⑤ ⑥
③ ④
① ②

**1996 and later
(coils on plugs)**

47017-SPECS HAYNES

**Cylinder and coil pack
terminal locations -
3.2L V6 engines**

Torque specifications (continued)

Ft-lbs (unless otherwise indicated)

Exhaust manifold nuts	42
Exhaust manifold heat shield bolts...	21
Exhaust pipe-to-manifold nuts	37
Fan pulley assembly mounting bolts	16
Crankshaft pulley-to-crankshaft bolt	123
Camshaft sprocket bolt	
SOHC engine	41
DOHC engine	72
Camshaft drive gear housing-to-cylinder head bolts (DOHC engine)	89 in-lbs
Camshaft tower bolts (SOHC engine)	
M6 bolts	69 in-lbs
M8 bolts	156 in-lbs
Camshaft bearing cap bolts (DOHC engine)	89 in-lbs
Timing belt covers	
SOHC engine	156 in-lbs
DOHC engine	168 in-lbs
Timing belt tensioner bolts	
SOHC engine	168 in-lbs
DOHC engine	18
Cylinder head bolts (in sequence - **see illustration 11.19**)	
SOHC engine	
M11 bolts (use new bolts only)	47
M8 bolts	180 in-lbs
DOHC engine	
Step 1...	21
Step 2...	47
Flywheel/driveplate mounting bolts...	40
Oil pan mounting bolts and nuts	
SOHC engine	84 in-lbs
DOHC engine	89 in-lbs
Oil pump assembly mounting bolts	
SOHC engine	156 in-lbs
DOHC engine	18
Oil pump relief plug cover..	69 in-lbs
Oil pick-up tube bolts	18
Oil pump cover screws	89 in-lbs
Valve cover bolts	
SOHC engine	69 in-lbs
DOHC engine	80 in-lbs

1   General information

This Part of Chapter 2 is devoted to in-vehicle repair procedures for the 3.2L V6 (Isuzu designates this as the 6VD1) engine. All information concerning engine removal and installation and engine block and cylinder head overhaul can be found in Part E of this Chapter.

The following repair procedures are based on the assumption that the engine is installed in the vehicle. If the engine has been removed from the vehicle and mounted on a stand, many of the steps outlined in this Part of Chapter 2 will not apply.

The Specifications included in this Part of Chapter 2 apply only to the procedures contained in this Part. Part E of Chapter 2 contains the Specifications necessary for cylinder head and engine block rebuilding.

The 75-degree V6 has an aluminum block with iron cylinder liners cast in, and aluminum heads. 1997 and earlier models use a Single Over-Head Camshaft (SOHC) design; each head has one camshaft which actuates two valves per cylinder through shaft-mounted rocker arms and hydraulic lifters. 1998 and later models use a Dual Over-Head Camshaft design; each head has two camshafts which actuate the valves directly, through solid lifters equipped with adjuster shims.

2   Repair operations possible with the engine in the vehicle

Many major repair operations can be accomplished without removing the engine from the vehicle.

Clean the engine compartment and the exterior of the engine with some type of degreaser before any work is done. It will make the job easier and help keep dirt out of the internal areas of the engine.

Depending on the components involved, it may be helpful to remove the hood to improve access to the engine as repairs are performed (refer to Chapter 11 if necessary). Cover the fenders to prevent damage to the paint. Special pads are available, but an old bedspread or blanket will also work.

If vacuum, exhaust, oil or coolant leaks develop, indicating a need for gasket or seal replacement, the repairs can generally be made with the engine in the vehicle. The intake and exhaust manifold gaskets, oil pan gasket, camshaft and crankshaft oil seals and cylinder head gaskets are all accessible with the engine in place.

Exterior engine components, such as the intake and exhaust manifolds, the oil pan (and the oil pump), the water pump, the starter motor, the alternator and the fuel sys-

3.5 On the left valve cover, pull out the PCV valve (left arrow) and disconnect the ground strap (right arrow)

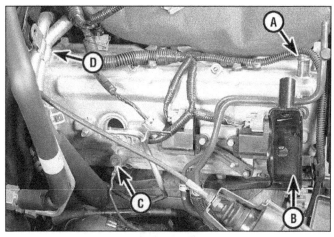

3.7 Remove the valve cover bolts (arrows) (1997 or earlier model shown; later models only have eight bolts)

3.10 On the right valve cover, remove the PCV hose (A), engine lifting bracket (B), ground strap (C), and move the heater hoses (D) out of the way

tem components can be removed for repair with the engine in place.

Since the cylinder heads can be removed without pulling the engine, camshaft and valve component servicing can also be accomplished with the engine in the vehicle. Replacement of the timing belt and sprockets is also possible with the engine in the vehicle.

In extreme cases caused by a lack of necessary equipment, repair or replacement of piston rings, pistons, connecting rods and rod bearings is possible with the engine in the vehicle. However, this practice is not recommended because of the cleaning and preparation work that must be done to the components involved.

3 Valve covers - removal and installation

Removal

1 Disconnect the negative cable from the battery.

Left valve cover

Refer to illustrations 3.5 and 3.7

2 Remove the air intake duct (see Chapter 4). On 1997 and earlier models, remove the EGR pipe (see Chapter 6).

3 Remove the ignition coil packs from the spark plugs (see Chapter 1) and lay them out of the way. Mark them clearly with pieces of masking tape to prevent confusion during installation.

4 Refer to Chapter 6 and disconnect the electrical connector from the camshaft position sensor.

5 Pull out the PCV valve and unbolt the ground cable at the front of the valve cover **(see illustration)**.

6 Detach the accelerator cable and unbolt the throttle body from the intake plenum (see Chapter 4B).

7 Remove the valve cover bolts and detach the valve cover **(see illustration)**. **Caution:** *If the cover is stuck to the head,*

3.14a Apply a bead of RTV sealant on each side (arrows) of the front and rear camshaft towers just before installing the valve cover

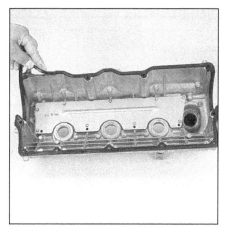

3.14b Install the new gasket firmly in the groove of the valve cover

bump one end with a block of wood and a hammer to jar it loose. If that doesn't work, try to slip a flexible putty knife between the head and cover to break the gasket seal. Don't pry at the cover-to-head joint or damage to the sealing surfaces may occur (leading to oil leaks in the future).

Right valve cover

Refer to illustration 3.10

8 If you're working on a 1997 or earlier model, refer to Chapter 4 and remove the intake plenum. If you're working on a 1998 or later model, remove the EGR pipe.

9 Refer to Chapter 1 for removal of the ignition coil packs.

10 Disconnect the PCV hose, ground strap, and engine lifting bracket **(see illustration)**. Move the two heater hoses if necessary to clear the valve cover.

11 Remove the valve cover bolts and remove the valve cover.

Installation

Refer to illustrations 3.14a and 3.14b

12 The mating surfaces of each cylinder

head and valve cover must be perfectly clean when the covers are installed. Use a gasket scraper to remove all traces of sealant and old gasket material, then clean the mating surfaces with lacquer thinner or acetone. If there's sealant or oil on the mating surfaces when the cover is installed, oil leaks may develop.

13 If necessary, clean the mounting bolt threads with a die to remove any corrosion and restore damaged threads. Make sure the threaded holes in the head are clean - run a tap into them to remove corrosion and restore damaged threads.

14 Apply a bead of RTV sealant to the front and rear camshaft towers in the areas indicated **(see illustration)**, then position the gasket firmly inside the groove around the cover **(see illustration)**.

15 Carefully position the cover on the head and install the bolts.

16 Tighten the bolts in three or four steps to the torque listed in this Chapter's Specifications.

17 The remaining installation steps are the reverse of removal.

18 Start the engine and check carefully for oil leaks as the engine warms up.

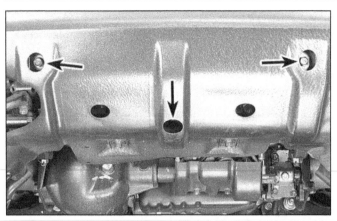

4.3 Remove the bolts (arrows) and take down the front splash apron

4.4 Use a strap wrench or large water-pump pliers over the pulley to keep the engine from turning while removing the center bolt with a breaker bar

4   Crankshaft front oil seal - replacement

Removal

Refer to illustrations 4.3, 4.4, 4.5, 4.7 and 4.8

1   Disconnect the negative cable from the battery.

2   Remove the drivebelt(s) (see Chapter 1) and refer to Chapter 3 for removal of the fan shroud and cooling fan.

3   Remove the bolts holding the front splash apron and remove the apron **(see illustration)**.

4   Remove the bolt **(see illustration)** attaching the crankshaft pulley and remove the pulley. **Note:** *Use a strap-type wrench over the pulley to keep it from turning while loosening the bolt.* If you don't have access to a strap-wrench, wrap a length of sandpaper over the pulley and grasp it with a large water-pump pliers. The sandpaper gives the pliers more grip.

5   You should be able to slide the crankshaft pulley off by hand **(see illustration)**. If the pulley is stuck, wedge two screwdrivers

behind it and carefully pry it off the crankshaft. Some sprockets are more difficult to remove because corrosion fuses them onto the nose of the crankshaft. Use some penetrating oil beforehand to loosen any corrosion.

6   Refer to Section 5 for removal of the timing belt covers and the timing belt.

7   Slip the crankshaft sprocket from the crankshaft **(see illustration)**. You may need two screwdrivers to pry behind it, but don't damage the front of the oil pump.

8   Use a seal pulley, or a screwdriver tip wrapped with tape, to pry the oil seal from the front of the oil pump **(see illustration)**. **Caution:** *Be careful not to nick the seal surface of the crankshaft.*

Installation

Refer to illustration 4.9

9   Apply a film of multi-purpose grease to the inside of the seal, and tap it into the bore in the oil pump, using a seal installation tool, an appropriate-size large socket, or a length of pipe **(see illustration)**.

10   Make sure the Woodruff key is in place in the crankshaft and slide the timing belt sprocket onto the crankshaft.

4.5 The crankshaft pulley should slide off the crankshaft

11   Installation of the remaining components is the reverse of removal. Be sure to refer to Section 5 for the timing belt installation and adjustment procedure. Tighten all bolts to the torque values listed in this Chapter's Specifications.

4.7 Pull the crankshaft sprocket off - use two screwdrivers to pry behind it if necessary

4.8 Use a screwdriver or seal puller to pry the crankshaft/oil pump seal out

4.9 Use an appropriate-size socket or section of pipe to tap the new seal in squarely - tap in only to the depth the original seal was installed

5.5 Remove the one bolt (top arrow) and two nuts (lower arrows), then take off the fan pulley support

5.7 Remove the bolts and remove the upper (A) and lower (B) timing belt covers

5 Timing belt - removal, installation and adjustment

Caution: *These engines are "interference" engines. This means that if the timing belt breaks, valve and/or piston damage will probably occur.*

Removal

★★ CAUTION ★★

The timing system is complex. Severe engine damage will occur if you make any mistakes. Do not attempt this procedure unless you are highly experienced with this type of repair. If you are at all unsure of your abilities, consult an expert. Double-check all your work and be sure everything is correct before you attempt to start the engine.

Refer to illustrations 5.5, 5.7, 5.8a, 5.8b and 5.9

1   Disconnect the cable from the negative terminal of the battery. Remove the air cleaner duct and on 1998 and later models, the air cleaner housing (see Chapter 4B).

2   Refer to Section 4 for removal of the crankshaft pulley.

3   Refer to Chapter 1 for removal of the drivebelt(s). On 1998 and later models also remove the belt tensioner. On all models, remove the spark plugs (see Chapter 1).

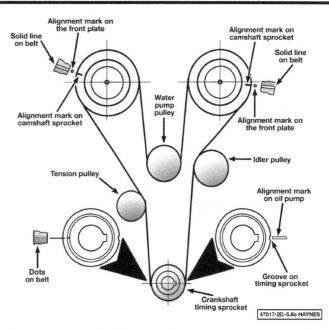

5.8a Timing belt and camshaft/crankshaft alignment mark details - 1997 and earlier SOHC engines

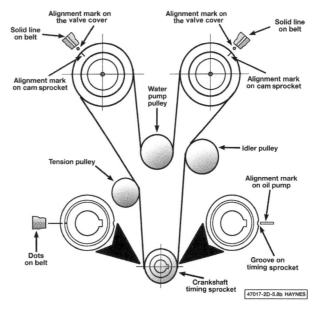

5.8b Timing belt and camshaft/crankshaft alignment mark details - 1998 and later DOHC engines

4   Remove the engine cooling fan and fan shroud (see Chapter 3).

5   Remove the bolt/nuts and take off the fan pulley support **(see illustration)**.

6   On 1997 and earlier models, disconnect the oil cooler hose and pull it out of the way. On 1998 and later models, remove the power steering pump (see Chapter 10).

7   Remove the timing belt covers **(see illustration)**. Note the various type and sizes of bolts by recording a diagram or making specific notes while the timing belt covers are being removed. There are four bolts in each upper cover and six bolts in the lower cover. The bolts must be reinstalled in their original locations. **Note:** *Remove the two upper covers first.*

8   Turn the engine (using the crankshaft pulley bolt temporarily installed in the front of the crankshaft) until the groove on the crankshaft sprocket aligns with the horizontal cast-in mark on the left (driver's) side of the oil pump body **(see illustrations)**. At the same time, check the camshaft sprockets to be sure that their marks align with the *dots* in the timing belt back covers (1997 and earlier models) or the marks on the valve covers (1998 and later models). If the marks on the camshaft sprockets don't line up

5.9 Unbolt the timing belt tensioner's hydraulic unit

5.11 Compress the hydraulic tensioner and insert a drill bit to hold it in the retracted position

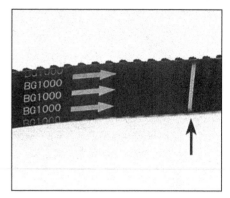

5.12 A factory Isuzu belt should be marked with writing, directional arrows and marks (arrow) to line up with the camshaft and crankshaft sprockets

with their corresponding marks, turn the crankshaft one complete revolution (360-degrees), realigning the marks on the crankshaft sprocket and the oil pump.

9   Relieve tension on the timing belt by unbolting the timing belt tensioner **(see illustration)**. Now remove the belt.

10   Inspect the timing belt. Look at the backside (the side without the teeth): If it's cracked or peeling, or it's hard, glossy and inflexible, and leaves no indent when pressed with your fingernail, replace the belt. Now look at the drive side: If any teeth are missing, cracked or excessively worn, replace the belt. **Caution:** *Do not nick the surface of the belt with tools or other objects, do not get oil or chemicals on it, and do not twist the belt.*

Installation

> ### ** CAUTION **
>
> Before starting the engine, carefully rotate the crankshaft by hand through at least two full revolutions (use a socket and breaker bar on the crankshaft pulley center bolt). If you feel any resistance, STOP! There is something wrong - most likely, valves are contacting the pistons. You must find the problem before proceeding. Check your work and see if any updated repair information is available.

Refer to illustrations 5.11 and 5.12
Caution: *The camshafts on 1998 and later models are driven by an idler gear. Because the ratio between the camshaft timing belt sprocket and camshafts is not 1:1, the camshafts might not be in the right positions if the timing belt has broken or if the camshafts or cylinder heads have been removed. To bring the camshafts into the proper phase, turn the crankshaft 1/8-turn (45-degrees) counterclockwise (so the pistons won't be in the way of the valves), then rotate the left (driver's side) camshaft timing belt sprocket clockwise until it "springs" into place; this could take up to nine turns of the sprocket. At this point the mark on the sprocket will still not be in the proper position - turn it a further 1/4 turn (90-degrees) and*

align the mark on the sprocket with the mark on the valve cover. To bring the right (passenger's side) cylinder bank camshafts into the proper phase rotate the camshaft sprocket until it "springs" into position; again, this could take up to nine revolutions. When this sprocket springs into position the timing marks on the sprocket and valve cover should be in alignment. Finally, turn the crankshaft 1/8-turn (45-degrees) clockwise and align the mark on the crankshaft timing belt sprocket with the mark on the oil pump. **Keep in mind that it is not necessary to do this during a routine timing belt replacement, since the alignment will not have been lost.**

11   Prepare to install the timing belt by compressing the hydraulic tensioner in a vise enough to insert a drill bit or pin into the hole **(see illustration)**.

12   Make certain the sprockets are all aligned with their marks **(see illustrations 5.8a and 5.8b)**. Install the belt on the crankshaft sprocket first, lining up the dotted line on the belt with the mark on the sprocket. Keeping the belt tight, slip it past the tensioner pulley and over the right (passenger's) side camshaft sprocket, lining up the solid line on the belt with the mark on the sprocket. **Note:** *The writing on the belt must face out and the arrows on the belt must point clockwise when viewed from the front* **(see illustration)**.

13   Route the belt inside the idler pulley, then onto the left camshaft sprocket and under the water pump sprocket. Make sure the solid line on the belt lines up with the mark on the camshaft sprocket. Be careful not to nudge the camshaft or crankshaft sprockets off their timing marks.

14   Make sure all three sets of timing marks are properly aligned **(see illustration 5.8)**. Install the hydraulic tensioner. Turn the crankshaft slightly clockwise to provide some slack in the belt between the crankshaft sprocket and the right (passenger's side) camshaft sprocket, then pull the pin or drill bit to allow the tensioner to take up the slack.

15   Slowly turn the crankshaft clockwise two full revolutions, returning the crankshaft sprocket mark to its original position. **Caution:** *If excessive resistance is felt while turn-*

ing the crankshaft, it's an indication that the pistons are coming into contact with the valves. Go back over the procedure to correct the situation before proceeding.

16   The remainder of installation is the reverse of removal.

---

6   Intake manifold - removal and installation

Removal

1   Relieve the fuel system pressure (see Chapter 4B).

2   Disconnect the cable from the negative terminal of the battery.

3   Remove the air cleaner assembly (see Chapter 4B).

4   Remove the throttle body, fuel lines and electrical connectors at each fuel injector (see Chapter 4B). **Note:** *If you're removing the upper and lower intake manifolds simply to replace manifold gaskets or to service some part of the upper engine, leave the throttle body attached to the air intake plenum and remove the plenum and throttle body as a single assembly.*

Air intake plenum (upper intake manifold)

Refer to illustration 6.8
Note: *The 3.2L engine has a separate air intake plenum (upper intake manifold) mounted above the lower intake manifold.*

5   Unplug the electrical connectors from the manifold absolute pressure (MAP) sensor, the power switch and the electronic vacuum sensing valve (see Chapter 6). If you didn't remove the throttle body in Step 4, unplug the electrical connectors from the idle air control (IAC) valve and the throttle position sensor (TPS) as well (see Chapter 6).

6   If you didn't remove the throttle body, disconnect the accelerator cable (see Chapter 4A) and (on models with an automatic transmission) the transmission linkage (see Chapter 7B) from the throttle body assembly. Detach the accelerator cable from the intake manifold.

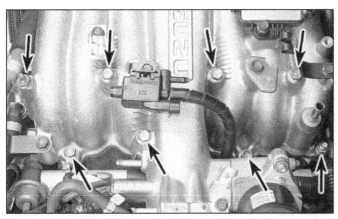

6.8 To detach the air intake plenum (upper intake manifold), remove the bolts on the backside of the plenum, the bolt at the engine hanger bracket and the bolt at the automatic transmission pipe bracket (not visible in this photo), then remove the plenum retaining bolts and nuts (1997 and earlier models)

6.13 Remove the mounting bolts/nuts (arrows) from the intake manifold - there are two nuts and one bolt on each side

7   Disconnect the brake booster vacuum hose from the pipe at the left rear corner of the plenum. Clearly label all other vacuum lines attached to the plenum and then detach them too. Label the EVAP canister purge lines and detach them. Also label any vacuum lines attached to the PCV valve, the EGR valve and the fuel pressure regulator and disconnect them. If you didn't remove the throttle body, clearly label and then detach any vacuum lines attached to it. If any of these hoses are attached to the plenum by clips or guides, detach them as well.

8   Remove the retaining bolts on the backside of the plenum, the retaining bolt at the engine hanger bracket and the bolt at the automatic transmission pipe bracket (if applicable). Remove the air intake plenum bolts and nuts **(see illustration)**.

9   Remove the air intake plenum (and the throttle body, if still attached) from the lower intake manifold.

10   Be sure to clean and inspect the mounting faces of the lower intake manifold and the air intake plenum before positioning the new gasket on the mating surface of the lower intake manifold.

Lower intake manifold

Refer to illustrations 6.13 and 6.14

11   Disconnect the fuel supply and return lines from the fuel rail (see Chapter 4) and detach the fuel line brackets from the engine cover.

12   Unplug the electrical connectors from each fuel injector (see Chapter 4).

13   Remove the two intake manifold retaining bolts and four nuts **(see illustration)**.

14   When the manifold is disconnected from all hoses and wires, pry at a corner of the manifold to break the gasket seal **(see illustration)**. **Caution:** *Do not pry under the sealing surfaces or leaks could develop.*

Installation

Refer to illustration 6.17

Note: *The mating surfaces of the cylinder heads and manifold must be perfectly clean when the manifold is installed. Gasket removal solvents in aerosol cans are available at most auto parts stores and may be helpful when removing old gasket material that's stuck to the heads and manifold (since they're made of aluminum, aggressive scraping can cause*

damage). Be sure to follow the directions printed on the container.

15   Use a gasket scraper to remove all traces of sealant and old gasket material, then clean the mating surfaces with lacquer thinner or acetone. If there's old sealant or oil on the mating surfaces when the manifold is installed, oil or vacuum leaks may develop. Use a vacuum cleaner to remove any material that falls into the intake ports in the heads.

16   Use a tap of the correct size to clean up the threads in the bolt holes, then use compressed air (if available) to remove the debris from the holes. **Warning:** *Wear safety glasses or a face shield to protect your eyes when using compressed air!*

17   Position the gaskets on the cylinder heads **(see illustration)**. No sealant is required, but make sure the beaded-sealant side of the gaskets are facing up.

18   Carefully set the manifold in place, being careful not to disturb the gaskets. Make sure all intake port openings and bolt holes are aligned correctly.

19   Install the bolts, washers and nuts and tighten them to the torque listed in this Chapter's Specifications. Work up to the final

6.14 Do not pry in the gasket sealing areas - pry at a casting protrusion to separate the manifold from the engine

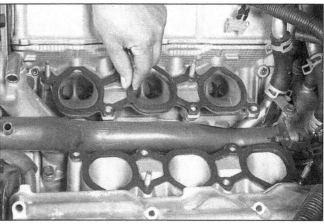

6.17 Install the new intake manifold gaskets to the cylinder heads, aligning them on the studs

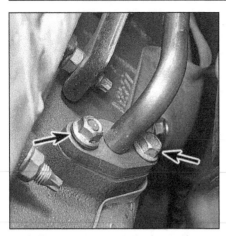

7.3 Remove the two bolts (arrows) and separate the EGR tube from the left exhaust manifold

7.5 Remove the three nuts (arrows) holding the exhaust pipe to the manifold

7.7 Remove the bolts (arrows) and take off the exhaust heat shield - left side shown, through fenderwell

torque in two steps.

20   Using a new gasket, install the air intake plenum (and throttle body assembly, if still attached to the plenum) onto the lower intake manifold. Make sure that the gasket remains in place. Install the plenum retaining bolts and tighten them in two or three steps to the torque listed in this Chapter's Specifications, starting with the center bolts first.

21   The remainder of installation is the reverse of removal.

22   Start the engine and check carefully for oil and coolant leaks at the intake manifold joints.

---

7   Exhaust manifolds - removal and installation

Warning: *Let the engine cool completely before this procedure is performed.*

Removal

Refer to illustrations 7.3, 7.5, 7.7 and 7.8

1   Disconnect the negative cable from the

battery. Raise the vehicle and support it securely on jackstands.

2   Spray penetrating oil on the exhaust manifold fasteners and allow it to soak in.

3   The procedure is virtually the same for either side, except that the air cleaner duct should be moved (see Chapter 4B) and the EGR pipe must be disconnected **(see illustration)** depending on which manifold you're removing.

4   Support the transmission with a floor-jack and remove the transmission crossmember and skid plate (see Chapter 7) to allow the exhaust pipes to be dropped down out of the way.

5   Remove the nuts that attach the flange(s) of the exhaust pipes to the manifold(s) to be removed **(see illustration)**.

6   Disconnect the electrical connector at the oxygen sensor (see Chapter 6). Leave the sensor in the manifold unless it is being replaced.

7   Remove the heat shield bolts **(see illustration)** and remove the shield.

8   Remove the exhaust manifold(s) and gasket(s) **(see illustration)**. **Note:** *The engine*

support brackets are retained by two of the exhaust manifold studs/nuts on each side. Remove the support brackets.

9   Carefully inspect the manifold(s) and fasteners for cracks and damage.

10   Use a scraper to remove all traces of old gasket material and carbon deposits from the manifold and cylinder head mating surfaces. If the gasket was leaking, have the manifold checked for warpage at an automotive machine shop and resurfaced if necessary.

Installation

11   Position the new gasket(s) over the cylinder head studs.

12   Install the manifold and thread the mounting nuts into place.

13   Working from the center out, tighten the nuts to the torque listed in this Chapter's Specifications in three or four equal steps.

14   Reinstall the remaining parts in the reverse order of removal.

15   Run the engine and check for exhaust leaks.

7.8 Remove the seven nuts (arrows indicate five here) on the exhaust manifold

8.4 Hold the sprocket from turning with a small, deep socket (arrow) pushed into the hole, and use a pry bar against it while using a breaker bar on the sprocket bolt

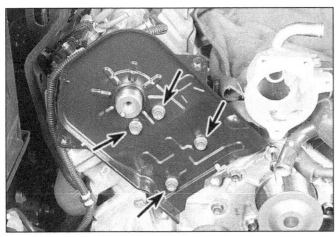

8.5 Remove the bolts (arrows) and take off the rear plate

8.6 Remove these bolts (arrows) holding the camshaft towers - two towers are shown here, with circles indicating the stamped numbers

8.7 Lift the camshaft from the head with the towers still on the camshaft

8.10 Pry the camshaft seal (arrow) out of the front camshaft tower and press in a new one, much like a crankshaft seal (see Section 4)

8 Camshaft(s) - removal, inspection and installation

1997 and earlier (SOHC) models

Removal

Refer to illustrations 8.4, 8.5, 8.6 and 8.7

1   Disconnect the negative battery cable.

2   Refer to Chapter 4 for intake air plenum removal, and Section 3 of this Chapter for removal of the valve covers.

3   Remove the timing belt (see Section 5).

4   Hold the camshaft sprocket with a strap-type wrench, and remove the camshaft sprocket bolt, then slip the sprocket off the camshaft. If you don't have access to a strap wrench or factory cam sprocket tool, select a small, deep socket that just fits into the hole in the face of the camshaft sprocket **(see illustration)**. Use a ratchet and socket on the sprocket bolt, while holding the sprocket from turning with a prybar against the other socket.

5   Remove the bolts and take off the plate behind the camshaft sprocket **(see illustration)**.

6   The camshaft can be removed without taking off the rocker arm assembly. Remove the bolts holding the five camshaft towers **(see illustration)**.

7   Lift the camshaft and its towers out of the cylinder head **(see illustration)**.

Inspection

8   Visually check the camshaft bearing surfaces for pitting, score marks, galling and abnormal wear. If the bearing surfaces are damaged, the head will have to be replaced. The camshaft journal oil clearance can be calculated by measuring the tower bores with an inside micrometer, measuring the camshaft journal with an outside micrometer, then subtracting the journal diameter from the bore diameter. If the clearances are excessive, the camshaft(s) and cylinder head(s) must be replaced.

9   Check the camshaft lobe height by measuring each lobe with a micrometer **(see illustrations in Chapter 2, Part E)**. Compare the measurement to the cam lobe height listed in this Chapter's Specifications. Then subtract the measured cam lobe height from the specified height to compute wear on the cam lobes. Compare it to the specified wear limit. If it's greater than the specified wear limit, replace the camshaft.

Installation

Refer to illustrations 8.10 and 8.12

10   Now is a good time to install a new camshaft seal in the front camshaft tower **(see illustration)**.

11   Lubricate the camshaft bearing journals and lobes with camshaft installation lube, then install it carefully in the head, with the camshaft towers in their original positions. **Note:** *Apply a bead of RTV sealant at the front and rear of the cylinder head where the front and rear camshaft towers bolt down.*

12   Tighten the camshaft tower bolts a little at a time, in sequence, to the torque listed in this

8.12 Tighten the camshaft tower bolts in this sequence

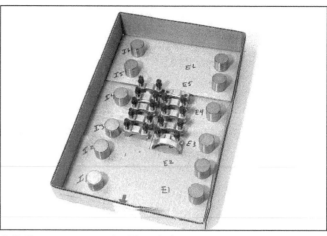

8.19 Be sure to store the lifters/shims and camshaft bearing caps in order

Chapter's Specifications **(see illustration)**.

13   Check to make sure the mark on the crankshaft sprocket is still aligned with its mark on the oil pump. Slide the camshaft sprockets onto the camshafts and align the marks on the sprockets with their corresponding marks on the cylinder heads. Use a strap wrench to hold the camshaft sprocket, while tightening the camshaft sprocket bolt to the specified torque.

14   Refer to Section 5 for installation of the timing belt. The remaining steps are the reverse of the removal procedure.

1998 and later (DOHC) models

Removal

Refer to illustration 8.19

15   Make sure the engine is positioned on TDC for number 1 piston (see Chapter 2A).

16   Remove the valve covers (see Section 4) and the timing belt (see Section 5).

17   Check the camshaft bearing caps for identification and installation markings. If none are present, use a sharp scribe and number them from one to five, from the front

of the head to the rear. Also make an arrow on each cap pointing to the front of the engine. Loosen the camshaft bearing cap bolts in 1/4-turn increments until they can be removed by hand. Follow the reverse of the recommended tightening sequence **(see illustration 8.32)**.

18   Remove the bearing caps and gently lift out the camshafts. Be sure to keep them level as this is done.

19   The lifters can be removed from the head now. Be sure to store them in order so they can be returned to their original bores **(see illustration)**. Also, don't lose or mix up the shims on top of the lifters.

20   If necessary, remove the camshaft sprocket, then unscrew the three bolts and remove the drive gear retainer from the front of the cylinder head.

Inspection

Refer to illustrations 8.21, 8.22, 8.23 and 8.24

21   Inspect each lifter for scuffing and score marks **(see illustration)**.

22   Visually examine the cam lobes and bearing journals for score marks, pitting,

galling and evidence of overheating (blue, discolored areas). Look for flaking away of the hardened surface layer of each lobe. Using a micrometer, measure the height of each camshaft lobe **(see illustration)**. Compare your measurements with this Chapter's Specifications. Replace the camshafts as a set if any undesirable conditions are found.

23   Using a micrometer, measure the diameter of each journal at several points **(see illustration)**. Compare your measurements with this Chapter's Specifications. If the diameter of any one journal is less than specified, replace the camshafts.

24   Check the oil clearance for each camshaft journal as follows:

a)   Clean the bearing caps and the camshaft journals with lacquer thinner or acetone.

b)   Carefully lay the camshaft(s) in place in the cylinder head. Don't install the lifters and don't use any lubrication.

c)   Lay a strip of Plastigage on each journal.

d)   Install the bearing caps in their proper locations with the arrows pointing toward the front of the engine.

8.21 Inspect each lifter for wear and scuffing

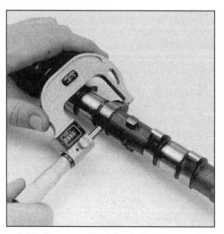

8.22 Measure the lobe height on each camshaft - if any lobe is worn more than the allowable limit, replace the camshafts

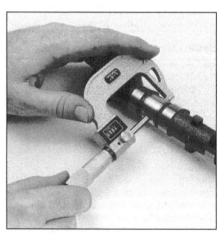

8.23 Measure each journal diameter with a micrometer (if any journal measures less than the specified limit, replace the camshafts)

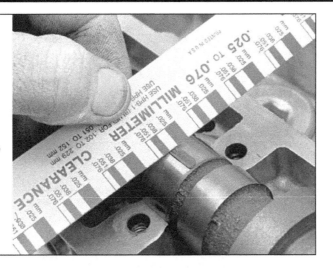

8.24 Compare the width of the crushed Plastigage to the scale on the envelope to determine oil clearance

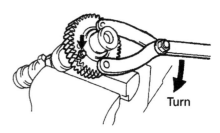

8.26 With the camshaft held in a vise, use a pin spanner to turn the sub-gear so a M5 x 0.5 mm bolt can be installed to lock the gears together

e) *Tighten the bolts to the torque listed in this Chapter's Specifications in 1/4-turn increments.* **Note:** *Don't turn the camshaft while the Plastigage is in place.*
f) *Remove the bolts and detach the caps.*
g) *Compare the width of the crushed Plastigage (at its widest point) to the scale on the Plastigage envelope* **(see illustration).**
h) *If the clearance is greater than specified, replace the camshaft and/or cylinder head (if the camshaft journal diameters are within specifications, the camshaft saddles in the cylinder head are excessively worn).*
i) *Scrape off the Plastigage with your fingernail or the edge of a credit card - don't scratch or nick the journals or bearing caps.*

Installation

Refer to illustrations 8.26, 8.30 and 8.32
25   If removed, install the drive gear retainer, tightening the bolts to the torque listed in this Chapter's Specifications.
25   Mount the camshaft in a padded vise.
26   Insert a M5 x 0.8 bolt into the hole in the camshaft sub-gear. Using a pin spanner, align the holes of the camshaft driven gear and sub-gear by turning the camshaft sub-gear clockwise **(see illustration)**. Tighten the bolt to clamp the gears together.
27   Apply engine assembly lube to the lifters, then install them in their original locations in the cylinder heads. Make sure the valve adjustment shims are in place in the lifters, and that all lifters are installed in their original bores.
28   Apply camshaft installation lube to the camshaft lobes, bearing journals and gear thrust faces.
29   Align the timing mark on the camshaft drive gear with the timing mark on the retainer (the mark on the left (driver's side) retainer is at approximately the 9 o'clock

position; the mark on the right [passenger's side] cylinder head is at the 12 o'clock position). When aligned, the dowel pin hole in the gear for the left (driver's side) cylinder head will be in the 9 o'clock position; the dowel pin hole in the gear for the right (passenger's side) cylinder head will be in the 12 o'clock position.
30   Set the camshafts in place in the cylinder head with the timing marks on the camshaft gears in alignment with the timing marks on the camshaft drive gear **(see illustration)**.
31   Install the bearing caps in numerical order with the arrows pointing toward the front of the engine.
32   Tighten the bearing cap bolts in 1/4-turn increments to the torque listed in this Chap-

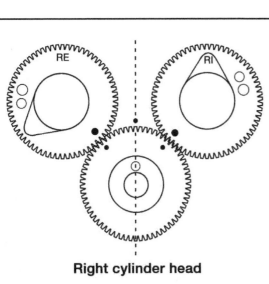

Right cylinder head

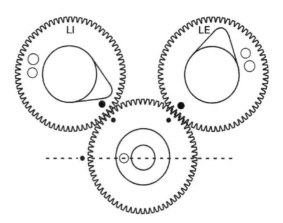

Left cylinder head

47017-2d-8.30 HAYNES

8.30 Camshaft gear and drive gear timing marks

ter's Specifications. Follow the recommended sequence **(see illustration)**.

33   Remove the bolt from the sub gear.

34   If it's necessary to replace the oil seal, refer to Section 4 (it's similar to replacing a crankshaft oil seal).

35   Install the timing belt sprocket and tighten the bolt to the torque listed in this Chapter's Specifications. Don't allow the sprocket to turn as the bolt is tightened.

36   Install the timing belt (see Section 5).

37   Remove the spark plugs (see Chapter 1). Using a socket and breaker bar on the crankshaft pulley bolt, turn the engine over in the normal direction of rotation (clockwise when viewed from the front) several times to verify proper cam timing. If resistance is felt at any point, STOP; the cam timing may not be correct, allowing the valves to contact the pistons. Go back and recheck all of the timing marks.

38   The remainder of installation is the reverse of removal.

39   Check and adjust, if necessary, the valve clearances (see Chapter 1).

40   Run the engine, then check for leaks and proper operation.

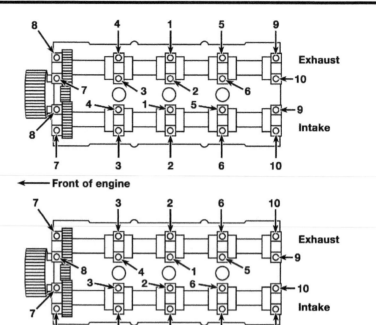

8.32 Camshaft bearing cap tightening sequence (DOHC engine)

9   Rocker arms, lash adjusters and shafts (SOHC) models - removal, inspection and installation

Refer to illustrations 9.3 and 9.4

1   Disconnect the negative battery cable.

2   Refer to Section 5 for timing belt removal and Section 8 for camshaft removal.

3   Prior to removal, scribe or paint identifying marks on the rockers to ensure they will be installed in their original locations. Wrap the ends of the rocker arms with tape to prevent the hydraulic lash adjusters from falling out when the rockers are removed **(see illustration)**.

4   Loosen the rocker arm shaft bolts **(see illustration)** in two or three stages, working your way from the ends toward the middle of the shafts. **Caution:** *Some of the valves will*

be open when you loosen the rocker arm shaft bolts and the rocker arm shafts will be under a certain amount of valve spring pressure. Therefore, the bolts must be loosened gradually. Loosening a bolt all at once near a rocker arm under spring pressure could bend or break the rocker arm shaft.

5   Lift off the rocker arm shaft assemblies and lay them down on a nearby workbench in the same relationship as when installed. They must be reinstalled on the same cylinder head.

6   Installation is the reverse of the removal procedure. Tighten the rocker arm shaft bolts, in several steps, to the torque listed in this Chapter's Specifications. Work from the ends of the shafts toward the middle.

Inspection

Refer to illustrations 9.7, 9.8 and 9.10

7   Inspect the contact and sliding surfaces of each hydraulic lash adjuster for scoring or damage **(see illustration)**. Replace any defective parts. **Note:** *Be sure to label each rocker arm and adjuster and place them in a partitioned box or something suitable to keep them from getting mixed up.*

8   Check the rocker arms and shafts for abnormal wear, pits, galling, score marks and rough spots **(see illustration)**. Don't attempt to restore rocker arms by grinding the pad surfaces. Replace any defective parts.

9   Lube the contact surfaces of the rocker arms with moly-based lubricant and coat the

9.3 Number the rocker arms and wrap some tape around the lash adjusters (arrows) where they fit into the ends of the rocker arms, to prevent them from falling out during rocker assembly removal

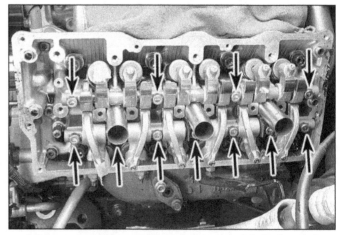

9.4 Loosen and remove the rocker arm shaft bolts (arrows)

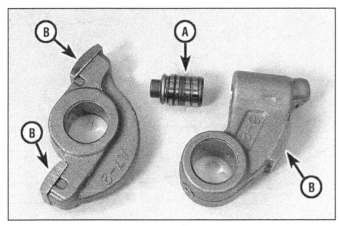

9.7 Examine the lash adjusters (A) for scoring or damage, and the wear points (B) on the rocker arms

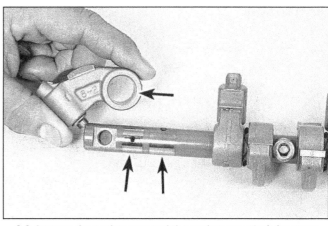

9.8 Inspect the rocker arm and the rocker arm shaft for wear where the rocker arms ride (arrows)

rocker arm shafts with clean engine oil before installation on the engine. **Caution:** *Keep all rocker arms and pedestals in their original order.*

10   Install the rocker arms/shaft assembly and tighten the bolts to torque listed in this Chapter's Specifications in the recommended sequence **(see illustration)**.

11   The remainder of installation is the reverse of removal.

---

10 Valve springs, retainers and seals - replacement

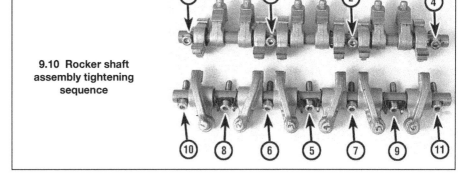

9.10 Rocker shaft assembly tightening sequence

SOHC models

This procedure is essentially the same as for the 2.3L/2.6L four-cylinder engines. Refer to Chapter 2, Part A, and follow the procedure outlined there.

DOHC models

Due to the design of the cylinder head on these models, it is not practical to attempt to replace the valve springs or seals with the cylinder head installed on the engine. Refer to Section 11 for cylinder head removal, then refer to Chapter 2 Part E for cylinder head servicing procedures.

---

11 Cylinder head(s) - removal and installation

Warning: *Allow the engine to cool completely before beginning this procedure.*

Removal

Refer to illustrations 11.5a, 11.5b and 11.7

1   Disconnect the negative battery cable and drain the engine coolant (see Chapter 1).
2   Remove the timing belt (see Section 5).
3   Remove the intake manifold (see Section 6).

4   Remove the exhaust manifold(s) as described in Section 7.
5   Behind the intake manifold area, unbolt and remove the coolant "manifold" that connects the cylinder heads on models so equipped **(see illustrations)**.
6   It is possible to remove the cylinder heads without removing the camshafts or rocker shaft assemblies, depending on the work to be done.
7   Remove the bolts and detach the power steering pump from the engine, tying it aside without disconnecting the hoses (see Chapter 10), then remove the bracket from the

11.5a Remove the bolts (arrows) and take out the large coolant pipe under the intake manifold

11.5b With the pipe out of the way, disconnect the hoses and senders attached to the coolant crossover manifold (arrow), then unbolt the manifold from the heads and remove it

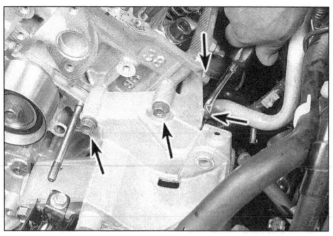

11.7 Remove the four power steering pump bracket bolts (arrows) from the left head

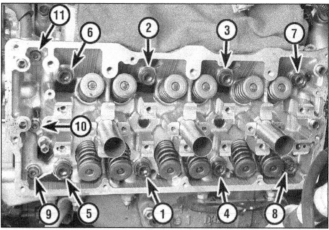

11.19 Cylinder head tightening sequence - 3.2L V6 (note that bolts 9, 10 and 11 [SOHC engine only] are smaller and require a different torque)

cylinder head **(see illustration)**.

8   Unbolt the oil dipstick tube from the right cylinder head. The bolt is accessible through the right fenderwell.

9   Loosen the cylinder head bolts 1/4-turn increments until they can be removed by hand. Be sure to follow the proper numerical sequence **(loosening sequence is the reverse of tightening sequence in illustration 11.19)**.

10   The large head bolts (M11) must be replaced with new ones, so discard the bolts after removing them (the three small ones at the front of the head can be reused).

11   Lift the head off the block. If resistance is felt, dislodge the head by striking it with a wood block and hammer. If prying is required, pry only on a casting protrusion - be very careful not to damage the aluminum head or block!

Installation

Refer to illustrations 11.19 and 11.20

12   Remove all traces of old gasket material from the cylinder heads and the engine block. The mating surfaces of the cylinder heads and block must be perfectly clean when the heads are installed.

13   Use a gasket scraper to remove all traces of carbon and old gasket material, then clean the mating surfaces with lacquer thinner or acetone. If there's oil on the mating surfaces when the heads are installed, the gaskets may not seal correctly and leaks may develop. Use a vacuum cleaner to remove any debris that falls into the cylinders. **Caution:** *Be careful not to gouge the soft aluminum of the head or block, which could cause coolant or combustion gas leakage.*

14   Check the block and head mating surfaces for nicks, deep scratches and other damage. If damage is slight, it can be removed with a file - if it's excessive, machining may be the only alternative.

15   Use a tap of the correct size to chase the threads in the head bolt holes. Dirt, corrosion, sealant and damaged threads will affect torque readings. Ensure that the threaded

holes in the block are clean and dry.

16   Position the new gaskets over the dowel pins on the block.

17   Carefully position the heads on the block without disturbing the gaskets.

18   Lightly oil the threads and install the bolts. Tighten them finger tight.

19   Follow the recommended sequence and tighten the bolts in three steps to the torque listed in this Chapter's Specifications **(see illustration)**.

20   The remaining installation steps are the reverse of removal. **Note:** *When reinstalling the water manifold and large coolant pipe, use a new O-ring where the pipe and manifold connect* **(see illustration)**.

21   Add coolant and change the engine oil and filter (see Chapter 1), then start the engine and check carefully for oil and coolant leaks.

---

12   Oil pan - removal and installation

Removal

1   Disconnect the negative cable from the battery.

2   Raise the vehicle and support it securely on jackstands.

3   Remove the under-vehicle splash pan.

4   Drain the engine oil and install a new oil filter (see Chapter 1).

5   Loosen the front wheel lug nuts, raise the vehicle and support it securely on jackstands. Place a floor jack under the front axle, and remove the front wheels.

6   If you're working on a 1997 or earlier model, refer to Chapter 10 and disconnect both the Pitman arm and idler arm, to allow the steering gear center link to drop down away from the front of the oil pan.

7   Refer to Chapter 7 and remove the flywheel inspection cover.

4WD models

Refer to illustration 12.8

8   Remove two bolts on each side of the under-engine crossmember, and remove the crossmember **(see illustration)**.

9   Refer to Chapter 8 and remove the front axle assembly.

All models

Refer to illustration 12.11

10   Remove the starter motor for access to the last few pan bolts on the left side of the engine (see Chapter 5).

11   Remove the oil pan bolts and nuts **(see illustration)**. **Note:** *Some of the fasteners are*

11.20 When replacing the water pipe to the water manifold at the back of the heads, use a new O-ring (arrow)

12.8 Unbolt the under-engine crossmember at each side (arrows)

12.11 Remove the oil pan bolts (arrows, not all bolts are shown here)

studs/nuts instead of bolts.

12   Detach the oil pan. Don't pry between the pan and block or damage to the sealing surfaces may result and oil leaks could develop. If the pan is stuck, dislodge it with a hammer and a block of wood.

13   Use a gasket scraper to remove all traces of old sealant from the block and pan. Clean the mating surfaces with lacquer thinner or acetone.

Installation

Refer to illustration 12.15

14   Ensure that the threaded holes in the block are clean (use a tap to remove any sealant or corrosion from the threads).

15   Apply a thin continuous bead of RTV sealant along the circumference of the oil pan flange **(see illustration)**. Where you encounter bolt holes, run the bead to the inside of the bolt holes. **Note:** *The pan should be installed on the block within five minutes of applying the sealant.*

16   Install the oil pan and tighten the bolts in three or four steps to the torque listed in this

Chapter's Specifications.

17   The remaining installation steps are the reverse of removal.

18   Allow at least 30 minutes for the sealant to dry. Fill the crankcase with oil (see Chapter 1), start the engine and check for oil pressure and leaks.

---

13   Oil pump - removal, inspection and installation

Removal

Refer to illustrations 13.2, 13.3, 13.4a, 13.4b and 13.5

1   Remove the timing belt and the crankshaft sprocket (see Sections 5 and 4). Remove the oil pan (see Section 12).

2   Unbolt the oil pump pickup from the oil gallery assembly **(see illustration)**.

3   Disconnect the oil pump "elbow" connection from the oil pump and the oil gallery assembly **(see illustration)**.

4   Refer to Chapter 3 and unbolt the air

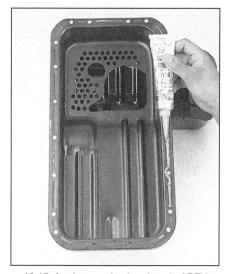

12.15 Apply an unbroken bead of RTV sealant to the oil pan and install it within five minutes (note how the sealant is applied to the *inner* sides of the bolt holes)

13.2 Remove the mounting bolts (arrows, two on the oil gallery assembly and two on the pump) then take down the oil pump pickup

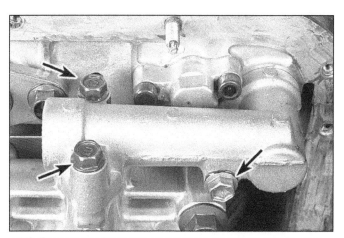

13.3 Remove the four bolts and disconnect the oil system elbow (arrow) from the pump and the oil gallery

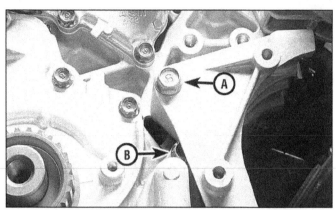

13.4a This drivebelt idler pulley must be unbolted to access one of the compressor mounting bracket bolts beneath it

13.4b Remove the hidden compressor mounting bolt (A), remove the remaining mount bolts and the mount, and there will be access to the oil pump bolt (B) hidden by the compressor mount

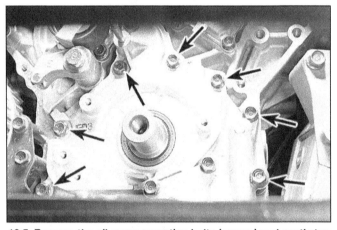

13.5 Remove the oil pump mounting bolts (arrows) and gently tap the pump loose with a hammer and block of wood (the upper bolt is not seen here)

13.9 Remove these screws and take the back cover from the oil pump

conditioning compressor and set it aside without disconnecting the hoses. Unbolt the compressor bracket from the engine **(see illustrations)**.

5   Remove the oil pump-to-engine block bolts from the front of the engine **(see illustration)**. **Note:** *Four of the bolts on the lower left of the pump hold the oil cooler adapter to the oil pump. Only the outer two need to be removed to take out the oil pump, with the cooler adapter attached by the other two bolts.*

6   Use a block of wood and a hammer to break the oil pump loose. **Note:** *The pump has dowels that locate in the block. Pull the oil pump straight off.*

7   Pull out on the oil pump to remove it from the engine block.

8   Use a scraper to remove old sealant from the oil pump and engine block mating surfaces. Clean the mating surfaces with lacquer thinner or acetone.

Inspection

Refer to illustrations 13.9, 13.11, 13.12a, 13.12b and 13.12c

9   Remove the screws holding the rear

cover to the oil pump **(see illustration)**.

10   Clean all components with solvent, then inspect them for wear and damage.

11   Remove the oil pressure relief valve cover, spring and valve (plunger) **(see illustration)**. Check the oil pressure relief valve sliding surface and valve spring. If the valve doesn't slide freely in it's bore, even after cleaning and oiling it, the pump must be replaced as an assembly.

12   Check the following clearances with a feeler gauge **(see illustrations)** and compare the measurements to the clearances listed in this Chapter's Specifications:

    Case-to-outer rotor
    Clearance above the rotors
    Clearance between the rotors

If any of the clearances are excessive, replace the entire oil pump assembly.

13   Pack the cavities of the oil pump with petroleum jelly to prime it. Assemble the oil pump and tighten the screws to the torque listed in this Chapter's Specifications. Oil and install the oil pressure relief valve and spring, then tighten the oil pressure relief valve cover to the torque listed in this Chapter's Specifications.

Installation

Refer to illustration 13.14

14   Apply a small bead of RTV sealant to the back of the oil pump **(see illustration)**.

15   Installation is the reverse of the removal procedure. Align the flats on the inner rotor of the oil pump with the flats on the crankshaft. Tighten all fasteners to the torque values listed in this Chapter's Specifications. **Note:** *New O-rings should be installed in the oil transfer "elbow" and the oil pump pickup tube.*

14   Flywheel/driveplate - removal and installation

This procedure is essentially the same for all engines. Refer to Part A and follow the procedure outlined there, but use the bolt torque listed in this Chapter's Specifications. **Note:** *Clean the threaded holes in the crankshaft, use only new bolts when installing the flywheel, and do not use thread-locking compound on the bolts.*

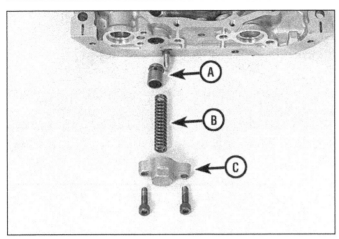

13.11 Relief valve assembly

A   Relief valve   C   Cover
B   Spring

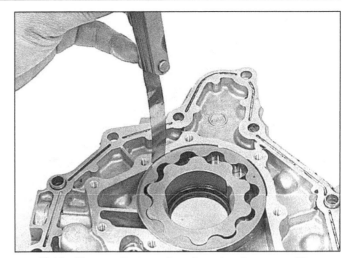

13.12a Measure the case-to-outer rotor clearance with a feeler gauge

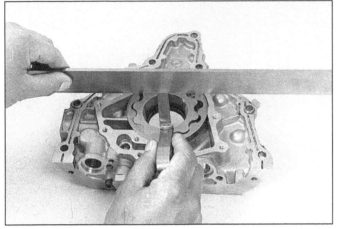

13.12b Put a precision straightedge across the back of the pump, and measure the clearance above the rotors

13.12c Use a feeler gauge to measure the clearance between the inner and outer rotors

15   Rear main oil seal - replacement

Refer to illustration 15.5

Note: *Although it's possible to pry out the old seal and install a new one without removing the seal retainer, we don't recommend it, because it's easy to nick or scratch the crankshaft or the seal bore. And if you're not extremely careful, you can damage the new seal lip during installation. However, if you don't feel like removing the oil pan and the retainer, use the procedure outlined in Chapter 2, Part A, (but be advised that the following method is the recommended way to change the rear main seal).*

1   Remove the transmission (see Chapter 7).

2   Remove the oil pan (See Section 12).

3   Remove the rear main oil seal retainer bolts and the retainer. **Note:** *The retainer has dowels that locate in the block. Pull the retainer plate straight off.*

4   Put the retainer in a bench vise and pry out the old seal.

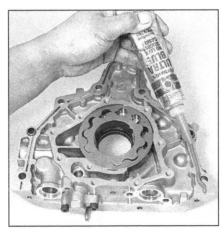

13.14 Apply a continuous bead of RTV sealant to the back of the oil pump and install it on the block right away

5   To install the new seal, lay the retainer on a clean, flat surface and carefully drive in

15.5 The new seal can be driven in squarely with a block of wood and hammer - drive it in only as far as the old seal was installed

the seal with a block of wood and a hammer **(see illustration).**

6   Coat the mating surface of the retainer with RTV sealant, apply a light coating of clean oil to the seal lip, install the retainer and tighten the retainer bolts to the torque listed in this Chapter's Specifications.

7   The remainder of installation is the reverse of removal.

16   Engine mounts - check and replacement

Refer to illustration 16.1

This procedure is essentially the same for all engines. See Part A of this Chapter and follow the procedure outlined there, but use the torque values listed in this Chapter's Specifications. Use the accompanying illustration of a 3.2L V6 mount for reference **(see illustration)**.

16.1 Engine mount details - 3.2L V6

A   *Engine bracket*
B   *Insulator*
C   *Insulator-to-chassis-bracket bolts*
D   *Chassis bracket*

Chapter 2 Part E
General engine overhaul procedures

Contents

Specifications

2.3L (4ZD1) and 2.6L (4ZE1) four-cylinder engines

General

Displacement	
4ZD1..	2.3L (138 cu. in)
4ZE1	2.6L (156.2 cu. in)
Oil pressure (at 3000 rpm)	56.9 psi
Compression pressure	
Standard..	142 to 170 psi
Minimum..	114 psi
Camshaft and rocker arms	See Part A
Cylinder head warpage limit	0.008 inch

Valves and related components

Valve stem diameter	
Intake..	0.3129 to 0.3134 inch
Exhaust..	0.3118 to 0.3124 inch
Valve margin width	
Intake..	0.0315 inch
Exhaust..	0.0393 inch
Valve stem-to-guide clearance	
Standard	
Intake	0.00098 to 0.0022 inch
Exhaust	0.0015 to 0.0028 inch
Limit	
Intake	0.0079 inch
Exhaust	0.0098 inch

2.3L (4ZD1) and 2.6L (4ZE1) four-cylinder engines (continued)

Valves and related components (continued)

Valve spring free length	
Standard	1.8307 inches
Limit	1.8937 inches
Valve spring out-of-square limit	0.0827 inch

Engine block

Deck distortion limit	0.0078 inch

Cylinder bore diameter

4ZD1	
Standard	3.516 to 3.517 inches
Limit	3.525 inches
4ZE1	
Grade A	3.6457 to 3.6461 inches
Grade B	3.6461 to 3.6464 inches
Grade C	3.6465 to 3.6468 inches
Grade D	3.6468 to 3.6472 inches
Cylinder bore taper/out-of-round limits	Less than 0.0006 inch
Piston-to-bore clearance	
4ZD1	0.0008 to 0.0016 inch
4ZE1	0.0010 to 0.0018 inch

Pistons and rings

Piston diameter	
4ZD1	
Standard	3.5168 to 3.5186 inch
1st oversize	3.5337 to 3.5352 inch
2nd oversize	3.5533 to 3.5549 inch
4ZE1	
Standard	3.6460 to 3.6470 inches
1st oversize	3.6640 to 3.6655 inches
2nd oversize	3.6836 to 3.6852 inches
Piston ring end gap	
Top compression ring	0.0118 to 0.0177 inch
Second compression ring	0.0236 to 0.0283 inch
Oil ring	0.0079 to 0.0276 inch
Piston ring side clearance	
Top compression ring	0.0010 to 0.0024 inch
Second compression ring	0.0008 to 0.0022 inch

Crankshaft, connecting rods and main bearings

Connecting rod side clearance	
Standard	0.0079 to 0.0013 inch
Service limit	0.0157 inch
Crankshaft endplay	
Standard	0.0024 to 0.0099 inch
Service limit	0.0118 inch
Main bearing journal diameter*	2.2016 to 2.2022 inches
Connecting rod journal diameter*	1.9262 to 1.9268 inches
Journal taper/out-of-round	
Standard	Less than 0.0012 inch
Service limit	0.0023 inch
Main bearing oil clearance	
Standard	0.0009 to 0.0020 inch
Service limit	0.0047 inch
Connecting rod bearing oil clearance	
Standard	0.0012 to 0.0024 inch
Service limit	0.0047 inch

*See charts in Section 20

Torque specifications*

	Ft-lbs
Main bearing cap bolts	72
Connecting rod cap nuts	43

* **Note:** Refer to Part A for additional torque specifications.

2.2L (X22SE) four-cylinder engine

General

Displacement	2.2L (134.1 cu. in.)
Oil pressure	
At idle	21.8 psi
At 3000 rpm	56 psi
Compression pressure	
Standard	153 psi
Minimum	128 psi
Camshaft and valve lifters	See Chapter 2B
Cylinder head warpage limit	0.008 inch

Valves and related components

Valve stem diameter	Not available
Valve margin width (intake and exhaust)	0.031 inch
Valve stem-to-guide clearance	
Standard	
Intake	0.001 inch
Exhaust	0.0012 inch
Limit (intake and exhaust)	0.008 inch
Valve spring free length	
Standard	1.756 inches
Limit	
Intake	1.7170 inches
Exhaust	1.7582 inches
Valve spring out-of-square limit	Not available

Engine block

Deck distortion limit	0.004 inch

Cylinder bore diameter

Standard	3.5036 inches
Service limit	3.5043 inches
Cylinder bore taper/out-of-round limits	0.001 inch
Piston-to-bore clearance	0.0017 inch

Pistons and rings

Piston diameter	
Grade number 99	3.3846 to 3.3850 inches
Grade number 00	3.3854 to 3.4043 inches
Oversize 0.5	3.4043 to 3.4051 inches
Piston ring end gap	
1998 through 2000	
Top and second compression rings	0.0118 to 0.0195 inch
Oil ring	0.0156 to 0.0546 inch
2001 and 2002	
Top compression ring	0.0078 to 0.0138 inch
Second compression ring	0.0118 to 0.0197 inch
Oil ring	0.0098 to 0.0299 inch
Piston ring side clearance	
Top and second compression rings	0.0008 to 0.0016 inch
Oil ring	0.0004 to 0.0012 inch

Crankshaft, connecting rods and main bearings

Connecting rod side clearance	Not available
Crankshaft endplay	
Standard	0.0004 to 0.0008 inch
Service limit	0.0118 inch
Main bearing journal diameter*	2.259 to 2.261 inches
Connecting rod journal diameter*	1.909 to 1.9105 inches
Journal taper/out-of-round	
Taper	0.00019 inch
Out-of-round	0.0012 inch
Main bearing oil clearance	
Standard	0.0007 to 0.0016 inch
Service limit	Not available

2.2L (X22SE) four-cylinder engine (continued)

Crankshaft, connecting rods and main bearings (continued)

Connecting rod bearing oil clearance
    Standard... 0.0008 to 0.0016 inch
    Service limit .... 0.0031 inch

See charts in Section 20

Torque specifications*   Ft-lbs

Balance unit retaining bolts
    Step 1 .... 168 in-lbs
    Step 2 .... 45 degrees rotation
Main bearing cap bolts
    Step 1 36
    Step 2 .... 45 degrees rotation
    Step 3 .... 15 degrees rotation
Connecting rod cap nuts
    Step 1 25
    Step 2 .... 45 degrees rotation
    Step 3 .... 15 degrees rotation

* **Note:** *Refer to Part B for additional torque specifications.*

3.1L V6 engine

General

Displacement .... 189 cu. in
Oil pressure
    At 500 rpm... 10 psi
    At 2000 rpm... 30 to 55 psi
Compression pressure
    Minimum.. 100 psi
    Maximum difference between cylinders .... 30 psi
Cylinder head warpage limit .... Less than 0.002 inch per 6 inches

Valves and related components

Valve stem-to-guide clearance.. 0.0010 to 0.0027 inch
Valve face angle (intake and exhaust) 45-degrees
Valve seat angle (intake and exhaust) 46-degrees
Valve spring
    Free length .... 1.9094 inches
    Installed height .... 1.5748 inches

Engine block

Cylinder bore diameter
    Standard... 3.5036 to 3.5067 inches
    Taper/out-of-round limit .... Less than 0.0005 inch

Pistons and rings

Piston diameter... 3.5029 to 3.5037 inches
Piston-to-bore clearance... 0.0009 to 0.0022 inch
Piston ring side clearance (top and second ring) .... 0.002 to 0.0035 inch
Piston ring end gap
    Top compression ring .... 0.010 to 0.020 inch
    Second compression ring .... 0.020 to 0.028 inch
    Oil control ring... 0.010 to 0.030 inch

Crankshaft, connecting rods and main bearings

Connecting rod endplay (side clearance) .... 0.014 to 0.027 inch
Crankshaft endplay... 0.0024 to 0.0083 inch
Crankshaft runout .... Less than 0.0002 inch
Main journal diameter .... 2.6473 to 2.6479 inches
Connecting rod journal diameter. .... 1.9987 to 1.9994 inches
Journal taper/out-of-round limits... Less than 0.0002 inch
Main bearing oil clearance.. 0.0013 to 0.0029 inch
Connecting rod bearing oil clearance... 0.0011 to 0.0033 inch

Torque specifications*   Ft-lbs

Main bearing cap bolts 73
Connecting rod cap nuts 39

* **Note:** *Refer to Part C for additional torque specifications.*

3.2L (6VD1) V6 engine

General

Displacement	193 cu. in
Oil pressure	
At 500 rpm	10 psi
At 2000 rpm	30 to 55 psi
Compression pressure	
Minimum	128 psi
Maximum difference between cylinders	30 psi
Cylinder head warpage limit	Less than 0.0079 inch

Valves and related components

Valve stem diameter	
Intake	
Standard	0.2346 to 0.2353 inch
Limit	0.2323 inch
Exhaust	
Standard	0.2343 to 0.2350 inch
Limit	0.2323 inch
Valve stem-to-guide clearance	
Intake	
Standard	0.0009 to 0.0022 inch
Limit	0.0079 inch
Exhaust	
Standard	0.0012 to 0.0025 inch
Limit	0.0079 inch
Valve spring	
SOHC engine	
Free length	1.97 inches
Installed height	1.54 inches
DOHC engine	
Free length	1.756 inches
Installed height	1.38 inches

Engine block

Cylinder bore diameter	
Grade A	3.6772 to 3.6776 inches
Grade B	3.6776 to 3.6779 inches
Grade C	3.6780 to 3.6783 inches
Taper/out-of-round limit	Less than 0.0005 inch

Pistons and rings

Piston diameter	
Grade A	3.6752 to 3.6756 inches
Grade B	3.6756 to 3.6760 inches
Grade C	3.6760 to 3.6764 inches
Piston-to-bore clearance	0.0009 to 0.0017 inch
Piston ring side clearance (top and second ring)	0.0006 to 0.0015 inch
Piston ring end gap	
Top compression ring	0.0138 to 0.0185 inch
Second compression ring	0.0177 to 0.0236 inch
Oil control ring	0.0059 to 0.0177 inch

Crankshaft, connecting rods and main bearings

Connecting rod endplay (side clearance)	0.006 to 0.014 inch
Crankshaft endplay	0.0024 to 0.0094 inch
Crankshaft runout	Less than 0.0002 inch
Main journal diameter*	2.5165 to 2.5170 inches
Connecting rod journal diameter*	2.1229 to 2.1235 inches
Journal taper/out-of-round limits	Less than 0.0002 inch
Main bearing oil clearance	0.0010 to 0.0019 inch
Connecting rod bearing oil clearance	0.0010 to 0.0023 inch

*See charts in Section 20

Torque specifications*

	Ft-lbs
Main bearing cap-to-block bolts	29
Oil gallery assembly to main caps/block	
Step 1	22
Step 2	Tighten an additional 60-degrees
Main cap side bolts	29
Connecting rod cap nuts	40

* **Note:** Refer to Part D for additional torque specifications.

2.3a An oil pressure gauge can be adapted to the oil pressure sending unit fitting for an accurate reading of oil pressure - (3.2L V6 engine shown)

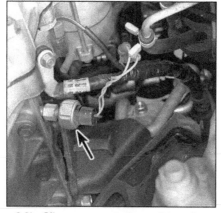

2.3b Oil pressure sending unit location (2.3L and 2.6L four-cylinder engines) - bottom of the block at the right rear

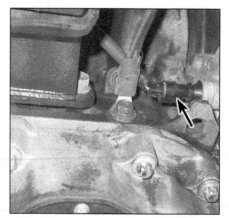

2.3c Oil pressure sending unit location (2.2L four-cylinder engine) - right rear corner of the oil pump housing

1 General information

Included in this portion of Chapter 2 are the general overhaul procedures for the cylinder head(s) and internal engine components.

The information ranges from advice concerning preparation for an overhaul and the purchase of replacement parts to detailed, step-by-step procedures covering removal and installation of internal engine components and the inspection of parts. The following procedures have been written based on the assumption the engine has been removed from the vehicle. For information concerning in-vehicle engine repair, as well as removal and installation of the external components necessary for the overhaul, see Part A, B, C or D of this Chapter and Section 8 of this Part.

The Specifications included here in Part E are only those necessary for the inspection and overhaul procedures which follow. Refer to Parts A, B, C or D for additional Specifications.

2 Engine overhaul - general information

Refer to illustrations 2.3a, 2.3b, 2.3c and 2.3d

It's not always easy to determine when, or if, an engine should be completely overhauled, as a number of factors must be considered. High mileage isn't necessarily an indication an overhaul is needed, while low mileage doesn't preclude the need for an overhaul. Frequency of servicing is probably the most important consideration. An engine that's had regular and frequent oil and filter changes, as well as other required maintenance, will most likely give many thousands of miles of reliable service. Conversely, a neglected engine may require an overhaul very early in its life.

Excessive oil consumption is an indication that piston rings, valve seals and/or valve guides are in need of attention. Make sure oil leaks aren't responsible before deciding the rings and/or guides are bad. Have a cylinder compression or leakdown test performed by an experienced tune-up mechanic to determine the extent of the work required.

If the engine is making obvious knocking or rumbling noises, the connecting rod and/or main bearings may be at fault. Check the oil pressure with a gauge installed in place of the oil pressure sending unit **(see illustrations)** and compare it to the Specifications. If it's extremely low, the bearings and/or oil pump are probably worn out.

Loss of power, rough running, excessive valve train noise and high fuel consumption rates may also point to the need for an overhaul, especially if they're all present at the same time. If a complete tune-up doesn't remedy the situation, major mechanical work is the only solution.

An engine overhaul involves restoring the internal parts to the specifications of a new engine. During an overhaul, the piston rings are replaced and the cylinder walls are reconditioned (rebored and/or honed). If a rebore is done, new pistons are required. The main and connecting rod bearings are generally replaced with new ones and, if necessary, the crankshaft may be reground to restore the journals. Generally, the valves are serviced as well, since they're usually in less-than-perfect condition at this point. While the engine is being overhauled, other components, such as the distributor, starter and alternator, can be rebuilt as well. The end result should be a like new engine that will give many trouble free miles. **Caution:** *Critical cooling system components such as the hoses, drivebelts, thermostat and water pump MUST be replaced with new parts when an engine is overhauled. The radiator should be checked carefully to ensure it isn't clogged or leaking; if in doubt, replace it with a new one. Some professional engine rebuilding shops will not guarantee their engines unless your radiator has been professionally cleaned and repaired. Also, we don't recom-*

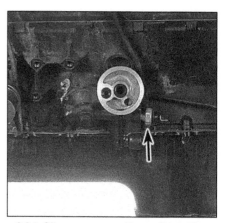

2.3d Oil pressure sending unit location (3.1L V6) - just to the rear of the oil filter on the left side

mend overhauling the oil pump - install a new one if and when the engine is rebuilt.

Before beginning the engine overhaul, read through the entire procedure to familiarize yourself with the scope and requirements of the job. Overhauling an engine isn't difficult, but it is time consuming. Plan on the vehicle being tied up for a minimum of two weeks, especially if parts must be taken to an automotive machine shop for repair or reconditioning. Check on availability of parts and make sure any necessary special tools and equipment are obtained in advance. Most work can be done with typical hand tools, although a number of precision measuring tools are required for inspecting parts to determine if they must be replaced. Often an automotive machine shop will handle the inspection of parts and offer advice concerning reconditioning and replacement. **Note:** *Always wait until the engine has been completely disassembled and all components, especially the engine block, have been inspected before deciding what service and repair operations must be performed by an automotive machine shop. Since the block's condition will be the major factor to consider when determining whether to overhaul the*

original engine or buy a rebuilt one, never purchase parts or have machine work done on other components until the block has been thoroughly inspected. As a general rule, time is the primary cost of an overhaul, so it doesn't pay to install worn or substandard parts.

As a final note, to ensure maximum life and minimum trouble from a rebuilt engine, everything must be assembled with care in a spotlessly-clean environment.

3 Vacuum gauge diagnostic checks

Refer to illustration 3.4

1 A vacuum gauge provides valuable information about what is going on in the engine at a low-cost. You can check for worn rings or cylinder walls, leaking head or intake manifold gaskets, incorrect carburetor adjustments, restricted exhaust, stuck or burned valves, weak valve springs, improper ignition or valve timing and ignition problems.

2 Unfortunately, vacuum gauge readings are easy to misinterpret, so they should be used in conjunction with other tests to confirm the diagnosis.

3 Both the absolute readings and the rate of needle movement are important for accurate interpretation. Most gauges measure vacuum in inches of mercury (in-Hg). The following references to vacuum assume the diagnosis is being performed at sea level. As elevation increases (or atmospheric pressure decreases), the reading will decrease. For every 1,000 foot increase in elevation above approximately 2,000 feet, the gauge readings will decrease about one inch of mercury.

4 Connect the vacuum gauge directly to intake manifold vacuum, not to ported (throttle body) vacuum **(see illustration)**. Be sure no hoses are left disconnected during the test or false readings will result.

5 Before you begin the test, allow the engine to warm up completely. Block the wheels and set the parking brake. With the transmission in Park, start the engine and allow it to run at normal idle speed. Carefully inspect the fan blades for cracks or damage before starting the engine. **Warning:** *Keep your hands and the vacuum gauge clear of the fan and do not stand in front of the vehicle or in line with the fan when the engine is running.*

6 Read the vacuum gauge; an average, healthy engine should normally produce about 17 to 22 inches of vacuum with a fairly steady needle. Refer to the following vacuum gauge readings and what they indicate about the engine's condition:

7 A low steady reading usually indicates a leaking gasket between the intake manifold and cylinder head(s) or throttle body, a leaky vacuum hose, late ignition timing or incorrect camshaft timing. Check ignition timing with a timing light and eliminate all other possible causes, utilizing the tests provided in this Chapter before you remove the timing chain cover to check the timing marks.

3.4 An inexpensive vacuum gauge can tell you a lot about an engine's condition and state of tune - here on a 3.2L V6, the gauge has been hooked to the PCV port on the intake manifold

8 If the reading is three to eight inches below normal and it fluctuates at that low reading, suspect an intake manifold gasket leak at an intake port or a faulty fuel injector.

9 If the needle has regular drops of about two-to-four inches at a steady rate, the valves are probably leaking. Perform a compression check or leak-down test to confirm this.

10 An irregular drop or down-flick of the needle can be caused by a sticking valve or an ignition misfire. Perform a compression check or leak-down test and read the spark plugs.

11 A rapid vibration of about four in-Hg vibration at idle combined with exhaust smoke indicates worn valve guides. Perform a leak-down test to confirm this. If the rapid vibration occurs with an increase in engine speed, check for a leaking intake manifold gasket or head gasket, weak valve springs, burned valves or ignition misfire.

12 A slight fluctuation, say one inch up and down, may mean ignition problems. Check all the usual tune-up items and, if necessary, run the engine on an ignition analyzer.

13 If there is a large fluctuation, perform a compression or leak-down test to look for a weak or dead cylinder or a blown head gasket.

14 If the needle moves slowly through a wide range, check for a clogged PCV system, incorrect idle fuel mixture, carburetor/throttle body or intake manifold gasket leaks.

15 Check for a slow return after revving the engine by quickly snapping the throttle open until the engine reaches about 2,500 rpm and let it shut. Normally the reading should drop to near zero, rise above normal idle reading (about 5 in-Hg over) and then return to the previous idle reading. If the vacuum returns slowly and doesn't peak when the throttle is snapped shut, the rings may be worn. If there is a long delay, look for a restricted exhaust system (often the muffler or catalytic converter). An easy way to check this is to temporarily disconnect the exhaust ahead of the suspected part and redo the test.

4.4 A compression gauge with a threaded fitting for the spark plug hole is preferred over the type that requires hand pressure to maintain the seal - be sure to open the throttle (and on carbureted models, the choke) as far as possible during the compression check!

4 Cylinder compression check

Refer to illustration 4.4

1 A compression check will tell you what mechanical condition the upper end (pistons, rings, valves, head gaskets) of your engine is in. Specifically, it can tell you if the compression is down due to leakage caused by worn piston rings, defective valves and seats or a blown head gasket. **Note:** *The engine must be at normal operating temperature and the battery must be fully charged for this check. Also, if the engine is equipped with a carburetor, the choke valve must be all the way open to get an accurate compression reading (if the engine's warm, the choke should be open).*

2 Begin by cleaning the area around the spark plugs before you remove them (compressed air should be used, if available, otherwise a small brush or even a bicycle tire pump will work). The idea is to prevent dirt from getting into the cylinders as the compression check is being done. Remove all of the spark plugs from the engine (Chapter 1).

3 Block the throttle wide open. Detach the coil wire from the distributor cap and ground it on the engine block. Use a heavy jumper wire with alligator clips at both ends to ensure a good ground. Also remove the fuel pump fuse. On engines that don't have distributors, disconnect the main relay to the PCM (see Chapter 12) to disable both the fuel system and ignition system.

4 With the compression gauge in the number one spark plug hole **(see illustration)**, depress the accelerator pedal all the way to the floor to open the throttle valve. Crank the engine over at least four compression strokes and watch the gauge. The compression should build up quickly in a healthy engine. Low compression on the first stroke, followed by gradually increasing pressure on successive strokes, indicates worn piston rings. A low compression reading on the first

stroke, which doesn't build up during successive strokes, indicates leaking valves or a blown head gasket (a cracked head could also be the cause). Record the highest gauge reading obtained.

5   Repeat the procedure for the remaining cylinders and compare the results to the Specifications.

6   Add some engine oil (about three squirts from a plunger-type oil can) to each cylinder, through the spark plug hole, and repeat the test.

7   If the compression increases after the oil is added, the piston rings are definitely worn. If the compression doesn't increase significantly, the leakage is occurring at the valves or head gasket. Leakage past the valves may be caused by burned valve seats and/or faces or warped, cracked or bent valves.

8   If two adjacent cylinders have equally low compression, there's a strong possibility that the head gasket between them is blown. The appearance of coolant in the combustion chambers or the crankcase would verify this condition.

9   If the compression is unusually high, the combustion chambers are probably coated with carbon deposits. If that's the case, the cylinder head(s) should be removed and decarbonized.

10   If compression is way down or varies greatly between cylinders, it would be a good idea to have a leak-down test performed by an automotive repair shop. This test will pinpoint exactly where the leakage is occurring and how severe it is.

5   Engine removal - methods and precautions

If you've decided the engine must be removed for overhaul or major repair work, several preliminary steps should be taken.

Locating a place to work is extremely important. Adequate work space, along with storage space for the vehicle, will be needed. If a shop or garage isn't available, at the very least a flat, level, clean work surface made of concrete or asphalt is required.

Cleaning the engine compartment and engine before beginning the removal procedure will help keep tools clean and organized.

An engine hoist or A-frame will also be needed. Make sure the equipment is rated in excess of the combined weight of the engine and transmission. Safety is of primary importance, considering the potential hazards involved in lifting the engine out of the vehicle.

If the engine is being removed by a novice, a helper should be available. Advice and aid from someone more experienced would also be helpful. There are many instances when one person cannot simultaneously perform all of the operations required when lifting the engine out of the vehicle.

Plan the operation ahead of time. Arrange for or obtain all of the tools and equipment you'll need prior to beginning the

job. Some of the equipment necessary to perform engine removal and installation safely and with relative ease are (in addition to an engine hoist) a heavy duty floor jack, complete sets of wrenches and sockets as described at the front of this manual, wooden blocks and plenty of rags and cleaning solvent for mopping up spilled oil, coolant and gasoline. If the hoist must be rented, make sure you arrange for it in advance and perform all of the operations possible without it ahead of time. This will save you money and time.

Plan for the vehicle to be out of use for quite a while. A machine shop will have to perform some of the work the do-it-yourselfer can't accomplish without special equipment. They often have a busy schedule, so it would be a good idea to consult them before removing the engine to accurately estimate the amount of time required to rebuild or repair components that need work.

Always be extremely careful when removing and installing the engine. Serious injury can result from careless actions. Plan ahead, take your time and a job of this nature, although major, can be accomplished successfully.

6   Engine - removal and installation

Refer to illustrations 6.9 and 6.29

Warning 1: *The engine is very heavy, so equipment designed specifically for lifting engines should be used. Never position any part of your body under the engine when it's supported by a hoist - it could shift or fall and cause serious injury or death!*

Warning 2: *The air conditioning system is under high pressure. DO NOT loosen any fittings or remove any components until after the system has been discharged. Air conditioning refrigerant should be properly discharged into an EPA-approved container at a dealer service department or an automotive air-conditioning repair facility. Always wear eye protection when disconnecting air conditioning system fittings.*

Warning 3: *Gasoline is extremely flammable, so take extra precautions when you work on any part of the fuel system. Don't smoke or allow open flames or bare light bulbs near the work area, and don't work in a garage where a gas-type appliance (such as a water heater or a clothes dryer) is present. Since gasoline is carcinogenic, wear latex gloves when there's a possibility of being exposed to fuel, and, if you spill any fuel on your skin, rinse it off immediately with soap and water. Mop up any spills immediately and do not store fuel-soaked rags where they could ignite. The fuel system on fuel-injected models is under constant pressure, so, if any fuel lines are to be disconnected, the fuel pressure in the system must be relieved first (see Chapter 4). When you perform any kind of work on the fuel system, wear safety glasses and have a Class B type fire extinguisher on hand.*

1   Read through the Section entitled

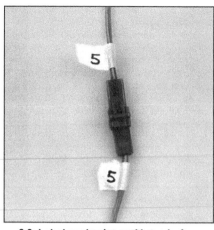

6.9 Label each wire and hose before unplugging them

Engine removal - methods and precautions before beginning any work.

2   Remove the hood (see Chapter 11) and store it in a safe place where it won't be damaged.

3   Remove the air cleaner assembly (see Chapter 4).

4   Disconnect the cables from the battery (negative first, then positive).

5   Drain the coolant from the radiator and engine block (refer to Chapter 1 if necessary).

6   Drain the engine oil.

7   Drain the oil from the transmission. Refer to Chapter 1 if necessary.

8   Remove the upper and lower radiator hoses.

9   Disconnect all wires and vacuum hoses running between the engine and components attached to the body. **Note:** *Before disconnecting or removing wires or hoses, mark them with pieces of tape to identify their installed locations. This will eliminate possible problems and confusion during the installation procedure* **(see illustration).**

10   Disconnect the fuel hoses from the mechanical fuel pump, if equipped, and plug them to prevent fuel leakage and the entry of dirt.

11   On fuel-injected models, relieve the fuel pressure (Chapter 4), then disconnect the fuel line from the fuel rail and return line and plug it.

12   If the vehicle is equipped with air conditioning, remove the compressor from the bracket and secure it out of the way. **Caution:** *Don't disconnect the hoses from the compressor unless the system has been discharged.*

13   If the vehicle is equipped with power steering, remove the power steering pump and secure it out of the way. Leave the hoses connected to it.

14   Disconnect the ignition coil wire from the distributor cap. On models with the 3.2L V6, mark with tape and disconnect the electrical connectors at each individual coil (one over each spark plug).

15   Detach the throttle cable from the carburetor or throttle body (see Chapter 4).

6.29 Attach a sturdy chain with grade-eight bolts to the front and rear engine lifting brackets in such a way that the engine will be lifted evenly

16 Disconnect the brake booster vacuum hose and position it out of the way.

17 Disconnect the heater hoses where they attach to the engine.

18 On carbureted engines, disconnect the rubber hose at the vacuum switch.

19 Raise the front of the vehicle and support it on jackstands. Apply the parking brake and block the rear wheels.

20 Disconnect the wires from the alternator and starter. Refer to Chapter 5 and remove the starter motor.

21 If the vehicle is equipped with an automatic transmission, disconnect the oil cooler hoses leading from the oil cooler located at the base of the radiator. Allow the fluid to drain into a container.

22 Remove the radiator and shroud. Refer to Chapter 3 if necessary.

23 Detach the speedometer cable from the transmission. Disconnect all wires attached to the transmission.

24 Disconnect the front exhaust pipe(s) from the exhaust manifold(s).

25 If the vehicle is equipped with a hydraulic clutch, detach the clutch operating cylinder from the transmission. Refer to Chapter 8 if necessary. If equipped with a clutch cable, take the tension off the cable by loosening the adjusting nut.

26 Disconnect the parking brake cable and position it out of the way.

27 Disconnect any remaining lines, hoses or wires leading to the engine. Mark them first to avoid confusion during installation.

28 Refer to Chapter 7 and remove the transmission crossmember and transmission, with the engine supported by an overhead hoist.

29 Attach an engine hoist to the two brackets mounted on the engine. They're located at the corners of the head. Make sure the chain is attached securely with hooks or high-strength nuts, bolts and washers. The hook on the hoist should be over the center of the engine with the two lengths of chain the same length so the engine can be lifted straight up **(see illustration)**.

30 Raise the engine just enough to remove all slack from the chains.

31 Remove the nuts that attach the front engine mount insulators to the block brackets (see the last Section in Part A, B, or C).

32 Raise the engine with the hoist until it clears the chassis and engine compartment. As you go, make sure there are no hidden hoses or wires (like ground straps) still connected to the engine.

33 When the engine is high enough to clear, slowly pull the hoist forward and, when clear of the front bumper, lower it until it can be attached to the engine stand. The flywheel/driveplate and rear main seal retainer plate should be removed before attaching the engine to the stand (see Part A, B or C).

34 Installation is the reverse of removal.

35 Tighten all fasteners to the torque listed in this Chapter's specifications.

36 Before starting the engine, make sure the oil and coolant levels are correct.

37 Run the engine and check for leaks and proper operation of all accessories, then install the hood and test drive the vehicle.

7 Engine rebuilding alternatives

The do-it-yourselfer is faced with a number of options when performing an engine overhaul. The decision to replace the engine block, piston/connecting rod assemblies and crankshaft depends on a number of factors, with the number one consideration being the condition of the block. Other considerations are cost, access to machine shop facilities, parts availability, time required to complete the project and the extent of prior mechanical experience on the part of the do-it-yourselfer.

Some of the rebuilding alternatives include:

Individual parts - If the inspection procedures reveal that the engine block and most engine components are in reusable condition, purchasing individual parts may be the most economical alternative. The block, crankshaft and piston/connecting rod assemblies should all be inspected carefully. Even if the block shows little wear, the cylinder bores should be surface honed.

Crankshaft kit - This rebuild package consists of a reground crankshaft and a matched set of pistons and connecting rods.

The pistons will already be installed on the connecting rods. Piston rings and the necessary bearings will be included in the kit. These kits are commonly available for standard cylinder bores, as well as for engine blocks which have been bored to a regular oversize.

Short block - A short block consists of an engine block with a crankshaft and piston/connecting rod assemblies already installed. All new bearings are incorporated and all clearances will be correct. The existing camshaft, valve train components, cylinder head(s) and external parts can be bolted to the short block with little or no machine shop work necessary.

Long block - A long block consists of a short block plus an oil pump, oil pan, cylinder head(s), rocker arm cover(s), camshaft and valve train components, timing sprockets and chain/belt. All components are installed with new bearings, seals and gaskets incorporated throughout. The installation of manifolds and external parts is all that's necessary.

Give careful thought to which alternative is best for you and discuss the situation with local automotive machine shops, auto parts dealers and experienced rebuilders before ordering or purchasing replacement parts.

8 Engine overhaul - disassembly sequence

1 It's much easier to disassemble and work on the engine if it's mounted on a portable engine stand. A stand can often be rented quite cheaply from an equipment rental yard. Before the engine is mounted on a stand, the flywheel/driveplate and rear main oil seal retainer plate should be removed from the engine.

2 If a stand isn't available, it's possible to disassemble the engine with it blocked up on a sturdy workbench or on the floor. Be extra careful not to tip or drop the engine when working without a stand.

3 If you're going to obtain a rebuilt engine, all external components must come off first, to be transferred to the replacement engine, just as they will if you're doing a complete engine overhaul yourself. These include:

Alternator and brackets
A/C compressor brackets
Power steering pump brackets
Emissions control components
Distributor (all except 2.2L four and 3.2L V6), spark plug wires and spark plugs
Thermostat and housing cover
Water pump
Fuel injection components or carburetor
Intake/exhaust manifolds
Oil filter
Engine mounts
Clutch and flywheel/driveplate
Engine rear plate

Note: *When removing the external components from the engine, pay close attention to details that may be helpful or important dur-*

9.2 Have several plastic bags ready (one for each valve) before disassembling the head - label each bag and put the entire contents of each valve assembly in one bag as shown

9.3a Use a valve spring compressor to compress the springs . . .

9.3b . . . compress the valve spring until the keepers can be removed from the valve with a small magnet . . .

ing installation. Note the installed position of gaskets, seals, spacers, pins, brackets, washers, bolts and other small items.

4   If you're obtaining a short block, which consists of the engine block, crankshaft, pistons and connecting rods all assembled, then the cylinder head(s), oil pan and oil pump will have to be removed as well. See Section 7 for additional information regarding the different possibilities to be considered.

5   If you're planning a complete overhaul, the engine must be disassembled and the internal components removed in the following general order:

Four-cylinder engines

> Valve cover
> Intake and exhaust manifolds
> Engine front cover/timing belt covers
> Timing belt and sprockets
> Rocker arm assembly and camshaft
> Cylinder head
> Oil pan
> Oil pan support (2.2L engines)
> Oil pump pickup tube
> Oil pump
> Balancer unit (2.2L engines)
> Piston/connecting rod assemblies
> Crankshaft and main bearings

V6 engines

> Valve covers
> Intake and exhaust manifolds
> Rocker arms and pushrods (3.1L V6)
> Timing belt covers and timing belt (3.2L V6)
> Camshafts and rocker arm/shaft assemblies (SOHC 3.2L V6)
> Camshafts and lifters (DOHC 3.2L V6)
> Cylinder heads
> Timing chain cover (3.1L V6)
> Timing chain and sprockets (3.1L V6)
> Camshaft (3.1L V6)
> Oil pan
> Oil pump
> Piston/connecting rod assemblies
> Crankshaft and main bearings

6   Before beginning the disassembly and overhaul procedures, make sure the following items are available:

> Common hand tools
> Small cardboard boxes or plastic bags for storing parts
> Gasket scraper
> Ridge reamer
> Vibration damper puller
> Micrometers
> Telescoping gauges
> Dial indicator set
> Valve spring compressor
> Cylinder surfacing hone
> Piston ring groove cleaning tool
> Electric drill
> Tap and die set
> Wire brushes
> Oil gallery brushes
> Cleaning solvent

---

9   Cylinder head - disassembly

Refer to illustrations 9.2, 9.3a, 9.3b, 9.3c, 9.3d, 9.3e, 9.3f and 9.3g
Note: *New and rebuilt cylinder heads are usually available for most engines at dealerships and auto parts stores. Due to the fact*

that some specialized tools are necessary for the disassembly and inspection procedures, and replacement parts may not be readily available, it may be more practical and economical for the home mechanic to purchase replacement head(s) rather than taking the time to disassemble, inspect and recondition the original(s).

1   Cylinder head disassembly involves removal of the intake and exhaust valves and related components. It's assumed the rocker arms (3.1L V6 engine), rocker arm shaft assembly and camshaft (four-cylinder and SOHC 3.2L V6 engines) or camshafts and lifters (DOHC 3.2L V6 engines) have been removed from the head.

2   Before the valves are removed, arrange to label and store them, along with their related components, so they can be kept separate and reinstalled in the same valve guides they're removed from **(see illustration)** .

3   Compress the springs on the first valve with a spring compressor and remove the keepers **(see illustrations)**. Carefully release the valve spring compressor and remove the retainer, the springs and the spring seat. Next, remove the oil seal from the guide, then pull the valve out of the head. If the valve

9.3c . . . or with a pair of needle nose pliers

9.3d Remove the compressor, then lift off the spring retainer . . .

9.3e ... and remove the valve spring

9.3f Pull the seal off the top of the valve guide ...

9.3g ... then remove the spring seat

binds in the guide (won't pull through), push it back into the head and deburr the area around the keeper groove with a fine file or whetstone.

4 Repeat the procedure for the remaining valves. Remember to keep all parts for each valve together so they can be reinstalled in the same locations.

5 Once the valves and related components have been removed and stored in an organized manner, the head should be thoroughly cleaned and inspected. If a complete engine overhaul is being done, finish the engine disassembly procedures before beginning the cylinder head cleaning and inspection process.

10 Cylinder head - cleaning and inspection

1 Thorough cleaning of the cylinder head(s) and related valve train components, followed by a detailed inspection, will enable you to decide how much valve service work must be done during the engine overhaul.

Cleaning

2 Scrape all traces of old gasket material and sealing compound off the head gasket, intake manifold and exhaust manifold sealing surfaces. Be very careful not to gouge the cylinder head. Special gasket removal solvents that soften gaskets and make removal much easier are available at auto parts stores.

3 Remove all built up scale from the coolant passages.

4 Run a stiff wire brush through the various holes to remove deposits that may have formed in them.

5 Run an appropriate-size tap into each of the threaded holes to remove corrosion and thread sealant that may be present. If compressed air is available, use it to clear the holes of debris produced by this operation. **Warning:** *Wear eye protection when using compressed air!*

6 Clean the exhaust and intake manifold stud threads with a wire brush.

7 Clean the cylinder head with solvent and dry it thoroughly. Compressed air will speed the drying process and ensure that all holes and recessed areas are clean. **Note:** *Decarbonizing chemicals are available and may prove very useful when cleaning cylinder heads and valve train components. They're very caustic and should be used with caution. Be sure to follow the instructions on the container.*

8 Clean the rocker arms with solvent and dry them thoroughly (don't mix them up during the cleaning process, they should be numbered with paint). Compressed air will speed the drying process and can be used to clean out the oil passages.

9 Clean all the valve springs, keepers and retainers with solvent and dry them thoroughly. Do the components from one valve at a time to avoid mixing up the parts.

10 Scrape off any heavy deposits that may have formed on the valves, then use a motorized wire brush to remove deposits from the valve heads and stems. Again, make sure the valves don't get mixed up.

Inspection

Cylinder head

Refer to illustrations 10.12, 10.14a and 10.14b

11 Inspect the head very carefully for cracks, evidence of coolant leakage and other damage. If cracks are found, a new cylinder head should be obtained. If in doubt, have the head checked for cracks at an automotive machine shop, where they have specialized equipment for crack detection.

12 Using a precision straightedge and feeler gauge, check the head gasket mating surface for warpage **(see illustration)**. If the warpage exceeds the specified limit, the head can be resurfaced at an automotive machine shop. **Note:** *If V6 cylinder heads are resurfaced, the intake manifold flanges may also require machining.*

13 Examine the valve seats in each of the combustion chambers. If they're pitted, cracked or burned, the head will require valve service that's beyond the scope of the home mechanic.

14 Check the valve stem deflection parallel to the rocker arms with a dial indicator

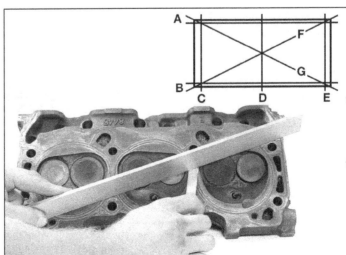

10.12 Check the cylinder head gasket for warpage by trying to slip a feeler gauge under the precision straightedge (see the Specifications for the maximum warpage allowed and use a feeler gauge of that thickness)

10.14a A dial indicator can be used to determine the valve stem-to-guide clearance (move the valve stem as indicated by the arrows)

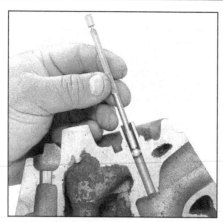

10.14b Machine shops can use a small-bore gauge to measure the stem-to-guide clearance at various points along the guide, to check for worn areas

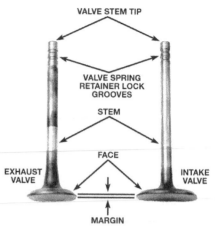

10.16 Check for valve wear at the points shown here

attached securely to the head **(see illustration)**. The valve must be in the guide and approximately 1/16-inch off the seat. The total valve stem movement indicated by the gauge needle must be noted, then divided by two to obtain the actual clearance. If it exceeds the specified limit, the valve stem-to-guide clearance should be checked by an automotive machine shop **(see illustration)**.

15 Inspect the camshaft bearing surfaces in the cylinder head for scoring.

Valves

Refer to illustrations 10.16 and 10.17

16 Carefully inspect each valve face for uneven wear, deformation, cracks, pits and burned areas. Check the valve stem for scuffing and galling and the neck for cracks. Rotate the valve and check for any obvious indication that it's bent. Look for pits and excessive wear on the end of the stem **(see illustration)**. The presence of any of these conditions indicates the need for valve service by an automotive machine shop.

17 Measure the margin width on each valve **(see illustration)**. Any valve with a margin narrower than 1/32-inch will have to be replaced with a new one.

Valve components

Refer to illustrations 10.18 and 10.19

18 Check each valve spring for wear (on the ends) and pits. Measure the free length and compare it to the specifications listed in this Chapter **(see illustration)**. Any springs that are shorter than specified have sagged and should not be reused. The tension of all springs should be checked with a special fixture before deciding they're suitable for use in a rebuilt engine (take the springs to an automotive machine shop for this check).

19 Stand each spring on a flat surface and check it for squareness **(see illustration)**. If any of the springs are distorted or sagged, replace all of them with new parts.

20 Check the spring retainers and keepers for obvious wear and cracks. Any questionable parts should be replaced with new ones, as extensive damage will occur if they fail during engine operation.

Rocker arm components

21 Refer to Part A (2.3L and 2.6L four-cylinder engines), C (3.1L V6 engine) or D (SOHC 3.2L V6) of this Chapter and inspect the rocker arms and related components.

22 If the inspection process indicates the valve components are in generally poor condition and worn beyond the limits specified, which is usually the case in an engine that's being overhauled, reassemble the valves in the cylinder head and refer to Section 11 for valve servicing recommendations.

23 If the inspection turns up no excessively worn parts, and if the valve faces and seats are in good condition, the valve train components can be reinstalled in the cylinder head without major servicing. Refer to the appropriate Section for the cylinder head reassembly procedure.

Valve lifters (2.2L engine only)

Refer to illustrations 10.24a and 10.24b

24 Inspect the contact and sliding surfaces of the valve lifters for signs of excessive wear, scratches and galling **(see illustrations)**. If the lifters look worn or damaged, replace them. If the camshaft contact surface on top of a lifter is excessively worn, inspect the corresponding camshaft lobe (see Section 16).

25 If aluminum from the cylinder head is adhering to the valve lifters, replace them. If any of the valve lifter bores are rough, scored or worn, replace the cylinder head.

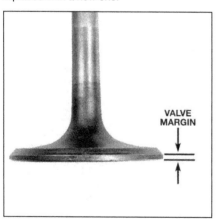

10.17 The margin width on each valve must be as specified (if no margin exists, the valve cannot be reused)

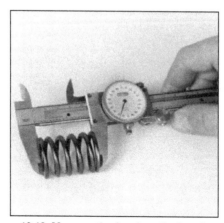

10.18 Measure the free length of each valve spring with a dial or vernier caliper

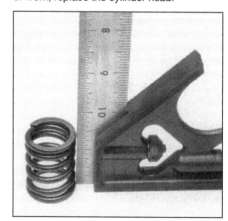

10.19 Check each valve spring for squareness

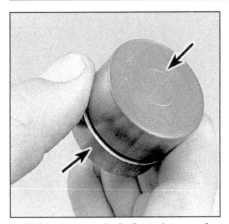

10.24a Inspect the indicated areas of each valve lifter

10.24b Also inspect the valve stem contact area of each lifter

12.3a Install the spring seat . . .

26   Verify that each lifter freely moves up and down in its bore in the cylinder head. If it doesn't, the valve might stick in the open position and cause internal engine damage.

11   Valves - servicing

1   Because of the complex nature of the job and the special tools and equipment needed, servicing of the valves, the valve seats and the valve guides, commonly known as a valve job, is best left to a professional.

2   The home mechanic can remove and disassemble the head, do the initial cleaning and inspection, then reassemble and deliver it to a dealer service department or an automotive machine shop for the actual valve servicing.

3   The dealer service department, or automotive machine shop, will remove the valves and springs, recondition or replace the valves and valve seats, recondition or replace the valve guides, check and replace the valve springs, spring retainers and keepers (as necessary), replace the valve seals with new ones, reassemble the valve components and make sure the installed spring height is cor-

rect. The cylinder head gasket surface will also be resurfaced if it's warped.

4   After the valve job has been performed by a professional, the head will be in like new condition. When it's returned, be sure to clean it again before installation on the engine to remove any metal particles and abrasive grit that may still be present from the valve service or head resurfacing operations. Use compressed air, if available, to blow out all the oil holes and passages.

12   Cylinder head - reassembly

Refer to illustrations 12.3a, 12.3b, 12.4a, 12.4b, 12.4c, 12.5, 12.6a, 12.6b, 12.7a, 12.7b and 12.8

1   Regardless of whether or not the head was sent to an automotive repair shop for valve servicing, make sure it's clean before beginning reassembly.

2   If the head was sent out for valve servicing, the valves and related components will already be in place. Begin the reassembly procedure with Step 8.

3   Install the valve spring seats (where applicable) prior to valve seal installation **(see**

12.3b . . . then install the seal protector (if applicable) to the valve

illustration). Some models may also be equipped with seal protectors **(see illustration)**.

4   Install new seals on each of the valve guides **(see illustration)**. Intake valve seals require a special tool or an appropriate-size deep socket **(see illustration)**. Gently tap each valve seal into place until it's seated on the guide **(see illustration)**. **Caution:** *Don't*

12.4a Install the new valve guide seal . . .

12.4b . . . and press it onto the guide with a suitable deep socket; there's a special tool for installing valve stem oil seals, but use a deep socket if you don't have the special tool

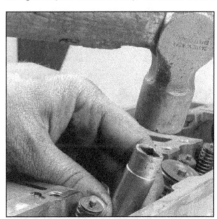

12.4c If a seal sticks during installation, gently tap it into place with a small hammer, but don't hammer on a seal once it's seated or you will ruin it

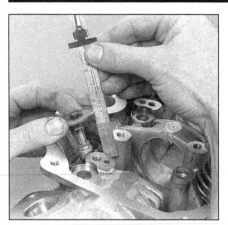

12.5 Check the distance from the top of the spring to the underside of the retainer to determine the installed valve spring height

12.6a Install the valve spring and the retainer

12.6b Make sure each valve spring is installed with the paint-marked end against the cylinder head; if the lower end of the spring is not marked by paint, install the spring with its tighter-wound end facing down, toward the head

hammer on the valve seals once they're seated or you may damage them. Don't twist or cock the seals during installation or they won't seal properly on the valve stems.

5   Apply moly-base grease or engine assembly lube to the first valve and install it in the head. Don't damage the new valve guide oil seal. Set the retainer and keepers in place. Check the installed spring height by lifting up on the retainer until the valve is seated. Measure the distance between the top of the spring seat(s) and the underside of the retainer **(see illustration)**. Compare your measurement to the specified installed height. Add shims, if necessary to obtain the specified height.

6   Once the correct height is established, remove the keepers and retainer and install the valve springs **(see illustration)**. **Note:** *The valve spring has a graduated pitch. Install it with the narrow-pitch end against the cylinder head* **(see illustration)**. *Some springs are marked with paint, indicating the end of the spring that should go towards the head.*

7   Compress the springs and retainer with a valve spring compressor and slip the keep-

ers into place **(see illustration)**. Release the compressor and make sure the keepers are seated completely in the valve stem groove. If necessary, grease can be used to hold the keepers in place as the compressor is released **(see illustration)**.

8   Double-check the installed valve spring height for each valve **(see illustration)**. If it was correct prior to reassembly, it should still be within the specified limits. If it isn't, you must install more shims until it's correct. **Caution:** *Do not, under any circumstances shim the springs to the point where the installed height is less than specified!*

9   If you're working on a four-cylinder engine, the valves should be adjusted cold (see Chapter 1).

10   Store the head in a clean plastic bag until you're ready to install it.

13   Piston/connecting rod assembly - removal

Refer to illustrations 13.1, 13.2, 13.3, 13.4 and 13.5
Note: *Prior to removing the piston/connect-*

ing rod assemblies, remove the cylinder head(s), the oil pan and the oil pump by referring to the appropriate Sections in Part A, B, C or D.

1   Completely remove the ridge at the top of each cylinder with a ridge reaming tool **(see illustration)**. Follow the manufacturer's instructions provided with the tool. Failure to remove the ridge before attempting to remove the piston/connecting rod assemblies may result in piston breakage.

2   After the cylinder ridges have been removed, turn the engine upside-down so the crankshaft is facing up. On 2.2L engines, the balancer unit must be removed before the connecting rods or crankshaft can be measured or removed. Unbolt the oil pump pickup tube (see Section 13 in Chapter 2B), unbolt the balancer unit (see Section 16 in Chapter 2B) and remove the balance unit spacer shim. Don't discard or lose the balancer unit shim; you'll need it for reassembly, or for reference if the balance unit backlash must be adjusted. On 3.2L V6 engines, the oil

12.7a Compress the valve spring and install the keepers with a small magnet or a pair of needle nose pliers

12.7b Keepers don't always stay in place, so apply a small dab of grease to each one as shown here before installation; the grease will hold them in place on the valve stem until the spring is released

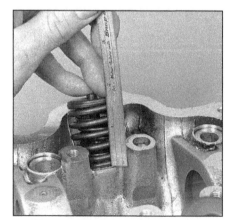

12.8 Double-check the height with the valve springs installed (do this for each valve)

13.1 A ridge reamer is required to remove the ridge from the top of the cylinder - do this before removing the pistons!

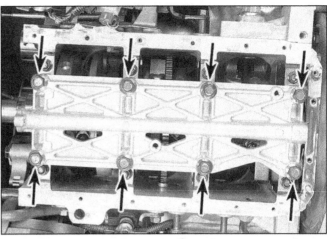

13.2 Remove the bolts and take off the oil gallery assembly on 3.2L V6 engines

gallery assembly must be removed before measuring or removing the connecting rod assemblies. Remove the bolts that go through the gallery into the crankshaft main caps and remove the gallery **(see illustration)**.

3   Before the connecting rods are removed, check the endplay with feeler gauges. Slide them between the connecting rod and the crankshaft throw until the play is removed **(see illustration)**. The endplay is equal to the thickness of the feeler gauge(s). If the endplay exceeds the service limit, new connecting rods will be required. If new rods (or a new crankshaft) are installed, the endplay may fall under the specified minimum (if it does, the rods will have to be machined to restore it - consult an automotive machine shop for advice if necessary). Repeat the procedure for the remaining connecting rods.

4   Check the connecting rods and caps for identification marks. If they aren't plainly marked, use a small center punch to make the appropriate number of indentations on

each rod and cap (1 through 4 or 6, depending on the engine type and cylinder they're associated with). On the 3.2L V6, the rods are marked on both sides. On one side, etched numbers indicate which cylinder they belong to **(see illustration)**. On the other side, stamped numbers indicate the factory "grading" of the rod size (see Section 25).

5   Loosen each of the connecting rod cap nuts 1/2-turn at a time until they can be removed by hand. Remove the number one connecting rod cap and bearing insert. Don't drop the bearing insert out of the cap. Slip a short length of plastic or rubber hose over each connecting rod cap bolt to protect the crankshaft journal and cylinder wall as the piston is removed **(see illustration)**. Push the connecting rod/piston assembly out through the top of the engine - use a wooden hammer handle to push on the upper bearing insert in the connecting rod. If resistance is felt, double-check to make sure all of the ridge was removed from the cylinder.

6   Repeat the procedure for the remaining

pistons. After removal, reassemble the connecting rod caps and bearing inserts in their respective connecting rods and install the cap nuts finger-tight. Leaving the old bearing inserts in place until reassembly will help prevent the connecting rod bearing surfaces from being accidentally nicked or gouged.

14   Crankshaft - removal

Refer to illustrations 14.1, 14.3 and 14.4
Note: *The crankshaft can be removed only after the engine has been removed from the vehicle. It's assumed the flywheel/driveplate, crankshaft damper/pulley, timing chain or belt, oil pan, oil pick-up, oil pump, front cover and piston/connecting rod assemblies have already been removed. The rear main oil seal retainer also must be unbolted and separated from the block before proceeding with crankshaft removal.*

1   Before the crankshaft is removed, check the endplay. Mount a dial indicator with the

13.3 Check the connecting rod side clearance with a feeler gauge as shown

13.4 The connecting rods and caps should be marked to indicate which cylinder they're installed in - if they aren't mark them with a center punch to avoid confusion during reassembly

13.5 To prevent damage to the crankshaft journal and cylinder, slip sections of hose over the rod bolts, then carefully drive the piston/connecting rod assembly out through the top of the engine with the end of a wooden hammer handle

14.1 Crankshaft endplay can be checked with a dial indicator, as shown here

14.3 Feeler gauges can also be used to check the crankshaft endplay

stem in-line with the crankshaft and just touching one of the crank throws **(see illustration)**.

2   Push the crankshaft all the way to the rear and zero the dial indicator. Next, pry the crankshaft to the front as far as possible and check the reading on the dial indicator. The distance that it moves is the endplay. If it's greater than specified, check the crankshaft thrust surfaces for wear. New main bearings or thrust bearings should correct the endplay.

3   If a dial indicator isn't available, feeler gauges can be used. Gently pry or push the crankshaft all the way to the front of the engine. Slip feeler gauges between the crankshaft and the front face of the thrust main bearing to determine the clearance **(see illustration)**. The thrust bearing on all engines is number three (counting from the front of the engine).

4   Check the main bearing caps to see if they're marked to indicate their locations. They should be numbered consecutively from the front of the engine to the rear. If they aren't, mark them with number stamping dies or a center punch. Main bearing caps gener-

ally have a cast-in arrow, which points to the front of the engine **(see illustration)**. Loosen the main bearing cap bolts 1/4-turn at a time until they can be removed by hand. On four-cylinder engines, loosen the caps in this order: number 5, 1, 4, 2 and 3. On V6 engines, loosen number 4, 1, 3 and 2. After the main cap bolts have been loosened on 3.2L V6 engines, loosen the main cap side bolts in the opposite order of the tightening sequence **(see illustration 25.27)**.

5   Gently tap the caps with a soft-face hammer, then separate them from the engine block. If necessary, use the bolts as levers to remove the caps.

6   Carefully lift the crankshaft out of the engine. With the bearing inserts in place in the engine block and main bearing caps or cap assembly, install the caps on the engine block and tighten the bolts finger-tight.

15   Engine block - cleaning

Refer to illustrations 15.1, 15.2, 15.7 and 15.9
1   Using a punch and hammer, drive one

edge of the core plug in toward the block, then pull the plug out with locking pliers **(see illustration)**. **Caution:** *Do not drive the whole plug into the block.*

2   Using a gasket scraper, remove all traces of gasket material from the engine block **(see illustration)**. **Caution:** *Be very careful not to nick or gouge the gasket sealing surfaces, especially on the aluminum block of the 3.2L V6.*

3   Remove the main bearing caps and separate the bearing inserts from the caps and the engine block. Tag the bearings, indicating which cylinder they were removed from and whether they were in the cap or the block, then set them aside.

4   If the engine is extremely dirty, it should be taken to an automotive machine shop to be steam cleaned or hot tanked.

5   After the block is returned, clean all oil holes and oil galleries one more time. Brushes specifically designed for this purpose are available at most auto parts stores. Flush the passages with warm water until the water runs clear, dry the block thoroughly and wipe all machined surfaces with a light,

14.4 The main bearing caps should be marked so they don't get mixed up - if they aren't, mark them with a center punch (the arrow indicates the front of the engine)

15.1 Drive one side of the core plug into the block, then pull the plug out with pliers

15.2 Use a gasket scraper to remove all traces of old gasket material and sealant from the block surfaces (gasket removal solvents are available and can make removal of stubborn gaskets much easier)

15.7 All bolt holes in the block - particularly the main bearing cap and head bolt holes - should be cleaned and restored with a tap (be sure to remove debris from the holes after this is done)

rust preventive oil. If you have access to compressed air, use it to speed the drying process and blow out all oil holes and galleries. **Warning:** *Wear eye protection when using compressed air!*

6   If the block isn't extremely dirty or sludged up, you can do an adequate cleaning job with hot soapy water and a stiff brush. Take plenty of time and do a thorough job. Regardless of the cleaning method used, be sure to clean all oil holes and galleries very thoroughly, dry the block completely and coat all machined surfaces with light oil.

7   The threaded holes in the block must be clean to ensure accurate torque readings during reassembly. Run the proper size tap into each of the holes to remove rust, corrosion, thread sealant or sludge and restore damaged threads **(see illustration)**. If possible, use compressed air to clear the holes of debris produced by this operation. Now is a good time to clean the threads on the head

bolts and the main bearing cap bolts as well.

8   Reinstall the main bearing caps or cap assembly and tighten the bolts finger-tight.

9   After coating the block-mating surfaces of the new core plugs with core plug sealant, install them in the engine block **(see illustration)**. Make sure they're driven in straight and seated properly or leakage could result. Special tools are available for this purpose, but equally good results can be obtained using a large socket, with an outside diameter that will just slip into the core plug, a 1/2-inch drive extension and a hammer.

10   If the engine isn't going to be reassembled right away, cover it with a large plastic trash bag to keep it clean.

---

16   Engine block, camshafts and camshaft bearings - inspection

Block

Refer to illustrations 16.4a, 16.4b, 16.4c, and 16.10

1   Before the block is inspected, it should be cleaned as described in Section 15. Double-check to make sure the ridge at the top of each cylinder has been completely removed.

2   Visually check the block for cracks, rust and corrosion. Look for stripped threads in the threaded holes. It's also a good idea to have the block checked for hidden cracks by an automotive machine shop that has the special equipment to do this type of work. If defects are found, have the block repaired, if possible, or replaced.

3   Check the cylinder bores for scuffing and scoring.

4   Check the cylinders for taper and out-of-round conditions as follows **(see illustrations)**:

5   Measure the diameter of each cylinder at the top (just under the ridge area), center and bottom of the cylinder bore, parallel to the crankshaft axis.

15.9 A large socket on an extension can be used to drive the new core plugs into the bores

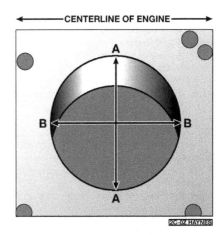

16.4a Measure the diameter of each cylinder at a right angle to the engine centerline (A), and parallel to engine centerline (B) - out-of-round is the difference between A and B; taper is the difference between A and B at the top of the cylinder and A and B at the bottom of the cylinder

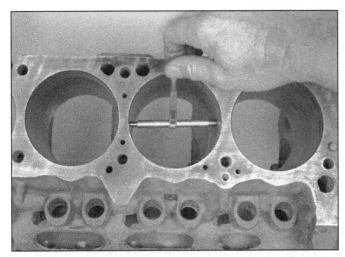

16.4b The ability to "feel" when the telescoping gauge is at the correct point will be developed over time, so work slowly and repeat the check until you're satisfied the bore measurement is accurate

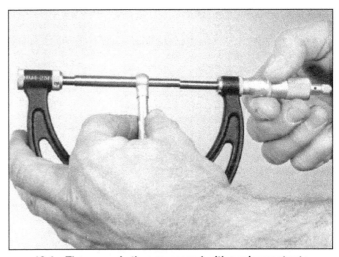

16.4c The gauge is then measured with a micrometer to determine the bore size

16.10 Check the block deck for distortion with a precision straightedge and feeler gauge

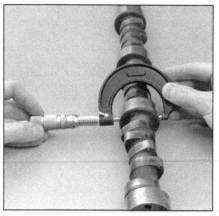

16.13 Measure the camshaft journals with a micrometer and compare them to the Specifications

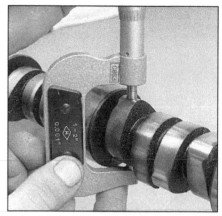

16.14a Measure the camshaft lobe at its greatest dimension . . .

6   Next measure each cylinder's diameter at the same three locations perpendicular to the crankshaft axis.

7   The taper of the cylinder is the difference between the bore diameter at the top of the cylinder and the diameter at the bottom. The out-of-round specification of the cylinder bore is the difference between the parallel and perpendicular readings. Compare your results to those listed in this Chapter's Specifications.

8   Repeat the procedure for the remaining pistons and cylinders.

9   If the cylinders are in reasonably good condition and not worn to the outside of the limits, and if the piston-to-cylinder clearances can be maintained properly, then they don't have to be rebored. Honing is all that's necessary (Section 17).

10   Using a precision straightedge and feeler gauge, check the block deck (the surface that mates with the cylinder head) for distortion **(see illustration)**. If it's distorted beyond the specified limit, it can be resurfaced by an automotive machine shop.

Camshafts and bearings

Refer to illustrations 16.13. 16.14a and 16.14b

11   Refer to Parts A, B, C or D of this Chapter for the removal of the camshaft(s).

12   After the camshaft has been removed from the engine, cleaned with solvent and dried, inspect the bearing journals for uneven wear, pitting and evidence of seizure. If the journals are damaged, the bearing inserts in the block (3.1L V6) or the camshaft towers in the head (four-cylinder and 3.2L V6 engines) are probably damaged as well. Both the camshaft and bearings will have to be replaced. **Note:** *Camshaft bearing replacement requires special tools and expertise that place it beyond the scope of the average home mechanic. The tools for bearing removal and installation are available at stores that carry automotive tools, possibly even found at a tool rental business. It is advisable though, if bearings are bad and the procedure is beyond your ability, remove the engine block and take it to an automotive machine*

shop to ensure that the job is done correctly.

13   Measure the bearing journals with a micrometer to determine if they are excessively worn or out-of-round **(see illustration)**.

14   Check the camshaft lobes for heat discoloration, score marks, chipped areas, pitting and uneven wear. If the lobes are in good condition and if the rocker arms or lifters are in good condition, the camshaft can be reused. Measure the lobes for evidence of wear **(see illustrations)**.

15   On 3.1L V6 engines, clean the lifters with solvent and dry them thoroughly without mixing them up. Check each lifter wall, pushrod seat and foot for scuffing, score marks and uneven wear. If the lifter walls are damaged or worn (which is not very likely), inspect the lifter bores in the engine block as well. If the pushrod seats are worn, check the pushrod ends. If new lifters are being installed, a new camshaft must also be installed. If a new camshaft is installed, then use new lifters as well. Never install used lifters unless the original camshaft is used and the lifters can be installed in their original locations.

16   On four-cylinder engines, there are no camshaft bearings. The camshaft(s) ride between the cylinder head and the camshaft caps. An accurate measurement of the camshaft clearance can be obtained by using strips of Plastigage. Install the camshaft in the head, apply the Plastigage, and torque the camshaft caps to Specifications. Remove the caps and compare the width of the crushed Plastigage to the scale, as performed in Section 23 for checking the crankshaft main bearing clearances. If the clearances are excessive, the camshaft(s) and cylinder head(s) must be replaced [see Chapter 2A (2.3L and 2.6L engines) or Chapter 2B (2.2L engine) for the Specifications].

17   On 3.2L V6 engines, there are no camshaft bearings either. The camshaft(s) on SOHC models ride inside camshaft "towers." The camshaft journal oil clearance can be calculated by measuring the tower bores with an inside micrometer, measuring the camshaft journal with an outside micrometer, then sub-

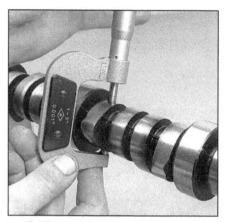

16.14b . . . then measure its smallest dimension - subtract the smallest from the largest to determine the lobe lift

tracting the journal diameter from the bore diameter. On DOHC models the oil clearance is measured with Plastigage (see Chapter 2D). If the clearances are excessive, the camshaft(s) and/or cylinder head(s) must be replaced (see Chapter 2D for the Specifications).

---

17 Cylinder honing

Refer to illustrations 17.3a and 17.3b

1   Prior to engine reassembly, the cylinder bores must be honed so the new piston rings will seat correctly and provide the best possible combustion chamber seal. If you don't have the tools or don't want to tackle the honing operation, most automotive machine shops will do it for a reasonable fee.

2   Before honing the cylinders, install the main bearing caps (without bearing inserts) and tighten the bolts to the specified torque.

3   Two types of cylinder hones are commonly available - the flex hone or "bottle brush" type and the more traditional surfacing hone with spring-loaded stones. Both will do the job, but for the less experienced mechanic the "bottle brush" hone will probably be easier to use. You'll also need plenty of light oil or honing oil, some rags and an

17.3a A "bottle brush" hone is the easiest type of hone to use

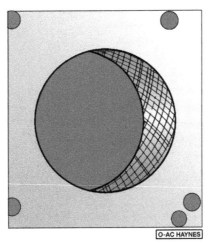

17.3b The cylinder hone should leave a smooth, crosshatch pattern with the lines intersecting at approximately a 60-degree angle

18.4a The piston ring grooves can be cleaned with a special tool, as shown here . . .

18.4b . . . or a section of a broken ring

18.5 Make sure the oil hole in the lower end of each connecting rod is clear - if the rods are separated from the pistons, make sure the oil hole is on the correct side when they're reassembled

electric drill motor. Proceed as follows:

a) *Mount the hone in the drill motor, compress the stones and slip it into the first cylinder* **(see illustration)**. *Be sure to wear safety goggles or a face shield!*

b) *Lubricate the cylinder with plenty of oil, turn on the drill and move the hone up-and-down in the cylinder at a pace that will produce a fine crosshatch pattern on the cylinder walls. Ideally, the crosshatch lines should intersect at approximately a 60-degree angle* **(see illustration)**. *Be sure to use plenty of lubricant and don't take off any more material than absolutely necessary to produce the desired finish.* **Note:** *Piston ring manufacturers may specify a smaller crosshatch angle than the traditional 60-degrees - read and follow any instructions included with the new rings.*

c) *Don't withdraw the hone from the cylinder while it's running. Instead, shut off the drill and continue moving the hone up-and-down in the cylinder until it comes to a complete stop, then compress the stones and withdraw the hone. If you're using a "bottle brush" type hone, stop the drill motor, then turn the chuck in the normal direction of rotation while withdrawing the hone from the cylinder.*

d) *Wipe the oil out of the cylinder and repeat the procedure for the remaining cylinders.*

4   After the honing job is complete, chamfer the top edges of the cylinder bores with a small file so the rings won't catch when the pistons are installed. Be very careful not to nick the cylinder walls with the end of the file!

5   The entire engine block must be washed again very thoroughly with warm, soapy water to remove all traces of abrasive grit produced during the honing operation. The bores can be considered clean when a white cloth - dampened with clean engine oil - used to wipe them down doesn't pick up any more honing residue, which will show up as gray areas on the cloth. Be sure to run a brush through all oil holes and galleries and flush them with running water.

6   After rinsing, dry the block and apply a coat of light rust preventive oil to all machined surfaces. Wrap the block in a plastic trash bag to keep it clean and set it aside until reassembly.

18 Piston/connecting rod assembly - inspection

Refer to illustrations 18.4a, 18.4b, 18.5, 18.10 and 18.11

1   Before the inspection process can be carried out, the piston/connecting rod assemblies must be cleaned and the original piston rings removed from the pistons. **Caution:** *Always use new piston rings when the engine is reassembled.*

2   Using a piston ring installation tool, carefully remove the rings from the pistons. Be careful not to nick or gouge the pistons in the process.

3   Scrape all traces of carbon from the top of the piston. A hand-held wire brush or a piece of fine emery cloth can be used once the majority of the deposits have been scraped away. Do not, under any circumstances, use a wire brush mounted in an electric drill to remove deposits from the pistons. The piston material is soft and will be eroded away by the wire brush.

4   Use a piston ring groove cleaning tool to remove carbon deposits from the ring grooves. If a tool isn't available, a piece broken off the old ring will do the job. Be very careful to remove only the carbon deposits - don't remove any metal and don't nick or scratch the sides of the ring grooves **(see illustrations)**.

5   Once the deposits have been removed, clean the piston/rod assemblies with solvent and dry them with compressed air (if available). Make sure the oil return holes in the back sides of the piston ring grooves and the oil hole in the lower end of each rod are clear

(see illustration).

6   If the pistons and cylinder walls aren't damaged or worn excessively, and if the engine block isn't rebored or replaced, new pistons won't be necessary. However, if new pistons are required, be sure to match them to each cylinder by referring to the grade numbers on the block. Normal piston wear appears as even vertical wear on the piston

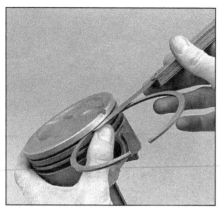

18.10 Check the ring side clearance with a feeler gauge at several points around the groove

18.11 Measure the piston diameter at a 90-degree angle to the piston pin

19.1 Chamfer the oil holes to remove sharp edges that might gouge or scratch the new bearings

thrust surfaces and slight looseness of the top ring in its groove. New piston rings, on the other hand, should always be used when an engine is rebuilt.

7   Carefully inspect each piston for cracks around the skirt, at the pin bosses and at the ring lands.

8   Look for scoring and scuffing on the thrust faces of the skirt, holes in the piston crown and burned areas at the edge of the crown. If the skirt is scored or scuffed, the engine may have been suffering from over-heating and/or abnormal combustion, which caused excessively high operating temperatures. The cooling and lubrication systems should be checked thoroughly. A hole in the piston crown is an indication that abnormal combustion (preignition) was occurring. Burned areas at the edge of the piston crown are usually evidence of spark knock (detonation). If any of the above problems exist, the causes must be corrected or the damage will occur again.

9   Corrosion of the piston, in the form of small pits, indicates coolant is leaking into the combustion chamber and/or the crankcase. Again, the cause must be corrected or the problem may persist in the rebuilt engine.

10   Measure the piston ring side clearance by laying a new piston ring in each ring groove and slipping a feeler gauge in beside it **(see illustration)**. Check the clearance at three or four locations around each groove. Be sure to use the correct ring for each groove, as they are different. If the side clearance is greater than specified, new pistons will have to be used.

11   Check the piston-to-bore clearance by measuring the bore (see Section 16) and the piston diameter. Make sure the pistons and bores are correctly matched. Measure the piston across the skirt, at a 90-degree angle to the piston pin and in-line with it **(see illustration)**. Subtract the piston diameter from the bore diameter to obtain the clearance. If it's greater than specified, the block will have to be rebored and new pistons and rings installed.

12   Check the piston-to-rod clearance by

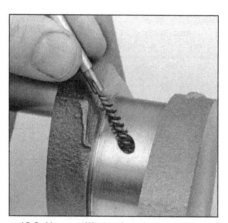

19.3 Use a stiff plastic bristle brush to clean the oil passages in the crankshaft

twisting the piston and rod in opposite directions. Any noticeable play indicates excessive wear, which must be corrected. The piston/connecting rod assemblies should be taken to an automotive machine shop to have the pistons and rods rebored and new pins installed.

13   If the pistons must be removed from the connecting rods for any reason, they should be taken to an automotive machine shop. While they're there, have the connecting rods checked for bend and twist, since automotive machine shops have special equipment for this purpose. Unless new pistons and/or connecting rods must be installed, do not disassemble the pistons from the connecting rods.

14   Check the connecting rods for cracks and other damage. Temporarily remove the rod caps, lift out the old bearing inserts, wipe the rod and cap bearing surfaces clean and inspect them for nicks, gouges and scratches. After checking the rods, replace the old bearings, slip the caps into place and tighten the nuts finger-tight.

19 Crankshaft - inspection

Refer to illustrations 19.1, 19.3, 19.5 and 19.7

1   Clean the crankshaft with solvent and dry it with compressed air (if available).

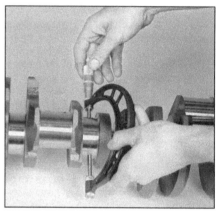

19.5 Measure the diameter of each crankshaft journal at several points to detect taper and out-of-round conditions

Warning: *Wear eye protection when using compressed air.* Remove all burrs from the crankshaft oil holes with a stone, file or scraper **(see illustration)**.

2   Check the main and connecting rod bearing journals for uneven wear, scoring, pits and cracks.

3   Be sure to clean the oil holes with a stiff brush and flush them with solvent **(see illustration)**.

4   Check the rest of the crankshaft for cracks and other damage. It should be magnafluxed to reveal hidden cracks - an automotive machine shop will handle the procedure.

5   Using a micrometer, measure the diameter of the main and connecting rod journals and compare the results to this Chapter's Specifications **(see illustration)**. By measuring the diameter at a number of points around each journal's circumference, you'll be able to determine whether or not the journal is out-of-round. Take the measurement at each end of the journal, near the crank throws, to determine if the journal is tapered.

6   If the crankshaft journals are damaged, tapered, out-of-round or worn beyond the limits given in the Specifications, have the crankshaft reground by an automotive machine shop. Be sure to use the correct size

19.7 If the seals have worn grooves in the crankshaft journals, or if the seal journals are nicked or scratched, the new seal(s) will leak

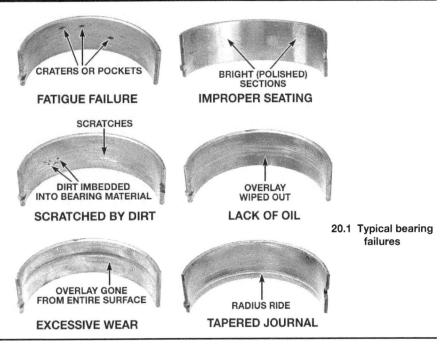

20.1 Typical bearing failures

bearing inserts if the crankshaft is reconditioned.

7   Check the oil seal journals at each end of the crankshaft for wear and damage. If the seal has worn a groove in the journal, or if it's nicked or scratched **(see illustration)**, the new seal may leak when the engine is reassembled. In some cases, an automotive machine shop may be able to repair the journal by pressing on a thin sleeve. If repair isn't feasible, a new or different crankshaft should be installed.

8   Refer to Section 20 and examine the main and rod bearing inserts.

20   Main and connecting rod bearings - inspection and bearing selection

Inspection

Refer to illustration 20.1

1   Even though the main and connecting rod bearings should be replaced with new ones during the engine overhaul, the old bearings should be retained for close examination, as they may reveal valuable information about the condition of the engine **(see illustration)**.

2   Bearing failure occurs because of lack of lubrication, the presence of dirt or other foreign particles, overloading the engine and corrosion. Regardless of the cause of bearing failure, it must be corrected before the engine is reassembled to prevent it from happening again.

3   When examining the bearings, remove them from the engine block, the main bearing caps, the connecting rods and the rod caps and lay them out on a clean surface in the same general position as their location in the engine. This will enable you to match any bearing problems with the corresponding crankshaft journal.

4   Dirt and other foreign particles get into the engine in a variety of ways. It may be left in the engine during assembly, or it may pass through filters or the PCV system. It may get into the oil, and from there into the bearings.

Metal chips from machining operations and normal engine wear are often present. Abrasives are sometimes left in engine components after reconditioning, especially when parts are not thoroughly cleaned using the proper cleaning methods. Whatever the source, these foreign objects often end up embedded in the soft bearing material and are easily recognized. Large particles will not embed in the bearing and will score or gouge the bearing and journal. The best prevention for this cause of bearing failure is to clean all parts thoroughly and keep everything spotlessly clean during engine assembly. Frequent and regular engine oil and filter changes are also recommended.

5   Lack of lubrication (or lubrication breakdown) has a number of interrelated causes. Excessive heat (which thins the oil), overloading (which squeezes the oil from the bearing face) and oil leakage or throw off (from excessive bearing clearances, worn oil pump or high engine speeds) all contribute to lubrication breakdown. Blocked oil passages, which usually are the result of misaligned oil holes in a bearing shell, will also oil starve a bearing and destroy it. When lack of lubrication is the cause of bearing failure, the bearing material is wiped or extruded from the steel backing of the bearing. Temperatures may increase to the point where the steel backing turns blue from overheating.

6   Driving habits can have a definite effect on bearing life. Low speed operation in too high a gear (lugging the engine) puts very high loads on bearings, which tends to squeeze out the oil film. These loads cause the bearings to flex, which produces fine cracks in the bearing face (fatigue failure). Eventually the bearing material will loosen in pieces and tear away from the steel backing. Short trip driving leads to corrosion of bearings because insufficient engine heat is produced to drive off the condensed water and corrosive gases. These

products collect in the engine oil, forming acid and sludge. As the oil is carried to the engine bearings, the acid attacks and corrodes the bearing material.

7   Incorrect bearing installation during engine assembly will lead to bearing failure as well. Tight fitting bearings leave insufficient bearing oil clearance and will result in oil starvation. Dirt or foreign particles trapped behind a bearing insert result in high spots on the bearing which lead to failure.

Bearing selection

8   If the original main or connecting rod bearings are worn or damaged, or if the oil clearances are incorrect (see Sections 23 and 25), the following procedure should be used to select the correct new bearings for engine reassembly. The following procedure applies to the four-cylinder engines and to the 3.2L V6. The 3.1L V6 does not have "graded" bearings, so bearings selected are either standard (for a crankshafts that measure within Specifications) or undersize, for crankshafts that have been machined. However, if the crankshaft in any engine has been reground, new undersize bearings must be installed - the automotive machine shop that reconditions the crankshaft will provide or help you select the correct size bearings. Regardless of how the bearing sizes are determined, use the oil clearance, measured with Plastigage (see Sections 23 and 25), as a guide to ensure the bearings are the right size. Remember, the oil clearance is the final judge when selecting new bearing sizes. If you have any questions or are unsure which bearings to use, get help from an Isuzu dealer parts or service department.

Main bearings

Refer to illustrations 20.9a and 20.9b

9   Locate the markings at the rear corner of the oil pan rail at the bottom of the block

that indicate the block main bearing journal bore sizes **(see illustration)**. The marks may be located along the oil pan rail, rather than just at one spot. Then look at the crankshaft for the markings that indicate the graded sizes of the main journals on the crankshaft. On the four-cylinder engines, the grade markings are on the "cheek" of each counterweight next to the main journal they are describing. On the 3.2L V6, the markings are all on the front counterweight of the crankshaft **(see illustration)**.

10   Locate the particular size marks on the charts and determine the color code for the crankshaft bearings.

Connecting rod bearings

11   Locate the size marking for the connecting rod big end diameter on the side of the connecting rod **(see illustration 25.2)**.

12   Locate the size mark on the chart and determine the color code for the connecting rod bearing for each rod.

20.9a Locate the marks along the oil pan rail (3.2L V6 shown, 4-cylinder marks are on the left rear corner) that indicate the sizes of the main bearing bores in the block

20.9b At the front of the first crank counterweight on the 3.2L V6, the marks indicate the crankshaft's main journal diameters, from #1 on the left (as seen from the front) to number 4

Bearing selection

2.2L four-cylinder engine

Crankshaft bearing journal diameter	2.259 to 2.261 inches
Standard size bearing colors	
White	2.260 to 2.261 inches
Green	2.261 to 2.2615 inches
Brown	2.2615 to 2.2618 inches
Undersize (0.0097 inch) bearing colors	
Green/blue	2.2515 to 2.2517 inches
Brown/blue	2.2517 to 2.252 inches

Undersize (0.0195 inch) bearing colors	
Green/white	2.2418 to 2.242 inches
Brown/white	2.242 to 2.2423 inches
Guide bearing standard size	1.012 to 1.014 inches
Undersize (0.0097 inch) guide bearing	1.019 to 1.021 inches
Undersize (0.0195 inch) guide bearing	1.027 to 1.029 inches

2.3L and 2.6L four-cylinder engines

Main bearing bore diameter	Marking on block	Diameter of main journal	Journal size marking	Correct bearing insert color	Oil clearance
2.3618 to 2.3622 inches	1	2.2016 to 2.2018 inches	--	Blue	0.001 to 0.002 inch
		2.2019 to 2.2022 inches	-	Black	0.001 to 0.002 inch
2.3614 to 2.3618 inches	2	2.2016 to 2.2018 inches	--	Black	0.0009 to 0.0018 inch
		2.2019 to 2.2022 inches	-	Brown	0.0009 to 0.0020 inch

3.2L V6 engine

Main bearing bore diameter	Marking on block	Diameter of main journal	Journal size marking	Correct bearing insert color	Oil clearance
2.7163 to 2.7165 inches	1	2.5165 to 2.5167 inches	--	Brown	0.0012 to 0.0019 inch
		2.5168 to 2.5170 inches	-	Green	0.0011 to 0.0019 inch
2.7160 to 2.7163 inches	2	2.5165 to 2.5167 inches	--		0.0011 to 0.0018 inch
		2.5168 to 2.5170 inches	-	Yellow	0.0011 to 0.0018 inch
2.7157 to 2.7160 inches	3	2.5165 to 2.5167 inches	--		0.0011 to 0.0019 inch
		2.5168 to 2.5170 inches	-	Pink	0.0010 to 0.0018 inch

Connecting rod bearings

2.2L four-cylinder engine

Connecting rod bearing journal diameter	Dimensions	Color code
Standard size	1.9098 to 1.9105 inches	No color code
Undersize (0.0097 inch)	1.9001 to 1.9008 inches	Blue
Undersize (0.0195 inch)	1.8903 to 1.891 inches	White
Connecting rod bearing width		
Standard size	1.0319 to 1.036 inches	No color code
Undersize (0.0097 inch)	1.0319 to 1.036 inches	Blue
Undersize (0.0195 inch)	1.0319 to 1.036 inches	White
Connecting rod width	1.0271 to 1.0292 inches	

2.3L and 2.6L four cylinder engines

Size mark	Big end bore diameter	Crankshaft pin diameter	Connecting rod bearing thickness	Color of size mark	Oil clearance
A	2.0486 to 2.0488 inches	1.9276 to 1.9282 inches	0.0594 to 0.0596 inch	Blue	0.0012 to 0.0022 inch
B	2.0483 to 2.0486 inches	1.9276 to 1.9282 inches	0.00593 to 0.0594 inch	Black	0.0012 to 0.0024 inch
C	2.0481 to 2.0482 inches	1.9276 to 1.9282 inches	0.0591 to 0.0593 inch	Brown	0.0013 to 0.0024 inch
Undersize (0.0098 inch)	2.0486 to 2.0488 inches	1.9163 to 1.9169 inches	0.0640 to 0.0645 inch		0.0007 to 0.0028 inch
Undersize (0.0197 inch)	2.0486 to 2.0488 inches	1.9065 to 1.9071 inches	0.0689 to 0.0684 inch		0.0007 to 0.0028 inch

3.2L V6 engine

Size mark	Big end bore diameter	Crankshaft pin diameter	Connecting rod bearing thickness	Color of size mark	Oil clearance
A	2.2439 to 2.2441 inches	2.1229 to 2.1235 inches	0.0595 to 0.0597 inch	Yellow	0.0010 to 0.0021 inch
B	2.2436 to 2.2439 inches	1.1229 to 1.1235 inches	0.00594 to 0.0595 inch	Green	0.0011 to 0.0022 inch
C	2.2434 to 2.2436 inches	1.1229 to 1.1235 inches	0.0592 to 0.0594 inch	Pink	0.0011 to 0.0023 inch

21   Engine overhaul - reassembly sequence

1   Before beginning engine reassembly, make sure you have all the necessary new parts, gaskets and seals as well as the following items on hand:

Common hand tools
A 1/2-inch drive torque wrench
Piston ring installation tool
Piston ring compressor
Short lengths of rubber or plastic hose to fit over connecting rod bolts
Plastigage
Feeler gauges
A fine-tooth file
New engine oil
Engine assembly lube or moly-base grease
RTV-type gasket sealant
Thread-locking compound

2   To save time and avoid problems, engine reassembly must be done in the following general order:

Four-cylinder engines

Piston rings
Crankshaft and main bearings
Piston/connecting rod assemblies
Rear main oil seal
Cylinder head (Part A)
Camshaft (Part A)
Rocker arm assembly (Part A)
Timing chain or belt and sprockets (Part A)
Front cover (Part A)
Oil strainer (Part A)
Oil pump (Part A)
Oil pan (Part A)
Intake and exhaust manifolds (Part A)
Rocker arm cover (Part A)
Flywheel/driveplate (Part A)

V6 engines

Piston rings
Crankshaft and main bearings
Piston/connecting rod assemblies
Oil gallery assembly (3.2L V6)
Rear main oil seal
Oil pump (Part B or C)
Oil pan (Part B or C)
Cylinder heads (Part B or C)
Camshafts and lifters (Part B or C)
Rocker arm assemblies (Part B or C)
Timing chain or belt (Part B or C)
Intake and exhaust manifolds (Part B or C)
Valve covers (Part B or C)
Engine rear plate
Flywheel/driveplate (Part B or C)

22   Piston rings - installation

Refer to illustrations 22.3, 22.4, 22.5, 22.9a, 22.9b and 22.12

1   Before installing the new piston rings, the ring end gaps must be checked. It's assumed the piston ring side clearance has been checked and verified correct (Section 18).

2   Lay out the piston/connecting rod assemblies and the new ring sets so the ring sets will be matched with the same piston and cylinder during the end gap measurement and engine assembly.

22.3 When checking piston ring end gap, the ring must be square in the cylinder bore - this is done by pushing the ring down with the top of a piston

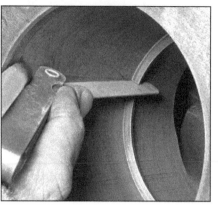

22.4 With the ring square in the cylinder, measure the end gap with a feeler gauge

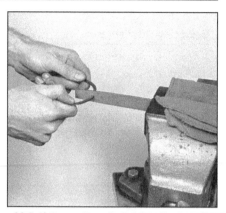

22.5 If the end gap is too small, clamp a file in a vise and file the ring ends (from the outside in only) to enlarge the gap slightly

3   Insert the top (number one) ring into the first cylinder and square it up with the cylinder walls by pushing it in with the top of the piston **(see illustration)**. The ring should be near the bottom of the cylinder, at the lower limit of ring travel.

4   To measure the end gap, slip feeler gauges between the ends of the ring until a gauge equal to the gap width is found **(see illustration)**. The feeler gauge should slide between the ring ends with a slight amount of drag. Compare the measurement to the Specifications. If the gap is larger or smaller than specified, double-check to make sure you have the correct rings before proceeding.

5   If the gap is too small, it must be enlarged or the ring ends may come in contact with each other during engine operation, which can cause serious damage to the engine. The end gap can be increased by filing the ring ends very carefully with a fine file. Mount the file in a vise equipped with soft jaws, slip the ring over the file with the ends contacting the file face and slowly move the ring to remove material from the ends. When performing this operation, file only from the outside in **(see illustration)**. After achieving the proper gap, deburr the filed ends of the rings with a fine whetstone.

6   Excess end gap isn't critical unless it's

greater than 0.040-inch. Again, double-check to make sure you have the correct rings for your engine.

7   Repeat the procedure for each ring that will be installed in the first cylinder and for each ring in the remaining cylinders. Remember to keep rings, pistons and cylinders matched up.

8   Once the ring end gaps have been checked/corrected, the rings can be installed on the pistons.

9   The oil control ring (lowest one on the piston) is installed first. On most models it's composed of three separate components. Slip the spacer/expander into the groove **(see illustration)**. If an anti-rotation tang is used, make sure it's inserted into the drilled hole in the ring groove. Next, install the lower side rail. Don't use a piston ring installation tool on the oil ring side rails, as they may be damaged. Instead, place one end of the side rail into the groove between the spacer/expander and the ring land, hold it firmly in place and slide a finger around the piston while pushing the rail into the groove **(see illustration)**. Next, install the upper side rail in the same manner.

10   After the three oil ring components have been installed, check to make sure both the upper and lower side rails can be turned smoothly in the ring groove.

11   The number two (middle) ring is installed next. It's stamped with a mark, usually a "2T"

which must face up, toward the top of the piston. Always follow the instructions printed on the ring package or box - different manufacturers may require different approaches. Do not mix up the top and middle rings, as they have different cross sections.

12   Use a piston ring installation tool and make sure the identification mark is facing the top of the piston, then slip the ring into the middle groove on the piston **(see illustration)**. Don't expand the ring any more than is necessary to slide it over the piston.

13   Install the number one (top) ring in the same manner. Make sure the mark, usually a "T", is facing up. Be careful not to confuse the number one and number two rings.

14   Repeat the procedure for the remaining pistons and rings.

23 Crankshaft - installation and main bearing oil clearance check

1   Crankshaft installation is the first step in engine reassembly. It's assumed at this point that the engine block and crankshaft have been cleaned, inspected and repaired or reconditioned.

2   Position the engine with the bottom facing up.

22.9a Install the three-piece oil control ring first, one part at a time, beginning with the spacer/expander . . .

22.9b . . . followed by the side rails - DO NOT use a piston ring installation tool to install the oil ring side rails

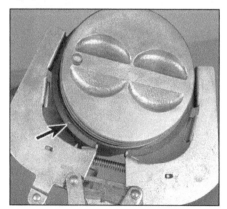

22.12 Install the compression rings with a ring expander (make sure the mark is facing up)

23.11 Lay the Plastigage strips on the main bearing journals, parallel to the crankshaft centerline

3   Remove the main bearing cap bolts and lift out the caps. Lay them out in the proper order to ensure correct installation.
4   If they're still in place, remove the original bearing inserts from the block and the main bearing caps. Wipe the bearing surfaces of the block and caps with a clean, lint-free cloth. They must be kept spotlessly clean.

Main bearing oil clearance check

Refer to illustrations 23.11 and 23.15
Caution: *Don't touch the faces of the new bearing inserts with your fingers. Oil and acids from your skin can etch the bearings.*
5   Clean the back sides of the new main bearing inserts and lay one in each main bearing saddle in the block. If one of the bearing inserts from each set has a large groove in it, make sure the grooved insert is installed in the block. Lay the other bearing from each set in the corresponding main bearing cap. Make sure the tab on the bearing insert fits into the recess in the block or cap. The oil holes in the block must line up with the oil holes in the bearing inserts. **Caution:** *Do not hammer the bearing into place and don't nick or gouge the bearing faces. No lubrication should be used at this time.*
6   If you're working on a 3.1L V6 engine, the flanged thrust bearing must be installed in the number three cap and saddle. On four-cylinder engines and the 3.2L V6, the thrust bearing inserts must be installed just ahead of the number 3 main with the oil grooves facing the crankshaft.
7   Clean the faces of the bearings in the block and the crankshaft main bearing journals with a clean, lint-free cloth.
8   Check or clean the oil holes in the crankshaft, as any dirt here can go only one way - straight through the new bearings.
9   Once you're certain the crankshaft is clean, carefully lay it in position in the main bearings.
10   Before the crankshaft can be permanently installed, the main bearing oil clearance must be checked.
11   Cut several pieces of the appropriate

23.15 Compare the width of the crushed Plastigage to the scale on the envelope to determine the main bearing oil clearance (always take the measurement at the widest point of the Plastigage); be sure to use the correct scale - inch and metric ones are included

size Plastigage (they should be slightly shorter than the width of the main bearings) and place one piece on each crankshaft main bearing journal, parallel with the journal axis **(see illustration)**.
12   Clean the faces of the bearings in the caps and install the caps in their original locations (don't mix them up) with the arrows pointing toward the front of the engine. Don't disturb the Plastigage.
13   Starting with the center main and working out toward the ends, tighten the main bearing cap bolts, in three steps, to the torque figure listed in this Chapter's Specifications. Don't rotate the crankshaft at any time during this operation.
14   Remove the bolts and carefully lift off the main bearing caps. Keep them in order. Don't disturb the Plastigage or rotate the crankshaft. If any of the main bearing caps are difficult to remove, tap them gently from side-to-side with a soft-face hammer to loosen them.
15   Compare the width of the crushed Plastigage on each journal to the scale printed on the Plastigage envelope to obtain the main bearing oil clearance **(see illustration)**. Check the Specifications to make sure

it's correct.
16   If the clearance is not as specified, the bearing inserts may be the wrong size (which means different ones will be required). Before deciding different inserts are needed, make sure no dirt or oil was between the bearing inserts and the caps or block when the clearance was measured. If the Plastigage was wider at one end than the other, the journal may be tapered (refer to Section 19).
17   Carefully scrape all traces of the Plastigage material off the main bearing journals and/or the bearing faces. Use your fingernail or the edge of a credit card - don't nick or scratch the bearing faces.

Final crankshaft installation

Refer to illustration 23.27
18   Carefully lift the crankshaft out of the engine.
19   Clean the bearing faces in the block, then apply a thin, uniform layer of moly-base grease or engine assembly lube to each of the bearing surfaces. Be sure to coat the thrust faces as well as the journal face of the thrust bearing. **Caution:** *Do not get any lubricant on the backside of the bearing shells.*
20   Make sure the crankshaft journals are clean, then lay the crankshaft back in place in the block.
21   Clean the faces of the bearings in the caps, then apply lubricant to them.
22   Install the caps in their original locations with the arrows pointing toward the front of the engine.
23   Install the bolts.
24   Tighten all except the thrust bearing cap bolts to the specified torque (work from the center out and approach the final torque in three steps).
25   Tighten the thrust bearing cap bolts to 10-to-12 ft-lbs.
26   Tap the ends of the crankshaft forward and backward with a lead or brass hammer to line up the main bearing and crankshaft thrust surfaces.
27   Retighten all main bearing cap bolts to the torque specified in this Chapter, starting with the center main and working out toward the ends (except on the 3.2L V6 engine). On 3.2L V6 engines, tighten the main caps in sequence, front to rear **(see illustration)** then

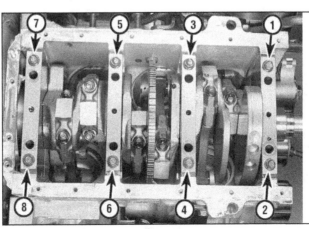

23.27 On 3.2L V6 engines, tighten the main caps in this sequence

24.2 Use a hammer and punch to drive the old seal out of the retainer plate - the plate must be well-supported by wood blocks

24.3 Drive the new seal into the retainer plate squarely with a block of wood and a hammer - only drive it in as far as the old seal was installed

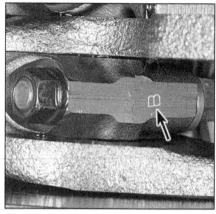

25.2 On graded rods (3.2L V6 shown), one side of each rod/cap should be stamped with a letter (here a B) to denote its size

install the main cap side bolts and tighten them to the specified torque. The number 5 main cap on four-cylinder engines must be installed so that its rear face aligns flush with the back of the block.

28   Rotate the crankshaft a number of times by hand to check for any obvious binding.

29   The final step is to check the crankshaft endplay with feeler gauges or a dial indicator as described in Section 14. The endplay should be correct if the crankshaft thrust faces aren't worn or damaged and new bearings have been installed.

30   Refer to Section 24 and install the new rear main oil seal.

24 Crankshaft rear oil seal - installation

Refer to illustrations 24.2 and 24.3

1   It is assumed that at this point you have installed the crankshaft and checked the final main bearing clearances. The rear main seal retainer plate should be off the engine.

2   Support the retainer plate over two equal-height blocks of wood and drive the old seal out with a punch and small hammer **(see illustration)**.

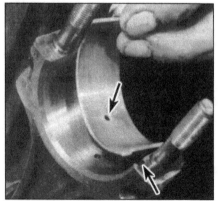

25.3 The tab on each bearing insert must fit into the recess in the rod or cap and the oil holes must line up

3   Lubricate the lip of the seal with engine oil and install it in the retainer plate with a hammer and a large block of wood **(see illustration)**.

4   Apply a thin bead of anaerobic sealant to the engine side of the retainer plate.

5   Lubricate the seal area of the crankshaft with engine oil and work the retainer and seal slowly onto the crankshaft, using as blunt tool to keep the lip of the seal pointing toward the front of the engine.

6   When the retainer plate is all the way on and close to the block, start the retainer plate bolts finger-tight. Make sure the bottom surface of the retainer plate is flush with the bottom of the engine block, then tighten the retainer bolts to Specifications.

25 Piston/connecting rod assembly - installation and rod bearing oil clearance check

Refer to illustrations 25.2, 25.3, 25.5, 25.6, 25.9, 25.11, 25.13, 25.17 and 25.27

1   Before installing the piston/connecting rod assemblies, the cylinder walls must be perfectly clean, the top edge of each cylinder must be chamfered, and the crankshaft must be in place.

2   Remove the cap from the end of the number one connecting rod (check the marks made during removal). Remove the original bearing inserts and wipe the bearing surfaces of the connecting rod and cap with a clean, lint-free cloth. They must be kept spotlessly clean. The rods should be marked to indicate factory rod bearing "grading" **(see illustration)**.

Connecting rod bearing oil clearance check

Caution: *Don't touch the faces of the new bearing inserts with your fingers. Oil and acids from your skin can etch the bearings.*

3   Clean the back side of the new upper bearing insert, then lay it in place in the connecting rod. Make sure the tab on the bearing fits into the recess in the rod **(see illustration)**. Don't hammer the bearing insert into place and be very careful not to nick or gouge the bearing face. Don't lubricate the bearing at this time.

4   Clean the back side of the other bearing insert and install it in the rod cap. Again, make sure the tab on the bearing fits into the recess in the cap, and don't apply any lubricant. It's critically important that the mating surfaces of the bearing and connecting rod are perfectly clean and oil free when they're assembled.

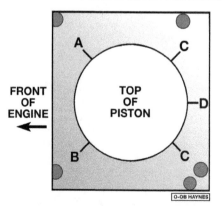

25.5 Ring end gap positions - align the oil ring spacer gap at D, the oil ring side rails at C (one inch either side of the pin centerline), and the compression rings at A and B, one inch either side of the pin centerline

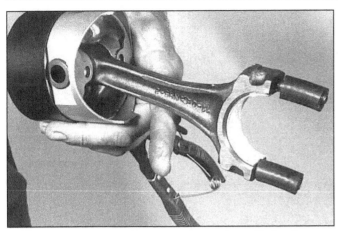

25.6 Slip sections of rubber hose over the rod bolts, then compress the rings with a ring compressor - leave the bottom of the piston sticking out so it'll slip into the cylinder

25.9 The mark or arrow on each piston must point towards the front of the engine

5   Position the piston ring gaps at intervals around the piston **(see illustration)**.

6   Slip a section of plastic or rubber hose over each connecting rod cap bolt **(see illustration)**.

7   Lubricate the piston and rings with clean engine oil and attach a piston ring compressor to the piston. Leave the skirt protruding at least 1/4-inch to guide the piston into the cylinder. The rings must be compressed until they're flush with the piston.

8   Rotate the crankshaft until the number one connecting rod journal is at BDC (bottom dead center) and apply a coat of engine oil to the cylinder walls.

9   With the mark or arrow on top of the piston facing the front of the engine **(see illustration)**, gently insert the piston/connecting rod assembly into the number one cylinder bore and rest the bottom edge of the ring compressor on the engine block.

10   Tap the top edge of the ring compressor to make sure it's contacting the block around its entire circumference.

11   Gently tap on the top of the piston with the end of a wooden or plastic hammer handle **(see illustration)** while guiding the end of the

connecting rod into place on the crankshaft journal. The piston rings may try to pop out of the ring compressor just before entering the cylinder bore, so keep some pressure on the ring compressor. Work slowly, and if any resistance is felt as the piston enters the cylinder, stop immediately. Find out what's hanging up and fix it before proceeding. Do not, for any reason, force the piston into the cylinder - you might break a ring and/or the piston.

12   Once the piston/connecting rod assembly is installed, the connecting rod bearing oil clearance must be checked before the rod cap is permanently bolted in place.

13   Cut a piece of the appropriate size Plastigage slightly shorter than the width of the connecting rod bearing and lay it in place on the number one connecting rod journal, parallel with the journal axis **(see illustration)**.

14   Clean the connecting rod cap bearing face, remove the protective hoses from the connecting rod bolts and install the rod cap. Make sure the mating mark on the cap is on the same side as the mark on the connecting rod.

15   Install the nuts and tighten them to the torque listed in this Chapter's Specifications.

Work up to it in three steps. Use a thin-wall socket to avoid erroneous torque readings that can result if the socket is wedged between the rod cap and nut. If the socket tends to wedge itself between the nut and the cap, lift up on it slightly until it no longer contacts the cap. Do not rotate the crankshaft at any time during this operation.

16   Remove the nuts and detach the rod cap, being very careful not to disturb the Plastigage.

17   Compare the width of the crushed Plastigage to the scale printed on the Plastigage envelope to obtain the oil clearance **(see illustration)**. Compare it to this Chapter's Specifications to make sure the clearance is correct.

18   If the clearance is not as specified, the bearing inserts may be the wrong size (which means different ones will be required). Before deciding different inserts are needed, make sure no dirt or oil was between the bearing inserts and the connecting rod or cap when the clearance was measured. Also, recheck the journal diameter. If the Plastigage was wider at one end than the other, the journal may be tapered (refer to Section 19).

25.11 Drive the piston gently into the cylinder bore with the end of a wooden or plastic hammer handle

25.13 Lay the Plastigage strips on each rod bearing journal, parallel to the crankshaft centerline

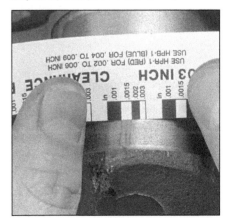

25.17 Measure the width of the crushed Plastigage to determine the rod bearing oil clearance (be sure to use the correct scale - inch and metric ones are included)

Final connecting rod installation

19   Carefully scrape all traces of the Plastigage material off the rod journal and/or bearing face. Be very careful not to scratch the bearing - use your fingernail or the edge of a credit card.

20   Make sure the bearing faces are perfectly clean, then apply a uniform layer of clean moly-base grease or engine assembly lube to both of them. You'll have to push the piston into the cylinder to expose the face of the bearing insert in the connecting rod - be sure to slip the protective hoses over the rod bolts first.

21   Slide the connecting rod back into place on the journal, remove the protective hoses from the rod cap bolts, install the rod cap and tighten the nuts to the torque specified in this Chapter. Again, work up to the torque in three steps.

22   Repeat the entire procedure for the remaining pistons/connecting rods.

23   The important points to remember are . . .
  a) Keep the back sides of the bearing inserts and the insides of the connecting rods and caps perfectly clean when assembling them.
  b) Make sure you have the correct piston/rod assembly for each cylinder.
  c) The arrow or mark on the piston must face the front (timing chain/belt end) of the engine.
  d) Lubricate the cylinder walls with clean oil.
  e) Lubricate the bearing faces when installing the rod caps after the oil clearance has been checked.

24   After all the piston/connecting rod assemblies have been properly installed, rotate the crankshaft a number of times by hand to check for any obvious binding.

25   As a final step, the connecting rod endplay must be checked. Refer to Section 13 for this procedure.

26   Compare the measured endplay to this Chapter's Specifications to make sure it's correct. If it was correct before disassembly and the original crankshaft and rods were reinstalled, it should still be right. If new rods or a new crankshaft were installed, the endplay may be inadequate. If so, the rods will have to be removed and taken to an automotive machine shop for resizing.

27   On 3.2L V6 engines, after the main caps and all connecting rod caps are tightened, install the oil gallery assembly and torque its bolts to Specifications, then tighten the main cap side bolts. All bolts, main, oil gallery and side bolts should be tightened in sequence **(see illustration)**.

Balancer unit adjustment and installation (2.2L four-cylinder engine)

Note: For more information regarding the removal, adjustment and installation of the balancer unit, see Section 16 in Chapter 2B.

28   A balancer unit **(see illustration 16.1 in**

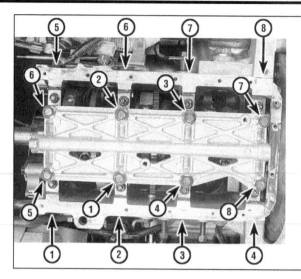

25.27 After the main and rod caps are torqued, install the oil gallery assembly and tighten its bolts in this sequence, then tighten the side bolts in sequence

Chapter 2B) is installed between the oil strainer and the crankshaft on this engine. The balancer unit consists of a pair of counter-rotating balancer shafts gear-driven by the crankshaft. The balancer unit is not rebuildable, but the backlash (between the gear on the front of the crankshaft and the smaller gears on each balancer shaft) is adjustable. As the gear teeth wear, the backlash increases. On high-mileage engines, it might become excessive enough to require adjustment. The balancer unit backlash is measured by installing a dial indicator and some special tools on the front end of the balancer unit **(see illustration 16.10 in Chapter 2B)**. These tools are difficult to obtain, so we recommend having the balancer unit backlash checked by a dealer service department with the right tools. Take the assembled short block, with the balance unit bolted on, to a dealer and have the service department do this job. This strategy is less expensive than buying the special tools (if you could even find them).

29   The crankshaft-to-balance unit gear backlash is adjusted by changing the spacer shim between the balancer unit and the engine block: A thicker shim moves the balancer unit down, away from the block, increasing backlash; a thinner shim moves the balancer unit up, toward the block, decreasing backlash. The shims are available in 11 thicknesses, each with a variation in thickness of up to 0.030 mm, beginning with code 55, which is 0.535 to 0.565 mm, code 58, which is 0.565 to 0.596 mm, etc., all the way up to code 85, which is 0.835 to 0.865 mm. The dealer service department will install the correct shim to bring the backlash within the correct range.

30   After the backlash has been measured and adjusted by the dealer, you can install the balance unit yourself (if it hasn't already been done by the dealer). Install the correct balance unit spacer shim, make sure that the mark on the crankshaft is at six o'clock, aligned with the stationary mark on the timing belt cover, and then install the balance unit with the flattened sides of both balancer

shafts facing downward **(see illustration 16.8 in Chapter 2B)**. Tighten the balance unit retaining bolts to the torque listed in this Chapter's Specifications.

26   Initial start-up and break-in after overhaul

1   Once the engine has been installed in the vehicle, double-check the engine oil and coolant levels.

2   With the spark plugs out of the engine and the ignition and fuel-injection systems disabled, crank the engine until oil pressure registers on the gauge.

3   Install the spark plugs, hook up the plug wires and restore the ignition and fuel system functions.

4   Start the engine. It may take a few moments for the gasoline to reach the carburetor or injectors, but the engine should start without a great deal of effort.

5   After the engine starts, it should be allowed to warm up to normal operating temperature. While the engine is warming up, make a thorough check for oil and coolant leaks.

6   Shut the engine off and recheck the engine oil and coolant levels.

7   Drive the vehicle to an area with no traffic, accelerate sharply from 30 to 50 mph, then allow the vehicle to slow to 30 mph with the throttle closed. Repeat the procedure 10 or 12 times. This will load the piston rings and cause them to seat properly against the cylinder walls. Check again for oil and coolant leaks.

8   Drive the vehicle gently for the first 500 miles (no sustained high speeds) and keep a constant check on the oil level. It's not unusual for an engine to use oil during the break-in period.

9   At approximately 500 to 600 miles, change the oil and filter.

10   For the next few hundred miles, drive the vehicle normally. Don't pamper it or abuse it.

11   After 2000 miles, change the oil and filter again and consider the engine fully broken in.

Chapter 3
Cooling, heating and air conditioning systems

Contents

Specifications

General

Radiator cap pressure rating	12.8 to 17.1 psi
Thermostat rating	
Four-cylinder engines	
2.2L	
Begins to open	197.6 degrees F
Fully open	224.6 degrees F
2.3L	179.6 degrees F
2.6L	
Begins to open	166.1 to 173.3 degrees F
Fully open	194 degrees F
3.1L V6 engine..	195-degrees F
3.2L V6 engine	
Begins to open..	166.1 to 173.3 degrees F
Fully open	194 degrees F

Torque specifications

Oil cooler center bolt (3.2L V6 engine)   **Ft-lbs** (unless otherwise indicated)

Oil cooler center bolt (3.2L V6 engine)	
1995 and earlier	20
1996 on	43
Thermostat housing bolts	
Four-cylinder engines	
2.2L	132 in-lbs
2.3L and 2.6L	18
3.1L V6 engine	18
3.2L V6 engine	168 in-lbs
Water pump bolts/nuts	
Four-cylinder engines	
2.2L	18
2.3L and 2.6L	
Bolts	168 in-lbs
Nut	20
3.1L V6 engine	22
3.2L V6 engine	156 in-lbs

1   General information

Engine cooling system

All vehicles covered by this manual employ a pressurized engine cooling system with thermostatically-controlled coolant circulation. An impeller type water pump mounted on the front of the block pumps coolant through the engine. The coolant flows around each cylinder and toward the rear of the engine. Cast-in coolant passages direct coolant around the intake and exhaust ports, near the spark plug areas and in proximity to the exhaust valve guides.

A wax-pellet type thermostat is located in the thermostat housing at the front of the engine. During warm up, the closed thermostat prevents coolant from circulating through the radiator. When the engine reaches normal operating temperature, the thermostat opens and allows hot coolant to travel through the radiator, where it is cooled before returning to the engine.

The cooling system is sealed by a pressure-type radiator cap. This raises the boiling point of the coolant, and the higher boiling point of the coolant increases the cooling efficiency of the radiator. If the system pressure exceeds the cap pressure-relief value, the excess pressure in the system forces the spring-loaded valve inside the cap off its seat and allows the coolant to escape through the overflow tube into a coolant reservoir. When the system cools, the excess coolant is automatically drawn from the reservoir back into the radiator.

The coolant reservoir serves as both the point at which fresh coolant is added to the cooling system to maintain the proper fluid level and as a holding tank for overheated coolant.

This type of cooling system is known as a closed design because coolant that escapes past the pressure cap is saved and reused.

Heating system

The heating system consists of a blower fan in a housing under the right end of the dashboard, a heater core in a housing under the center of the dash, the inlet and outlet hoses connecting the heater core to the engine cooling system and the heater/air conditioning control head on the dashboard. Hot engine coolant is circulated through the heater core. When the heater mode is activated, a flap door opens to expose the heater box to the passenger compartment. A fan switch on the control head activates the blower motor, which forces air through the core, heating the air.

Air conditioning system

The air conditioning system consists of a condenser mounted in front of the radiator, an evaporator mounted adjacent to the heater core, a compressor mounted on the engine, a receiver-drier which contains a high-pressure relief valve and the plumbing connecting all of the above.

A blower fan forces the warmer air of the passenger compartment through the evaporator core (sort of a radiator-in-reverse), transferring the heat from the air to the refrigerant. The liquid refrigerant boils off into low pressure vapor, taking the heat with it when it leaves the evaporator. The compressor keeps refrigerant circulating through the system, pumping the warmed coolant through the condenser where it is cooled and then circulated back to the evaporator.

2   Antifreeze - general information

Refer to illustration 2.4

Warning: *Do not allow antifreeze to come in contact with your skin or painted surfaces of the vehicle. Rinse off spills immediately with plenty of water. Antifreeze is highly toxic if ingested. Never leave antifreeze lying around in an open container or in puddles on the floor; children and pets are attracted by it's sweet smell and may drink it. Check with local authorities about disposing of used antifreeze. Many communities have collection centers which will see that antifreeze is disposed of safely. Never dump used antifreeze on the ground or into drains.*

Note: *Non-toxic antifreeze is now manufactured and available at local auto parts stores, but even these types should be disposed of properly.*

The cooling system should be filled with a water/ethylene-glycol based antifreeze solution, which will prevent freezing down to at least -20 degrees F, or lower if local climate requires it. It also provides protection

2.4 An inexpensive hydrometer can be used to test the condition of your coolant

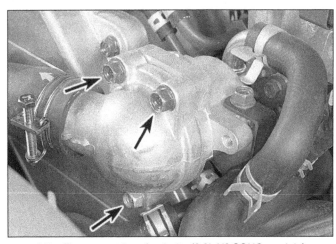

3.7a Thermostat housing bolts (3.2L V6 SOHC models)

against corrosion and increases the coolant boiling point.

The cooling system should be drained, flushed and refilled every 30,000 miles or every two years (see Chapter 1). The use of antifreeze solutions for periods of longer than two years is likely to cause damage and encourage the formation of rust and scale in the system. If your tap water is "hard", i.e. contains a lot of dissolved minerals, use distilled water with the antifreeze.

Before adding antifreeze to the system, check all hose connections, because antifreeze tends to leak through very minute openings. Engines do not normally consume coolant. Therefore, if the level goes down, find the cause and correct it.

The exact mixture of antifreeze-to-water you should use depends on the relative weather conditions. The mixture should contain at least 50-percent antifreeze, but should never contain more than 70-percent antifreeze. Consult the mixture ratio chart on the antifreeze container before adding coolant. Hydrometers are available at most auto parts stores to test the ratio of antifreeze to water

(see illustration). Use antifreeze which meets the vehicle manufacturer's specifications.

3 Thermostat - check and replacement

Warning: *Do not attempt to remove the radiator cap, coolant or thermostat until the engine has cooled completely.*

Check

1   Before assuming the thermostat is responsible for a cooling system problem, check the coolant level (Chapter 1), drivebelt tension (Chapter 1) and temperature gauge (or light) operation.
2   If the engine takes a long time to warm up (as indicated by the temperature gauge or heater operation), the thermostat is probably stuck open. Replace the thermostat with a new one.
3   If the engine runs hot, use your hand to check the temperature of the lower radiator hose. If the hose is not hot, but the engine is, the thermostat is probably stuck in the closed

position, preventing the coolant inside the engine from traveling through the radiator. Replace the thermostat. **Caution:** *Do not drive the vehicle without a thermostat. The computer may stay in open loop and emissions and fuel economy will suffer.*
4   If the lower radiator hose is hot, it means that the coolant is flowing and the thermostat is open. Consult the *Troubleshooting* section at the front of this manual for further diagnosis.

Replacement

Refer to illustrations 3.7a, 3.7b, 3.7c, 3.8a, 3.8b and 3.9

Note: *On 3.2L V6 DOHC engines, remove the intake manifold assembly for access to the thermostat housing (see Chapter 2D).*

5   Disconnect the negative cable from the battery.
6   Drain the coolant from the radiator (see Chapter 1).
7   Detach the housing from the engine. Be prepared for some coolant to spill as the gasket seal is broken. The radiator hose can be left attached to the housing, unless the housing itself is to be replaced **(see illustrations)**.

3.7b On 3.1L V6 models, the thermostat housing is secured by two bolts; one of the bolts may have a stud on some models

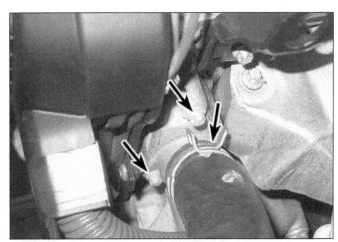

3.7c To remove the thermostat housing on 2.2L four-cylinder models, loosen the hose clamp, disconnect the upper radiator hose and remove these two bolts

8   Remove the thermostat, noting the direction in which it was installed in the housing, and thoroughly clean the sealing surfaces **(see illustrations)**.

9   Fit a new gasket or RTV sealant onto the thermostat housing **(see illustration)**. Make sure it is evenly fitted all the way around. Most new thermostats come with a new gasket, which should be coated with a thin application of RTV sealant for installation.

10   Install the thermostat and housing, positioning the jiggle pin at the highest point.

11   Tighten the housing fasteners to the torque listed in this Chapter's Specifications and reinstall the remaining components in the reverse order of removal.

12   Refill the cooling system, run the engine and check for leaks and proper operation.

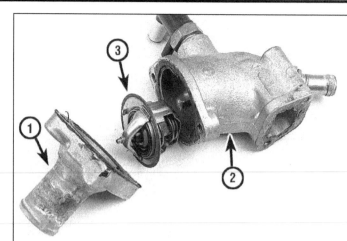

3.8a On 2.3L and 2.6L four-cylinder engines, note the installed order of the thermostat and related components:

1   Thermostat housing cover
2   Thermostat housing
3   Thermostat

---

4   Engine cooling fan and clutch - check and replacement

Warning: *Keep hands, tools and clothing away from the fan. To avoid injury or damage DO NOT operate the engine with a damaged fan. Do not attempt to repair fan blades - replace a damaged fan with a new one.*

Belt-driven fans

Check

1   Symptoms of failure of the fan clutch are continuous noisy operation, looseness leading to vibration and evidence of silicone fluid leaks.

2   Rock the fan back and forth by hand to check for excessive bearing play.

3   With the engine cold, turn the blades by hand. The fan should turn freely.

4   Visually inspect for substantial fluid leakage from the clutch assembly, a deformed bi-metal spring or grease leakage from the cooling fan bearing. If any of these conditions exist, replace the fan clutch.

5   When the engine is warmed up, turn off the ignition switch and disconnect the cable from the negative battery terminal. Turn the fan by hand. Some resistance should be felt. If the fan turns easily, replace the fan clutch.

3.8b The thermostat is installed with the spring end towards the engine and (on models so equipped) the jiggle pin at the top (3.2L V6 shown, 2.2L four-cylinder models similar)

Removal and installation

Refer to illustrations 4.7a, 4.7b, 4.8 and 4.9

6   Loosen the drivebelt tension (see Chapter 1).

7   On some models, the fan shroud is one piece, and it should be disconnected from the radiator and pulled back over the engine for access to the fan mounting bolts/nuts. On

3.9 A 1/8-inch bead of RTV sealant can be applied to the thermostat cover (3.1L V6 shown) in place of a gasket - allow the sealant to "skin over" before installation and don't overtighten the bolts

later models, the fan shroud is two-piece. To access the fan assembly, unbolt the upper shroud from the radiator and remove the clips holding the two shroud halves together **(see illustrations)**. Remove the upper shroud. If necessary for other procedures, the lower shroud half is tabbed into two slots in the bottom of the radiator and pulls out easily.

4.7a Remove the bolts holding the fan shroud to the radiator

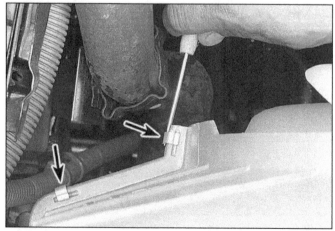

4.7b To separate upper and lower shrouds on two-piece models, remove the two clips on each side and lift out the upper half

4.8 Remove the four nuts and remove the fan assembly from the water pump or mount (3.2L V6 shown)

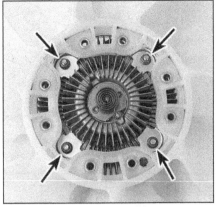

4.9 Remove the four nuts retaining the fan clutch to the fan

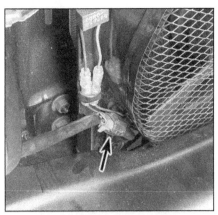

4.12a To remove an electric condenser fan, disconnect the electrical connector . . .

8   Remove the fan assembly-to-water pump hub bolts **(see illustration)**. On the 3.2L V6, the fan bolts to a hub/mount on the front of the timing cover, not the water pump.

9   The fan clutch (if equipped) can be unbolted from the fan blade assembly for replacement **(see illustration)**. **Caution:** *To prevent silicone fluid from draining from the clutch assembly into the fan drive bearing and ruining the lubricant, DON'T place the drive unit in a position with the rear of the shaft pointing down. Store the clutch either front down, or prop it up vertically as it would be on an engine.*

10   Installation is the reverse of removal.

Electric fans

Condenser fan

Refer to illustrations 4.12a and 4.12b

11   V6 models with air conditioning are equipped with a condenser fan mounted in front of the air conditioning condenser, behind the grille. Refer to Chapter 11 for removal of the grille.

12   To replace the electric fan, disconnect the negative battery cable, remove the four mounting screws and unplug the electrical connector **(see illustrations)**. To test the fan, unplug the electrical connector and apply a

4.12b . . . and remove the four fan shroud retaining screws (the two screws at the bottom can be accessed through the openings in the bumper cover)

fused source of 12 volts to the power side of the fan and ground to the other side. If the fan doesn't operate, replace it. If it does operate, trace the circuit to find the source of the problem (check the fuse first). Other possible causes are a defective condenser fan relay, a faulty air conditioning pressure switch or a bad control head. Refer to the wiring diagrams at the back of this manual.

4.14 Unplug the electrical connector from the cooling fan motor

Engine cooling system fan (2.2L models)

Refer to illustrations 4.14, 4.15a, 4.15b, 4.16, 4.17 and 4.18

13   Disconnect the negative battery cable.

14   Unplug the electrical connector from the cooling fan motor **(see illustration)**.

15   Remove the upper fan shroud retaining bracket bolts **(see illustration)** and then disengage the lower edge of the fan shroud from the crossmember **(see illustration)**.

4.15a Remove the upper fan shroud retaining bracket bolts . . .

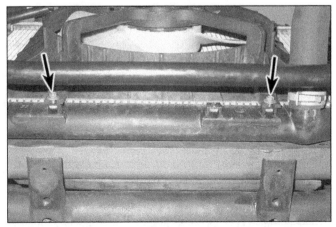

4.15b . . . and remove the lower fan shroud retaining bracket bolts

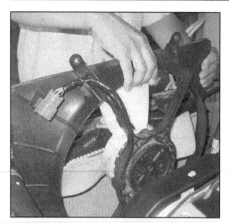

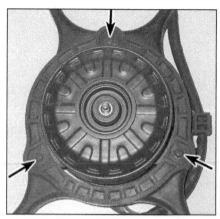

4.16 Removing the engine cooling system fan assembly from a 2.2L model (note that it's unnecessary to disconnect the upper radiator hose to remove the fan assembly)

4.17 To remove the fan from the fan motor shaft, remove this nut

4.18 To detach the fan motor from the fan shroud, remove these three screws

16   Remove the fan and fan shroud as a single assembly **(see illustration)**.
17   Remove the fan from the fan motor shaft **(see illustration)**.
18   Remove the fan motor from the fan shroud **(see illustration)**.
19   Installation is the reverse of removal.

---

5   Radiator and coolant reservoir - removal and installation

---

Warning: *Do not start this procedure until the engine is completely cool.*

Radiator

Removal

Refer to illustrations 5.3a, 5.3b, 5.4a, 5.4b, 5.6, 5.7a, 5.7b and 5.11

1   Disconnect the negative battery cable.
2   Drain the coolant into a container (see Chapter 1).
3   Remove both the upper and lower radia-

5.3a To detach the upper radiator hose, loosen the hose clamp and pull off the hose (2.2L Amigo model shown, other models similar)

tor hoses **(see illustrations)**.
4   Disconnect the reservoir hose from the radiator filler neck **(see illustrations)**.
5   Remove the cooling fan and shroud (see Section 4).

5.3b To detach the lower radiator hose, loosen the hose clamp and pull off the hose (2.2L Amigo model shown, other models similar)

6   If equipped with an automatic transmission, disconnect the cooler lines from the radiator **(see illustration)**, then plug the hoses and cap the fittings. Place a drip pan underneath to catch the fluid.

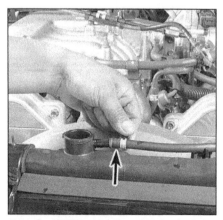

5.4a On 1997 and earlier models, the hose to the coolant reservoir is routed to the left because the coolant reservoir is located on the left side of the engine compartment (3.2L V6 Rodeo model shown, other models similar)

5.4b On 1998 and later models, the hose to the coolant reservoir is routed to the right, because the coolant reservoir is located at the right front corner of the engine compartment (2.2L Amigo model shown, other models similar)

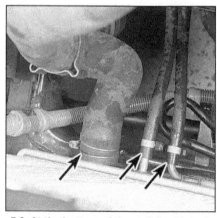

5.6 At the bottom of the radiator, detach the automatic transmission cooler lines and (if you haven't already done so) the lower radiator hose (1997 and earlier Rodeo model shown, later models similar)

5.7a To detach the upper radiator mounting brackets from 1997 and earlier models, remove these bolts (some models, such as the 3.2L V6 Rodeo shown here, have only one clamp near the center of the radiator)

5.7b To detach the upper radiator mounting brackets from 1998 and later models, remove this bolt (2.2L Amigo model shown, other models similar)

5.11 Inspect the two rubber insulators for cracks, tears and deterioration; if they're damaged or worn, replace them

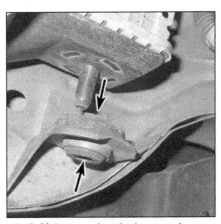

5.12 Make sure that the locator pins on the bottom of the radiator seat in the insulators

7   Remove the two upper radiator mounting brackets **(see illustrations)**. (The bottom of the radiator is positioned and cushioned by rubber insulators that fit into holes in the body.)

8   Lift out the radiator. Be aware of dripping fluids and the cooling fins, which are razor sharp.

9   Inspect the radiator for leaks, damage and internal blockage. If it needs to be repaired, take it to a radiator shop or dealer service department.

10   Clean bugs and dirt from the radiator with compressed air and a soft brush, but be careful not to bend the cooling fins. **Warning:** *Wear eye protection.*

11   Inspect the two lower insulators **(see illustration)** for cracks, tears and deterioration. If the insulators are worn or damaged, replace them.

Installation

Refer to illustration 5.12

12   When installing the radiator, make sure that the lower insulators are in place and make sure that the locator pins on the radiator are correctly engaged with the insulators **(see illustration)**. Installation is otherwise the reverse of removal.

13   After installation, fill the cooling system with the proper mixture of antifreeze and water. On models with an automatic transmission, check and add automatic transmission fluid as needed. Refer to Chapter 1 if necessary.

14   Start the engine and check for leaks. Allow the engine to reach normal operating temperature, indicated by both radiator hoses becoming hot. Recheck the coolant level and add more if required.

Coolant reservoir

Refer to illustrations 5.15a and 5.15b

15   On some models, the coolant reservoir simply pulls up and out of its mounting bracket. On most models, it is bolted to the fenderwell **(see illustrations)**.

16   Pour the coolant into a container. Wash out and inspect the reservoir for cracks and chafing. Replace it if damaged. Wash the outside and inside of the reservoir with soapy water and a brush to remove any dirt or scale, to make reading the coolant level easier.

17   Installation is the reverse of removal.

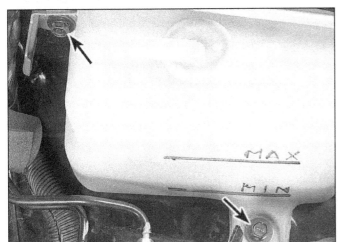

5.15a On 1997 and earlier models, locate the coolant reservoir on the left fenderwell; to detach the reservoir from the vehicle, remove these two retaining bolts

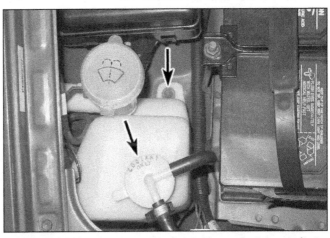

5.15b On 1998 and later models, locate the coolant reservoir at the right front corner of the engine compartment; to detach the reservoir, remove the cap and set it aside, then remove the retaining bolt

6.2 Check the weep hole for signs of leakage: a small amount of gray discoloration is normal, but large brown stains indicate seal failure (2.2L four-cylinder pump shown)

7.5 Before removing the water pump on 2.2L engines, remove the lower rear timing belt cover; to detach the cover, remove this bolt

7.6a Remove the five bolts retaining the water pump to the engine block (2.3L/2/6L four-cylinder engines)

6   Water pump - check

Refer to illustration 6.2

1   A failure in the water pump can cause serious engine damage due to overheating.

2   Water pumps are equipped with weep or vent holes **(see illustration)**. If a failure occurs in the pump seal, coolant will leak from this hole. An inspection mirror and flashlight can be used to look at the underside of the pump. A small amount of black discoloration is normal around the weep hole, but the presence of brown residue or dripping water indicates pump replacement is required. If the weep hole is leaking, shaft bearing failure will follow. Replace the water pump immediately.

3   If the water pump shaft bearings fail there may be a howling sound at the front of the engine while it is running. Bearing wear can be felt if the water pump pulley is rocked up and down. Do not mistake drivebelt slippage, which causes a squealing sound, for water pump failure. Spray automotive drivebelt dressing on the belts to eliminate the belt as a possible cause of the noise.

7   Water pump - replacement

Removal

Refer to illustrations 7.5, 7.6a, 7.6b, 7.6c, 7.6d, 7.6e and 7.8

Warning: *Do not start this procedure until the engine is completely cool.*

1   Disconnect the negative battery cable and drain the cooling system (see Chapter 1).

2   Remove the power steering belt, alternator/water pump belt and air conditioning belt (see Chapter 1).

3   Remove the fan shroud and the fan (see Chapter 4).

4   On four cylinder and 3.2L V6 engines, remove the upper and lower timing belt covers (four-cylinder engines, see Chapter 2A or 2B; 3.2L V6 engine, see Chapter 2D). On 3.1L V6 engines, remove the power steering pump (see Chapter 10) and air conditioning compressor and brackets, if necessary (see Section 15). On 2.3L and 2.6L four-cylinder engines, loosen the timing belt tensioner and move it out of the way of the water pump bolts (see Chapter 2A). Be careful not to disturb the timing belt or sprockets or the valve

timing could be altered.

5   If you're working on a 2.2L four-cylinder engine or a 3.2L V6 engine, remove the timing belt (2.2L engine, see Chapter 2B; 3.2L engine, see Chapter 2D). On 2.2L engines, remove the lower rear timing belt cover **(see illustration)**.

6   Remove the water pump mounting bolts **(see illustrations)**. On 2.2L engines, note the alignment lugs (one on the pump and one on the pump housing); when installing the pump, these lugs must be aligned exactly as they were before the pump was removed.

7   Remove the water pump from the engine block and scrape all gasket and sealant material from the mounting surfaces.

8   On 2.2L engines, remove and discard the old water pump O-ring **(see illustration)**. If you're installing a new pump, make sure that it includes a new O-ring. Even if you're going to install the old pump, install a new O-ring.

Installation

9   On all pumps except the one used on 2.2L engines, apply RTV sealant to both sides of the new gasket and install it and the water pump on the engine block. On 2.2L engines, make sure that the new O-ring is installed on the pump **(see illustra-**

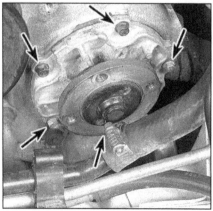

7.6b On the 3.1L V6, the mounting bolts are located around the perimeter of the pump

7.6c On 3.2L V6 engines, unbolt the timing belt idler pulley for access to some of the water pump bolts . . .

7.6d . . . then remove the water pump bolts and detach the pump from the block

7.6e Water pump bolts (2.2L four-cylinder engine); when installing the pump, make sure that the lugs (one on the pump and one on the pump housing) are aligned

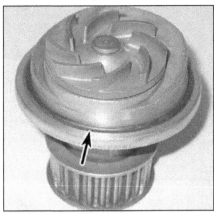

7.8 On 2.2L engines, remove and discard the old O-ring and install a new O-ring on the pump before installing it again; if you're going to install a new pump, make sure that it includes a new O-ring

8.1 Components of the oil cooler assembly include:

A   Oil filter
B   Oil cooler
C   Oil cooler/filter adapter
D   Oil pressure gauge sending unit
E   Oil pressure warning light switch

Check around the water pump and radiator for any leaks.

15   Recheck the coolant level and add more if necessary.

8   Engine oil cooler - replacement

Removal

Refer to illustrations 8.1 and 8.5

Warning: *The engine must be completely cool for this procedure.*

1   Models with the 3.2L V6 have an engine oil cooler, which is sandwiched between the oil filter and the engine block **(see illustration)**. Coolant flows through the cooler from two hoses connected to the engine.

2   To replace the oil cooler, refer to Chapter 1 for removal of the oil filter and draining of the cooling system.

3   Disconnect the two hoses from the cooler.

4   Disconnect the electrical connector from the oil pressure sending unit on the oil filter adapter. (There are two electrical units. The larger one is the sender for the oil pressure gauge, the smaller one is a switch that turns on a warning light on the dash if oil pressure is too low.)

5   Use a deep socket to remove the threaded adapter holding the oil cooler housing to the oil filter adapter housing **(see illustration)**.

6   Pull the oil cooler from the adapter on the block.

7   If only the oil cooler is to be replaced, this is as far as you need to disassemble components. If the oil cooler adapter is to be removed, remove the bolts holding it to the block and remove it.

Installation

Refer to illustration 8.9

8   Clean the block and back of the oil cooler adapter of any old gasket material. Install the adapter to the block with a new gasket and tighten the bolts securely.

9   Clean the coolant sealing surfaces of the adapter and oil cooler, and install a O-ring **(see illustration)**. Clean the threads of the threaded tube and apply thread-locking compound to the end that goes into the oil cooler adapter. Insert it through the oil cooler and tighten it to this Chapter's Specifications.

10   The remainder of installation is the reverse of removal. Install a new oil filter, refill and bleed the cooling system (see Chapter 1), then run the engine to check for oil or coolant leaks.

tion 7.8).

10   Install the pump. On 2.2L engines, make sure that the lugs on the pump and on the pump housing are aligned **(see illustration 7.6e)**.

11   Tighten all the water pump bolts a little at a time to the torque listed in this Chapter's Specifications.

12   The remainder of installation is the

reverse of the removal procedure.

13   Adjust the drivebelts to the proper tension. Refer to Chapter 1 if necessary.

14   Fill the cooling system with the proper mixture of antifreeze and water, again referring to Chapter 1 if necessary. Then start the engine and allow it to idle until it reaches normal operating temperature. This is indicated by the upper radiator hose becoming hot.

9   Coolant temperature sending unit - check and replacement

Check

Refer to illustrations 9.3a and 9.3b

Warning: *Do not start this procedure until the engine is completely cool.*

8.5 Use a large deep socket to remove the oil cooler adapter center bolt

8.9 When reinstalling the oil cooler, use a new O-ring in the groove

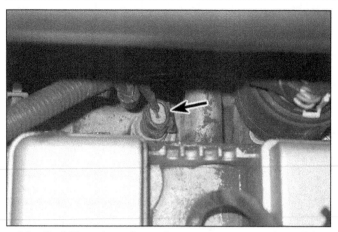

9.3a On 2.2L engines, the coolant temperature sending unit is located at the rear of the cylinder head, behind the end of the valve cover

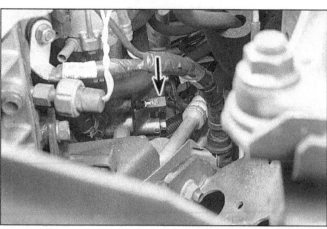

9.3b On 2.6L engines, the coolant temperature sending unit is located on the right side of the engine, in front of the starter motor (shown from the bottom right, looking up)

1   If the coolant temperature gauge is inoperative, check the fuses first (see Chapter 12).
2   If the temperature gauge indicates excessive temperature after running awhile, see the *Troubleshooting* section in the front of the manual.
3   If the temperature gauge indicates Hot as soon as the engine is started cold, disconnect the wire at the coolant temperature sender **(see illustrations)**. If the gauge reading drops, replace the sending unit. If the reading remains high, the wire to the gauge may be shorted to ground or the gauge is faulty. On 2.6L four-cylinder engines, the sending unit is located on the right side of the engine, below the intake manifold. On the 3.1L V6 engine it's located at the left front of the engine, near the power steering pump. On the 3.2L V6 engine it's located at the rear of the engine, in the coolant manifold (pipe) behind the right (passenger-side) cylinder head.
4   If the coolant temperature gauge fails to show any indication after the engine has been warmed up, (approx. 10 minutes) and the fuses checked out OK, shut off the engine. Disconnect the wire at the sending

unit and, using a jumper wire, connect the wire to a clean ground on the engine. Briefly turn on the ignition without starting the engine. If the gauge now indicates Hot, replace the sending unit.
5   If the gauge fails to respond, the circuit may be open or the gauge may be faulty.
6   If the temperature sending unit is removed, you can test it in a pan of water on the stove. Have an ohmmeter connected to the sending unit body and the boss for the electrical connector. When the temperature of the water reaches around 140-degrees F, the electrical resistance should be about 160 ohms. If not, replace the sending unit.

Replacement

7   Drain the coolant (see Chapter 1).
8   Disconnect the wiring connector from the sending unit.
9   Using a deep socket or a wrench, remove the sending unit.
10   Install the new unit and tighten it securely. Do not use thread sealant as it may electrically insulate the sending unit.
11   Reconnect the wiring connector, refill

the cooling system and check for coolant leakage and proper gauge function.

---

10 Heating and air conditioning blower motor and circuit - check and component replacement

---

Refer to illustrations 10.3, 10.6, 10.9a and 10.9b

Warning: *Some models covered by this manual are equipped with airbags. Always disable the airbag system (see Chapter 12) before working in the vicinity of the impact sensors, steering column or instrument panel. Failure to follow these procedures may cause accidental deployment of the airbag, which could cause personal injury. The airbag circuits are easily identified by yellow insulation covering the entire wiring harness. Do not use electrical test equipment on any of these wires or tamper with them in any way.*

1   Disconnect the negative cable from the battery.
2   The blower unit is located in the passenger compartment above the right front

10.3 The connector at the blower motor should have 12 volts at the terminal for the blue/white wire and ground through the terminal for the blue/black wire

10.6 Blower motor resistor location: the resistor is retained by two screws, in a housing just to the left of the blower motor housing

10.9a Remove the four screws and pull down the blower motor for replacement (one screw not visible in this photo)

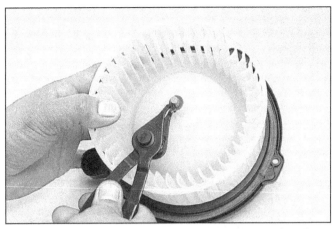

10.9b Use snap-ring pliers to remove the small snap-ring, then slide the fan from the blower motor shaft

footwell, ahead of the glovebox. If the blower doesn't work, check the fuse and all connections in the circuit for looseness and corrosion.

3   If the blower motor does not operate, remove the glovebox (see Chapter 11), disconnect the electrical connector at the blower motor, turn the ignition key On (engine not running) and check for battery voltage on the blue/white stripe wire **(see illustration)**. If battery voltage is not present, there is a problem in the ignition feed circuit.

4   If battery voltage is present, reconnect the terminal to the blower motor and back-probe the blue/black stripe wire with a jumper wire connected to ground. If the motor still does not operate, the motor is faulty.

5   If the motor is good, but doesn't operate at any speed, the heater/air conditioning control switch is probably faulty. Remove the control assembly (see Section 12) and check for continuity through the switch in each position.

6   If the blower motor operates at High speed, but not at one or more of the lower speeds, check the blower motor resistor, located in the housing just to the left of the blower housing **(see illustration)**.

7   Disconnect the electrical connector from

the blower motor resistor. Remove the screws and withdraw the resistor from the housing.

8   Inspect the blower motor resistor for damaged or broken resistance coils. Check for continuity across each resistor coil.

9   If the blower motor must be replaced, remove the bracket holding the wiring connector to the blower, then remove the four mounting screws and drop the blower assembly down out of the housing **(see illustration)**. The fan can be removed and used again on the new blower motor **(see illustration)**.

10   Installation is the reverse of removal. Check for proper operation.

11   Heater core - removal and installation

Refer to illustrations 11.3a, 11.3b, 11.7, 11.8, 11.9a, 11.9b and 11.10

Warning 1: *Some models covered by this manual are equipped with airbags. Always disable the airbag system (see Chapter 12) before working in the vicinity of the impact sensors, steering column or instrument panel.*

Failure to follow these procedures may cause accidental deployment of the airbag, which could cause personal injury. The airbag circuits are easily identified by yellow insulation covering the entire wiring harness. Do not use electrical test equipment on any of these wires or tamper with them in any way.

Warning 2: *The air conditioning system is under high pressure. Do not loosen any hose fittings or remove any components until the system has been discharged. Air conditioning refrigerant should be properly discharged into an EPA-approved recovery/recycling unit by a dealer service department or an automotive air conditioning repair facility. Always wear eye protection when disconnecting air conditioning system fittings.*

1   Disconnect the negative cable from the battery.

2   Drain the cooling system (see Chapter 1).

3   Working in the engine compartment, disconnect the heater hoses where they enter the firewall and the two evaporator lines at the firewall **(see illustrations)**. Push the rubber seal around the hoses toward the inside of the vehicle, releasing it from the sheet metal.

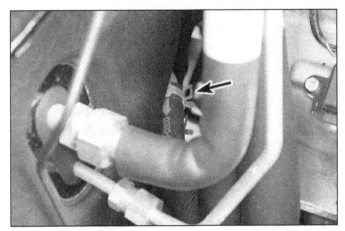

11.3a Disconnect the heater hoses at the firewall (hoses are somewhat hidden by the air conditioning lines in the foreground)

11.3b Use two wrenches when disconnecting air conditioning lines at the firewall

11.7 Remove the screws and take out the lower duct below the heater core housing

11.8 There are four retaining nuts, one top and bottom on each side of the heater core housing - when they are removed, pull straight back to remove the housing

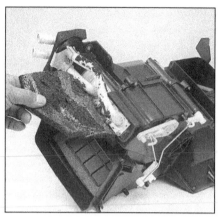

11.9a Pull back the foam insulation around the heater core pipes . . .

4   Refer to Section 12 of this Chapter to remove the heat/air conditioning controls, and refer to Chapter 12 to remove the instrument panel.

5   There are three separate housings under the dash. At right is the blower motor housing, to the left of it is the housing for the evaporator core, and the third unit near the center of the dash area is the heater core housing.

6   To remove the heater core housing, the evaporator core housing must be removed first (see Section 17).

7   Remove the lower duct from below the heater core housing **(see illustration)**.

8   Remove the four nuts from the studs retaining the heater core housing (they're on the engine compartment side of the firewall) and pull the housing back and out **(see illustration)**. Wrap duct tape around the heater core tubes to prevent coolant from leaking on the carpeting.

9   Once the housing is out of the vehicle, remove the insulation around the heater core pipes and remove the clamp over the pipes **(see illustrations)**. Try not to tear the insulation, as it will be reused later.

10   Remove the screws around the perimeter of the heater core housing and separate

the two halves to remove the heater core **(see illustration)**.

11   Installation is the reverse order of removal.

12   Refill the cooling system, reconnect the battery and run the engine. Check for leaks and proper system operation. Have the air conditioning system evacuated and charged by the shop that discharged it.

---

12   Heating and air conditioning control assembly - check, removal and installation

---

Warning: *Some models covered by this manual are equipped with airbags. Always disable the airbag system (see Chapter 12) before working in the vicinity of the impact sensors, steering column or instrument panel. Failure to follow these procedures may cause accidental deployment of the airbag, which could cause personal injury. The airbag circuits are easily identified by yellow insulation covering the entire wiring harness. Do not use electrical test equipment on any of these wires or tamper with them in any way.*

Removal and installation

Refer to illustrations 12.3a, 12.3b, 12.4a, 12.4b, 12.4c and 12.4d

1   Disconnect the negative cable from the battery.

2   Remove the center cluster trim panels (see Chapter 11).

3   Remove the mounting screws located on the front of the control assembly **(see illustrations)**.

4   Pull out the heating and air conditioning control assembly far enough to disconnect all electrical connectors and cables from the control assembly. Pry back the clips on the control cable mountings and remove the cables from the control. Disconnect the electrical connectors **(see illustrations)**.

5   Installation is the reverse of the removal procedure.

6   Run the engine and check for proper functioning of the heater (and air conditioning, if equipped).

Cable adjustment

Refer to illustration 12.8

7   With the cables attached at the control end and the control assembly installed in the dash, adjust the cables at their ends. There are three cables. The leftmost cable attaches

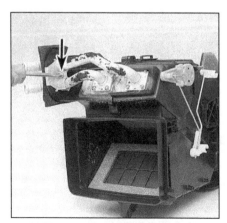

11.9b . . . and remove the screw on the pipe clamp

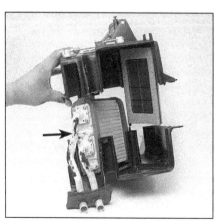

11.10 With the screws removed, separate the two halves of the heater core housing and pull out the core

12.3a Remove these screws to release the heating and air conditioning control assembly on 1997 and earlier models

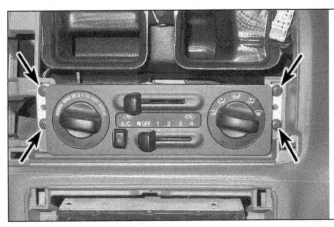

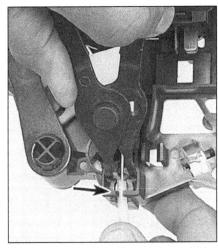

12.3b Remove these screws to detach the heating and air conditioning control assembly on 1998 and later models

to the temperature control lever on the left side of the heater core housing (see Section 11). The center cable is the air selector control which connects to linkage at the bottom-right side of the heater core housing, and the right-hand cable (air source or "mode" control) connects to the blower housing.

8   To adjust the air source control cable, move the dash control to the left (RECIRC position), set the damper lever to its recirculation position (toward the firewall), install the cable and clamp it in place **(see illustration)**.

9   To adjust the temperature control cable, set the dash control to COLD, move the lever on the heater box to the Cold position (up) install the cable and lock the clamp while applying slight pressure on the outer cable.

10   To adjust the air selector control cable, set the dash control to the right (DEFROST), move the lever on the lower right of the heater box to the defrost position (far right), hook the cable end on and tighten the clamp.

Electrical checks

Refer to illustrations 12.11a, 12.11b, 12.12a, 12.12b and 12.12c

11   Check the air conditioning switch continuity. On 1997 and earlier models, with the switch off, there should be continuity between

12.4b Disconnect the electrical connectors by squeezing the tabs in and pulling the connector off (1997 and earlier models)

terminals 2 and 3 only **(see illustration)**; when the switch is on, there should be continuity between terminals 1, 2 and 3. On 1998 and later models, with the switch off, there should be continuity between terminals 1 and 3 only **(see illustration)**; when the switch is on, there should be continuity between terminals 1, 2 and 3.

12.4a Use snap-ring pliers to "spread" the clips retaining each of the three cables to release it from the control unit, then lift the cable end "eye" off the control lever pin (1997 and earlier models)

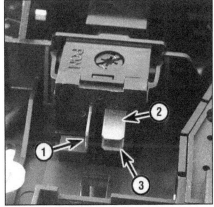

12.4c On 1998 and later models, pull out the heating and air conditioning control assembly far enough to disconnect the control cables from the lever arms and from the cable guides on top of the assembly . . .

12.4d . . . and disconnect the electrical connectors from the underside of the assembly

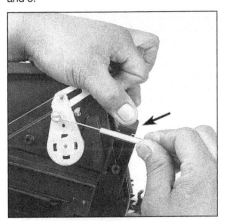

12.8 On top of the blower motor housing, adjust the air source cable by setting the dash control to RECIRC, push the air source door arm (with cable attached) toward the engine, and place the cable in the clip

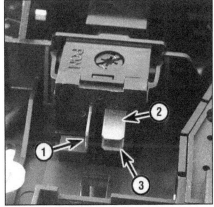

12.11a Air conditioning switch terminal guide (1997 and earlier models)

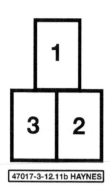

47017-3-12.11b HAYNES

12.11b Air conditioning switch terminal guide (1998 and later models)

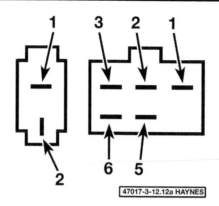

47017-3-12.12a HAYNES

12.12a Blower speed selector switch terminal guide (1989 through 1992 Amigo and 1991 and 1992 Rodeo models)

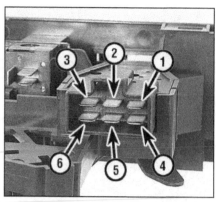

12.12b Blower speed selector switch terminal guide (1993 and 1994 Amigo, 1993 through 1997 Rodeo and 1995 through 1997 Passport models)

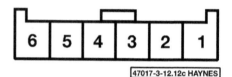

47017-3-12.12c HAYNES

12.12c Blower speed selector switch terminal guide (all 1998 and later models)

12   Test the blower speed control switch for continuity **(see illustrations)**.

 a) *On 1989 through 1992 Amigo and 1991 and 1992 Rodeo models: With the switch in position 1, there should be continuity between terminals 1 and 2; in position 2, there should be continuity between 3 and 2; in position 3, there should be continuity between 2 and 2; in position 4, there should be continuity between 1 and 2.*

 b) *On 1993 through 1997 Rodeo, 1993 and 1994 Amigo and 1995 through 1997 Passport models: With the switch in position 1, there should be continuity between terminals 1 and 4; in position 2, there should be continuity between 1 and 5; in position 3, there should be continuity between 1 and 6; and in position 4, there should be continuity between 1 and 3.*

 c) *On 1998 and later Amigo, Rodeo and Passport models: With the switch in position 1, there should be continuity between terminals 1 and 2; in position 2, there should be continuity between 1 and 3; in position 3, there should be continuity between 1 and 4; in position 4, there should be continuity between 1 and 5.*

13   If the switch fails any of these tests, replace it.

13 Air conditioning and heating system - check and maintenance

Air conditioning system

Warning: *The air conditioning system is under high pressure. Do not loosen any hose fittings or remove any components until the system has been discharged. Air conditioning refrigerant should be properly discharged into an EPA-approved recovery/recycling unit by a dealer service department or an automotive air conditioning repair facility. Always wear eye protection when disconnecting air conditioning system fittings.*

1   The following maintenance checks should be performed on a regular basis to ensure that the air conditioner continues to operate at peak efficiency:

 a) *Inspect the condition of the compressor drivebelt. If it is worn or deteriorated, replace it (see Chapter 1).*

 b) *Check the drivebelt tension and, if necessary, adjust it (see Chapter 1).*

 c) *Inspect the system hoses. Look for cracks, bubbles, hardening and deterioration. Inspect the hoses and all fittings for oil bubbles or seepage. If there is any evidence of wear, damage or leakage, replace the hose(s).*

 d) *Inspect the condenser fins for leaves, bugs and any other foreign material that may have embedded itself in the fins. Use a "fin comb" or compressed air to remove debris from the condenser.*

 e) *Make sure the system has the correct refrigerant charge.*

2   It's a good idea to operate the system for about ten minutes at least once a month. This is particularly important during the winter months because long term non-use can cause hardening, and subsequent failure, of the seals.

3   Leaks in the air conditioning system are best spotted when the system is brought up to operating temperature and pressure, by running the engine with the air conditioning ON for five minutes. Shut the engine off and inspect the air conditioning hoses and connections. Traces of oil usually indicate refrigerant leaks.

4   Because of the complexity of the air conditioning system and the special equipment required to effectively work on it, accurate troubleshooting of the system should be left to a professional technician.

5   If the air conditioning system doesn't operate at all, check the fuse panel and the air conditioning relay, located in the fuse/relay box in the engine compartment. Refer to Sections 4, 10 and 12 for electrical checks of heating/air conditioning system components.

6   The most common cause of poor cooling is simply a low system refrigerant charge. If a noticeable drop in cool air output occurs, the following quick check will help you determine if the refrigerant level is low. For more

complete information on the air conditioning system, refer to the *Haynes Automotive Heating and Air Conditioning Manual.*

Checking the refrigerant charge

Refer to illustrations 13.9 and 13.11

7   Warm the engine up to normal operating temperature.

8   Place the air conditioning temperature selector at the coldest setting and put the blower at the highest setting. Open the doors (to make sure the air conditioning system doesn't cycle off as soon as it cools the passenger compartment).

9   With the compressor engaged - the clutch will make an audible click and the center of the clutch will rotate. After the system reaches operating temperature, feel the two pipes connected to the evaporator at the firewall **(see illustration)**.

10   The pipe (thinner tubing) leading from the condenser outlet to the evaporator should be cold, and the evaporator outlet line (the thicker tubing that leads back to the compressor) should be slightly colder (3 to 10-degrees F). If the evaporator outlet is considerably warmer than the inlet, the system needs a charge. Insert a thermometer in the center air distribution duct while operating the air conditioning system - the temperature of the output air should be 35 to 40-degrees F below the ambient air temperature (down to approximately 40-degrees F). If the ambient (outside) air temperature is very high, say 110-degrees F, the duct air temperature may be as high as 60-degrees F, but generally the air conditioning is 30 to 50-degrees F cooler than the ambient

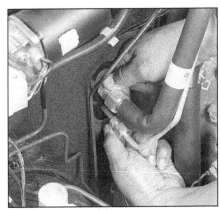

13.9 Feel the two air conditioning pipes where they enter the firewall - if the thicker tube feels warmer than the thinner (inlet) pipe, the system may need a charge

13.11 The sight glass is located on the top of the receiver/drier in the left front corner ahead of the radiator (grille removed here for clarity)

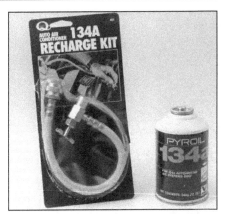

13.12 A basic charging kit is available at most auto parts stores - it must say R-134a (1994 and later models) and so must the cans of refrigerant you buy

air. If the air isn't as cold as it used to be, the system probably needs a charge. Further inspection or testing of the system is beyond the scope of the home mechanic and should be left to a professional.

11   Inspect the sight glass If the refrigerant looks foamy when running, it's low **(see illustration)**. When ambient temperatures are very hot, bubbles may show in the sight glass even with the proper amount of refrigerant. With the proper amount of refrigerant, when the air conditioning is turned off, the sight glass should show refrigerant that foams, then clears.

Adding refrigerant

Refer to illustrations 13.12 and 13.15

Caution 1: *Refrigerant has changed from the use of R-12 (1989 through 1993 models) to R-134a (1994 and later models). The two refrigerants are NOT compatible. Special fittings and manifold gauge sets are used on the different refrigerant types so that you can't accidentally hook up R-12 charging equipment to R-134a, or vice versa.*

Caution 2: *When replacing air conditioning components, additional refrigerant oil - equal to the amount removed with the component being replaced - should be added. Mineral-based refrigerant oils for R-12 and synthetic refrigerant oils for R-134a are incompatible. Be sure to read the can before adding any oil to the system, to make sure it is compatible with the type of system being repaired.*

Note: *Because of EPA regulations, R-12 refrigerant is no longer available to home mechanics. However, R-134 refrigerant is available in auto parts stores. Models with R-12 systems will have to be serviced at a dealer service department or air conditioning shop. Check prices for having your R-12 system converted to R-134a - you might find it a more economical alternative over time.*

12   Buy an automotive charging kit at an auto parts store. A charging kit includes a 12-ounce can of R-134a refrigerant, a tap valve and a short section of hose that can be attached between the tap valve and the system low side service valve **(see illustration)**.

Because one can of refrigerant may not be sufficient to bring the system charge up to the proper level, it's a good idea to buy a couple of additional cans. Try to find at least one can that contains red refrigerant dye. If the system is leaking, the red dye will leak out with the refrigerant and help you pinpoint the location of the leak.

13   Connect the charging kit by following the manufacturer's instructions.

14   Back off the valve handle on the charging kit and screw the kit onto the refrigerant can, making sure first that the O-ring or rubber seal inside the threaded portion of the kit is in place. **Warning:** *Wear protective eye wear when dealing with pressurized refrigerant cans.*

15   Remove the dust cap from the low-side charging port and attach the quick-connect fitting on the kit hose **(see illustration)**. **Warning:** *DO NOT hook the charging kit hose to the system high side! The fittings on the charging kit are designed to fit **only** on the low side of the system.*

16   Warm the engine to normal operating temperature and turn on the air conditioner. Keep the charging kit hose away from the fan and other moving parts.

17   Turn the valve handle on the kit until the stem pierces the can, then back the handle out to release the refrigerant. You should be able to hear the rush of gas. Add refrigerant to the low side of the system until both the outlet and the evaporator inlet pipe feel about the same temperature. Allow stabilization time between each addition. **Warning:** *Never add more than two cans of refrigerant to the system.* The can may tend to frost up, slowing the procedure. Wrap a shop towel wet with hot water around the bottom of the can to keep it from frosting.

18   If you have an accurate thermometer, you can place it in the center air conditioning duct inside the vehicle to monitor the air temperature. A charged system that is working properly, should output air down to approximately 40-degrees F.

19   When the can is empty, turn the valve handle to the closed position and release the

connection from the low-side port. Replace the dust cap.

20   Remove the charging kit from the can and store the kit for future use with the piercing valve in the UP position, to prevent inadvertently piercing the can on the next use.

Heating systems

21   If the air coming out of the heater vents isn't hot, the problem could stem from any of the following causes:

a) *The thermostat is stuck open, preventing the engine coolant from warming up enough to carry heat to the heater core. Replace the thermostat (see Section 3).*

b) *A heater hose is blocked, preventing the flow of coolant through the heater core. Feel both heater hoses at the firewall. They should be hot. If one of them is cold, there is an obstruction in one of the hoses or in the heater core, or the heater control valve is shut. Detach the hoses and back flush the heater core with a water hose. If the heater core is clear but circulation is impeded, remove the two hoses and flush them out with a water hose.*

13.15 Add R-134a refrigerant to the low-side port only - the procedure will go faster if you wrap the can with a warm, wet towel to prevent icing

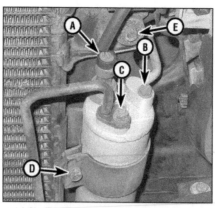

14.3a To remove the receiver-drier on 1997 and earlier vehicles, disconnect the electrical connector (A), unbolt the line (B) to the evaporator, unbolt the line (C) to the condenser and remove the clamp bolt (D)

14.3b To remove the receiver-drier on 1998 and later models, disconnect the electrical connector (A), unbolt the line (B) to the evaporator, unbolt the line (C) to the condenser (the other end of this line is welded to the condenser) and remove the clamp bolt (D); to detach the left end of the condenser, remove this bolt (E)

15.4a Typical 1997 and earlier compressor: Disconnect the electrical connector (A), unbolt the refrigerant line flanges (arrow B indicates one flange bolt; other refrigerant line flange bolt not shown) and remove the compressor mounting bolts (C indicates upper mounting bolts; lower bolts not shown)

c) *If flushing fails to remove the blockage from the heater core, the core must be replaced.* (see Section 11).

22   If the blower motor speed does not correspond to the setting selected on the blower switch, the problem could be a bad fuse, circuit, switch, blower motor resistor or motor (see Sections 10 and 12).

23   If there isn't any air coming out of the vents:

a) *Turn the ignition ON and activate the fan control. Place your ear at the heating/air conditioning register (vent) and listen. Most motors are audible. Can you hear the motor running?*

b) *If you can't (and have already verified that the blower switch and the blower motor resistor are good), the blower motor itself is probably bad (see Section 10).*

24   If the carpet under the heater core is damp, or if antifreeze vapor or steam is coming through the vents, the heater core is leaking. Remove it (see Section 11) and install a new unit (most radiator shops will not repair a leaking heater core).

25   Inspect the drain hose from the evaporator case, make sure it is not clogged. It is located at the firewall, just above and to the right of the bellhousing.

Eliminating air conditioner odors

26   Unpleasant odors that often develop in air conditioning systems are caused by the growth of a fungus, usually on the surface of the evaporator core. The warm, humid environment there is a perfect breeding ground for mildew to develop.

27   The evaporator core on most vehicles is difficult to access, and factory dealerships have a lengthy, expensive process for eliminating the fungus by opening up the evaporator case and using a powerful disinfectant and rinse on the core until the fungus is gone. You can service your own system at home, but it takes something much stronger than basic household germ-killers or deodorizers.

28   Aerosol disinfectants for automotive air conditioning systems are available in most auto parts stores, but remember when shopping for them that the most effective treatments are also the most expensive. The basic procedure for using these sprays is to start by running the system in the RECIRC mode for ten minutes with the blower on its highest speed. Use the highest heat mode to dry out the system and keep the compressor from engaging by disconnecting the wiring plug at the compressor (see Section 15).

29   The disinfectant can usually comes with a long spray hose. Remove the blower motor resistor (see Section 10), point the nozzle inside the hole and spray, according to the manufacturer's recommendations. Try to cover the whole surface of the evaporator core, by aiming the spray up, down and sideways. Follow the manufacturer's recommendations for the length of spray and waiting time between applications.

30   Once the evaporator has been cleaned, the best way to prevent the mildew from coming back again is to make sure your evaporator housing drain tube is clear.

14   Air conditioning receiver/drier - removal and installation

Refer to illustration 14.3a and 14.3b
Warning: *The air conditioning system is under high pressure. Do not loosen any hose fittings or remove any components until the system has been discharged. Air conditioning refrigerant should be properly discharged into an EPA-approved recovery/recycling unit by a dealer service department or an automotive air conditioning repair facility. Always wear eye protection when disconnecting air conditioning system fittings.*

1   Have the refrigerant discharged and recovered by an air conditioning technician.

2   Remove the grille (see Chapter 11).

3   Disconnect the electrical connector and the refrigerant lines **(see illustrations)** from the receiver/drier and cap the open fittings to prevent entry of moisture.

4   Loosen the clamp bolt and slip the receiver/drier out of the bracket.

5   Installation is the reverse of removal.

6   Have the system evacuated, charged and leak tested by the shop that discharged it. If the receiver was replaced, have them add new refrigeration oil to the compressor, about 20 cc (0.7 oz.) for R-12 systems, or 10 cc (.34-oz.) for R-134a systems. Use only the refrigerant oil compatible with the refrigerant of your system (R-12 or R-134a).

15   Air conditioning compressor - removal and installation

Refer to illustrations 15.4a, 15.4b and 15.4c
Warning: *The air conditioning system is under high pressure. Do not loosen any hose fittings or remove any components until the system has been discharged. Air conditioning refrigerant should be properly discharged into an EPA-approved recovery/recycling unit by a dealer service department or an automotive air conditioning repair facility. Always wear eye protection when disconnecting air conditioning system fittings.*

1   Have the refrigerant discharged and recovered by an automotive air conditioning technician.

2   Disconnect the negative cable from the battery.

3   Remove the drivebelt from the compressor (see Chapter 1).

4   Disconnect the electrical connector and disconnect the refrigerant lines **(see illustrations)**. Remove and discard the old O-rings.

5   Unbolt the compressor and lower it from the vehicle.

6   If a new or rebuilt compressor is being installed, follow the directions which come

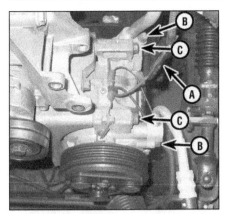

15.4b Typical 1998 and later compressor (power steering pump removed for clarity): Disconnect the electrical connector (A), remove the refrigerant line bolts (B), remove the upper mounting bolts (C) . . .

with it regarding the proper level of oil prior to installation.

7   Installation is the reverse of removal. Be sure to replace all O-rings with new ones specifically made for the type of refrigerant in your system and lubricate them with refrigerant oil, also designed specifically for your refrigerant.

8   Have the system evacuated, recharged and leak tested by the shop that discharged it.

16 Air conditioning condenser - removal and installation

Refer to illustrations 16.3a, 16.3b, 16.4, 16.5a, 16.5b and 16.5c

Warning 1: *The air conditioning system is under high pressure. Do not loosen any hose fittings or remove any components until the system has been discharged. Air conditioning refrigerant should be properly discharged into an EPA-approved recovery/recycling unit by a dealer service department or an automotive air conditioning repair facility. Always wear eye protection when disconnecting air conditioning system fittings.*

Warning 2: *Some models covered by this manual are equipped with airbags. Always disable the airbag system (see Chapter 12) before working in the vicinity of the impact*

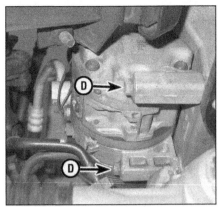

15.4c . . . and remove the lower mounting bolts (D)

sensors, steering column or instrument panel. Failure to follow these procedures may cause accidental deployment of the airbag, which could cause personal injury. The airbag circuits are easily identified by yellow insulation covering the entire wiring harness. Do not use electrical test equipment on any of these wires or tamper with them in any way.

1   Have the refrigerant discharged by an air conditioning technician.

2   Remove the grille (see Chapter 11). If equipped with air conditioning, remove the condenser cooling fan (see Section 4).

3   Disconnect the inlet and outlet fittings **(see illustrations)**. The left-side fitting is at the receiver/drier (see Section 14). Cap the open fittings immediately to keep moisture and dirt out of the system. Remove and discard all old O-rings. The O-rings must NOT be reused, or the system will leak.

4   On 1997 and earlier models, remove the two bolts through the radiator support, and the hood latch and its support bracket **(see illustration)**. The bottom of the condenser is mounted in rubber "doughnuts" on the body, and the top of the condenser is "captured" by the radiator core support. To remove the condenser once the lines and fasteners are off, pull the condenser straight up (inside the core support sheet metal) until the bottom is out of its rubber mounts. Then swing the bottom of the condenser forward until it clears the body and can be dropped down. When the condenser top is free of the radiator support, pull it up and out of the vehicle.

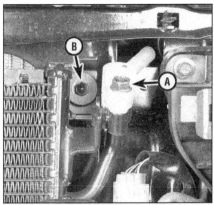

16.3a On 1997 and earlier models, disconnect the refrigerant line on the left side by unbolting this flange (A) (another line is attached to the condenser in this manner on the right side); to detach the left end of the condenser, remove the left mounting bolt (B)

16.3b On1998 and later models, disconnect the compressor discharge line from the condenser by removing this bolt. The other refrigerant line is welded to the bottom left corner of the condenser; disconnect it from the receiver-drier (see illustration 14.3b)

5   On 1998 and later models, remove the hood latch and the vertical support **(see illustrations)**, then remove the left and right

16.4 To detach the condenser on 1997 and earlier models, remove the hood latch and vertical support bracket and then remove the left and right condenser retaining bolts

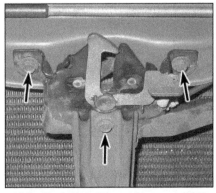

16.5a To detach the condenser on 1998 and later models, remove the hood latch bolts and the hood latch assembly, remove the upper bolt from the vertical support bracket . . .

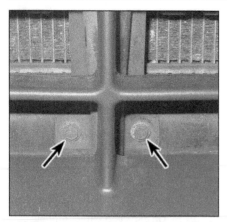

16.5b ... remove the two lower vertical support bracket bolts and remove the vertical support bracket ...

16.5c ... and remove the right condenser mounting bracket bolt (the left mounting bracket bolt is shown in illustration 14.3b)

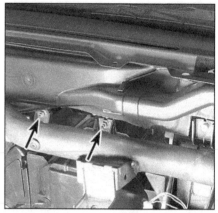

17.3a Remove the upper mounting nut (A) for the evaporator case (B is the upper right nut for the heater core housing) ...

condenser mounting bolts **(left bolt, see illustration 14.3b; right bolt, see illustration)**. Then tilt the upper edge of the condenser forward and lift the condenser assembly out of the vehicle.

6   Inspect the rubber insulators that position and cushion the lower edge of the condenser. If they're cracked, torn or deteriorated, replace them.

7   When installing the condenser, make sure that the condenser locator pins are fully inserted into the insulator cushions.

8   Using new O-rings, reconnect the refrigerant lines,. If a new condenser has been installed, add about 1.4 to 1.7 ounces of new refrigerant oil.

9   Installation is otherwise the reverse of removal.

10   Have the system evacuated, charged and leak tested by the shop that discharged it.

---

17   Air conditioning evaporator - removal and installation

Refer to illustrations 17.3a, 17.3b, 17.4 and 17.6

Warning 1: *The air conditioning system is under high pressure. Do not loosen any hose fittings or remove any components until the system has been discharged. Air conditioning refrigerant should be properly discharged into an EPA-approved recovery/recycling unit by a dealer service department or an automotive air conditioning repair facility. Always wear eye protection when disconnecting air conditioning system fittings.*

Warning 2: *Some models covered by this manual are equipped with airbags. Always disable the airbag system (see Chapter 12) before working in the vicinity of the impact sensors, steering column or instrument panel. Failure to follow these procedures may cause accidental deployment of the airbag, which could cause personal injury. The airbag circuits are easily identified by yellow insulation*

17.3b ... and the lower nut for the evaporator case

covering the entire wiring harness. Do not use electrical test equipment on any of these wires or tamper with them in any way.

1   After the system has been discharged and the refrigerant recovered, disconnect the two refrigerant lines at the firewall.

2   Refer to Chapter 11 for removal of the instrument panel.

3   Remove two nuts holding the evaporator unit to the firewall and pull the unit out of the vehicle **(see illustrations)**.

4   With the evaporator unit on the bench, remove the screws and clips and separate the top and bottom halves of the case and pull out the evaporator **(see illustration)**.

5   Pull the thermistor sensor probe from the evaporator core and unbolt the expansion valve and the two short refrigerant lines **(see illustration 17.4)**.

6   The evaporator core can be cleaned with a "fin comb" and blown off with compressed air **(see illustration)**.

7   If the evaporator core is replaced with a new unit, add 1.4 ounces of new refrigerant oil (make sure it is the oil compatible with your type of refrigerant) to the system.

8   The remainder of the installation is the reverse of the removal process. Be sure to

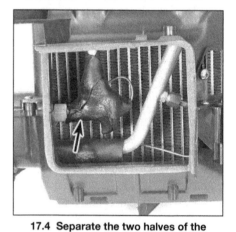

17.4 Separate the two halves of the evaporator housing and remove the core - arrow indicates the expansion valve

use new O-rings, and new gaskets on the expansion valve.

9   Have the system evacuated, charged and leak tested by the shop that discharged it.

17.6 The fins of the core can be cleaned with a "fin comb" and blown off with compressed air - use the side of the tool with the correct fins-per-inch spacing for your core

Chapter 4 Part A
Fuel and exhaust systems - carbureted engines

Contents

Specifications

Fuel pressure	2 to 8 psi
Curb idle speed	See Chapter 1
Choke valve opening clearance	0.031 to 0.050 inch
Vacuum break clearance	0.105 to 0.109 inch
Fast idle clearance	0.048 to 0.058 inch
Valve plate interlock clearance	0.272 to 0.331 inch

1 General information

The fuel system on all carbureted models consists of a fuel tank, a mechanically operated fuel pump and a carburetor. A combination of metal and rubber fuel hoses is used to connect these components.

The electric assist choke system consists of a thermostatic spring and cover (choke assembly), temperature sensing switch and a ceramic (positive temperature coefficient) heater. At temperatures below 60 degrees, the switch is open and no current is supplied to the ceramic heater located within the thermostatic spring. At temperatures above 60 degrees F, the temperature sensing switch closes and current is supplied to the ceramic heater. As the heater warms, it causes the spring to pull the choke plate open within 1 to 1-1/2 minutes.

The carburetor is either a Hitachi or Nikki two-barrel type, depending on engine displacement and year of production. USA models are equipped with electronic feedback carburetors. These carburetors are linked with a variety of sensors and output actuators to help control the emissions output.

The fuel system is interrelated with the emissions control systems on all vehicles produced for sale in the United States. Components of the emissions control systems are described in Chapter 6.

2 Fuel pump/fuel pressure - check

Warning: *Gasoline is extremely flammable, so take extra precautions when you work on any part of the fuel system. Don't smoke or allow open flames or bare light bulbs near the work area, and don't work in a garage where a gas-type appliance (such as a water heater or a clothes dryer) is present. Since gasoline is carcinogenic, wear latex gloves when there's a possibility of being exposed to fuel, and, if you spill any fuel on your skin, rinse it off immediately with soap and water. Mop up any spills immediately and do not store fuel-soaked rags where they could ignite. When you perform any kind of work on the fuel sys-*

tem, wear safety glasses and have a Class B type fire extinguisher on hand.

Note: It is a good idea to check the fuel pump and lines for any obvious damage or fuel leakage. Also check all hoses from the tank to the pump, particularly the suction hoses at the fuel tank and pump which, if they have cracked, may not allow fuel to the fuel pump. It is also possible for the fuel pump diaphragm to rupture internally and leak fuel into the crankcase. If you have excess fuel consumption or fuel smell, and you can't find any external leaks, check the condition of the engine oil for any signs of fuel mixing with the engine oil; the oil level will usually be abnormally high and the oil will be thinned out and have a fuel smell.

1   Disconnect the fuel line from the carburetor and install a T-fitting. Connect a fuel pressure gauge to the T-fitting with a section of fuel hose that is no longer than six inches.

2   Start the engine and allow it to idle. The pressure on the gauge should be 2 to 8 psi. It should remain constant and return to zero slowly when the engine is shut off. **Note:** If the engine will not start, crank the engine until you get a reading on the gauge.

3   An instant pressure drop indicates a faulty outlet valve and the fuel pump must be replaced.

4   If the pressure is too high, check the air vent in the carburetor float bowl to see if it is plugged before replacing the pump.

5   If the pressure is too low, be sure the fuel hoses and lines are in good shape and not plugged, then replace the pump.

---

3   Fuel lines and fittings - repair and replacement

Warning: Gasoline is extremely flammable, so take extra precautions when you work on any part of the fuel system. See the **Warning** in Section 2.

Inspection

Refer to illustrations 3.2a and 3.2b

1   Once in a while, you will have to raise the vehicle to service or replace some component (an exhaust pipe hanger, for example). Whenever you work under the vehicle, always inspect the fuel lines and fittings for possible damage or deterioration.

2   Check all hoses and pipes for cracks, kinks, deformation or obstructions **(see illustrations)**.

3   Make sure all hose and pipe clips attach their associated hoses or pipes securely to the underside of the vehicle.

4   Verify all hose clamps attaching rubber hoses to metal fuel lines or pipes are snug enough to assure a tight fit between the hoses and the metal pipes.

Replacement

5   If you must replace any damaged sections, use hoses approved for use in fuel systems or pipes made from steel only (it's best

3.2a Check carefully for damaged fuel lines at the carburetor inlet

to use an original-type pipe from a dealer that's already flared and pre-bent). Do not install substitutes constructed from inferior or inappropriate material, as this could result in a fuel leak and a fire.

6   Always, before detaching or disassembling any part of the fuel line system, note the routing of all hoses and pipes and the orientation of all clamps and clips to ensure that replacement sections are installed in exactly the same manner.

7   Before detaching any part of the fuel system, be sure to relieve the fuel tank pressure by removing the fuel filler cap.

8   Always use new hose clamps if they have lost their tension after loosening or removing them. It's a good idea to replace spring-type clamps with screw-type clamps.

9   While you're under the vehicle, it's a good idea to check the following related components:

a) Check the condition of the fuel filter - make sure that it's not clogged or damaged (see Chapter 1).

b) Inspect the evaporative emission control (EVAP) system. Verify that all hoses are attached and in good condition (see Chapter 6).

---

4   Fuel tank - removal and installation

Refer to illustrations 4.5, 4.6, 4.9a, 4.9b, 4.11a, 4.11b, 4.11c and 4.12

Warning: Gasoline is extremely flammable, so take extra precautions when you work on any part of the fuel system. See the **Warning** in Section 2.

Note: The following procedure is much easier to perform if the fuel tank is empty. Some tanks have a drain plug for this purpose. If the tank does not have a drain plug, drain the fuel into an approved fuel container using a commercially available siphoning kit (NEVER start a siphoning action by mouth) or wait until the fuel tank is nearly empty, if possible.

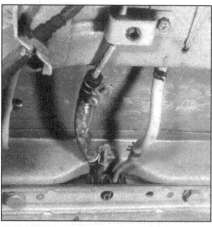

3.2b Carefully inspect the fuel lines that extend from the fuel pump to the fuel tank

1   Remove the fuel tank filler cap to relieve fuel tank pressure.

2   Detach the cable from the negative terminal of the battery.

3   If the tank still has fuel in it, you can drain it at the fuel feed line after raising the vehicle. If the tank has a drain plug remove it and allow the fuel to collect in an approved gasoline container.

4   Raise the vehicle and place it securely on jackstands.

5   Disconnect the fuel filler tube, overflow tube and vent hoses from the fuel filler neck **(see illustration)**.

6   Disconnect the fuel lines and the vapor return line **(see illustration)**. **Note:** The fuel feed and return lines and the vapor return line are three different diameters, so reattachment is simplified. If you have any doubts, however, clearly label the three lines and the fittings. Be sure to plug the hoses to prevent leakage and contamination of the fuel system.

7   If there is still fuel in the tank, siphon it out from the fuel feed port into an approved fuel container. Remember - NEVER start the

4.5 Remove the clamps from the fuel filler tube, overflow tube and vent hoses, then separate the tubes and hoses at the filler neck

4.6 Disconnect the fuel lines and vapor line clamps and detach the hoses from the metal lines

4.9a Remove the fuel tank brackets (arrows)

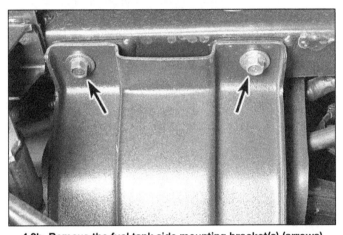

4.9b Remove the fuel tank side mounting bracket(s) (arrows)
(1997 Rodeo/Passport shown)

4.11a Remove the fuel tank mounting nuts (arrows)

siphoning action by mouth! Use a siphoning kit, which can be purchased at most auto parts stores.

8   Support the fuel tank with a floor jack. Position a wood block between the jack head and the fuel tank to protect the tank.

9   If equipped, remove the bolts from the fuel tank brackets located on the side of the fuel tank **(see illustrations)**.

10   Disconnect the electrical connectors for the fuel pump and fuel level sending unit located under the rear seat (see Chapter 4B for additional details on fuel injected models).

11   Remove the bolts that retain the fuel tank to the chassis **(see illustrations)**.

12   Remove the tank from the vehicle. **Note:** *On Rodeo/Passport models, allow the left side of the fuel tank to drop slightly then slide*

the tank sideways to allow the right side to separate from the bracket assembly on the chassis. Carefully drop the tank and move the electrical harness and hoses around any obstructions while lowering the fuel tank **(see illustration)**. **Note:** *On early Amigo models, the fuel tank brackets are arranged slightly different.*

13   Installation is the reverse of removal.

4.11b Some of the fuel tank mounting nuts must be accessed from the top on some models

4.11c Location of the fuel tank mounting fasteners (arrows)

4.12 On Rodeo/Passport models, slide the filler tube through the space between the frame and chassis when lowering the fuel tank

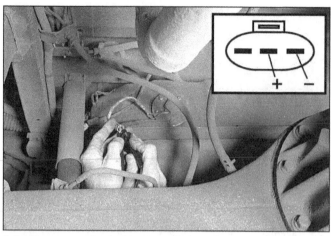

7.4a Disconnect the fuel level sending unit harness and check the resistance (early systems). It should read approximately 120 ohms EMPTY and to 17 ohms FULL

5   Fuel tank cleaning and repair - general information

1   All repairs to the fuel tank or filler neck should be carried out by a professional who has experience in this critical and potentially dangerous work. Even after cleaning and flushing of the fuel system, explosive fumes can remain and ignite during repair of the tank.

2   If the fuel tank is removed from the vehicle, it should not be placed in an area where sparks or open flames could ignite the fumes coming out of the tank. Be especially careful inside garages where a gas-type appliance is located, because it could cause an explosion.

6   Fuel pump - removal and installation

Warning: *Gasoline is extremely flammable, so take extra precautions when you work on*

any part of the fuel system. See the **Warning** *in Section 2.*

1   The fuel pump is located under the intake manifold, near the bottom of the engine block next to the engine mounts.

2   Place rags under the fuel pump to catch any gasoline which may be spilled during removal.

3   Carefully unscrew the fuel line fittings and detach the lines from the pump. A flare-nut wrench along with a backup wrench should be used on the pressure-side fitting to prevent damage to the line fittings.

4   Unbolt and remove the fuel pump.

5   Before installation, coat both sides of the gasket surface with gasket sealant, position the gasket and fuel pump against the block and install the bolts, tightening them securely.

6   Attach the lines to the pump and tighten the pressure fitting securely (use a flare-nut wrench, if one is available, to prevent damage to the fittings).

7   Run the engine and check for leaks.

7   Fuel level sending unit - check and replacement

Warning: *Gasoline is extremely flammable, so take extra precautions when you work on any part of the fuel system. See the* **Warning** *in Section 2.*

Check

Refer to illustrations 7.4a, 7.4b, 7.6, 7.10 and 7.11

1   Before performing any tests on the fuel level sending unit, completely fill the tank with fuel.

2   Raise the vehicle and support it securely with jackstands.

3   Working under the vehicle near the fuel tank, disconnect the fuel level sending unit electrical connector.

4   Position the ohmmeter probes into the electrical connector and check the resistance **(see illustrations)**. Use the 200-ohm or low scale on the ohmmeter.

5   With the fuel tank completely full, the

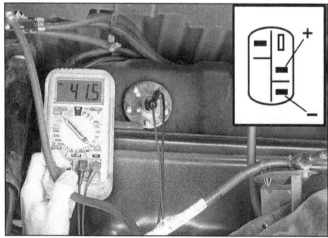

7.4b Checking the sending unit on late systems. It should range from 200 ohms EMPTY to 17 ohms FULL

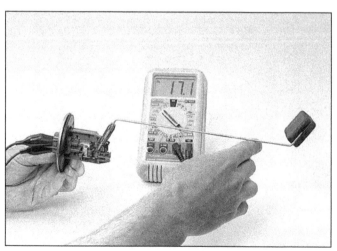

7.6 Raise the float level and observe the resistance

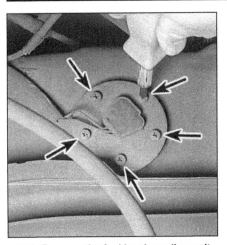

7.10 Remove the fuel level sending unit mounting screws

7.11 Carefully angle the float assembly out of the tank without damaging the assembly

9.1a Remove the accelerator cable and bracket (arrow) from the valve cover

resistance should be approximately 17 ohms. With the tank half full, the resistance should be approximately 60 to 100 ohms. With the fuel tank empty, the resistance of the sending unit should be approximately 120 ohms (early models) or 200 ohms (late models).

6   If the readings are incorrect, replace the sending unit. **Note:** *A more accurate check of the sending unit can be made by removing it from the fuel tank and checking its resistance while manually operating the float arm* **(see illustration)**.

7   If the readings are incorrect, replace the sending unit.

Replacement

8   Raise the vehicle and support it securely with jackstands.

9   Working under the vehicle near the fuel tank, disconnect the fuel level sending unit electrical connector.

10   Remove the mounting screws **(see illustration)**.

11   Lift the sending unit from the tank **(see illustration)**. Carefully angle the sending unit out of the opening without damaging the fuel level float located at the bottom of the assembly.

12   Installation is the reverse of removal. Be sure to use a new gasket.

8   Air cleaner assembly - removal and installation

1   Remove the air filter element from the air filter housing (see Chapter 1).

2   Disconnect the PCV hose, the TCA vacuum hose, electrical connectors and mounting bolts that are attached to the housing and mark them with paint or marked pieces of tape for reassembly purposes.

3   Lift the air filter housing from the engine compartment.

4   Installation is the reverse of removal.

9   Accelerator cable - removal, installation and adjustment

Removal

Refer to illustrations 9.1a, 9.1b and 9.2

1   Remove the accelerator cable mounting

bracket from the valve cover and detach the cable from any other brackets **(see illustrations)**.

2   Rotate the throttle valve and remove the cable end from the slot **(see illustration)**.

3   Detach the screws and the clips retaining the lower instrument trim panel on the driver's side and remove the trim piece (see Chapter 11).

4   Pull the cable end out and then up from the accelerator pedal recess.

5   To disconnect the cable at the firewall, push the cable assembly through the firewall from inside the passenger compartment.

Installation

6   Installation is the reverse of removal. **Note:** *To prevent possible interference, flexible components (hoses, wires, etc.) must not be routed within two inches of moving parts, unless routing is controlled.*

7   Operate the accelerator pedal and check for any binding condition by completely opening and closing the throttle.

8   If necessary, at the engine compartment side of the firewall, apply sealant around the accelerator cable to prevent water from entering the passenger compartment.

9.1b Loosen the large nut and slide it away from the bracket. Rotate the throttle and pass the cable through the recess in the bracket

9.2 Rotate the throttle and pass the cable through the slot

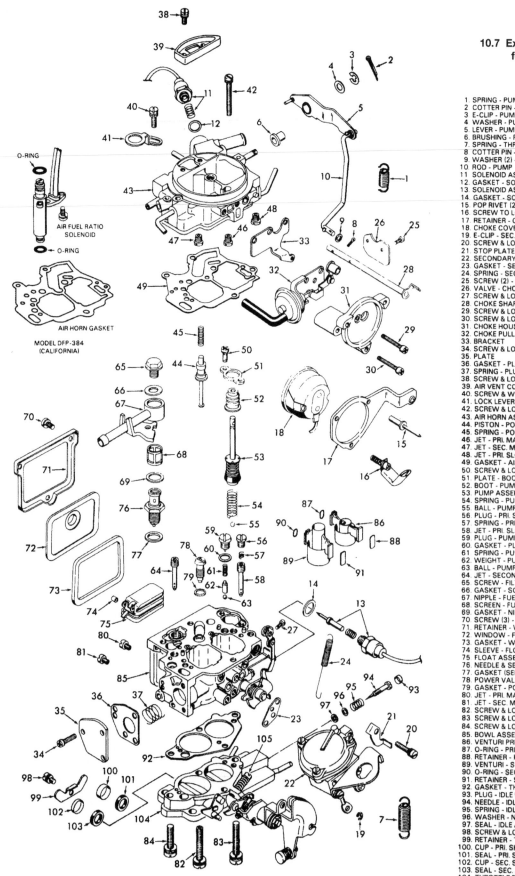

10.7 Exploded view of the typical feedback carburetor

1. SPRING - PUMP LEVER RETURN
2. COTTER PIN - PUMP LEVER
3. E-CLIP - PUMP LEVER
4. WASHER - PUMP LEVER
5. LEVER - PUMP
6. BRUSHING - PUMP LEVER
7. SPRING - THROTTLE RETURN
8. COTTER PIN - PUMP ROD (LOWER)
9. WASHER (2) - PUMP ROD
10. ROD - PUMP
11. SOLENOID ASSY. - BOWL VENT (SINGLE LEAD)
12. GASKET - SOLENOID
13. SOLENOID ASSY. - IDLE SHUT OFF (TWO LEAD)
14. GASKET - SOLENOID
15. POP RIVET (2) - CHOKE RETAINER
16. SCREW TO LOCKWASHER - CHOKE RETAINER
17. RETAINER - CHOKE COVER
18. CHOKE COVER ASSY. - (ELECTRIC)
19. E-CLIP - SEC. DIAPHRAGM STEM
20. SCREW & LOCKWASHER (3) - SEC. DIAPHRAGM ASSY.
21. STOP PLATE - FAST IDLE CAM
22. SECONDARY DIAPHRAGM ASSY.
23. GASKET - SEC. DIAPHRAGM ASSY
24. SPRING - SEC. THROTTLE RETURN
25. SCREW (2) - CHOKE VALVE
26. VALVE - CHOKE
27. SCREW & LOCKWASHER - FAST IDLE LINK LEVER
28. CHOKE SHAFT
29. SCREW & LOCKWASHER - CHOKE HOUSING
30. SCREW & LOCKWASHER - CHOKE HOUSING
31. CHOKE HOUSING
32. CHOKE PULL-OFF & BRACKET ASSY.
33. BRACKET
34. SCREW & LOCKWASHER (3) - PLATE
35. PLATE
36. GASKET - PLATE
37. SPRING - PLUG
38. SCREW & LOCKWASHER (2) - VENT COVER
39. AIR VENT COVER
40. SCREW & WASHER - LOCK LEVER
41. LOCK LEVER - FILTER SET SCREW
42. SCREW & LOCKWASHER (4) - AIR HORN (DIFF. LENGTHS)
43. AIR HORN ASSY.
44. PISTON - POWER VALVE
45. SPRING - POWER PISTON
46. JET - PRI. MAIN AIR BLEED
47. JET - SEC. MAIN AIR BLEED
48. JET - PRI. SLOW AIR BLEED
49. GASKET - AIR HORN
50. SCREW & LOCKWASHER (2) - PUMP BOOT PLATE
51. PLATE - BOOT
52. BOOT - PUMP
53. PUMP ASSEMBLY
54. SPRING - PUMP RETURN
55. BALL - PUMP INTAKE CHECK
56. PLUG - PRI. SLOW JET
57. SPRING - PRI. SLOW JET
58. JET - PRI. SLOW
59. PLUG - PUMP CHECK
60. GASKET - PLUG
61. SPRING - PUMP BALL WEIGHT
62. WEIGHT - PUMP BALL
63. BALL - PUMP DISC. CHECK
64. JET - SECONDARY SLOW
65. SCREW - FILTER SET
66. GASKET - SCREW
67. NIPPLE - FUEL
68. SCREEN - FUEL FILTER
69. GASKET - NIPPLE
70. SCREW (3) - RETAINER
71. RETAINER - WINDOW
72. WINDOW - FUEL BOWL
73. GASKET - WINDOW
74. SLEEVE - FLOAT RETAINER
75. FLOAT ASSEMBLY
76. NEEDLE & SEAT ASSEMBLY
77. GASKET (SERVICE) - NEEDLE SEAT
78. POWER VALVE ASSEMBLY
79. GASKET - POWER VALVE
80. JET - PRI. MAIN
81. JET - SEC. MAIN
82. SCREW & LOCKWASHER (1) - THROTTLE BODY (HOLLOW)
83. SCREW & LOCKWASHER (1) - THROTTLE BODY (LONG)
84. SCREW & LOCKWASHER (2) - THROTTLE BODY
85. BOWL ASSEMBLY
86. VENTURI PRIMARY
87. O-RING - PRI. VENTURI SEAL
88. RETAINER - PRI. VENTURI
89. VENTURI - SECONDARY
90. O-RING - SEC. VENTURI SEAL
91. RETAINER - SEC. VENTURI
92. GASKET - THROTTLE BODY
93. PLUG - IDLE NEEDLE
94. NEEDLE - IDLE ADJUSTING
95. SPRING - IDLE NEEDLE
96. WASHER - NEEDLE SEAL
97. SEAL - IDLE ADJUSTING NEEDLE
98. SCREW & LOCKWASHER - RETAINER
99. RETAINER - THROTTLE SHAFT SEALS
100. CUP - PRI. SHAFT SEAL
101. SEAL - PRI. SHAFT
102. CUP - SEC. SHAFT SEAL
103. SEAL - SEC. SHAFT
104. THROTTLE BODY ASSEMBLY
105. IDLE SPEED SCREW

O-RING

AIR FUEL RATIO
SOLENOID

O-RING

AIR HORN GASKET

MODEL DFP-384
(CALIFORNIA)

Adjustment

9   The cable housing can be adjusted to allow for freeplay in the cable. Loosen the cable bracket screw **(see Illustration 9.1)** and move the cable housing up or down to allow for more or less cable freeplay (slack). The throttle valve should be able to completely close (idle) without hindrance.

---

10   Carburetor - diagnosis and overhaul

Warning: *Gasoline is extremely flammable, so take extra precautions when you work on any part of the fuel system. See the* **Warning** *in Section 2.*

Diagnosis

1   A thorough road test and check of carburetor adjustments should be done before any major carburetor service. Specifications for some adjustments are listed on the *Vehicle Emissions Control Information* (VECI) label found in the engine compartment.

2   Carburetor problems usually show up as flooding, hard starting, stalling, severe backfiring and poor acceleration. A carburetor that's leaking fuel and/or covered with wet looking deposits definitely needs attention.

3   Some performance complaints directed at the carburetor are actually a result of loose, out-of-adjustment or malfunctioning engine or electrical components. Others develop when vacuum hoses leak, are disconnected or are incorrectly routed. The proper approach to analyzing carburetor problems should include the following items:

a)   *Inspect all vacuum hoses and actuators for leaks and correct installation (see Chapters 1 and 6).*

b)   *Tighten the intake manifold and carburetor mounting nuts/bolts evenly and securely.*

c)   *Perform a compression test and vacuum test (see Chapter 2B).*

d)   *Clean or replace the spark plugs as necessary (see Chapter 1).*

e)   *Check the spark plug wires (see Chapter 1).*

f)   *Inspect the ignition primary wires.*

g)   *Check the ignition timing (follow the instructions printed on the* Emissions Control Information *label).*

h)   *Check the fuel pump pressure/volume (see Section 2).*

i)   *Check the heat control valve in the air filter assembly for proper operation (see Chapter 1).*

j)   *Check/replace the air filter element (see Chapter 1).*

k)   *Check the PCV system (see Chapter 6).*

l)   *Check/replace the fuel filter (see Chapter 1). Also, the strainer in the tank could be restricted.*

m)   *Check for a plugged exhaust system.*

n)   *Check EGR valve operation (see Chapter 6).*

o)   *Check the choke - it should be completely open at normal engine operating temperature (see Chapter 1).*

p)   *Check for fuel leaks and kinked or dented fuel lines (see Chapters 1 and 4)*

q)   *Check accelerator pump operation with the engine off (remove the air filter assembly cover and operate the throttle as you look into the carburetor throat - you should see a stream of gasoline enter the carburetor).*

r)   *Check for incorrect fuel or bad gasoline.*

s)   *Check the valve clearances (if applicable) and camshaft lobe lift (see Chapters 1 and 2)*

t)   *Have a dealer service department or repair shop check the electronic engine and carburetor controls.*

4   Diagnosing carburetor problems may require that the engine be started and run with the air filter assembly off. While running the engine without the air filter assembly, backfires are possible. This situation is likely to occur if the carburetor is malfunctioning, but just the removal of the air filter assembly can lean the fuel/air mixture enough to produce an engine backfire. **Warning:** *Do not position any part of your body, especially your face, directly over the carburetor during inspection and servicing procedures. Wear eye protection!*

Overhaul

Refer to illustration 10.7

5   Once it's determined that the carburetor needs an overhaul, several options are available. If you're going to attempt to overhaul the carburetor yourself, first obtain a good-quality carburetor rebuild kit (which will include all necessary gaskets, internal parts, instructions and a parts list). You'll also need some special solvent and a means of blowing out the internal passages of the carburetor with compressed air.

6   An alternative is to obtain a new or rebuilt carburetor. They are readily available from dealers and auto parts stores. Make absolutely sure the exchange carburetor is identical to the original. A tag is usually attached to the top of the carburetor or a number is stamped on the float bowl. It will help determine the exact type of carburetor you have. When obtaining a rebuilt carburetor or a rebuild kit, make sure the kit or carburetor matches your application exactly. Seemingly insignificant differences can make a large difference in engine performance.

7   If you choose to overhaul your own carburetor, allow enough time to disassemble it carefully, soak the necessary parts in the cleaning solvent (usually for at least one-half day or according to the instructions listed on the carburetor cleaner) and reassemble it, which will usually take much longer than disassembly **(see illustration)**. When disassembling the carburetor, match each part with the illustration in the carburetor kit and lay the parts out in order on a clean work surface. Overhauls by inexperienced mechanics can

result in an engine which runs poorly or not at all. To avoid this, use care and patience when disassembling the carburetor so you can reassemble it correctly.

8   Because carburetor designs are constantly modified by the manufacturer in order to meet increasingly more stringent emissions regulations, it isn't feasible to include a step-by-step overhaul of each type. You'll receive a detailed, well illustrated set of instructions with the carburetor overhaul kit.

---

11   Electric choke heater - testing

Caution: *If there is any loss of electrical current to the choke heater, operation of any type, including idling, should be avoided. Loss of power to the choke will cause the choke to remain partly closed during engine operation. A very rich air-to-fuel mixture will be created and result in abnormally high exhaust system temperatures, which may cause damage to the catalytic converter or other underbody parts of the vehicle.*

1   With the engine cold, connect a jumper wire from the choke heater wire to the positive battery terminal. The choke heater housing should begin becoming hot and the choke plate in the carburetor should slowly open after some time. Be sure it opens completely in about five minutes or replacement is necessary. **Note:** *As the heater is heating the choke coil, occasionally tap the accelerator to allow the choke to open.* **Note:** *The electric assist choke system consists of a thermostatic spring and cover (choke assembly). As the heater warms, it causes the spring to pull the choke plate open within 1 to 1-1/2 minutes.*

2   Working with the temperature of the choke assembly at 68 degrees F, install the positive probe (+) of an ohmmeter to the choke positive terminal (+) and the negative probe (-) of the ohmmeter to the choke body. The resistance should be 1.1 to 3.1 ohms. If the resistance is incorrect, replace the choke assembly.

3   Start the vehicle and check for battery voltage to the coil as the engine warms up. If there is no battery voltage available, check the harness for a short or blown fuse. Refer to the wiring diagrams at the end of Chapter 12 for additional information.

---

12   Idle shut-off solenoid - check and replacement

Check

1   The idle shut-off solenoid prevents excess engine idle and surges after the ignition switch has been turned off. The solenoid extends a plunger into the idle circuit within the carburetor when the ignition is turned OFF, immediately shutting off fuel to the idle circuit.

2   Turn the ignition key ON (engine not run-

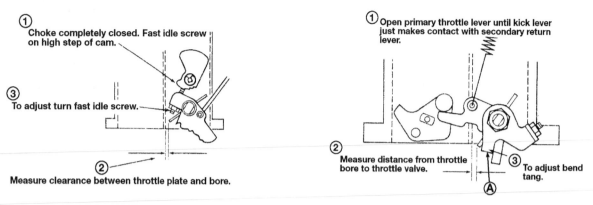

13.10 Checking the fast idle adjustment   **13.11 Checking the valve plate interlock adjustment**

ning) and using a voltmeter, check for battery voltage at the idle shut-off solenoid. There should be approximately 12 volts.

3   Touch the electrical connector to the solenoid and listen for a distinct "click" from the solenoid. If necessary, remove the solenoid from the carburetor and watch the solenoid plunger move as voltage is applied. If there is no movement from the plunger, replace it with a new part (be sure to use a new gasket). Refer to illustration 10.7, number 13 for the exact location of the idle shut-off solenoid.

Replacement

4   Disconnect the electrical connector from the idle shut-off solenoid.
5   Remove the solenoid from the carburetor using an open-end wrench.
6   Installation is the reverse of removal.

13   Carburetor - adjustments

Note: *These carburetor adjustments are strictly in-vehicle adjustments. During overhaul, refer to the instructions included in the overhaul kit for complete procedures and any additional adjustments that are required.*

Choke valve opening

1   The choke valve opening is a critical carburetor adjustment directly involved with cold-running conditions and choke enrichment during cranking. The choke unloader is a mechanical device that partially opens the choke valve at wide open throttle to eliminate choke enrichment during hard acceleration.
2   Engines which have been stalled or flooded by excessive choke enrichment can be cleared by the use of the choke unloader. With the throttle valve at wide open throttle, the choke valve should be slightly open to allow a sufficient amount of intake air into the carburetor venturi.
3   Turn the throttle lever until the primary throttle valve is fully open. If it is not, bend the tongue on the unloader to compensate for the necessary adjustment between the choke

valve and the choke valve chamber wall. Refer to the Specifications listed in this Chapter.

Vacuum break

4   The choke vacuum break is a diaphragm which opens the choke valve plate immediately after cold starting to create a suitable air/fuel mixture during warm-up. With the engine cold, remove the air filter assembly and close the choke plate manually.
5   Fully open the primary throttle valve by rotating the throttle. Measure the clearance between the edge of the valve plate and the wall of the carburetor.
6   To adjust the vacuum break, bend the adjusting lever tang. Refer to the Specifications listed in this Chapter.

Fast idle speed

Refer to illustration 13.10
7   Make sure the choke valve is fully closed.
8   Position the fast idle speed screw onto the first step on the fast idle cam.
9   Measure the clearance between the primary throttle valve and the throttle valve chamber wall. Refer to the Specifications listed in this Chapter.
10   Turn the fast idle screw IN or OUT to obtain the correct setting **(see illustration)**.

Valve plate interlock

Refer to illustration 13.11
11   Manually turn the throttle valve until the throttle arm contacts the lock lever at point A **(see illustration)**.
12   Measure the clearance between the throttle valve and throttle bore. If not as specified, bend the throttle arm tongue to adjust.

Float height

13   Start the engine and allow it to idle. Carefully observe through the inspection glass. The float chamber fuel level should be even with the level mark scribed onto the glass. The indicator mark is round with a

smaller circle in the middle.
14   If the fuel level is higher or lower than the prescribed mark, adjust the float height by bending the float tang.

14   Carburetor - removal and installation

Warning: *Gasoline is extremely flammable, so take extra precautions when you work on any part of the fuel system. See the* **Warning** *in Section 2.*

Removal

1   Remove the fuel filler cap to relieve fuel tank pressure.
2   Remove the air filter assembly from the carburetor. Be sure to label all vacuum hoses attached to the air filter housing.
3   Disconnect the accelerator cable from the throttle valve (see Section 9).
4   If the vehicle is equipped with an automatic transmission, disconnect the kickdown cable from the throttle valve.
5   Clearly label all vacuum hoses and fittings, then disconnect the hoses.
6   Disconnect the fuel line from the carburetor.
7   Label the wires and terminals, then unplug all the electrical connectors.
8   Remove the mounting fasteners and detach the carburetor from the intake manifold. Remove the carburetor mounting gasket. Stuff a rag into the intake manifold openings.

Installation

9   Use a gasket scraper to remove all traces of gasket material and sealant from the intake manifold (and the carburetor, if it's being reinstalled), then remove the shop rag from the manifold openings. Clean the mating surfaces with lacquer thinner or acetone.
10   Place a new gasket on the intake manifold.
11   Position the carburetor on the gasket and install the mounting fasteners.
12   To prevent carburetor distortion or dam-

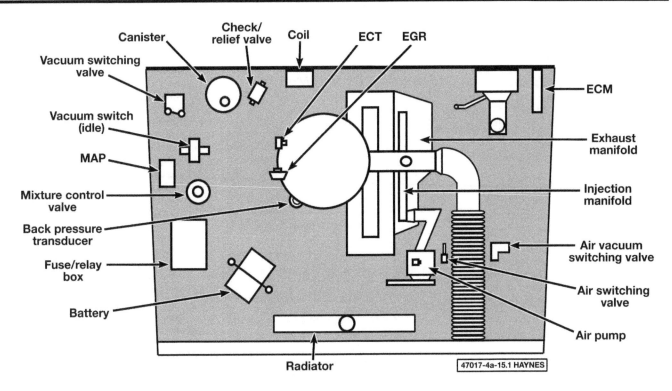

15.1 Typical feedback carburetor system component locations

age, tighten the fasteners to approximately 16 ft-lbs in a criss-cross pattern, 1/4-turn at a time.

13 The remaining installation steps are the reverse of removal.

14 Check and, if necessary, adjust the idle speed (see Chapter 1).

15 If the vehicle is equipped with an automatic transmission, refer to Chapter 7B for the kickdown cable adjustment procedure.

16 Start the engine and check carefully for fuel leaks.

15 Electronic feedback carburetor systems general information and sensors/output actuators - check and replacement

Refer to illustration 15.1

Note 1: *This section covers only a select few of the output actuators incorporated onto the carburetor. Refer to Chapter 6 for additional information on the information sensors such as coolant temperature sensor, oxygen sensor, etc.* **Note 2:** *Refer to Chapter 6 for additional information and accessing trouble codes for the feedback carburetor systems.*

1 The electronically-controlled carburetor emission system (most commonly called the electronic feedback carburetor system) relies on an electronic signal, which is generated by an exhaust gas oxygen sensor, to control a variety of devices and keep emissions within limits **(see illustration)**. The system works in conjunction with a three-way catalyst to con-

trol the levels of carbon monoxide, hydrocarbons and oxides of nitrogen. The feedback carburetor system also works in conjunction with the computer. The two systems share certain sensors and output actuators; therefore, diagnosing the feedback carburetor system will require a thorough check of all the feedback carburetor components (refer to Chapter 6 for additional information).

2 The system operates in two modes: open loop and closed loop. When the engine is cold, the air/fuel mixture is controlled by the computer in accordance with a program designed at the time of production. The air/fuel mixture during this time will be richer to allow for proper engine warm-up. When the engine is at operating temperature, the system operates in closed loop and the air/fuel mixture is varied depending on the information supplied by the exhaust gas oxygen sensor.

3 Here is a list of the various sensors and output actuators involved with these feedback carburetor systems:

> *Coolant temperature sensor*
> *Oxygen sensor*
> *Air/fuel ratio solenoid*
> *ECM*
> *Vacuum switching valve (MAP, ignition, canister purge)*
> *Vacuum switching valve (AIR)*
> *Air Injection Reactor (AIR) system)*
> *Backpressure transducer (see Chapter 6)*
> *Check and relief valve*
> *EGR valve (see Chapter 6)*

Air/fuel ratio solenoid

Check

4 The function of the air/fuel ratio solenoid is to provide limited regulation of the air/fuel ratio of a feedback carburetor in response to the electronic signals sent by the computer. This is accomplished by metering the main fuel jets in the carburetor with the use of a fuel metering pintle. The pintle extends into the fuel passage and shuts the fuel off (lean) and on (rich), allowing the fuel/air mixture ratio to change. By controlling the duration of this voltage signal, the ratio of power ON-time versus the power OFF-time is called the duty cycle.

5 Maintain an engine speed of 1,500 rpm. Disconnect the air/fuel ratio solenoid connector from the carburetor. Average engine speed should increase a minimum of 50 rpm. Reconnect the air/fuel ratio solenoid connector. The engine speed should slowly return to 1,500 rpm. If rpm does not change as specified, there's a problem in the feedback carburetor system, although the oxygen sensor is usually the cause. First check for battery voltage to the air/fuel ratio solenoid electrical connector, then check the oxygen sensor. If the oxygen sensor is OK and all connections are OK, the air/fuel ratio solenoid is bad.

6 Another quick check on the air/fuel ratio solenoid is to measure the resistance on the electrical connector terminals. The resistance should be approximately 34 to 47 ohms.

Replacement

7 Disconnect the cable from the negative

battery terminal.

8   Disconnect the electrical connector from the air/fuel ratio solenoid.

9   Remove the carburetor air cleaner housing from the carburetor.

10   Remove the mounting screw for the solenoid.

11   Install a new air/fuel ratio solenoid grommet onto the solenoid and apply silicon grease to the O-ring.

12   Tighten the mounting screw.

13   The remainder of installation is the reverse of removal.

Manifold Absolute Pressure (MAP) sensor

General description

14   The MAP sensor monitors the intake manifold pressure changes resulting from changes in engine load and speed. The vacuum sensor operates the same as the MAP sensor in later models. Refer to Chapter 6 for the diagnostic procedures.

Coolant temperature sensor (ECT) sensor

General description

15   The coolant temperature sensor monitors the cooling system temperature as the engine operates from cold to fully warm temperatures. Refer to Chapter 6 for the diagnostic procedures.

Oxygen sensor

General description

16   The O2 sensor monitors exhaust gasses released from the combustion chamber. These readings are sent to the ECM in the form of fluctuating voltage signals. The ECM uses this information to control the air/fuel solenoid (RICH/LEAN limits). Refer to Chapter 6 for the diagnostic procedures.

Electronic Control Module (ECM)

General description

17   The ECM is a microprocessor for the sensors and output actuators used in the feedback control system. Any problems with the ECM should be handled by a dealer service department.

---

16   Exhaust system - servicing and general information

Warning: *Inspection and repair of exhaust system components should be done only after enough time has elapsed after driving the vehicle to allow the system components to cool completely. Also, when working under the vehicle, make sure it is securely supported on jackstands.*

1   The exhaust system consists of the exhaust manifold(s), the catalytic converter(s), the muffler, the tailpipe and all connecting pipes, brackets, hangers and clamps. The exhaust system is attached to the body with mounting brackets and rubber hangers. If any of the parts are improperly installed, excessive noise and vibration will be transmitted to the body.

2   Conduct regular inspections of the exhaust system to keep it safe and quiet. Look for any damaged or bent parts, open seams, holes, loose connections, excessive corrosion or other defects which could allow exhaust fumes to enter the vehicle. Deteriorated exhaust system components should not be repaired; they should be replaced with new parts.

3   If the exhaust system components are extremely corroded or rusted together, welding equipment will probably be required to remove them. The convenient way to accomplish this is to have a muffler repair shop remove the corroded sections with a cutting torch. If, however, you want to save money by doing it yourself (and you don't have a welding outfit with a cutting torch), simply cut off the old components with a hacksaw. If you have compressed air, special pneumatic cutting chisels can also be used. If you do decide to tackle the job at home, be sure to wear safety goggles to protect your eyes from metal chips and work gloves to protect your hands.

4   Here are some simple guidelines to follow when repairing the exhaust system:

a) *Work from the back to the front when removing exhaust system components.*

b) *Apply penetrating oil to the exhaust system component fasteners to make them easier to remove.*

c) *Use new gaskets, hangers and clamps when installing exhaust system components.*

d) *Apply anti-seize compound to the threads of all exhaust system fasteners during reassembly.*

e) *Be sure to allow sufficient clearance between newly installed parts and all points on the underbody to avoid overheating the floor pan and possibly damaging the interior carpet and insulation. Pay particularly close attention to the catalytic converter and heat shield.*

Chapter 4 Part B
Fuel and exhaust systems - fuel-injected engines

Contents

Specifications

Fuel pressure

Fuel system pressure (at idle)

Through 1995
  2.6L
    Amigo
      Vacuum hose attached .... 36 psi
      Vacuum hose detached .... 43 psi
    Rodeo
      Vacuum hose attached .... 35 psi
      Vacuum hose detached .... 42 psi
  3.1L .... 9 to 13 psi
  3.2L
    Vacuum hose attached .... 25 to 30 psi
    Vacuum hose detached .... 41 to 46 psi
1996 on
  2.2L .... 41 to 55 psi
  3.2L .... 42 to 55 psi

Injector resistance

TBI system .... 1.16 to 1.36 ohms
MPFI system
    2.2L engine .... 11.8 to 12.6 ohms
    2.6L engine .... 13.8 ohms
    3.2L engine .... 11.8 to 12.6 ohms

Torque specifications

Ft-lbs (unless otherwise indicated)

Throttle body mounting bolts (use cross pattern)
    2.2L and 2.6L engines .... 120 in-lbs
    3.2L engine 17
Fuel pressure regulator mounting screws
    2.2L engine (clamp screw) .... N/A
    2.6L engine .... 60 in-lbs
    3.2L engine .... 26 in-lbs
Air intake plenum (upper intake manifold) mounting bolts
    2.6L engine 20
    3.2L engine 17
Fuel rail mounting bolts/nuts
    2.2L (nuts) .... 75 in-lbs
    2.6L engine (bolts) .... 75 in-lbs
    3.2L engine (bolts) .... N/A
Exhaust pipe-to-exhaust manifold flange bolts 15

1   General information

The fuel system on fuel-injected engines is either Throttle Body Injection (TBI) or Multi-Port Fuel Injection (MPFI). 3.1L engines are equipped with TBI and 2.2L, 2.6L and 3.2L engines are equipped with the Isuzu I-TEC fuel injection system, which is a multi-port system. Both types of fuel injection systems consist of a fuel tank, an electric fuel pump (located in the fuel tank), a fuel pump relay, fuel injectors, an air induction system and a throttle body unit. On TBI systems, the fuel injectors are integral with the throttle body. MPFI systems use separate injectors located in the intake ports. 1996 and later models are equipped with On Board Diagnostic (OBD) II. OBD II requires a special SCAN tool to access trouble codes and engine data. Earlier models use the OBD I system of self diagnosis. Codes on OBD I models can be read by counting the number of flashes on the CHECK ENGINE light on the dash. Refer to Chapter 6 for OBD information and diagnostic codes.

Throttle Body Injection (TBI)

3.1L V6 engines are equipped with the Throttle Body Injection (TBI) system. The throttle body system utilizes two injectors, centrally mounted in a carburetor-like housing. The injector is an electrical solenoid, with fuel delivered to the injector at a constant pressure level. To maintain the fuel pressure at a constant level, fuel is supplied at a greater pressure than required and controlled by a fuel pressure regulator, which returns excess fuel to the fuel tank.

A signal from the ECM opens the solenoid, allowing fuel to spray through the injector into the throttle body. The amount of time the injector is held open by the ECM determines the fuel/air mixture ratio.

Multi-Port Fuel Injection (MPFI) system

Multi-Port Fuel Injection uses timed impulses to inject the fuel directly into the intake port of each cylinder. The injectors are controlled by the Engine Control Module (ECM). The ECM monitors various engine parameters and delivers the exact amount of fuel required into the intake ports. The throttle body serves only to control the amount of air passing into the system. Because each cylinder is equipped with its own injector, much better control of the fuel/air mixture ratio is possible.

Fuel pump and lines

Fuel is circulated from the fuel tank to the fuel injection system, and back to the fuel tank, through a pair of metal lines running along the underside of the vehicle. An electric fuel pump is located inside the fuel tank. A vapor return system routes all vapors back to the fuel tank through a separate return line.

The fuel pump will operate as long as the engine is cranking or running and the ECM is receiving ignition reference pulses from the electronic ignition system. If there are no reference pulses, the fuel pump will shut off after two or three seconds.

Exhaust system

The exhaust system includes an exhaust manifold fitted with an exhaust oxygen sensor, a catalytic converter, an exhaust pipe, and a muffler.

The catalytic converter is an emission control device added to the exhaust system to reduce pollutants. A single-bed converter is used in combination with a three-way (reduction) catalyst. Refer to Chapter 6 for more information regarding the catalytic converter.

2   Fuel pressure relief procedure

Warning: *Gasoline is extremely flammable, so take extra precautions when you work on any part of the fuel system. Don't smoke or allow open flames or bare light bulbs near the work area, and don't work in a garage where a gas-type appliance (such as a water heater or a clothes dryer) is present. Since gasoline is carcinogenic, wear latex gloves when there's a possibility of being exposed to fuel, and, if you spill any fuel on your skin, rinse it off immediately with soap and water. Mop up any spills immediately and do not store fuel-soaked rags where they could ignite. The fuel system is under constant pressure, so, if any fuel lines are to be disconnected, the fuel pressure in the system must be relieved first. When you perform any kind of work on the fuel system, wear safety glasses and have a Class B type fire extinguisher on hand.*
Note: *After the fuel pressure has been relieved. it's a good idea to lay a shop towel over any fuel connection to be disassembled, to absorb the residual fuel that may leak out when servicing the fuel system.*

TBI systems

Note: *Although the Model 220 TBI units have an automatic internal bleed system, follow the procedure described below to relieve the fuel pressure on all TBI engines.*
1   TBI systems have an internal bleed feature that allows the fuel system to automatically depressurize while not running (ignition key OFF). It is a good idea to make sure that any built-up pressure has been relieved from the fuel tank by removing the fuel tank cap. When working on the fuel system, place shop rags or towels under the fuel line to be disconnected to soak up any fuel that may spill out.

2.2 Location of the fuel pump fuse in the fuse and relay control box (2.6L engine shown; on 2.2L four-cylinder and on 3.2L V6 engines, refer to fuse guide on cover for location of fuel pump fuse)

3.3a Remove the fuel pump relay from the fuse and relay control box (2.6L engine shown; on 2.2L four-cylinder and on 3.2L V6 engines, refer to relay guide on cover for location of fuel pump relay)

3.3b Check for battery voltage on the fuel pump relay connector

MFI systems

Refer to illustration 2.2

2   Remove the fuel pump fuse from the fuse panel **(see illustration)**.

3   Start the engine and allow it to run until it stops. This should take only a few seconds. Crank the engine over for about 30 more seconds. Disconnect the cable from the negative terminal of the battery before working on the fuel system.

4   The fuel system pressure is now relieved. When you're finished working on the fuel system, simply install the fuel pump fuse back into the fuse panel and connect the negative cable to the battery.

5   It is a good idea to cover any fuel line that will be disconnected using a shop rag to catch any fuel that might spill out.

3   Fuel pump/fuel pressure - check

Warning: *Gasoline is extremely flammable, so take extra precautions when you work on any part of the fuel system. See the* **Warning**

in Section 2.
Note: *To perform the fuel pressure test, you will need to obtain a fuel pressure gauge and adapter set (fuel line fittings).*

Preliminary inspection

Refer to illustrations 3.3a, 3.3b and 3.4

1   Should the fuel system fail to deliver the proper amount of fuel, or any fuel at all, inspect it as follows. Remove the fuel filler cap. Have an assistant turn the ignition key to the ON position (engine not running) while you listen through the fuel filler opening for the sound of the fuel pump priming itself. You should hear a whirring sound that lasts for a couple of seconds.

2   If you don't hear anything, check the fuel pump fuse (see Chapter 12). If the fuse is blown, replace it and see if it blows again. If it does, trace the fuel pump circuit for a short. If it isn't blown, check the fuel pump relay.

3   Remove the relay and check for battery voltage to the fuel pump relay connector **(see illustrations)**. If there is battery voltage present, check the relay for proper operation

(see Chapter 12).

4   If battery voltage is present and the relay is good, reinstall the relay and check for battery voltage at the fuel pump electrical connector **(see illustration)**. If there is no voltage, check the fuel pump circuit. If there is voltage present and the ground circuit is good, replace the pump (see Section 4).

Operating pressure check

Refer to illustrations 3.6a, 3.6b, 3.6c, 3.9, 3.11a, 3.11b and 3.13
Note: *It is not possible to attach a hand-held vacuum pump to the TBI unit on the 3.1L engines to check the operation of the fuel pressure regulator. The regulator can be checked to an extent by pinching off the fuel return line (see Step 9).*

5   Relieve the fuel system pressure (see Section 2). Detach the cable from the negative battery terminal.

6   Attach a fuel pressure gauge to the fuel rail.

a)   *On 2.6L engines, disconnect the inlet line directly at the air intake plenum* **(see illustration)** *and using a T-fitting, install the fuel pressure gauge between the rubber fuel line and the metal fuel line.*

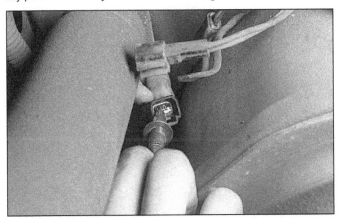

3.4 Disconnect the fuel pump harness connector at the fuel tank and check for battery voltage to the fuel pump. Be sure the vehicle is secured properly and the transmission is in neutral (manual transmission) or park (automatic transmission) while the assistant is cranking the engine over

3.6a Using a T-fitting, install the fuel pressure gauge between the fuel inlet line and the fuel rail (2.6L engine shown)

3.6b Location of the fuel pressure test port on the 3.2L engine

3.6c Location of the fuel pressure test port on the 2.2L engine

3.9 If the fuel pressure is low, squeeze the fuel return line attached to the bottom of the fuel pressure regulator (arrow) - if the fuel pressure increases, the pressure regulator is defective. The tool shown here is a pair of flat-nosed pliers with no sharp edges (it's a good idea to wrap a rag around the hose to prevent damage)

b) On 3.1L engines, install the fuel pressure gauge between the fuel inlet line and the throttle body using a T-fitting. Tighten the hose clamps securely.
c) On 3.2L engines, connect the fuel pressure gauge to the Schrader valve test port fitting on the fuel rail **(see illustration)**.
d) On 2.2L engines, hook up the fuel pressure gauge to the Schrader valve test port fitting on the fuel rail **(see illustration)**.

7   Attach the cable to the negative battery terminal. Start the engine.
8   Check the fuel pressure at idle, comparing your reading with the value listed in this Chapter's Specifications. On 2.2L, 2.6L and 3.2L engines, disconnect the vacuum hose and watch the gauge - the pressure should jump up considerably as soon as the hose is disconnected. If it doesn't, check for a vacuum signal to the fuel pressure regulator (see Step 13).
9   If the fuel pressure is low, pinch the fuel return line shut **(see illustration)** and watch the gauge. If the pressure doesn't rise, the fuel pump is defective or there is a restriction in the fuel feed line (possibly a clogged fuel filter). If the pressure rises sharply, replace

the pressure regulator.
10   If the fuel pressure is too high, turn the engine off. Disconnect the fuel return line and blow through it to check for a blockage. If there is no blockage, replace the fuel pressure regulator.
11   On 2.2L, 2.6L and 3.2L engines, hook up a hand-held vacuum pump to the port on the fuel pressure regulator **(see illustrations)**.
12   Read the fuel pressure gauge with vacuum applied to the pressure regulator and also with no vacuum applied. The fuel pressure should decrease as vacuum increases (and increase as vacuum decreases).
13   Connect a vacuum gauge to the pressure regulator vacuum hose. Start the engine and check for vacuum **(see illustration)**. If there isn't vacuum present, check for a clogged hose or vacuum port or a vacuum leak. If the amount of vacuum is adequate, replace the fuel pressure regulator. **Note:** Some models are equipped with a vacuum switching valve to regulate fuel pressure regulator vacuum. Check for battery voltage to the VSV with the ignition key ON (engine not running). If battery voltage is available, have the VSV checked at a dealer service department or other qualified repair facility.
14   On 2.2L, 2.6L and 3.2L engines, turn the

ignition switch to OFF and recheck the pressure on the gauge. The pressure should drop slightly then hold steady. If the pressure continues to drop to zero psi:

a) The fuel lines may be leaking.
b) The fuel pressure regulator may be allowing the fuel pressure to bleed through to the return line
c) A fuel injector (or injectors) may be leaking.
d) The fuel pump may be defective.

---

4   Fuel pump/fuel level sending unit - replacement

Removal

Refer to illustrations 4.5, 4.7a, 4.7b and 4.7c
Warning: Gasoline is extremely flammable, so take extra precautions when you work on any part of the fuel system. See the **Warning** in Section 2.

3.11a Connect a hand-held vacuum pump to the fuel pressure regulator, apply vacuum and confirm that fuel pressure drops as vacuum is applied

3.11b Release the vacuum - fuel pressure should increase (2.6L engine shown)

3.13 Detach the vacuum line from the fuel pressure regulator and see if vacuum is present when the engine is running

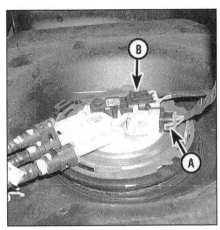

4.5 After lowering the fuel tank, disconnect the electrical connectors from the fuel pump/fuel level sending unit and disconnect the fuel supply and return lines and the EVAP line (1998 and later unit with locking tangs shown; other units similar)

1   Unless the vehicle has been driven far enough to completely empty the tank, it's a good idea to siphon the residual fuel out before removing the fuel pump from the vehicle. **Warning:** *DO NOT start the siphoning action by mouth! Use a siphoning kit (available at most auto parts stores).*

2   Relieve the fuel pressure (see to Section 2).

3   Detach the cable from the negative terminal of the battery.

4   Lower the fuel tank far enough to allow access to the fuel pump/fuel level sending unit electrical connectors and to the fuel supply and return lines (see Chapter 4A).

5   Disconnect the electrical connectors from the fuel pump/fuel level sending unit **(see illustration).**

6   Disconnect the fuel pressure and return lines from the fuel pump **(see illustration 4.5).**

7   On 1989 through 1997 models, remove the fuel pump cover mounting screws **(see illustration).** On some 1998 and later models, the pump is secured to the fuel tank by

locking tangs. To release the fuel pump on these models, turn it counterclockwise with special tool J-39763. If you don't have this special tool, use a *brass* drift and a plastic-tipped hammer to loosen the pump tangs **(see illustration). Warning:** *Do NOT use a steel drift to loosen the pump. Using a steel drift to loosen the pump could produce sparks, which could cause an explosion.* On other 1998 and later models, remove the pump snap-ring **(see illustration).**

Disassembly

1997 and earlier models

Refer to illustrations 4.8, 4.9, 4.10, 4.11a and 4.11b

8   Lift the fuel pump assembly from the fuel tank **(see illustration). Caution:** *The fuel level float and sending unit are delicate. Do not bump them against the tank during removal or the accuracy of the sending unit may be affected.*

9   Disconnect the fuel line from the fuel

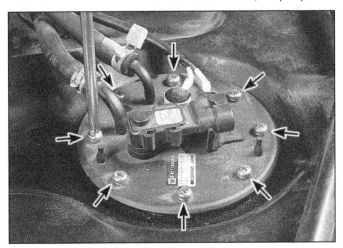

4.7a On 1989 through 1997 models, remove the fuel pump cover screws (arrows)

4.7b On some 1998 and later models, release the fuel pump from the fuel tank by turning the pump's locking tangs counterclockwise a 1/4-turn with either special tool J-39763, with a pair of large water pump pliers (shown) or with a *brass* punch and a plastic-tipped hammer

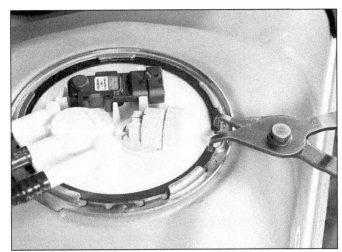

4.7c On other 1998 and later models, remove the fuel pump snap-ring to release the fuel pump from the fuel tank

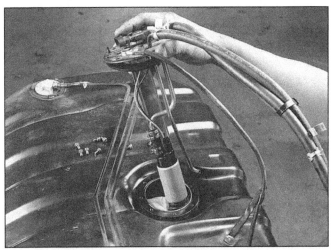

4.8 Lift the fuel pump from the fuel tank (1997 and earlier models)

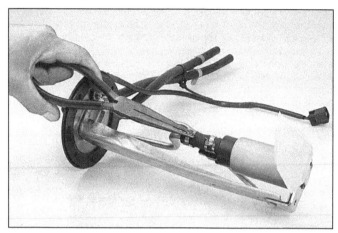

4.9 Using a needle-nose pliers, remove the fuel line clamp from the fuel pump (1997 and earlier models)

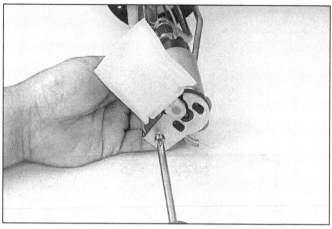

4.10 Remove the fuel pump bracket mounting bolt (1997 and earlier models)

4.11a Lift the clip and remove the fuel filter (sock) from the base of the fuel pump (1997 and earlier models)

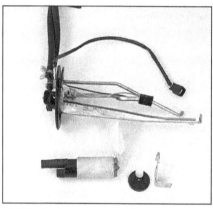

4.11b Fuel pump assembly components (1997 and earlier models)

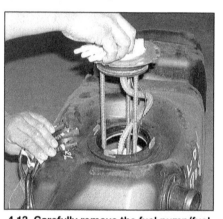

4.13 Carefully remove the fuel pump/fuel level sending unit assembly from the tank and drain the fuel from the reservoir (1998 and later models with locking-tang-type mounting flange)

pump **(see illustration)**.
10   Remove the mounting bolt from the fuel pump bracket **(see illustration)**. Remove the fuel pump from the bracket.
11   Remove the retaining clip **(see illustration)** and separate the filter sock from the fuel pump **(see illustration)**.
12   Reassembly is the reverse of disassembly.

1998 and later models

Locking tang type
Refer to illustrations 4.13, 4.14, 4.15 and 4.16
13   Lift the fuel pump assembly from the fuel tank **(see illustration)**. **Caution:** *The fuel level float and sending unit are delicate. Do not bump them against the tank during removal or the accuracy of the sending unit*

may be affected.
14   Disconnect the fuel level sending unit electrical connector from the module cover **(see illustration)**.
15   Remove the sending unit retaining clip **(see illustration)**.
16   Pinch the tabs together and slide the

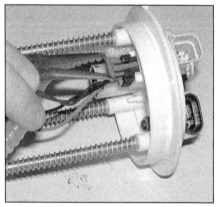

4.14 Disconnect the fuel pump/fuel level sending unit electrical connector from the fuel pump module (1998 and later models with locking-tang-type mounting flange)

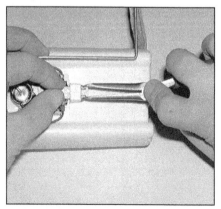

4.15 Remove the sending unit from the retaining clip (1998 and later models with locking-tang-type mounting flange)

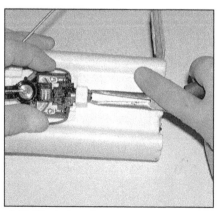

4.16 Pinch the tabs together and remove the fuel level sending unit from the fuel pump module (1998 and later models with locking-tang-type mounting flange)

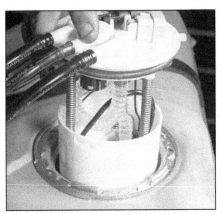

4.18 Lift the fuel pump/fuel level sending unit assembly from the fuel tank (1998 and later models with snap-ring-type mounting flange)

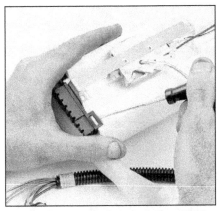

4.20a To remove the protective shield from the foot of the fuel pump/fuel level sending unit assembly, pry open the plastic tab (1998 and later models with snap-ring-type mounting flange)

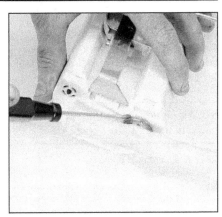

4.20b Carefully pry the fuel strainer from the inlet pipe (1998 and later models with snap-ring-type mounting flange)

fuel level sending unit off the module **(see illustration)**. Note the routing of the wiring for installation.

17 Reassembly is the reverse of disassembly.

Snap-ring type

Refer to illustrations 4.18, 4.20a, 4.20b, 4.22, 4.23, 4.24a and 4.24b

18 Lift the fuel pump/fuel level sending unit assembly from the fuel tank **(see illustration)**. **Caution:** *The fuel level float and sending unit are delicate. Do not bump them against the tank during removal or the accuracy of the sending unit may be affected.*

19 Inspect the condition of the O-ring around the opening of the tank. If it is dried, cracked or deteriorated, replace it.

20 Remove the strainer from the lower end of the fuel pump **(see illustrations)**. If it is dirty, clean it with a suitable solvent and blow it out with compressed air. If it is too dirty to be cleaned, replace it.

21 Remove the fuel pressure sensor mounting bolts and separate the sensor from the top of the fuel pump assembly.

22 Disconnect the fuel line from the pump

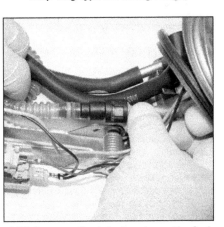

4.22 Squeeze the tabs to release the fuel line from the collar (1998 and later models with snap-ring-type mounting flange)

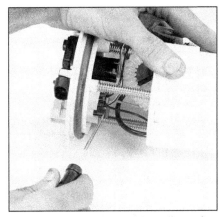

4.23 Detach the fuel level sending unit electrical connector from the fuel pump (1998 and later models with snap-ring-type mounting flange)

(see illustration).

23 Disconnect the sending unit electrical connector from the assembly **(see illustration)**.

24 Carefully separate the sending unit

bracket from the base of the fuel pump assembly **(see illustrations)**.

25 Reassembly is the reverse of disassembly.

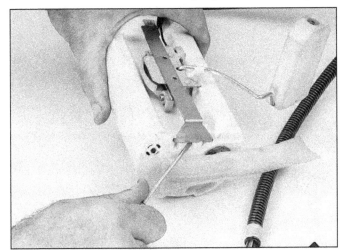

4.24a Pry the bracket from the base of the fuel pump (1998 and later models with snap-ring-type mounting flange)

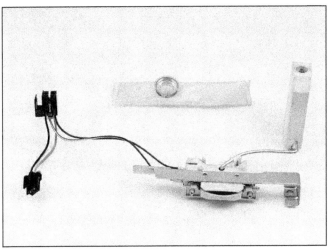

4.24b Typical fuel level sending unit and fuel strainer

Installation

26   Installation is the reverse of removal. If the fuel pump/fuel level sending unit used a gasket between the unit and the fuel tank, install a new gasket.

5   Air cleaner assembly - removal and installation

Refer to illustrations 5.2, 5.3a and 5.3b

1   Remove the air cleaner cover and the air filter element from the air cleaner housing (see Chapter 1).
2   Disconnect the air intake duct from the air cleaner assembly **(see illustration)**.
3   Remove the mounting bolts from the air cleaner housing **(see illustrations)**.
4   Loosen the clamp on the air intake duct and separate the duct from the housing.
5   Installation is the reverse of removal.

6   Electronic fuel injection system - general information

General Information

The fuel system on fuel injected engines is either Throttle Body Injection (TBI) or Multi-Port Fuel Injection (MPFI). 3.1L engines are equipped with TBI and 2.2L, 2.6L and 3.2L engines are equipped with Isuzu MPFI systems (I-TEC). Both types of fuel injection systems consist of a fuel tank, an electric fuel pump (located in the fuel tank), a fuel pump relay, fuel injectors, an air induction system and a throttle body unit. TBI systems incorporate the fuel injectors into the throttle body while MPFI systems use separate injectors positioned over the intake valves, mounted in the intake manifold. 1996 and later models are equipped with On Board Diagnostic (OBD) II. OBD II requires a special SCAN tool to access trouble codes and engine data. Earlier models use the OBD I system of self

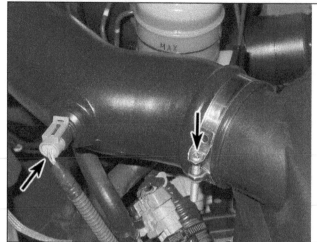

5.2 To disconnect the air intake duct from the air cleaner, loosen this hose clamp; before pushing the intake duct out of the way, you'll also need to unplug the electrical connector from the Intake Air Temperature (IAT) sensor on some models (such as this 2.2L engine)

diagnosis. Codes can be read by counting the number of flashes on the CHECK ENGINE light on the dash. Refer to Chapter 6 for OBD information and diagnostic codes.

Throttle Body Injection (TBI) systems

The main component of the TBI system is the Throttle Body Injection (TBI) unit, which is mounted on the intake manifold just like a carburetor. The TBI unit is made up of two major assemblies: the throttle body and the fuel metering assembly.

The throttle body contains two throttle plates, controlled by the accelerator pedal, similar to a carburetor. Attached to the exterior of the body are the Throttle Position Sensor (TPS), which sends throttle position information to the ECM, and the Idle Air Control (IAC) assembly, which is used by the ECM to maintain a constant idle speed during normal engine operation.

The fuel metering assembly contains the fuel pressure regulator and the two fuel injectors. The regulator dampens the pulsations of the fuel pump and maintains a steady pressure at the injectors. The fuel injectors are controlled by the ECM through electrically

operated solenoids. The amount of fuel injected into the intake manifold is varied by the length of time the injector plungers are held open.

Multi-port Fuel Injection (MPFI)

Multi-port Fuel Injection (MPFI) consists of an air intake plenum, an intake manifold, the throttle body, the injectors, the fuel rail assembly, an electric fuel pump and associated plumbing. The Multi-Port Fuel Injection (MPFI) system is a multi-point, pulse timed, speed density controlled fuel injection system. On the MPFI system, fuel is metered into each intake port in accordance with engine demand through injectors mounted on the intake manifold.

Air is drawn through the air cleaner and throttle body. A Manifold Absolute Pressure (MAP) sensor informs the ECM of temperature and pressure variations.

While the engine is running, the fuel constantly circulates through the fuel rail, which removes vapors and keeps the fuel cool while maintaining sufficient pressure to the injectors under all running conditions.

As with TBI, the operation of the MPFI

5.3a Air cleaner assembly mounting bolts (arrows) (2.6L engine)

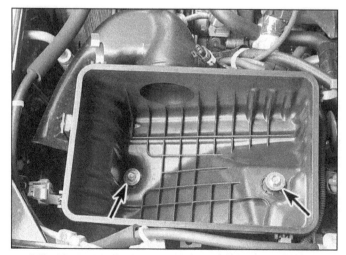

5.3b Air cleaner assembly mounting bolts (arrows) (3.2L V6 engine shown; 2.2L four-cylinder engine identical)

injection system is controlled by the ECM so that it works in conjunction with the rest of the vehicle functions to provide optimum driveability and emissions control. This system incorporates an on-board Engine Control Module (ECM) (called the Powertrain Control Module, or PCM, on later models) computer that accepts inputs from various engine sensors to compute the required fuel flow rate necessary to maintain a prescribed air/fuel ratio throughout the entire engine operational range. The computer then outputs a command to the fuel injectors to meter the correct quantity of fuel. The system automatically senses and compensates for changes in altitude, load and speed.

Because the MPFI system meters fuel and air precisely, it is important to the proper operation of the vehicle that the fuel and air filters be changed at the specified intervals.

7   Electronic fuel injection system - check

Warning: *Gasoline is extremely flammable, so take extra precautions when you work on any part of the fuel system. See the* **Warning** *in Section 2.*
Note: *The following procedure is based on the assumption that the fuel pump is working and the fuel pressure is adequate (see Section 3).*

Preliminary checks

1   Check all electrical connectors that are related to the system. Loose electrical connectors and poor grounds can cause many problems that resemble more serious malfunctions.
2   Check to see that the battery is fully charged, as the control unit and sensors depend on an accurate supply voltage in order to properly meter the fuel.
3   Check the air filter element - a dirty or partially blocked filter will severely impede performance and economy (see Chapter 1).
4   If a blown fuse is found, replace it and see if it blows again. If it does, search for a grounded wire in the harness to the fuel pump (see Chapter 12).

System checks

Refer to illustrations 7.7, 7.8 and 7.9
5   Check the condition of the vacuum hoses connected to the intake manifold.
6   Remove the air intake duct from the throttle body and check for dirt, carbon or other residue build-up in the throttle body, particularly around the throttle plate. If it's dirty, clean it with aerosol carburetor cleaner, a rag and a toothbrush, if necessary.
7   With the engine running, place an automotive stethoscope against each injector, one at a time, and listen for a clicking sound, indicating operation **(see illustration)**. If you don't have a stethoscope, you can place the tip of a long screwdriver against the injector and listen through the handle.

7.7 Use a stethoscope to determine if the injectors are working properly - they should make a steady clicking sound that rises and falls with engine speed changes

8   If an injector isn't functioning (not clicking), purchase a special injector test light (sometimes called a "noid" light) and install it into the injector electrical connector **(see illustration)**. Start the engine and check to see if the noid light flashes. If it does, the injector is receiving proper voltage. If it doesn't flash, further diagnosis should be performed by a dealer service department or other repair shop.
9   With the engine OFF and the fuel injector electrical connectors disconnected, measure the resistance of each injector **(see illustration)**. Compare your measurements with the injector resistance listed in this Chapter's Specifications. If any injector is open or has an abnormally high resistance, replace it with a new one.

8   Model 220 Throttle Body Injection (TBI) - removal, overhaul and installation

Warning: *Gasoline is extremely flammable, so take extra precautions when you work on any part of the fuel system. See the* **Warning** *in Section 2.*
Note: *Because of its relative simplicity, the throttle body assembly does not need to be removed from the intake manifold or disassembled for component replacement. However, for the sake of clarity, the following procedures are shown with the TBI assembly removed from the vehicle.*

1   Relieve system fuel pressure (see Section 2).
2   Detach the cable from the negative terminal of the battery.
3   Remove the air cleaner housing assembly, adapter and gaskets.

Fuel meter cover/fuel pressure regulator assembly

Refer to illustrations 8.6 and 8.7
Note: *The fuel pressure regulator is housed in the fuel meter cover. Whether you are replacing the meter cover or the regulator itself, the entire assembly must be replaced. The regulator must not be removed from the cover.*

7.8 Install a special injector test light ("noid light") into the electrical connector and confirm that it blinks when the engine is cranked

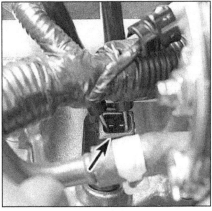

7.9 Measure the resistance of each injector (arrow). Some injectors may be very difficult to reach with ohmmeter probes and it will be necessary to remove the air intake plenum to gain access to the fuel injectors

4   Unplug the electrical connectors to the fuel injectors.
5   Remove the long and short fuel meter cover screws and remove the fuel meter cover.

8.6 Carefully peel away the old fuel meter outlet passage gasket and fuel meter cover gasket with a razor blade

8.7 Never remove the four pressure regulator screws (arrows) from the fuel meter cover

8.14 To remove either injector electrical connector, depress the two tabs on the front and rear of each connector and lift straight up

6   Remove the fuel meter outlet passage gasket, cover gasket and pressure regulator seal. Carefully remove any old gasket material that is stuck with a razor blade **(see illustration)**. **Caution:** *Do not attempt to re-use either of these gaskets.*

7   Inspect the cover for dirt, foreign material and casting warpage. If it is dirty, clean it with a clean shop rag soaked in solvent. Do not immerse the fuel meter cover in cleaning solvent - it could damage the pressure regulator diaphragm and gasket. **Warning:** *Do not remove the four screws* **(see illustration)** *securing the pressure regulator to the fuel meter cover. The regulator contains a large spring under compression which, if accidentally released, could cause injury. Disassembly might also result in a fuel leak between the diaphragm and the regulator housing. The new fuel meter cover assembly will include a new pressure regulator.*

8   Install the new pressure regulator seal, fuel meter outlet passage gasket and cover gasket.

9   Install the fuel meter cover using Loctite 262 or equivalent on the screws. **Note:** *The short screws go next to the injectors.*

10   Attach the electrical connectors to both injectors.

11   Attach the cable to the negative terminal of the battery.

12   With the engine off and the ignition on, check for leaks around the gasket and fuel line couplings.

13   Install the air cleaner, adapter and gaskets.

Fuel injector(s)

Refer to illustrations 8.14, 8.16, 8.21, 8.22, 8.23, 8.24 and 8.25

Note: *When replacing a fuel injector, be sure to use one having the identical part number. Injectors from other models are calibrated with different flow rates but can be interchanged with other Model 220 TBI units. Check with a dealer parts department for correct identification.*

14   To unplug the electrical connectors from the fuel injectors, squeeze the plastic tabs

and pull straight up **(see illustration)**.

15   Remove the fuel meter cover/pressure regulator assembly. **Note:** *Do not remove the fuel meter cover assembly gasket - leave it in place to protect the casting from damage during injector removal.*

16   Use two screwdrivers to pry out the injector(s) **(see illustration)**.

17   Remove the upper (larger) and lower (smaller) O-rings and filter from the injector(s).

18   Remove the steel backup washer from the top of each injector cavity.

19   Inspect the fuel injector filters for evidence of dirt and contamination. If present, check for the presence of dirt in the fuel lines and fuel tank.

20   Be sure to replace the fuel injector with an identical part. Injectors from other models can fit in the Model 220 TBI assembly, but

8.16 To remove an injector, slip the tip of a flat-bladed screwdriver under the lip of the lug on top of the injector and, using another screwdriver as a fulcrum, carefully pry the injector up and out

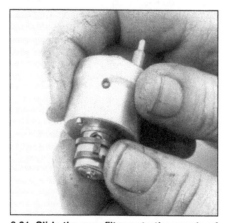

8.21 Slide the new filter onto the nozzle of the fuel injector

8.22 Lubricate the lower O-ring with transmission fluid then place it on the shoulder in the bottom of the injector cavity

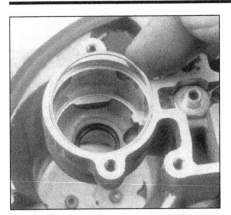

8.23 Place the steel back-up washer on the shoulder near the top of the injector cavity

8.24 Lubricate the upper O-ring with transmission fluid then install it on top of the steel washer

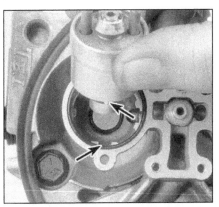

8.25 Make sure that the lug is aligned with the groove in the bottom of the fuel injector cavity (arrows)

are calibrated for different flow rates.

21   Slide the new filter into place on the nozzle of the injector **(see illustration)**.

22   Lubricate the new lower (smaller) O-ring with automatic transmission fluid and place it on the small shoulder at the bottom of the fuel injector cavity in the fuel meter body **(see illustration)**.

23   Install the steel back-up washer in the injector cavity **(see illustration)**.

24   Lubricate the new upper (larger) O-ring with automatic transmission fluid and install it on top of the steel back-up washer **(see illustration)**. **Note:** *The backup washer and the large O-ring must be installed before the injector. If they aren't, improper seating of the large O-ring could cause fuel leakage.*

25   To install an injector, align the raised lug on the injector base with the notch in the fuel meter body cavity **(see illustration)**. Push down on the injector until it is fully seated in the fuel meter body. **Note:** *The electrical terminals should be parallel with the throttle shaft.*

26   Install the fuel meter cover assembly and gasket.

27   Attach the cable to the negative terminal of the battery.

28   With the engine off and the ignition on, check for fuel leaks.

29   Attach the electrical connectors to the fuel injectors.

30   Install the air cleaner housing assembly, adapter and gaskets.

Throttle Position Sensor (TPS)

31   Remove the two TPS attaching screws and retainers and remove the TPS from the throttle body.

32   If you intend to re-use the same TPS, do not attempt to clean it by soaking it in any liquid cleaner or solvent. The TPS is a delicate electrical component and can be damaged by solvents.

33   Install the TPS on the throttle body while lining up the TPS lever with the TPS drive lever.

34   Install the two TPS attaching screws and retainers.

35   Install the air cleaner housing assembly,

adapter and gaskets.

36   Attach the cable to the negative terminal of the battery.

Idle Air Control (IAC) valve

Refer to illustration 8.37

37   Unplug the electrical connector from the IAC valve and remove the IAC valve **(see illustration)**.

38   Remove and discard the old IAC valve gasket. Clean any old gasket material from the surface of the throttle body assembly to insure proper sealing of the new gasket.

39   All pintles in IAC valves on Model 220 TBI units have the same dual taper. However, the pintles on some units have a 12 mm diameter and the pintles on others have a 10 mm diameter. A replacement IAC valve must have the appropriate pintle taper and diameter for proper seating of the valve in the throttle body.

40   Measure the distance between the tip of the pintle and the housing mounting surface. If the distance from the pintle tip to the gasket mounting surface is greater than 1-1/8 inches, it must be reduced to prevent damage to the valve.

41   To adjust the pintle of an IAC valve, grasp the valve and exert firm pressure on the pintle with the thumb. Use a slight side-

to-side movement on the pintle as you press it in with your thumb.

42   Install the IAC valve and tighten it to the specified torque. Attach the electrical connector.

43   Install the air cleaner housing assembly, adapter and gaskets.

44   Attach the cable to the negative terminal of the battery.

45   Start the engine and allow it to reach operating temperature, then turn it off. No adjustment of the IAC valve is required after installation. The IAC valve is reset by the ECM when the engine is turned off.

Fuel meter body assembly

Refer to illustrations 8.50 and 8.52

46   Unplug the electrical connectors from the fuel injectors.

47   Remove the fuel meter cover/pressure regulator assembly, fuel meter cover gasket, fuel meter outlet gasket and pressure regulator seal.

48   Remove the fuel injectors.

49   Unscrew the fuel inlet and return line threaded fittings **(see illustration 8.65)**, detach the lines and remove the O-rings.

50   Remove the fuel inlet and outlet nuts and gaskets from the fuel meter body assembly **(see illustration)**. Note the locations of

8.37 The IAC valve can be removed with an adjustable wrench (shown) or a 1-1/4 inch wrench

8.50 Remove the fuel inlet and outlet nuts from the fuel meter body

the nuts to ensure proper reassembly. The inlet nut has a larger passage than the outlet nut.

51 Remove the gasket from the inner end of each fuel nut.

52 Remove the fuel meter body-to-throttle body attaching screws and remove the fuel meter body from the throttle body (see illustration).

53 Install the new throttle body-to-fuel meter body gasket. Match the cut-out portions in the gasket with the openings in the throttle body.

54 Install the fuel meter body on the throttle body. Coat the fuel meter body-to-throttle body attaching screws with thread locking compound before installing them.

55 Install the fuel inlet and outlet nuts, with new gaskets, in the fuel meter body and tighten the nuts to the specified torque. Install the fuel inlet and return line threaded fittings with new O-rings. Use a backup wrench to prevent the nuts from turning.

56 Install the fuel injectors.

57 Install the fuel meter cover/pressure regulator assembly.

58 Attach the cable to the negative terminal of the battery.

59 Attach the electrical connectors to the fuel injectors.

60 With the engine off and the ignition on, check for leaks around the fuel meter body, the gasket and around the fuel line nuts and threaded fittings.

61 Install the air cleaner housing assembly, adapters and gaskets.

Throttle body assembly

Refer to illustrations 8.65 and 8.66

62 Unplug all electrical connectors - the IAC valve, TPS and fuel injectors. Detach the grommet with the wires from the throttle body.

63 Detach the throttle linkage, return spring(s), transmission control cable (automatics) and, if equipped, cruise control.

64 Clearly label, then detach, all vacuum hoses.

65 Using a backup wrench, detach the inlet and outlet fuel line nuts (see illustration). Remove the fuel line O-rings from the nuts and discard them.

66 Remove the TBI mounting bolts (see illustration) and lift the TBI unit from the intake manifold. Remove and discard the TBI manifold gasket.

67 Place the TBI unit on a holding fixture. **Note:** *If you don't have a holding fixture, and decide to place the TBI directly on a work bench surface, be extremely careful when servicing it. The throttle valve can be easily damaged.*

68 Remove the fuel meter body-to-throttle body attaching screws and separate the fuel meter body from the throttle body.

69 Remove the throttle body-to-fuel meter body gasket and discard it.

70 Remove the TPS.

71 Invert the throttle body on a flat surface for greater stability and remove the IAC valve.

72 Clean the throttle body assembly in a cold immersion cleaner. Clean the metal parts thoroughly and blow dry with compressed air. Be sure that all fuel and air passages are free of dirt or burrs. **Caution:** *Do not place the TPS, IAC valve, pressure regulator diaphragm, fuel injectors or other components containing rubber in the solvent or cleaning bath. If the throttle body requires cleaning, soaking time in the cleaner should be kept to a minimum. Some models have throttle shaft dust seals that could lose their effectiveness by extended soaking.*

73 Inspect the mating surfaces for damage that could affect gasket sealing. Inspect the throttle lever and valve for dirt, binds, nicks and other damage.

74 Invert the throttle body on a flat surface for stability and install the IAC valve and the TPS.

75 Install a new throttle body-to-fuel meter body gasket and place the fuel meter body assembly on the throttle body assembly. Coat the fuel meter body-to-throttle body attaching screws with thread locking compound and tighten them securely.

8.52 Once the fuel inlet and outlet nuts are off, pull the fuel meter body straight up to separate it from the throttle body

76 Install the TBI unit and tighten the mounting bolts to the specified torque. Use a new TBI-to-manifold gasket.

77 Install new O-rings on the fuel line nuts. Install the fuel line and outlet nuts by hand to prevent stripping the threads. Using a backup wrench, tighten the nuts to the specified torque once they have been correctly threaded into the TBI unit.

78 Attach the vacuum hoses, throttle linkage, return spring(s), transmission control cable (automatics) and, if equipped, cruise control cable. Attach the grommet, with wire harness, to the throttle body.

79 Plug in all electrical connectors, making sure that the connectors are fully seated and latched.

80 Check to see if the accelerator pedal is free by depressing the pedal to the floor and releasing it with the engine off.

81 Connect the negative battery cable, and, with the engine off and the ignition on, check for leaks around the fuel line nuts.

82 Check the TPS output (see Chapter 6).

83 Install the air cleaner housing assembly, adapter and gaskets.

8.65 When disconnecting the fuel feed and return lines from the fuel inlet and outlet nuts, be sure to use a backup wrench to prevent damage to the lines

8.66 To remove the Model 220 throttle body from the intake manifold, remove the three bolts (arrows) (third bolt not visible)

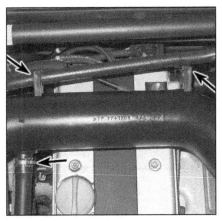

9.2a Loosen the clamp that attaches the PCV fresh air intake hose to the air intake duct, detach the PCV hose from the duct, detach any other hoses connected to the duct (such as the power brake booster hose on this 2.2L engine) . . .

9.2b . . . loosen the hose clamp at the throttle body and remove the intake air duct (2.2L engine shown, others similar)

9.3 Unplug the electrical connectors from the Idle Air Control (IAC) valve and from the Throttle Position (TP) sensor (2.2L engine shown, others similar)

9 Throttle body (MPFI system) - removal and installation

Refer to illustration 9.2a, 9.2b, 9.3, 9.4, 9.5a, 9.5b, 9.6, 9.7a, 9.7b and 9.7c

Warning: *Gasoline is extremely flammable, so take extra precautions when you work on any part of the fuel system. See the* **Warning** *in Section 2.*

Note: *The following procedure assumes that you are removing the throttle body in order to replace the throttle body gasket or the throttle body itself or the intake plenum. If you are simply removing the throttle body in order to get to some other component, it's not necessary to detach the throttle body from the air intake plenum (2.6L and 3.2L engines) or from the intake manifold (2.2L engine). Instead, follow this procedure up to the point at which everything has been removed or disconnected from the throttle body, and then proceed to the appropriate "Intake manifold - removal and installation" Section in Chapter 2.*

1   Detach the cable from the negative battery terminal.

9.4 Loosen the hose clamps and detach the coolant hoses (arrows) from the throttle body; plug or pinch off the hoses to prevent coolant from spilling

2   Remove the air intake duct **(see illustration)**.
3   Unplug the electrical connectors from the Throttle Position (TP) sensor and the Idle Air Control (IAC) valve **(see illustration)**.
4   Detach the coolant hoses **(see illustra-**

9.5a To detach the accelerator cable from the cable bracket, back off the locknut and the adjusting nut and slide the cable out of its slot in the bracket . . .

tion) from the throttle body and plug them or pinch them off to prevent coolant from spilling.
5   Disconnect the accelerator cable from the throttle cam **(see illustrations)**.
6   Carefully label and then remove all vacuum hoses from the throttle body **(see illustration)**.

9.5b . . . then disengage the plug on the end of the cable from its slot in the throttle cam

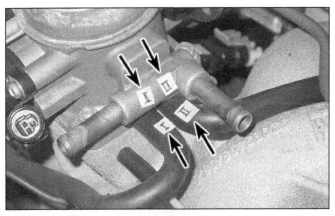

9.6 Before disconnecting any vacuum hoses, label everything to ensure that it will be correctly reconnected

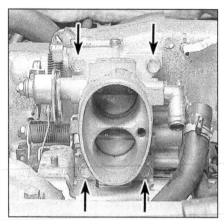

9.7a To detach the throttle body from the air intake plenum (upper intake manifold) on a 2.6L four-cylinder engine, remove these four bolts (arrows)

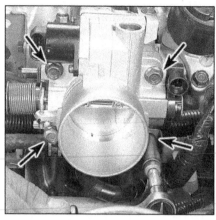

9.7b To detach the throttle body from the air intake plenum (upper intake manifold) on a 3.2L V6 engine, remove these four bolts (arrows)

9.7c To detach the throttle body from the air intake plenum (upper intake manifold) on a 2.2L four-cylinder engine, remove these four nuts (arrows)

7　Unscrew the four throttle body mounting bolts or nuts **(see illustrations)** and remove the throttle body. Remove and discard the gasket between the throttle body and air intake plenum or intake manifold

8　Clean the gasket mating surfaces. If scraping is necessary, be careful not to damage the gasket surfaces or allow material to fall into the air intake plenum or intake manifold.

9　Installation is otherwise the reverse of removal. Be sure to use a new gasket and tighten the throttle body mounting nuts to the torque listed in this Chapter's Specifications.

10　Fuel pressure regulator (MPFI system) - check and replacement

Check

Note: *This procedure assumes the fuel filter is in good condition.*

1　Start the engine and check for leakage around the fuel rail, the fuel lines and the fuel pressure regulator.

2　Check the fuel pressure and verify that the pressure regulator is functioning correctly (see Section 3).

3　If the fuel pressure regulator is faulty or leaking, replace it.

Replacement

Refer to illustrations 10.7 and 10.10

4　Relieve the fuel pressure from the system (see Section 2).

5　Disconnect the cable from the negative terminal of the battery.

6　Wipe off all dirt and oil from the area on the fuel rail around the fuel pressure regulator.

7　Disconnect the vacuum line from the fuel pressure regulator **(see illustration)**.

8　On 2.6L four-cylinder engines and on 3.2L V6 engines, loosen the hose clamp and detach the fuel return line from the bottom of the fuel pressure regulator. On 2.2L engines, remove the banjo bolt and sealing washers and disconnect the fuel return line from the regulator **(see illustration 10.7)**. Discard the

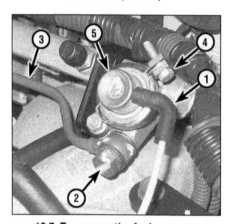

10.7 To remove the fuel pressure regulator from the fuel rail on a 2.2L engine, detach the vacuum line (1) from the regulator, remove the banjo bolt (2) and sealing washers and disconnect the fuel return line (3), loosen the clamp screw (4), remove the clamp and remove the regulator (5)

old sealing washers.

9　On 2.6L four-cylinder engines and on 3.2L V6 engines, remove the two retaining bolts from the fuel pressure regulator and detach the regulator from the fuel rail. On 2.2L four-cylinder engines, loosen the clamp screw **(see illustration 10.7)**, remove the clamp and remove the pressure regulator from the fuel rail.

10　Remove and discard the old pressure regulator O-ring **(see illustration)**. Install a new O-ring on the pressure regulator and lubricate it with a light coat of oil.

11　Place the pressure regulator in position on the fuel rail. On 2.6L four-cylinder engines and on 3.2L V6 engines, install the pressure regulator mounting bolts and tighten them securely. On 2.2L four-cylinder engines, install the pressure regulator clamp and tighten the screw securely.

12　On 2.6L four-cylinder engines and on 3.2L V6 engines, reattach the fuel return line, install the hose clamp and make sure that it's

10.10 Fuel pressure regulator details (2.2L engine)

1　Clamp
2　Fuel pressure regulator
3　O-ring
4　Fuel rail

tight. On 2.2L four-cylinder engines, reconnect the fuel return line with the banjo bolt and with new sealing washers and tighten the banjo bolt securely.

13　Reconnect the vacuum hose to the regulator. Make sure that the hose fits tightly and is in good condition.

14　Installation is otherwise the reverse of removal.

15　When you're done, start up the engine and check the fuel pressure regulator for leaks. (Starting the engine might take a moment because of air trapped in the fuel line).

11　Fuel rail and injectors (MPFI system) - removal and installation

Removal

Refer to illustrations 11.4, 11.7a, 11.7b and 11.7c

1　Relieve the fuel pressure (see Section 2).

2　Detach the cable from the negative ter-

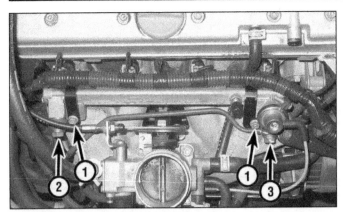

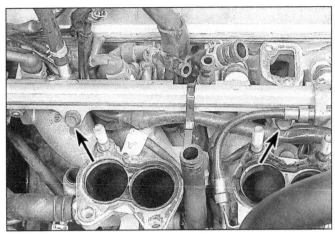

11.4 To detach the fuel injector harness from the fuel rail on a 2.2L four-cylinder engine, remove the two harness retaining bracket bolts (1); then remove the banjo bolts from the fuel supply line (2) and from the fuel return line (3), disconnect the fuel lines from the fuel rail and discard the old sealing washers

11.7a To detach the fuel rail from a 2.6L four-cylinder engine, remove these bolts

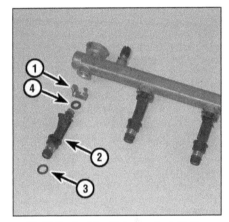

11.7b To detach the fuel rail from a 3.2L V6 engine, remove these bolts

11.7c To detach the fuel rail from a 2.2L four-cylinder engine, remove these nuts

11.15 Typical fuel injector installation details (2.2L engine shown, others similar)

1   Injector retainer clip
2   Fuel injector
3   Lower injector O-ring
4   Upper injector O-ring

minal of the battery.

3   On 2.2L engines, remove the throttle body assembly from the intake manifold (see Section 9). On 2.6L and 3.2L engines, disconnect everything from the throttle body (see Section 9) and then remove the air intake plenum and throttle body as a single assembly (2.6L engine, see Section 5 in Chapter 2A; 3.2L engine, see Section 6 in Chapter 2D).

4   Detach the fuel injector harness from the fuel rail. On some models, the fuel injector harness is attached to the fuel rail by cable ties. On 2.2L models, the harness is attached to the fuel rail by small brackets (see illustration).

5   Disconnect the fuel inlet and return lines from the fuel rail assembly (see illustration 11.4) and then cap the openings to prevent dirt from entering the system. On 3.2L V6 engines, the fuel inlet line is staked to the fuel rail, so it cannot be disconnected. Removing the fuel inlet line will damage the fuel rail and/or the O-ring seal. If you're going to replace the fuel rail on a 3.2L engine, the new fuel rail will include a fuel inlet line.

6   Disconnect the fuel injector electrical connectors and set the fuel injector harness aside. (If the injector connectors are difficult

to disconnect at this stage, leave them connected for now and wait until you have detached the fuel rail and injector assembly from the engine before unplugging them.)

7   Remove the fuel rail retaining nuts or bolts (see illustrations).

8   Carefully lift the fuel rail and fuel injectors from the (lower) intake manifold as a single assembly. Take your time and gently work the injectors free of the manifold. Be extremely careful not to damage the injector electrical terminals and the injector spray nozzles. Unless you're planning to replace the injectors or the injector O-rings, make sure that you don't pull an injector out of the fuel rail. If an injector is separated from the fuel rail, the O-rings and retainer clip must be replaced.

9   If you're replacing the fuel rail, remove the fuel pressure regulator (see Section 10) and the injectors (go to Step 13). If you're going to replace an injector or if you're replacing the injector O-rings, go to Step 13. Otherwise, go to the next step.

Installation

10   If the fuel pressure regulator was removed, install it (see Section 10). If the injectors were removed, install them (see

Step 19). If the holes for the injectors in the intake manifold are sealed with rubber grommets, pry out the old rubber grommets and install new ones.

11   Install the fuel rail and injector assembly onto the engine and install the retaining bolts or nuts. Tighten the bolts or nuts to the torque listed in this Chapter's Specifications.

12   The remainder of installation is the reverse of removal.

Fuel injectors

Removal

Refer to illustration 11.15

13   Relieve the system fuel pressure (see Section 2).

14   Remove the fuel rail (see Steps 1 through 8).

15   Remove the injector retainer clip from each injector you're replacing or servicing (see illustration). Discard the old retainer clip.

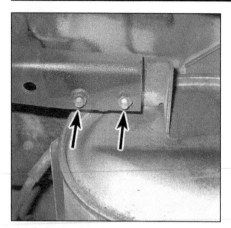

12.1a Typical muffler bracket; make sure that all exhaust brackets are in good shape and all fasteners are securely tightened

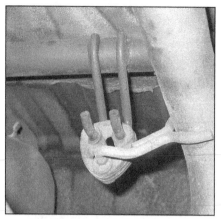

12.1b When inspecting the exhaust system, pay particularly close attention to all rubber exhaust hangers; because they're subjected to constant exhaust heat, rubber hangers dry out, break down, develop cracks and tear apart

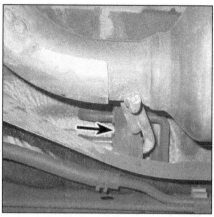

12.1c Some rubber exhaust hangers are difficult to inspect and difficult to replace, but if they break, they can allow the exhaust system to fall down onto heat shields, crossmembers or the axle, which produces an annoying "buzzy" vibration

16   Carefully pry the injector loose from the fuel rail with a flat-bladed screwdriver and then gently rock it from side-to-side and pull it out of the fuel rail.

17   Remove both O-rings from the injector **(see illustration 11.15)** and install new ones.

18   Inspect the injector around the sealing area at each end of the injector. Look for signs of damage or cracking that may cause fuel leaks. Replace the injector if you note any damage.

Installation

19   Lubricate the new O-rings with light grade oil and install them on the injector. **Caution:** *Do not use silicone grease. It will clog the injectors.*

20   Using a light twisting motion, install the injector.

21   Install a new retainer clip on the injector.

22   The remainder of installation is the reverse of removal.

---

12   Exhaust system - servicing

Refer to illustrations 12.1a, 12.1b, 12.1c and 12.4

Warning: *Inspection and repair of exhaust system components should be done only after the system components have cooled completely.*

1   The exhaust system consists of the exhaust manifold(s), the catalytic converter, the muffler, the tailpipe and all connecting pipes, brackets, hangers and clamps. The exhaust system is attached to the body with mounting brackets and rubber hangers **(see illustrations)**. If any of these parts are damaged or deteriorated, excessive noise and vibration will be transmitted to the body.

2   Conducting regular inspections of the exhaust system will keep it safe and quiet.

Look for any damaged or bent parts, open seams, holes, loose connections, excessive corrosion or other defects which could allow exhaust fumes to enter the vehicle. Pay particular attention to rubber exhaust hangers. Because they're subjected to constant exhaust heat, they dry out, break down, develop cracks and then tear apart, allowing the exhaust system to hang unsupported. Sometimes an unsupported exhaust system "clunks" when the vehicle is driven over bumps and dips in the road; sometimes the exhaust touches the axle or other metal parts, which produces an annoying vibration at certain speeds.

3   If the exhaust system components are extremely corroded or rusted together, they will probably have to be cut from the exhaust system. The convenient way to accomplish this is to have a muffler repair shop remove the corroded sections with a cutting torch. If, however, you want to save money by doing it yourself and you don't have an oxy/acetylene welding outfit with a cutting torch, simply cut off the old components with a hack-saw. If you have compressed air, special pneumatic cutting chisels can also be used. If you do decide to tackle the job at home, be sure to wear eye protection to protect your eyes from metal chips and work gloves to protect your hands.

4   Here are some simple guidelines to apply when repairing the exhaust system:

a) *Work from the back to the front when removing exhaust system components.*

b) *To facilitate disassembly of the exhaust system, always apply penetrating oil to nuts and bolts* **(see illustration)**.

c) *Use new gaskets, hangers and clamps when installing exhaust system components.*

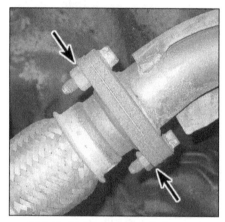

12.4 To facilitate removal of rusted exhaust fasteners, always apply penetrating oil to nuts and bolts, and then allow it to soak in for awhile before trying to loosen them

d) *Apply anti-seize compound to the threads of all exhaust system fasteners during reassembly. Be sure to allow sufficient clearance between newly installed parts and all points on the underbody to avoid overheating the floor pan and possibly damaging the interior carpet and insulation. Pay particularly close attention to the catalytic converter and its heat shield.* **Warning:** *The catalytic converter operates at very high temperatures and takes a long time to cool. Wait until it's completely cool before attempting to remove the converter. Failure to do so could result in serious burns.*

Chapter 5
Engine electrical systems

Contents

Specifications

General

Battery voltage
    Engine off .... 12 volts
    Engine running .... 14 to 15 volts

Ignition system

Ignition coil resistance
    2.3L four-cylinder
        Primary resistance .... 1.2 to 1.4 ohms
        Secondary resistance .... 8,600 to 13,000 ohms
    2.6L four-cylinder
        Primary resistance .... 0.81 to 0.99 ohms
        Secondary resistance .... 7,500 to 11,300 ohms
    3.1L V6
        Primary resistance .... 0.2 to 1.5 ohms
        Secondary resistance .... 5,000 to 25,000 ohms
    2.2L four-cylinder engine and 3.2L V6 .... Not available
Primary winding-to-case resistance (all ignition coils) .... Infinite
Ignition coil-to-distributor cap wire resistance .... 5,000 ohms per foot

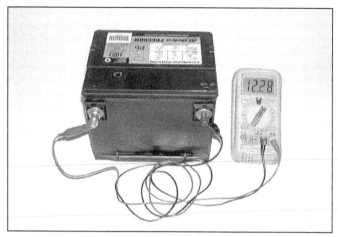

3.2 To test the open circuit voltage of the battery, connect a voltmeter to the battery; a fully charged battery should measure at least 12.4 volts (depending on outside air temperature)

3.3 Connect a battery load tester to the battery and check the battery condition under load following the tool manufacturer's instructions

1 General information and precautions

General information

The engine electrical systems include all ignition, charging and starting components. Because of their engine-related functions, these components are discussed separately from body electrical devices such as the lights, the instruments, etc. (which are included in Chapter 12).

Precautions

Always observe the following precautions when working on the electrical system:

a) *Be extremely careful when servicing engine electrical components. They are easily damaged if checked, connected or handled improperly.*

b) *Never leave the ignition switched on for long periods of time when the engine is not running.*

c) *Never disconnect the battery cables while the engine is running.*

d) *Maintain correct polarity when connecting battery cables from another vehicle during jump starting - see the "Booster battery (jump) starting" section at the front of this manual.*

e) *Always disconnect the negative battery cable before working on the electrical system.*

It's also a good idea to review the safety-related information regarding the engine electrical systems located in the *"Safety first!"* section at the front of this manual, before beginning any operation included in this Chapter.

Battery disconnection

Several systems on the vehicle require battery power to be available at all times, either to ensure their continued operation (such as the clock) or to maintain control unit memories (such as that in the engine management system's Powertrain Control Module [PCM]) which would be wiped out if the battery were to be disconnected. Therefore, whenever the battery is to be disconnected, first note the following to ensure that there are no unforeseen consequences of this action:

a) *First, on any vehicle with power door locks, it is a wise precaution to remove the key from the ignition and to keep it with you, so that it does not get locked inside if the power door locks should engage accidentally when the battery is reconnected!*

b) *The engine management system's PCM will lose the information stored in its memory when the battery is disconnected. This includes idling and operating values, and any fault codes detected (see Chapter 6). Whenever the battery is disconnected, the information relating to idle speed control and other operating values will have to be re-programmed into the unit's memory. The PCM does this by itself, but until then, there may be surging, hesitation, erratic idle and a generally inferior level of performance. To allow the PCM to relearn these values, start the engine and run it as close to idle speed as possible until it reaches its normal operating temperature, then run it for approximately two minutes at 1200 rpm. Next, drive the vehicle as far as necessary - approximately 5 miles of varied driving conditions is usually sufficient - to complete the relearning process.*

Devices known as "memory-savers" can be used to avoid some of the above problems. Precise details vary according to the device used. Typically, it is plugged into the cigarette lighter, and is connected by its own wires to a spare battery; the vehicle's own battery is then disconnected from the electrical system, leaving the "memory-saver" to pass sufficient current to maintain audio unit security codes and PCM memory values, and also to run permanently live circuits such as the clock, all the while isolating the battery in the event of a short-circuit occurring while work is carried out. **Warning:** *Some of these devices allow a considerable amount of current to pass, which can mean that many of the vehicle's systems are still operational when the main battery is disconnected. If a "memory-saver" is used, ensure that the circuit concerned is actually "dead" before carrying out any work on it!*

2 Battery - emergency jump starting

Refer to the *Booster battery (jump) starting* procedure at the front of this manual.

3 Battery - check and replacement

Warning: *Hydrogen gas is produced by the battery, so keep open flames and lighted cigarettes away from it at all times. Always wear eye protection when working around a battery. Rinse off spilled electrolyte immediately with large amounts of water.*

Check

Refer to illustrations 3.2 and 3.3

1   The battery's surface charge must be removed before accurate voltage measurements can be made. Turn On the high beams for ten seconds, then turn them Off, let the vehicle stand for two minutes. Remove the battery from the vehicle (see Steps 4 through 10).

2   Check the battery state of charge. Visually inspect the indicator eye on the top of the battery, if the indicator eye is clear, charge the battery as described in Chapter 1. Next

3.6 Detach the nuts from the battery hold-down clamps, then remove the cable from the negative battery terminal, followed by the positive battery cable

4.4 The negative battery cable is grounded to the right side of the block, right in front of the right engine mount

perform an open voltage circuit test using a digital voltmeter **(see illustration)**. With the engine and all accessories Off, connect the negative probe of the voltmeter to the negative terminal of the battery and the positive probe to the positive terminal of the battery. The battery voltage should be 12.4 volts or more. If the battery is less than the specified voltage, charge the battery before proceeding to the next test. Do not proceed with the battery load test unless the battery charge is correct.

3   Perform a battery load test. An accurate check of the battery condition can only be performed with a load tester (available at most auto parts stores). This test evaluates the ability of the battery to operate the starter and other accessories during periods of heavy amperage draw (load). Install a special battery load testing tool onto the terminals **(see illustration)**. Load test the battery according to the tool manufacturer's instructions. This tool utilizes a carbon pile to increase the load demand (amperage draw) on the battery. Maintain the load on the battery for 15 seconds or less and observe that the battery voltage does not drop below 9.6 volts. If the battery condition is weak or defective, the tool will indicate this condition immediately. **Note:** *Cold temperatures will cause the minimum voltage requirements to drop slightly. Follow the chart given in the tool manufacturer's instructions to compensate for cold climates. Minimum load voltage for freezing temperatures (32 degrees F) should be approximately 9.1 volts.*

Replacement

Refer to illustration 3.6

4   Disconnect the cable from the negative battery terminal.
5   Disconnect the positive battery cable.
6   Remove the battery retainer bolt and retainer **(see illustration)**.
7   Lift the battery off the battery tray and remove the battery insulator. **Note:** *Battery*

handling tools are available at most auto parts stores for a reasonable price. They make it easier to remove and carry the battery.
8   While the battery is removed, inspect the tray, retainer brackets and related fasteners for corrosion or damage.
9   If corrosion is evident, remove the battery tray and use a baking soda/water solution to clean the corroded area to prevent further oxidation. Repaint the area as necessary using rust resistant paint.
10   Clean and service the battery and cables (see Chapter 1).
11   If you are replacing the battery, make sure you purchase one that is identical to yours, with the same dimensions, amperage rating, cold cranking amps rating, etc. Make sure it is fully charged prior to installation in the vehicle.
12   Installation is the reverse of removal. Connect the positive cable first and the negative cable last.
13   After connecting the cables to the battery, apply a light coating of petroleum jelly or grease to the connections to help prevent corrosion.

---

4   Battery cables - check and replacement

Refer to illustration 4.4
1   Periodically inspect the entire length of each battery cable for damage, cracked or burned insulation and corrosion. Poor battery cable connections can cause starting problems and decreased engine performance.
2   Check the cable-to-terminal connections at the ends of the cables for cracks, loose wire strands and corrosion. The presence of white, fluffy deposits under the insulation at the cable terminal connection is a sign that the cable is corroded and should be replaced. Check the terminals for distortion, missing mounting bolts and corrosion.

3   When removing the cables, always disconnect the negative cable first and hook it up last or the battery may be shorted by the tool used to loosen the cable clamps. Even if only the positive cable is being replaced, be sure to disconnect the negative cable first (see Chapter 1 for further information regarding battery cable maintenance).
4   Disconnect the old cables from the battery, then disconnect them at the opposite end. The positive battery cable is connected to the starter solenoid (see Section 19) and the negative battery cable is connected to the right side of the engine block **(see illustration)**. Note the routing of each cable to ensure correct installation.
5   If you are replacing either or both of the battery cables, take them with you when buying new cables. It is vitally important that you replace the cables with identical parts. Cables have characteristics that make them easy to identify: positive cables are usually red and larger in cross-section; ground cables are usually black and smaller in cross-section.
6   Clean the threads of the starter solenoid or ground connection with a wire brush to remove rust and corrosion. Apply a light coat of battery terminal corrosion inhibitor or petroleum jelly to the threads to prevent future corrosion.
7   Attach the cable to the terminal and tighten the mounting nut/bolt securely.
8   Before connecting a new cable to the battery, make sure that it reaches the battery without having to be stretched.

---

5   Ignition system - general information

1   The ignition system is designed to ignite the fuel/air charge entering each cylinder at just the right moment. It does this by producing a high voltage spark between the elec-

trodes of each spark plug.

2 The vehicles covered by this manual are equipped with an electronic ignition system. 2.3L, 2.6L and 3.1L engines are equipped with a conventional distributor-type ignition system. 2.2L and 3.2L engines are equipped with a distributorless ignition system. All ignition systems have an ignition module, ignition coil(s) and primary and secondary wiring circuits, but they differ in other ways.

3 2.3L engines are equipped with a vacuum advance unit mounted on the distributor. All other models use an electronic spark timing system in which the Powertrain Control Module (PCM) controls the spark timing advance characteristics.

4 The distributor on the 2.6L engine cannot be repaired. In the event of failure, the distributor must be removed as a complete unit and exchanged for a new or rebuilt unit. 2.3L and 3.1L engines are equipped with a distributor that can be repaired. The pick-up coil, reluctor and other internal parts are available at a dealer or auto parts store.

5 Distributorless systems rely on s crankshaft position sensor and (on 2.2L engines) a camshaft position sensor for ignition and timing control. In the event of a no-start problem, check the operation of these sensors (see Chapter 6).

Distributor-type ignition systems

6 3.1L V6 models use a special distributor that is mounted at the rear of the engine. All spark timing changes are carried out by the Electronic Control Module (ECM) which monitors data from various engine sensors, computes the desired spark timing and signals the distributor to change the timing accordingly. No vacuum or mechanical advance is used.

7 2.3L and 2.6L four-cylinder models are equipped with an electronic ignition system. This ignition system consists of the ignition module, distributor housing, cap and rotor, distributor drive shaft, an ignition relay, an ignition coil, primary and secondary wiring, spark plugs and the necessary control circuits (wiring harness) for the entire system. Early systems (2.3L engine) use a vacuum advance to mechanically advance the timing. Later models (2.6L engine) use the ECM to control the ignition timing. The ECM monitors the information from various engine sensors and computes the correct spark timing via a direct line from the computer.

Distributorless ignition systems

8 Newer engines use a distributorless ignition system. This system uses a "waste spark" method of spark distribution. Each cylinder is paired with its opposing cylinder in the firing order (1-4, 3-2 on the 2.2L four-cylinder engine, 1-4, 6-3, 2-5 on the 3.2L V6) so that one cylinder on compression fires simultaneously with its opposing cylinder on the exhaust stroke. Since the cylinder on the exhaust stroke requires very little of the avail-

able voltage to fire its plug, most of the voltage is used to fire the cylinder on the compression stroke.

3.2L V6 engine

Note: *For more information regarding the information sensors described below, refer to Section 6.*

9 The DIS system includes six coil packs, an ignition control module (ICM) (through 1997), a crankshaft position sensor (CKP) and the Powertrain Control Module (PCM).

10 The PCM, which is located in the left side of the engine compartment, controls ignition timing in accordance with crankshaft position, engine rpm, engine temperature and manifold absolute pressure, all of which are monitored by information sensors (see Chapter 6).

11 On 1993 through 1995 systems, the ignition control module (ICM) is located under the coil packs and is connected to the PCM. On 1996 and 1997 systems, the ICM is located on the air intake plenum. On 1998 and later models, the ICM is an integral part of the PCM; there is no separate ignition module.

12 The crankshaft position (CKP) sensor is located on the lower side of the engine block. The crankshaft sensor is positioned within 0.050-inch of the crankshaft reluctor ring. The reluctor ring is a special disc cast into the crankshaft, which acts as a signal generator for the ignition timing.

13 On 1996 and later models, the ignition coil packs are mounted directly over the spark plugs on the valve covers. This eliminates the secondary wires used on the early systems.

14 The DIS on 2000 models is equipped with an "ion sensing module" which monitors combustion quality in order to detect cylinder misfires. The ion sensing module is located on top of the intake plenum.

2.2L four-cylinder engine

Note: *For more information regarding the information sensors described below, refer to Section 6.*

15 The DIS system includes two coil packs, an ignition control module (ICM), a crankshaft position sensor (CKP), a camshaft position sensor and the Powertrain Control Module (PCM). The ignition control module is located at the right rear of the engine.

16 The PCM, which is located below the center of the dash, controls ignition timing in accordance with crankshaft position, engine rpm, engine temperature and manifold absolute pressure, all of which are monitored by information sensors (for more information on the PCM and information sensors, see Chapter 6).

17 The ignition control module (ICM) is an integral part of the ignition coil assembly, which is located at the right rear of the engine. The ICM is an electronic switching device that turns the primary voltage ground for the coils on and off. The actual timing, duration and strength of the spark is determined by the PCM.

18 The camshaft position (CMP) sensor is located at the top front center of the engine.

The CMP sends a voltage signal to the PCM, which uses this signal as the timing pulse to trigger the injectors in the correct sequence The PCM also uses the CMP signal to indicate the position of the number one piston on the power stroke, information which it needs to initiate the sequential firing of the injectors when the engine is started. If the PCM detects an incorrect CMP signal while the engine is running, it will set a trouble code (see Chapter 6).

19 The crankshaft position (CKP) sensor is located at the lower front of the engine. The CKP sends a voltage signal to the PCM which uses this signal to calculate the ignition sequence. The CKP sensor wheel has 58 teeth on it and a spot where one tooth is missing. The CKP sensor responds to this "missing tooth" by sending a voltage pulse to the PCM, which uses it to determine engine speed and crankshaft position.

20 The ignition coil assembly is located at the right rear of the engine. There are two coils, one for each two cylinders. Radio frequency interference produced by the coils is controlled by a condenser mounted near the coils.

---

6   Ignition system - check

Warning: *Because of the very high voltage generated by the ignition system, extreme care should be taken whenever an operation is performed involving ignition components. This not only includes the coils, control module and spark plug wires, but related items connected to the system as well, such as the plug connections, tachometer and any test equipment.*

General checks

Refer to illustration 6.3a and 6.3b

1 Check all ignition wiring connections for tightness, cuts, corrosion or any other signs of a bad connection. A faulty or poor connection at a spark plug could also result in a misfire. Also check for carbon deposits inside the spark plug boots. Remove the spark plugs, if necessary, and check for fouling.

2 Check for ignition and battery supply to the ECM. Check the ignition fuses (see Chapter 12). **Note:** *The ECM fuse controls the computer and the fuel injection system while the IGN fuse controls power to the ignition system.*

3 Use a calibrated ignition tester to verify adequate available secondary voltage (25,000 volts) at the spark plug **(see illustrations)**. Using an ohmmeter, check the resistance of the spark plug wires. Each wire should measure less than 30,000 ohms.

4 On fuel injected engines, check to see if the fuel pump and relay are operating properly (see Chapter 4B). The fuel pump should activate for two seconds when the ignition key is cycled ON. Install an injector test light and monitor the blinks as the injector voltage signal pulses (see Chapter 4B, *Fuel injection system - check*).

6.3a To use a calibrated ignition tester, simply disconnect a spark plug wire and connect it to the tester, clip the tester to a convenient ground (like a valve cover bolt) and operate the starter - if there is enough power to fire the plug, sparks will be visible between the electrode tip and the tester body

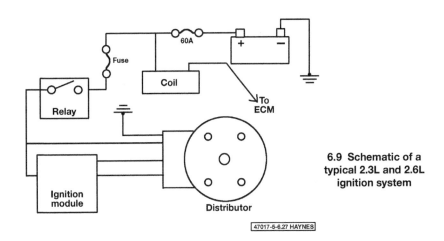

6.9 Schematic of a typical 2.3L and 2.6L ignition system

6.3b Checking the spark on a Distributorless Ignition System (DIS)

Ignition systems with a distributor

2.3L and 2.6L four-cylinder models

No start condition

Refer to illustration 6.9

5   Using a calibrated ignition tester **(see illustration 6.3a)**, verify that each cylinder is receiving a clean (bright blue) and consistent spark. Also, check the condition of each spark plug.

6   At least two plug wires are checked to make sure there is the proper output from the distributor ignition system. Ensure that an open is not present in a spark plug wire.

7   Install a spark tester into the ignition coil wire and check for a clean and consistent spark. A spark indicates that the problem must be the distributor unit. This test separates the ignition wires from the ignition coil. **Note:** *A few sparks followed by no spark is the same condition as no spark at all.*

8   Check for battery voltage to the ignition coil with the key ON (engine not running). Check the condition of the coil (see Section 8).

9   Disconnect the ignition module connector and check for battery voltage **(see illustration)** to the ignition module from the ignition system relay. If battery voltage is not available check the relay and the ignition system fuse. Refer to Chapter 12 for additional information on the electrical system and the location of the relays and fuses

10   If all the tests results are correct and there still is a NO START condition, it will be very difficult to distinguish a faulty ECM from a faulty ignition module. In most cases, the ignition module, when defective, will not produce a pulsing voltage signal from the computer with symptoms of a definite start or no-start condition while a faulty module might also be defective. Take the vehicle to a dealer service department or other qualified repair shop for diagnosis.

3.1L V6 models

Refer to illustration 6.11

11   Refer to the illustration for terminal pin designations for testing the ignition module **(see illustration)**.

12   Check for spark at the spark plugs with

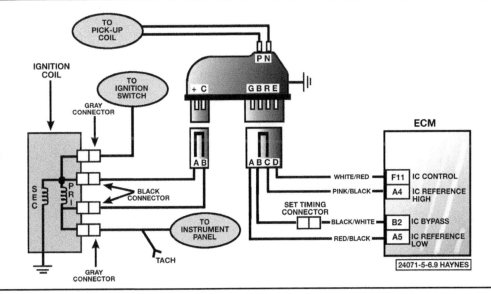

6.11 Schematic of the 3.1L V6 ignition system

a calibrated ignition tester **(see illustration 6.3a)**. Check at least two or more spark plug wires. On a NO START condition, check the fuel pump relay and fuel pump systems for fuel delivery problems in the event the ignition spark is available.

13   Unplug the four-wire connector from the ignition module at the distributor and check for spark at the coil wire. If the fuel system is working properly and the ignition system will not start the engine, a spark indicates that the problem must be the distributor cap or rotor. **Note:** *A few sparks followed by no spark is the same condition as no spark at all.*

14   Check the ignition coil (see Section 8). Next, disconnect the two terminal plug on the module and check for voltage on the "C" and the "+" terminals. Normally, there should be battery voltage at the "C" and "+" terminals. Low voltage indicates an open or high resistance circuit from the distributor to the coil or ignition switch. If the "C" terminal voltage is low, but the "+" terminal voltage is 10 volts or more, the circuit from "C" terminal to the ignition coil or ignition coil primary winding is open.

15   Reconnect the "C,+" connector onto the ignition module and with the ignition key ON (engine not running), check for voltage at the TACH terminal. The TACH terminal is taped back in the harness near the distributor. If there is less than 10 volts available, check the TACH circuit from the distributor. This test, checks for a shorted module or a grounded circuit from the ignition coil to the module. The distributor module should be turned off, so normal voltage should be about 12 volts. If the module is turned ON, the voltage will be low, but above one volt. This could cause the ignition coil to fail from excessive heat.

16   Check the module itself. Construct a battery pack that is more than 1.5 volts and less than 8 volts (penlight batteries). Disconnect the ignition module four-wire connector from the distributor. Disconnect the pick-up coil two-wire connector and remove the distributor cap and rotor. Connect a voltmeter from TACH to ground. Connect a test light from the battery pack (1.5 to 8.0 volts) and touch the test light probe to terminal P on the ignition module and watch the voltmeter. Voltage should drop. Applying a voltage (1.5V) to module terminal "P" should turn the module on and the TACH terminal voltage should drop to about 7 to 9 volts. **Note:** *Remember, the TACH terminal is usually taped to the harness near the distributor.*

17   This test will determine whether the module or coil is faulty or if the pick-up coil is not generating the proper signal to turn the module on. This test can be performed by using a DC battery with a rating of 1.5 volts. Monitor the change in voltage with a voltmeter connected to the TACH terminal. Some digital multimeters can also be used to trigger the module by selecting ohms, usually the diode position. In this position, the meter may have a voltage across its terminals which can be used to trigger the module. The voltage in the ohms position can be checked by using a second meter or by checking the manufacturer's specifications for the tool being used.

18   Install the calibrated spark tester **(see illustration 6.3a)**. Repeat the previous test but instead observe spark at the spark plug when the test light tip is removed from terminal P. There should be a spark. This test checks the ignition module's ability to cause a spark. If no spark occurs, the fault is most likely in the ignition coil or pick-up coil because most module problems would have been found before this point in the procedure. A GM HEI module tester (Kent Moore J-24642-F or equivalent) can determine which is at fault. If you cannot obtain the module tester, take the vehicle to a dealer service department or other qualified repair shop at this point. Also, check the resistance of the pick-up coil and if the specifications are incorrect, replace it with a new part (see Section 11).

Distributorless ignition systems

3.2L V6 models

Engine runs but misfires

19   Disconnect the electrical connector from the IAC valve and with the engine idling, disconnect, one-by-one, each spark plug wire using insulated pliers. Be sure to ground the disconnected plug wire to the engine block using a jumper wire. Listen carefully for a drop in engine rpm. The engine rpm can be observed on a tachometer also. If the engine rpm does not decrease when the spark plug wire is removed from the cylinder, then there is a problem with that cylinder. Refer to the electrical schematic of the ignition system including the module, coil packs and ECM in the wiring diagram section at the end of Chapter 12. Also, check the engine compression (see Chapter 2E). **Note:** *This test is difficult to perform on 1996 and later models because the coil pack must be removed from the cylinder head in order to insert a spark tester. However, the system is equipped with an ignition misfire detection system built within the self diagnosis system. Use a special SCAN tool or have the system diagnosed by a dealer service department.*

20   Using a calibrated ignition tester **(see illustration 6.3b)**, verify that each cylinder is receiving a clean (bright blue) and consistent spark. The spark tester verifies that there is adequate secondary voltage at the spark plug. Also, inspect the condition of each spark plug (refer to the chart inside the back cover of this manual).

21   Ground the opposite plug wire (companion cylinder) using a jumper wire and observe

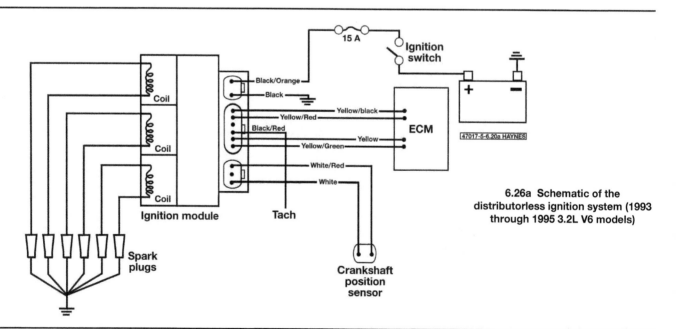

6.26a Schematic of the distributorless ignition system (1993 through 1995 3.2L V6 models)

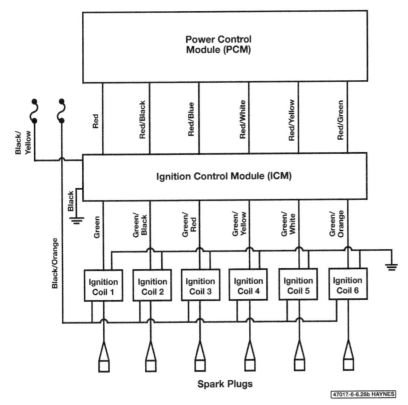

6.26b Schematic of the distributorless ignition system (1996 and 1997 3.2L V6 models)

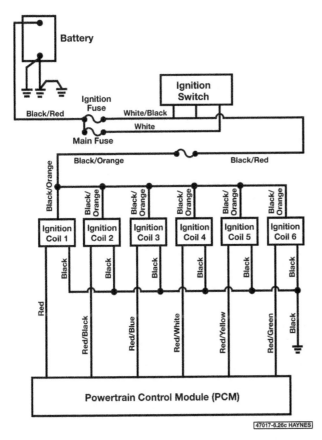

6.26c Schematic of the distributorless ignition system (1998 and 1999 3.2L V6 models)

the spark on the spark tester. The spark should jump the gap easily. If the spark jumps the test gap after grounding the opposite plug wire, it indicates excessive resistance in the plug which was bypassed. A faulty or poor connection at that plug could also result in a misfire condition. Also check for carbon deposits inside the spark plug boot.

22   On 1993 through 1995 models, check the resistance of the plug wires. It should be less than 30,000 ohms resistance. On all models, remove the coil pack and check for carbon tracking. If carbon tracking is evident, replace the coil and be sure that the plug wires relating to that coil are clean and tight. Excessive wire resistance or faulty connections could cause damage to the coil. **Note:** *This test is difficult to perform on 1996 and later models because the coil pack must be removed from the cylinder head in order to insert a spark tester. However, the system is equipped with an ignition misfire detection system built within the self diagnosis system. Use a special SCAN tool or have the system diagnosed by a dealer service department.*

23   If possible, switch the coil pack with a known good unit and check the ignition spark. If a no spark condition disappears when the coil is switched for another coil, the original coil is faulty. If not, the ignition module is the cause of the no spark condition. This test can also be performed by substituting a known good coil for the one causing the no spark condition.

No start condition

Refer to illustrations 6.26a, 6.26b, 6.26c and 6.26d

24   Using a calibrated ignition tester **(see illustration 6.3b)**, verify that each cylinder is receiving a clean (bright blue) and consistent spark. Also, check the condition of each spark plug.

25   On 1993 through 1995 models, check the resistance of the plug wires. It should be less than 30,000 ohms resistance. No spark at one cylinder may be caused by an open plug wire or secondary winding. Both wires related to a coil and the secondary winding resistance should therefore be checked. Resistance readings over the upper limit, but not infinite, will probably not cause a no start but may cause an engine miss under certain conditions.

26   Disconnect the ignition coil harness connectors at the coil pack and install a test light to the battery positive terminal and touch the tip of the test light to the coil module terminals while an assistant cranks the engine over **(see illustrations)**. This tests the triggering circuit in the ignition module. A blinking test light indicates the module is triggering (see Section 8). **Note 1:** *On 1993 through 1995 systems, it's impossible to separate the ignition module from the coil packs to check the trigger signal. On 1996 and later systems, you can check the primary wires.* **Note 2:** *Remove the fuel pump fuse and disable the ignition system to prevent the other cylinders from starting and running. The easi-*

est method to disable the ignition system partially is to remove each plug wire and ground it to the engine with a jumper wire. This prevents the other cylinders from firing but still allows the trigger signal to reach the coils.

27   A slowly blinking light, at this point, indicates the ECM is not seeing a crank sensor signal (see Chapter 6). At this point, the problem is in the camshaft or crankshaft sensor(s), sensor circuits or the ignition module. **Note:** *Refer to Chapter 6 for additional information and testing procedures for the camshaft sensor and the crankshaft sensors and Section 12 for testing procedures for the ignition module.*

28   Check for battery voltage to the ignition module (black/orange [1993 through 1995] or black/yellow [1996 and 1997]).

2.2L four-cylinder models

Refer to illustrations 6.31a and 6.31b

29   Using a calibrated ignition tester **(see illustration 6.3a)**, verify that each cylinder is receiving a clean (bright blue) and consistent spark. Also, inspect the condition of each spark plug (refer to the chart inside the back cover of this manual).

30   Check the resistance of the plug wires. It should be less than 30,000 ohms resistance. No spark at a cylinder may be caused by an open plug wire or by an open in the coil secondary winding. Be sure to check both wires for each coil. Resistance readings over the upper limit, but not infinite, won't necessarily cause a no start condition, but might cause an engine to misfire. If the plug wires check out okay, have the coil/module unit checked by a dealer service department. (There are no secondary winding resistance specs available for these units.)

31   Disconnect the electrical connector at the coil/module unit **(see illustration)**. Hook up a test light between the terminal on the harness side of the connector for the black/orange wire **(see illustration)** and ground. While an assistant cranks over the engine, note whether the test light flashes on

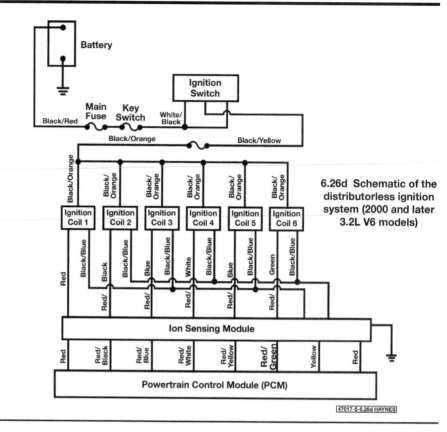

6.26d Schematic of the distributorless ignition system (2000 and later 3.2L V6 models)

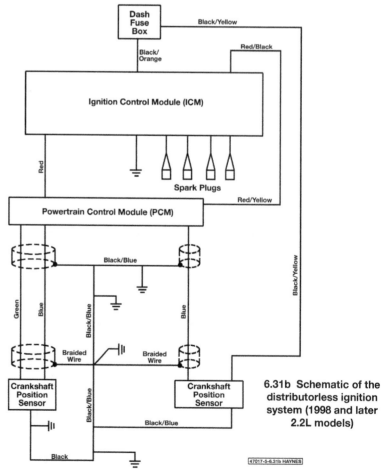

6.31b Schematic of the distributorless ignition system (1998 and later 2.2L models)

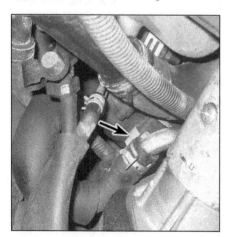

6.31a To check primary voltage to the coil and the trigger signal from the ignition module, unplug the coil/module electrical connector

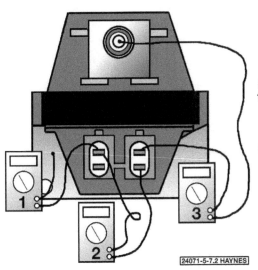

8.2 Perform these three checks; first, check the coil grounds (1), then the primary resistance (2), and the secondary resistance (3) (3.1L V6 model)

24071-5-7.2 HAYNES

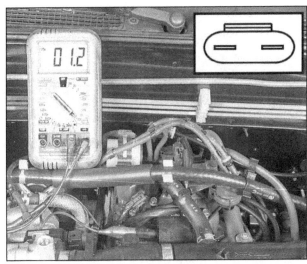

8.11 Checking the coil primary resistance on the 2.6L engine

and off. This verifies that the coil is receiving primary voltage. If it isn't, check the wiring and connectors between the PCM and the coil. If the wiring and connectors are okay, the PCM might be defective. Have the PCM checked out by a dealer service department.

32 At the same coil/module connector, hook up the test light between the terminal on the harness side of the connector for the red wire and ground. While an assistant cranks over the engine, note whether the light flashes on and off. This verifies that the ignition control module is turning the primary voltage on and off. **Note:** *Remove the fuel pump fuse and disable the ignition system to prevent the other cylinders from starting and running. The easiest method to disable the ignition system partially is to remove each plug wire and ground it to the engine with a jumper wire. This prevents the other cylinders from firing but still allows the trigger signal to reach the coils.*

33 An intermittently blinking light or a light that doesn't blink at all indicates that the PCM is probably not receiving a good signal, or any signal at all, from the camshaft and/or crankshaft sensor (see Chapter 6). There is a problem in the camshaft or crankshaft sensor, the sensor circuit or the ignition module. Refer to Chapter 6 for additional information and testing procedures for the camshaft sensor and the crankshaft sensors and Section 12 for testing procedures for the ignition module.

7 Air gap (2.3L four-cylinder models only) - check and adjustment

Note: *2.3L models are equipped with a replaceable pick-up coil unit that is mounted on the breaker plate. The air gap (distance between the reluctor and the pick-up coil) must be adjusted periodically.*

1 Remove the distributor cap from the dis-

tributor (see Chapter 1).

2 Position a brass feeler gauge between the reluctor and pick-up coil and measure the clearance. It should be between 0.012 to 0.020 inch.

3 If the measurement is incorrect, loosen the stator mounting screws and adjust the air gap.

8 Ignition coil - check and replacement

1 Disconnect the cable from the negative terminal of the battery.

3.1L V6 models

Check

Refer to illustration 8.2

2 Check the coil for opens and grounds by performing the following three tests with an ohmmeter **(see illustration)**.

3 Using the ohmmeter's high scale, hook up the ohmmeter leads as illustrated (see test 1 in **illustration 8.2**). The ohmmeter should indicate a very high, or infinite, resistance value. If it doesn't, replace the coil. Test 1 checks to make sure that the coil is not grounding against the case.

4 Check the coil primary resistance. Using the low scale, hook up the leads as illustrated (see test 2 in **illustration 8.2**). The ohmmeter should indicate a very low, or zero, resistance value. If it doesn't, replace the coil. Check the Specifications listed in the beginning of this Chapter.

5 Check the coil secondary resistance. Using the high scale, hook up the leads as illustrated (see test 3 in **illustration 7.2**). The ohmmeter should not indicate an infinite resistance. If it does, replace the coil.

Replacement

6 Unplug the coil high tension wire and both electrical leads from the coil.

7 Remove both mounting nuts and remove the coil from the engine.

8 Installation of the coil is the reverse of the removal procedure.

2.3L and 2.6L four-cylinder models

Check

Refer to illustrations 8.11 and 8.12

9 Detach the primary electrical connector from the coil.

10 Check to make sure the coil is receiving battery voltage with the ignition key ON (engine not running).

11 Using the ohmmeter's low scale, hook up the ohmmeter leads to the primary terminals on the ignition coil **(see illustration)**. The ohmmeter should indicate a very low resistance value. If it doesn't, replace the coil. Refer to the Specifications listed in the beginning of this Chapter.

12 Using the high scale, hook up one lead to the ignition coil primary terminal and the other lead to the secondary terminal **(see illustration)**. Refer to the Specifications listed

8.12 Checking the coil secondary resistance on the 2.6L engine

8.25 To detach an ignition coil from a 1996 and later 3.2L V6 engine, disconnect the electrical connector (A) and remove the two retaining screws (B)

8.26 To remove an ignition coil from a 1996 and later 3.2L V6 engine, pull it straight up

8.29 Use a 5 mm Allen wrench to remove the spark plug cable cover bolts

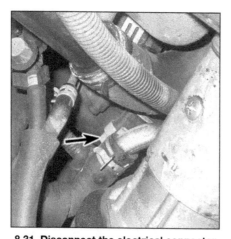

8.31 Disconnect the electrical connector from the ignition coil/ignition control module assembly (connector shown from below; connector is not visible from above, but you can unplug it from above)

8.32a To detach the ignition coil/ignition control module assembly from the mounting bracket, remove this bolt on top . . .

in the beginning of this Chapter. The ohmmeter should not indicate an infinite resistance. If it does, replace the coil.

13   Using the ohmmeter's low scale, hook up the negative probe (-) of the ohmmeter to the coil body and the positive probe (+) of the ohmmeter to the coil primary terminal on the ignition coil. The ohmmeter should indicate infinite resistance. If it doesn't, replace the coil. Refer to the Specifications listed in the beginning of this Chapter.

Replacement

14   Detach the cable from the negative terminal of the battery.
15   Unplug the coil high tension wire and both electrical leads from the coil.
16   Remove both mounting nuts and remove the coil from the engine.
17   Installation of the coil is the reverse of the removal procedure.

Distributorless Ignition System (DIS)

Check

18   These systems require special testing equipment. Have the vehicle checked by a dealer service department.

Replacement

1993 through 1995 3.2L V6 models

19   Disconnect the negative battery cable.
20   Disconnect the spark plug wires.
21   Unplug the electrical connectors from the PCM, power supply and crank angle sensor.
22   Remove the three ignition coil mounting bolts and remove the ignition coil assembly.
23   Installation is the reverse of removal.

1996 and later 3.2L V6 models

Refer to illustrations 8.25 and 8.26

24   Disconnect the negative battery cable.
25   Disconnect the electrical connector and remove the coil pack retaining screws **(see illustration)**.
26   To remove the coil pack, pull straight up on it **(see illustration)**.
27   Installation is the reverse of removal.

2.2L four-cylinder models

Refer to illustrations 8.29, 8.31, 8.32a, 8.32b, 8.33a and 8.33b

28   Disconnect the negative battery cable.

29   Remove the spark plug cable cover bolts **(see illustration)** with a 5 mm Allen wrench and remove the cover.
30   Unplug the spark plug wires from the spark plugs.
31   Disconnect the electrical connector

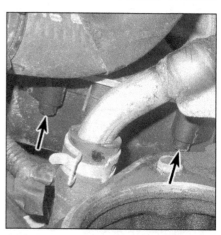

8.32b . . . and these two bolts underneath

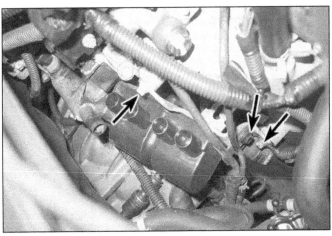

8.33a If you're unable to remove the three ignition coil/ignition control module mounting bolts, then remove the four mounting bracket bolts (far left bolt not visible in this photo) . . .

8.33b . . . remove the ignition coil/ignition control module and mounting bracket as a single assembly and then remove the three ignition coil/ignition control module bolts to detach the coil/ignition module from the bracket

from the ignition coil/ignition control module assembly **(see illustration)**.

32   Remove the three ignition coil/ignition control module retaining bolts **(see illustrations)**.

33   If you're unable to remove the three ignition coil/ignition control module retaining bolts, remove the ignition coil/ignition control module bracket **(see illustration)** and then remove the ignition coil/ignition control module from the bracket **(see illustration)**.

34   Remove the ignition coil/ignition control module assembly.

35   Installation is the reverse of removal.

9   Distributor - removal and installation

Note: *The distributor on the 2.6L engine cannot be repaired. In the event of failure, the distributor must be removed as a complete unit and exchanged for a new or rebuilt unit. 2.3L models are equipped with a distributor that can be repaired. The pick-up coil, reluctor and other internal parts are available at a dealer or auto parts stores.*

Removal

Refer to illustrations 9.5a and 9.5b

1   Detach the cable from the negative terminal of the battery.

2   Disconnect the electrical connector from the distributor. Follow the wires as they exit the distributor to find the connector.

3   Note the raised "1" on the distributor cap. This marks the location for the number one cylinder spark plug wire terminal. **Note:** *Some distributor caps may have been replaced with aftermarket units that are not marked.*

4   Remove the distributor cap (see Chapter 1). Using a socket and breaker bar on the crankshaft pulley bolt, rotate the engine until the rotor is pointing toward the number one spark plug terminal (see the TDC locating procedure in Chapter 2).

9.5a Before removing the distributor, be sure to mark the relationship of the rotor to the distributor body to ensure that the two components are correctly aligned when reinstalled

5   Make a mark on the edge of the distributor base directly below the rotor tip and inline with it **(see illustration)**. Also, mark the distributor base and the engine block to ensure that the distributor can be reinstalled correctly **(see illustration)**.

6   Remove the distributor hold-down bolt and clamp, then pull the distributor straight out to remove it. **Caution:** *DO NOT turn the engine while the distributor is removed, or the alignment marks will be useless.*

Installation

7   Insert the distributor into the engine in exactly the same relationship to the block that it was in when removed.

8   To mesh the helical gears on the camshaft and the distributor, it may be necessary to turn the rotor slightly. Make sure the distributor is seated completely and the alignment marks made previously are aligned. If not, remove the distributor and reposition it. **Note:** *If the crankshaft has been moved while the distributor is out, locate Top*

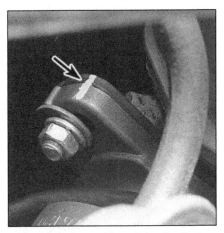

9.5b Mark the position of the distributor in relation to the block or the cylinder head (arrow) before loosening the distributor hold down clamp

Dead Center (TDC) for the number one piston (see Chapter 2A) and position the distributor and rotor accordingly.

9   Place the hold-down clamp in position and loosely install the bolt.

10   Install the distributor cap and tighten the screws securely.

11   Plug in the electrical connector.

12   Reattach the spark plug wires to the plugs (if removed).

13   Connect the cable to the negative terminal of the battery.

14   Check and, if necessary, adjust the ignition timing (see Chapter 1) and tighten the distributor hold-down bolt securely.

10   Distributor centrifugal and vacuum advance system (2.3L and 2.6L models) - check

Note: *The distributor on the 2.6L engine cannot be repaired. In the event of failure, the distributor must be removed as a complete unit and exchanged for a new or rebuilt unit.*

11.4 Before testing or removing the pick-up coil, unplug the lead from the ignition module

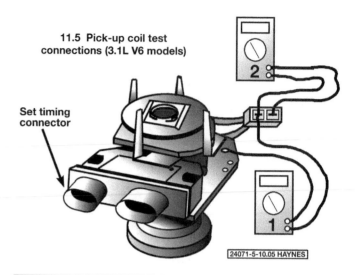

11.5 Pick-up coil test connections (3.1L V6 models)

Set timing connector

24071-5-10.05 HAYNES

2.3L models are equipped with a distributor that can be repaired. The pick-up coil, reluctor and other internal parts are available at a dealer or auto parts store.

General information

1   To provide the proper spark advance or retard during different speeds and throttle openings, 2.3L and 2.6L models are equipped with a vacuum advance and a centrifugal advance mechanism in the distributor. These parts may stick, wear out or become defective after years of service. **Note:** *The vacuum advance unit governs ignition timing according to engine load, while the centrifugal advance unit governs ignition timing according to engine speed. Some models are equipped with dual diaphragm vacuum advance units that provide additional timing retard during engine cranking to aid in starting and also reduce emissions during idle and during coast down.*
2   Both the vacuum advance and the centrifugal advance systems can be tested using the timing light method. After the initial test to determine its working capabilities, the different systems must be checked separately.

Timing light test

3   Disconnect the vacuum hose to the vacuum advance unit, plug the hose and run the engine at approximately 1,200 rpm.
4   Install a timing light and observe the timing mark on the pulley while slowly raising the engine speed to 4,000 rpm. The timing mark should move smoothly and without any hesitations or sudden movements. If the timing mark jumps around, it will be necessary to remove the distributor cap and check the centrifugal weights for binding, missing parts or defective parts (see Section 10).
5   Next, check the operation of the vacuum advance unit. Run the engine at a steady 1,200 rpm and point the timing light at the timing mark.
6   Note the exact position of the timing

mark and raise the rpm while observing the mark. Connect the vacuum hose to the vacuum advance unit, when the vacuum hose is connected the timing mark should advance quickly and completely. Most systems will advance the timing approximately 35 degrees when opened to wide open throttle.

Centrifugal advance check

7   With the engine OFF and the distributor cap removed from the distributor, advance the timing plate manually using the rotor and release the mechanism. It should snap back to its original position very quickly and without hesitation. Try oiling the weights and springs with a spray lubricant and repeat the test several times to loosen the springs and weights if they show signs of sticking or binding.
8   Repeat the test several more times until the timing can be advanced without trouble.

Vacuum advance check

9   The diaphragm assembly is attached to the distributor breaker plate. A vacuum line attaches the diaphragm housing to a ported vacuum source. As vacuum changes from idle to acceleration through deceleration and back to idle, the timing changes accordingly.
10   With the engine OFF check the condition of the vacuum hose from the distributor to the vacuum source. Make sure the hose is connected and the ends sealed.
11   Disconnect and plug the vacuum hose to the distributor. Connect a vacuum pump to the vacuum diaphragm and apply between 15 to 20 in-Hg to the advance unit.
12   Make sure the vacuum advance diaphragm holds vacuum. If the vacuum bleeds down, replace the unit with a new vacuum advance.

Vacuum advance removal

13   Remove the distributor cap and the vacuum hose to the advance unit.

14   Remove the two screws that retain the vacuum advance unit to the distributor body.
15   Remove the clip that holds the vacuum advance plate to the vacuum advance link. Be very careful not to drop it into the distributor.
16   Remove the old vacuum advance unit and replace it with a new part.
17   Installation is the reverse of removal.

---

11   Ignition pick-up coil (2.3L and 3.1L models) - check and replacement

---

3.1L V6 models

Refer to illustration 11.4
1   Detach the cable from the negative terminal of the battery.
2   Remove the distributor cap and rotor.
3   Remove the distributor from the engine (only do this if you have difficulty in accessing the terminals of the pick-up coil) (see Section 8).
4   Detach the pick-up coil leads from the module **(see illustration)**.

Check

Refer to illustration 11.5
5   Connect one lead of an ohmmeter to the terminal of the pick-up coil lead and the other to ground as shown in test 1 **(see illustration)**. Flex the leads by hand to check for intermittent opens. The ohmmeter should indicate infinite resistance at all times. If it doesn't, the pick-up coil is defective and must be replaced.
6   Connect the ohmmeter leads to both terminals of the pick-up coil lead (see test 2 in **illustration 11.5**). Flex the leads by hand to check for intermittent opens. The ohmmeter should read one steady value between 500 and 1,500 ohms as the leads are flexed by hand. If it doesn't, the pick-up coil is defective and must be replaced.

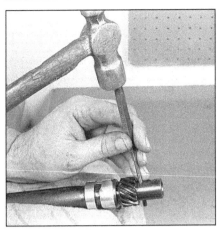

11.8a Mount the distributor shaft in a soft-jawed vise and, using a drift punch and hammer, drive out the roll pin

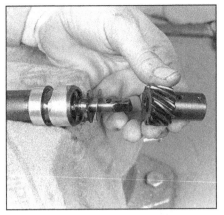

11.8b Remove the driven gear and spacer washers from the end of the shaft, making sure to note the order in which you remove any spacers

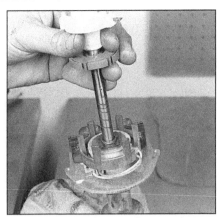

11.9 Remove the distributor shaft

11.10 To remove the pick-up coil from the 3.1L engine distributor, remove the retaining clip

Replacement

Refer to illustrations 11.8a, 11.8b, 11.9, and 11.10

7   Remove the distributor, if not already done. Mark the distributor tang drive and shaft so they can be reassembled in the same position.

8   Carefully set the distributor in a vise with a shop rag to protect the shaft. Using a hammer and punch, remove the roll pin from the distributor shaft and gear **(see illustrations)**.

9   Remove the distributor shaft **(see illustration)**.

10   Remove the spring clip from the distributor shaft **(see illustration)**.

11   Lift the pick-up coil assembly straight up and remove it from the distributor. Note the order in which you remove the pieces.

12   Reassembly is the reverse of disassembly.

2.3L four-cylinder models

Note: *The distributor on the 2.6L engine cannot be repaired. In the event of failure, the distributor must be removed as a complete unit and exchanged for a new or rebuilt unit. 2.3L models are equipped with a distributor that can be repaired. The pick-up coil, reluctor and other internal parts are available at a dealer or auto parts store.*

13   Detach the cable from the negative terminal of the battery.

14   Remove the distributor cap and rotor.

15   Remove the distributor from the engine (only do this if you have difficulty in accessing the terminals of the pick-up coil) (see Section 8).

Check

16   The manufacturer does not supply specifications or testing procedures for checking the pick-up coil. Have the distributor tested at a dealer service department or other qualified electronics repair facility.

Replacement

17   Remove the distributor, if not already done. Remove the rotor from the distributor

shaft (see Chapter 1).

18   Remove the reluctor from the distributor shaft. Do not lose the locking pin.

19   Carefully set the distributor in a vise with a shop rag to protect the shaft. Remove the screws that mount the pick-up coil to the breaker plate.

20   Reassembly is the reverse of disassembly.

---

12   Ignition control module - check and replacement

Caution: *The ignition module is a delicate and relatively expensive electrical component. Failure to follow the step-by-step procedures could result in damage to the module and/or other electronic devices, including the PCM. Additionally, all PCM controlled devices are protected by a Federally mandated emissions warranty. Check with the dealer concerning this warranty before attempting to diagnose and replace this unit yourself.*

3.1L V6 models

Note 1: *It is not necessary to remove the distributor to check or replace the module.*
Note 2: *Refer to illustration 6.9 for the terminal designations on the ignition module.*

Check

1   Disconnect the four-terminal connector from the distributor.

2   Check for a spark at the coil using a spark tester connected to the coil wire (see Section 6). If there is spark, check the cap, rotor and coil wire for opens or other damage.

3   If there is no spark, remove the distributor cap. Reconnect the four-terminal connector to the ignition module (distributor).

4   Unplug the two-wire connector from the distributor. With the ignition switch turned ON, check for voltage at the module positive (+) terminal of the two-wire connector.

5   If the reading is less than ten volts, there is a fault in the wire between the module positive (+) terminal and the ignition coil positive

connector or the ignition coil and primary circuit-to-ignition switch.

6   If the reading is ten volts or more, check the "C" terminal on the module (two-wire connector).

7   If the reading is less than one volt, there is an open or grounded lead in the distributor-to-coil "C" terminal connection or ignition coil or an open primary circuit in the coil itself.

8   If the reading is one to ten volts, replace the module with a new one and check for a spark (see Section 6). If there is a spark the module was faulty and the system is now operating properly. If there is no spark, there is a fault in the ignition coil.

9   If the reading in Step 4 is 10 volts or more, unplug the pick-up coil connector (two-wire) from the module (terminals P and N). Locate the TACH harness terminal (taped to the harness near the coil) and install a voltmeter positive probe (+). With the ignition key ON (engine not running), use a 1.5 volt power source (flashlight battery) and jumper wires and carefully touch the positive probe to terminal P on the module **(see illustration 6.11)**. Voltage at TACH should momentarily drop then return to normal.

10   If there is no drop in voltage, check the module ground and, if it is good, replace the module with a new one.

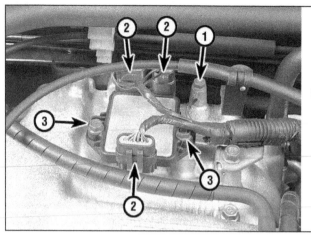

12.29 On 1996 and 1997 3.2L V6 models, check for battery voltage to the ignition module on the black/yellow wire; to remove the module, disconnect the ground wire bolt (1), unplug the three electrical connectors (2) and remove the mounting bolts (3)

11   If the voltage drops, check for spark at the coil wire (spark tester hooked to coil tower wire) as the test light is removed from the module terminal. If there is no spark, the module is faulty and should be replaced with a new one. If there is a spark, the pick-up coil or connections are faulty or not grounded.

Replacement

12   Detach the cable from the negative terminal of the battery.
13   Remove the distributor cap and rotor (see Chapter 1).
14   Disconnect the electrical connectors from the module **(see illustration 11.4)**. Note that the connectors cannot be interchanged.
15   Remove both module attaching screws and lift the module up and away from the distributor.
16   Do not wipe the grease from the module or the distributor base if the same module is to be reinstalled. If a new module is to be installed, a package of silicone grease will be included with it. Wipe the distributor base and the new module clean, then apply the silicone grease on the face of the module and on the distributor base where the module seats. This grease is necessary for heat dissipation.
17   Install the module and attach both electrical leads.
18   Install the distributor rotor and cap (see Chapter 1).
19   Attach the cable to the negative terminal of the battery.

2.3L and 2.6L four-cylinder models

Check

20   First, check the ignition system as outlined in Steps 5 through 9 in Section 6).
21   Check for battery voltage to the coil and to the ignition module (see the wiring diagrams at the end of Chapter 12).

Replacement

22   Detach the cable from the negative terminal of the battery.
23   Disconnect the ignition module electrical connector.

24   Remove the module attaching screws and lift the module up and away from its mount.
25   Install the module and tighten the screws securely.
26   Plug in the electrical connector.
27   Attach the cable to the negative terminal of the battery.

Distributorless Ignition Systems (DIS)

Note: *The ignition module on 1998 and later 3.2L V6 models is an integral part of the PCM. Do not attempt to check it. If a check of the ignition system (see Section 6) indicates that the ignition module might be defective, have it checked out by a dealer service department. If the ignition module is bad on these models, the PCM must be replaced.*

Check

Refer to illustration 12.29

28   First, check the ignition system as outlined in Steps 19 through 28 (3.2L V6 models) or Steps 29 through 33 (2.2L four-cylinder models) in Section 6.
29   Disconnect the electrical connector from the ignition control module and check for battery voltage **(3.2L V6, see illustration; 2.2L four, see illustration 6.31a)**. On 1993 through 1995 3.2L V6 models, the battery voltage feed wire to the module is black/orange **(see illustration 6.26a)**; on 1996 and 1997 3.2L models, it's black/yellow **(see illustration 6.26b)**. On 2.2L four-cylinder models, the battery voltage feed wire to the module is black/orange **(see illustration 6.31b)**.
30   If no voltage is present, check the circuit from the ignition key switch to the ignition module (refer to the wiring diagrams at the end of Chapter 12). **Note:** *If battery voltage is not present, check the IGNITION fuse. This might be a simple solution to a no start condition.*
31   Check the operation of the crankshaft sensor and the camshaft sensor (see Chapter 6).
32   If all these tests are correct and there is no spark output at any of the coils, the ignition control module is defective.

Replacement

33   Detach the cable from the negative terminal of the battery.

1997 and earlier 3.2L V6 models

34   Remove the ground wire bolt **(see illustration 12.29)**.
35   Unplug the three electrical connectors from the ignition control module **(see illustration 12.29)**.
36   Remove the ignition control module mounting bolts **(see illustration 12.29)**.
37   Remove the ignition control module.

2.2L four-cylinder models

38   Remove the spark plug cable cover bolts with a 5mm Allen wrench and remove the cover.
39   Clearly label, then disconnect, the spark plug wires from the spark plugs.
40   Unplug the electrical connector at the module **(see illustration 8.31)**.
41   Remove the three ignition coil/ignition control module retaining bolts **(see illustrations 8.32a and 8.32b)**.
42   Remove the ignition coil/ignition control module unit.

All models

43   Installation is the reverse of removal. Be sure to attach the wires of the new module to the coil assembly spade terminals in exactly the same order in which they were removed.

---

13 Charging system - general information and precautions

The charging system includes the alternator, an internal voltage regulator, a charge indicator or warning light, the battery and the wiring between all the components. The charging system supplies electrical power for the ignition system, the lights, the radio, etc. The alternator is driven by a drivebelt at the front of the engine. Refer to the wiring diagrams at the end of Chapter 12 for additional information.

The charging system on the 3.1L engine is equipped with the CS-130 alternator. This unit is not serviceable; if it's defective, exchange it as a core for a new or rebuilt unit. The first alternator overhaul covered in this manual applies to the 2.3L and 2.6L four-cylinder engines and to the 3.2L V6 engine; the second one applies to the 2.2L four-cylinder engine.

The purpose of the voltage regulator is to limit the alternator's voltage to a preset value. This prevents power surges, circuit overloads, etc., during peak voltage output.

The charging system doesn't ordinarily require periodic maintenance. However, the drivebelt, battery and wires and connections should be inspected at the intervals outlined in Chapter 1.

Be very careful when making electrical circuit connections to a vehicle equipped with an alternator and note the following:

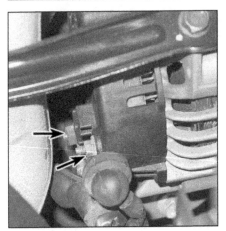

15.2 Disconnect the electrical wires from the terminals on the back of the alternator (alternator on 2.2L engine shown; other alternators similar)

a) *When reconnecting wires to the alternator from the battery, be sure to note the polarity.*

b) *Before using arc welding equipment to repair any part of the vehicle, disconnect the wires from the alternator and the battery terminals.*

c) *Never start the engine with a battery charger connected.*

d) *Always disconnect both battery cables before using a battery charger (negative cable first, positive cable last).*

14 Charging system - check

1   If a malfunction occurs in the charging circuit, do not immediately assume that the alternator is causing the problem. First check the following items:

a) *The battery cables where they connect to the battery. Make sure the connections are clean and tight.*

b) *The battery electrolyte specific gravity. If it is low, charge the battery.*

c) *Check the external alternator wiring and connections.*

d) *Check the fuses and fusible links (see Chapter 12).*

e) *Check the drivebelt condition and tension (see Chapter 1).*

f) *Check the alternator mounting bolts for tightness.*

g) *Run the engine and check the alternator for abnormal noise.*

2   Using a voltmeter, check the battery voltage with the engine off. It should be approximately 12-volts.

3   Start the engine and check the battery voltage again. It should now be approximately 14 to 15-volts.

4   If the indicated voltage reading is less or more than the specified charging voltage, the problem may be within the alternator.

5   Due to the special equipment necessary to test or service the alternator, it is recom-

15.5 Using a chain or strap wrench to hold the pulley, loosen the pulley retaining nut, and then use a small puller, if necessary, to remove the pulley from the shaft; be sure to use a strip of old inner tube to protect the pulley from the chain

mended that if a fault is suspected the vehicle be taken to a shop with the proper equipment. But if the home mechanic feels confident in the use of an ohmmeter, and in some cases a soldering iron, the component check and replacement procedures for the most common alternator type are included in Section 14.

6   Some models are equipped with an ammeter on the instrument panel that indicates charge or discharge - current passing in or out of the battery. With all electrical equipment switched ON, and the engine idling, the gauge needle may show a discharge condition. At fast idle or normal driving speeds the needle should stay on the charge side of the gauge, with the charged state of the battery determining just how far over (the lower the battery state of charge, the farther the needle should swing toward the charge side).

7   Some models are equipped with a voltmeter on the instrument panel that indicates battery voltage with the key ON (engine not

running), and alternator output when the engine is running.

8   The charge light on the instrument panel illuminates with the key ON and engine not running, and should go out when the engine runs.

9   If the gauge does not show a charge when it should or the alternator light (if equipped) remains on, there is a fault in the system. Before inspecting the brushes or replacing the alternator, the battery condition, alternator belt tension and electrical cable connections should be checked.

15 Alternator - removal and installation

Refer to illustrations 15.2, 15.5 and 15.7

1   Detach the cable from the negative terminal of the battery.

2   Clearly label and then disconnect the electrical connectors from the alternator **(see illustration)**.

3   On 2.3L and 2.6L four-cylinder and on 3.1L V6 models, loosen the alternator adjustment bolt(s) and detach the accessory drivebelt from the alternator drive pulley.

4   On 2.2L four-cylinder and 3.2L V6 and engines, loosen the serpentine drivebelt (see Section 25 in Chapter 1) and disengage it from the alternator drive pulley.

5   If you're going to swap the existing alternator unit for a new or rebuilt unit, remove the alternator pulley **(see illustration)**. (You'll need to install the pulley on the new or rebuilt unit.)

6   On all models except 2.2L four-cylinder models, remove the adjustment and pivot bolts and separate the alternator from its mounting brackets.

7   On 2.2L models, remove the alternator mounting brackets and the lower pivot bolt **(see illustration)**.

8   Installation is the reverse of removal.

9   After the alternator is installed, adjust the drivebelt tension (see Section 25 in Chapter 1).

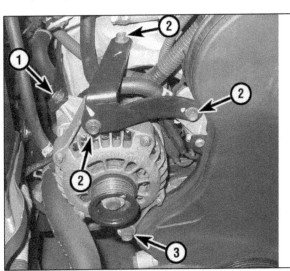

15.7 To detach the alternator from a 2.2L engine:

A   Remove the front bolt (1) from the smaller bracket, loosen the bolt (not shown) at the other end of this bracket and rotate the bracket out of the way

B   Remove all three bolts (2) from the bigger bracket and remove the bracket

C   Remove the lower pivot bolt (3)

16.2 Paint a mark across the front cover, stator and rear cover to aid in reassembly

16.3 Remove the pulley nut, then remove the pulley

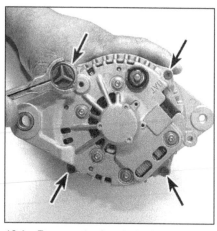

16.4a Remove the four bolts from the rear cover (arrows)

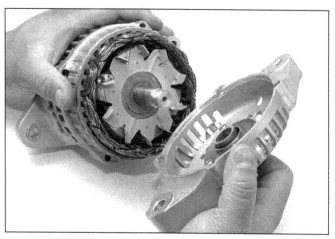

16.4b Separate the front cover from the alternator body

16.4c Remove the rotor from the alternator assembly

16   Alternator components - check and replacement

Note: *The alternator overhaul covered in this manual applies to the original equipment on 2.3L, 2.6L and 1997 and earlier 3.2L engines. The charging system on the 3.1L engine is equipped with the CS-130 alternator. These alternators should be considered non-serviceable and, if found to be defective, should be exchanged as cores for new or rebuilt units. These alternators cannot be disassembled and repaired. The alternators used on 1998 and later 3.2L V6 and 2.2L four-cylinder engines should also be considered non-serviceable.*

2.3L and 2.6L four-cylinder and 1997 and earlier 3.2L V6 models

Disassembly

Refer to illustrations 16.2, 16.3, 16.4a, 16.4b, 16.4c, 16.5a, 16.5b, 16.6a, 16.6b, 16.6c and 16.7

1   Remove the alternator from the vehicle (Section 15).

2   Scribe or paint marks on the front and rear end frame housings of the alternator to facilitate reassembly **(see illustration)**.

3   Remove the nut retaining the pulley to the rotor shaft and remove the pulley **(see illustration)**.

4   Remove the four through-bolts holding the front and rear covers together, then sepa-

rate the rear cover assembly from the front cover. Remove the rotor **(see illustrations)**.

5   Remove the nuts retaining the stator, then separate the stator and diode assembly from the end frame **(see illustrations)**.

6   Remove the screws attaching the regulator and the brush holder to the diode assembly and remove the brush holder and

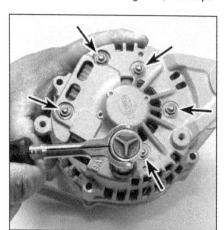

16.5a Remove the mounting nuts (arrows)

16.5b Separate the rear cover from the stator and diode assembly

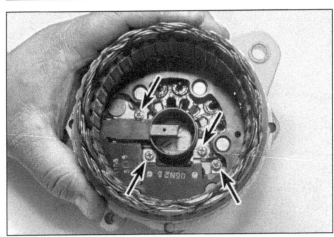

16.6a Remove the diode assembly mounting screws (arrows)

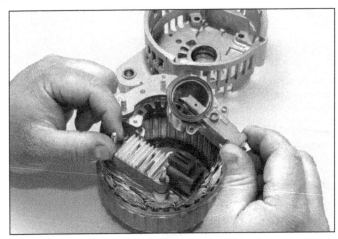

16.6b Remove the brush assembly holder from the diode assembly

regulator **(see illustrations)**.

7   Remove the diode assembly from the stator **(see illustration)**. On some types of alternators it will be necessary to use a solder gun and heat sink to melt the solder joints that connect the stator leads to the diode assembly.

Component checks

Refer to illustrations 16.8a, 16.8b, 16.9 and 16.10

8   Check the rotor for an open between the two slip rings **(see illustration)**. There should be continuity between the slip rings. Check for grounds between each slip ring and the rotor shaft **(see illustration)**. There should be no continuity (infinite resistance) between the rotor shaft and either slip ring. If the rotor fails either test, or if the slip rings are excessively worn, the rotor is defective.

9   Check for opens between the center terminal and each end terminal of the stator windings **(see illustration)**. If either reading is high (infinite resistance), the stator is defective. Check for a grounded stator winding

16.6c Remove the voltage regulator

between each stator terminal and the frame. If there's continuity between any stator winding and the frame, the stator is defective.

10   Start the checks on the diode assembly

by touching one probe (+) of the ohmmeter on the diode plate to the other probe on one of the other designated diode terminals (-)

16.7 Remove the four set screws and lift the diode assembly from the stator

16.8a Continuity should exist between the rotor slip rings

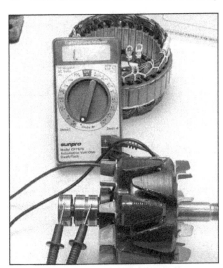

16.8b No continuity should exist between the slip ring(s) and rotor shaft

16.9 Check for continuity on each stator lead. There should be no breaks in the windings, therefore continuity should exist between each terminal

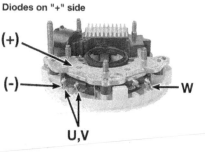

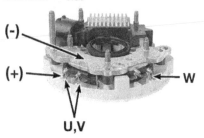

16.10 Check for continuity between the diode terminals U, V, W and the rectifier body then reverse the probes and check once again. Continuity should exist in only ONE direction

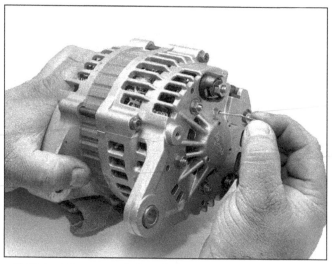

16.13 Install a paper clip into the backside of the alternator to hold the brushes in place during installation - after installation, simply pull the paper clip out

(see illustration). Then reverse the probes and check again. Follow the chart carefully. The diode should have continuity one way and no continuity when the probes are reversed. Check each of the terminals in this manner. If any of the diodes fail the test, the diode assembly is defective. **Note:** *Select the diode function on the ohmmeter.*

11   Replace the alternator brushes. **Note:** *On some models the brush leads are soldered onto the voltage regulator assembly.*

Reassembly

Refer to illustration 16.13

12   Install the components in the reverse order of removal, noting the following:

13   Before installing the brush holder, push the brushes into the holder and slip a straightened paper clip or other suitable pin through the hole in the brush holder to hold the brushes in a retracted position. After the front and rear end frames have been bolted together, remove the paper clip **(see illustration)**.

2.2L four-cylinder and 1998 and later 3.2L V6 models

14   The alternators on these models are not serviceable.

17   Starting system - general information and precautions

1   The starting system is composed of the starter motor, starter solenoid (some models), battery, ignition switch, clutch start switch (manual transmission models), neutral start/back-up light switch (automatic transmission models), and connecting wires.

2   Turning the ignition key to the Start position actuates the starter relay through the starter control circuit. The starter solenoid then connects the battery to the starter. The battery supplies the electrical energy to the starter motor, which does the actual work of cranking the engine. Refer to the starting system wiring schematics at the end of Chapter 12 for addition information.

3   All vehicles are equipped with a starter assembly that is mounted to the transmission bellhousing.

4   All vehicles are equipped with a clutch start switch or a neutral start switch in the starter control circuit, which prevents operation of the starter unless the shift lever is in Neutral or Park (automatic) or the clutch is depressed (manual).

5   Never operate the starter motor for more than 15 seconds at a time without pausing to allow it to cool for at least two minutes. Excessive cranking can cause overheating, which can seriously damage the starter.

18   Starter motor and circuit - in-vehicle check

Note: *Before diagnosing starter problems, make sure the battery is fully charged.*

General check

1   If the starter motor doesn't turn at all when the switch is operated, make sure the shift lever is in Neutral or Park or the clutch is fully depressed.

2   Make sure the battery is charged and that all cables at the battery and starter solenoid terminals are secure.

3   If the starter motor spins but the engine doesn't turn over, the drive assembly in the starter motor is slipping and the starter motor must be replaced (see Section 17).

4   If, when the switch is actuated, the starter motor doesn't operate at all but the starter solenoid operates (clicks), the problem lies with either the battery, the starter solenoid contacts or the starter motor connections.

5   If the starter solenoid doesn't click when the ignition switch is actuated, either the starter solenoid circuit is open or the solenoid itself is defective. Check the starter solenoid circuit (see the wiring diagrams at the end of Chapter 12) or replace the solenoid (see Section 20).

6   To check the starter solenoid circuit, remove the push-on connector from the solenoid wire. Make sure that the connection is clean and secure and the relay bracket is grounded. If the connections are good, check the operation of the solenoid with a jumper wire. To do this, place the transmission in Park or Neutral and apply the parking brake. Remove the push-on connector from the solenoid. Connect a jumper wire between the battery positive terminal and the exposed terminal on the solenoid. If the starter motor now operates, the starter solenoid is okay. The problem is in the ignition switch, Neutral start switch or in the starting circuit wiring (look for open or loose connections).

7   If the starter motor still doesn't operate, replace the starter motor or starter solenoid (see Section 19 or 20).

8   If the starter motor cranks the engine at an abnormally slow speed, first make sure the battery is fully charged and all terminal connections are clean and tight. Also check the connections at the starter solenoid and battery ground. Eyelet terminals should not be easily rotated by hand. Also check for a short to ground. If the engine is partially seized, or has the wrong viscosity oil in it, it will crank slowly.

19.4a To remove the starter from the 3.1L V6 engine, detach the electrical wires from the solenoid terminals, then remove both mounting bolts

19.4b Starter mounting bolts on the 3.2L V6 engine

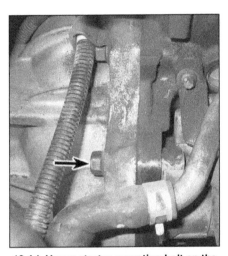

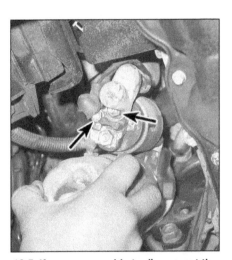

19.4c Lower starter mounting bolt on the 2.2L four-cylinder engine

19.4d Upper starter mounting bolt on the 2.2L four-cylinder engine

19.5 If you were unable to disconnect the battery cable and the solenoid lead while the starter motor was still bolted in place, pull the starter out far enough to access the terminals and disconnect both wires

Starter cranking circuit test

Note: *To determine the location of excessive resistance in the starter circuit, perform the following simple series of tests.*

9   Disconnect the ignition coil wire from the distributor cap and ground it on the engine.

10   Connect a remote control starter switch from the battery terminal of the starter solenoid to the S terminal of the solenoid.

11   Connect a voltmeter positive lead to the starter motor terminal of the starter solenoid, then connect the negative lead to ground.

12   Actuate the ignition switch and take the voltmeter readings as soon as a steady figure is indicated. Do not allow the starter motor to turn for more than 15 seconds at a time. A reading of 9-volts or more, with the starter motor turning at normal cranking speed, is normal. If the reading is 9-volts or more but the cranking speed is slow, the motor is faulty. If the reading is less than 9-volts and the cranking speed is slow, the solenoid contacts are probably burned.

19   Starter motor - removal and installation

Refer to illustrations 19.4a, 19.4b and 19.4c and 19.5

1   Detach the cable from the negative terminal of the battery.

2   Raise the vehicle and support it securely on jackstands.

3   If you can reach them, disconnect the battery cable and the solenoid connector from the starter motor solenoid. (On some engines, such as the 2.2L four-cylinder engine, the solenoid terminals are very difficult to reach while the starter is installed; it's actually easier to unbolt the starter, pull it out and then disconnect the starter cables.)

4   Remove the starter motor mounting bolts and detach the starter from the engine **(see illustrations)**. Depending upon the year and engine type, the starter/solenoid assembly may be mounted above the transmission or below the intake manifold. **Note:** *On some*

models it may be necessary to remove the exhaust system to make room for the starter assembly to pass between the engine and chassis. Make sure the engine is completely cool before working on the exhaust system.

5   If you haven't yet disconnected the battery cable and solenoid lead, do so now **(see illustration)**.

6   Installation is the reverse of removal.

20   Starter solenoid - replacement

Refer to illustration 20.2

Note: *Some engines are equipped with a starter solenoid that is mounted on the starter assembly. All other engines use an overrunning clutch, idle pinion gear and a starter relay that is mounted separately from the starter.*

1   Remove the nut and disconnect the wire from the solenoid electrical terminal.

2   Remove the two bolts from the solenoid and separate the solenoid from the starter assembly **(see illustration)**.

3   Installation is the reverse of removal.

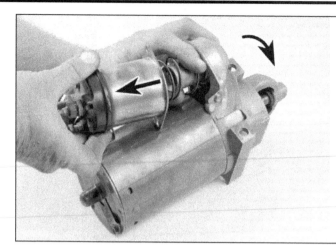

20.2 To remove the starter solenoid from the starter motor, remove the solenoid-to-starter frame mounting screws, turn the solenoid in a clockwise direction and pull (starter for 3.1L V6 engine shown, other units similar)

Chapter 6
Emissions and engine control systems

Contents

Specifications

Torque specifications

Ft-lbs (unless otherwise noted)

Camshaft Position (CMP) sensor retaining bolt	78 in-lbs
Crankshaft Position (CKP) Sensor retaining bolt	78 in-lbs
Engine Coolant Temperature (ECT) sensor	22
Knock sensor retaining bolt	177 in-lbs
Oxygen sensor	40
Vehicle Speed Sensor (VSS) hold-down bolt	144 in-lbs

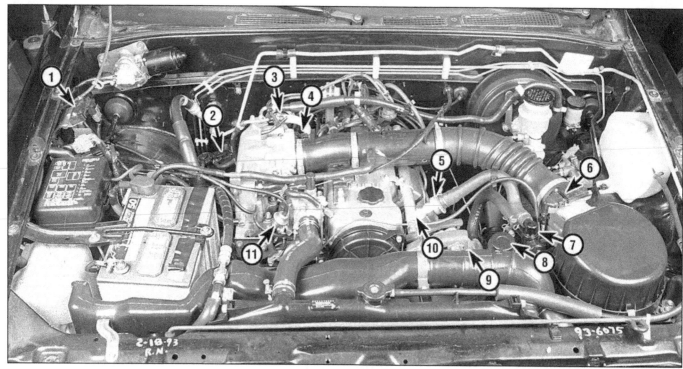

1.1a Emission and engine control system component locations on the 2.6L four-cylinder engine

1	Manifold Absolute Pressure (MAP)	4	Throttle Position Sensor (TPS)	8	Backpressure transducer
	sensor	5	Air injection manifold check valve	9	Air pump
2	Charcoal canister	6	Mass Airflow (MAF) sensor	10	Injection manifold
3	Exhaust Gas Recirculation (EGR) valve	7	Air diverter valve	11	Fuel pressure regulator

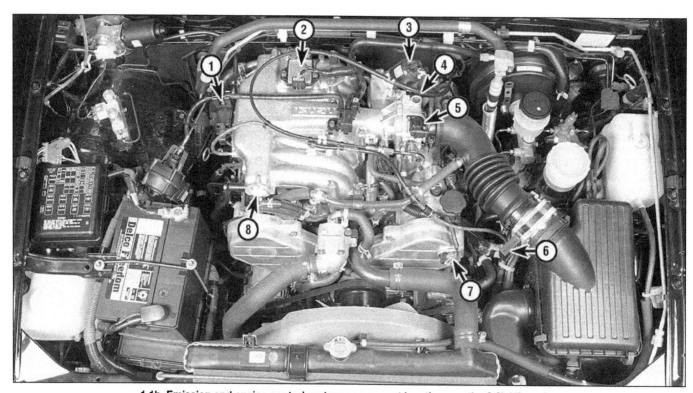

1.1b Emission and engine control system component locations on the 3.2L V6 engine

1	Manifold Absolute Pressure (MAP)	3	Linear EGR valve	6	Mass Airflow (MAF) sensor
	sensor	4	Throttle Position Sensor (TPS)	7	Camshaft position sensor
2	Ignition control module	5	Idle Air Control (IAC) valve	8	Fuel pressure test port

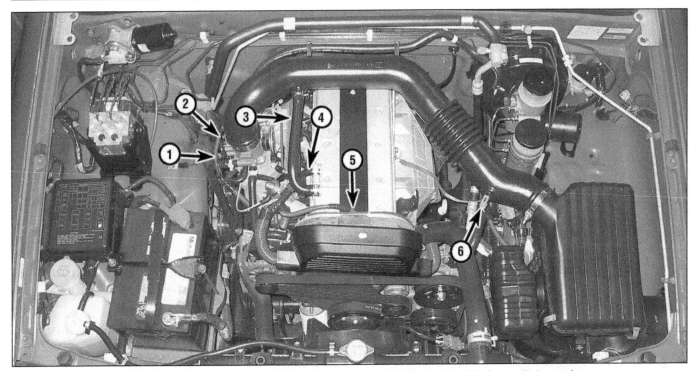

1.1c Emission and engine control system component locations on the 2.2L four-cylinder engine

1   Throttle Position (TP) sensor (not
    visible, on right side of throttle body)
2   Idle Air Control (IAC) valve (not visible,
    on right side of throttle body)

3   Positive Crankcase Ventilation (PCV)
    system, crankcase fresh air intake
    hose
4   Positive Crankcase Ventilation (PCV)
    system, PCV hose to intake manifold

5   Camshaft Position (CMP) sensor (not
    visible, located between camshaft
    drive sprockets)
6   Intake Air Temperature (IAT) sensor

1   General information

Refer to illustrations 1.1a, 1.1b, 1.1c and 1.6

To prevent pollution of the atmosphere from incompletely burned and evaporating gases, and to maintain good driveability and fuel economy, a number of emission control systems are incorporated **(see illustrations)**. They include the:

Air Injection Reactor (AIR) system
Air regulator (2.6L engine)
Fuel cut-off system
Mixture Control valve system
High Altitude Emission control system
Idle air control valve
Automatic choke system
*Thermostatically Controlled Air Cleaner
    (THERMAC) system*
Feedback carburetors (2.3L engine)*
*Evaporative Emission Control (EVAP)
    system*
*Throttle Body Injection (TBI) system
    (3.1L engine)*
*Multi Port Fuel Injection (MPFI) system
    (2.2L, 2.6L and 3.2L engines)*
*Positive Crankcase Ventilation (PCV)
    system*
*Exhaust Gas Recirculation (EGR) system
    (electronic)*
Catalytic converter

*Refer to Chapter 4A for information on the components and locations for the 2.3L feedback carburetor.

The Sections in this Chapter include general descriptions, checking procedures within the scope of the home mechanic and component replacement procedures (when possible) for each of the systems listed above.

Before assuming that an emissions control system is malfunctioning, check the fuel and ignition systems carefully. The diagnosis of some emission control devices requires specialized tools, equipment and training. If checking and servicing become too difficult or if a procedure is beyond your ability, consult a dealer service department or other qualified repair facility. Remember, the most frequent cause of emissions problems is simply a loose or broken vacuum hose or wire, so always check the hose and wiring connections first.

This doesn't mean, however, that emission control systems are particularly difficult to maintain and repair. You can quickly and easily perform many checks and do most of the regular maintenance at home with common tune-up and hand tools. **Note:** *Because of a Federally mandated extended warranty which covers the emission control system components, check with your dealer about warranty coverage before working on any emissions-related systems. Once the warranty has expired, you may wish to perform some of the component checks and/or replacement procedures in this Chapter to save money.*

Pay close attention to any special precautions outlined in this Chapter. It should be noted that the illustrations of the various systems may not exactly match the system installed on the vehicle you're working on because of changes made by the manufacturer during production or from year-to-year.

A Vehicle Emissions Control Information (VECI) label is located in the engine compartment **(see illustration)**. This label contains important emissions specifications and adjustment information, as well as a vacuum

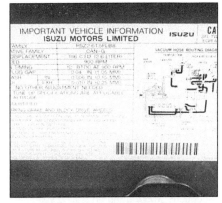

1.6 The Vehicle Emission Control Information (VECI) label on some models is located in the engine compartment under the hood and contains information on the emission devices on your vehicle, vacuum line routing, etc.

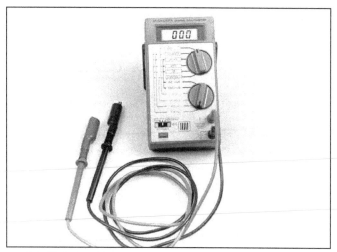

2.1 Digital multimeters can be used for testing all types of circuits; because of their high impedance, they are much more accurate than analog meters for measuring low-voltage computer circuits

2.2 Scanners like the Actron Scantool and the AutoXray XP240 are powerful diagnostic aids; programmed with comprehensive diagnostic information, they can tell you jus about anything you want to know about your engine management system

hose schematic with emissions components identified. When servicing the engine or emissions systems, the VECI label in your particular vehicle should always be checked for up-to-date information.

---

2   On Board Diagnostic (OBD) systems and trouble codes

Diagnostic tool information

Refer to illustrations 2.1 and 2.2

1   A digital multimeter is necessary for checking fuel injection and emission related components **(see illustration)**. A digital volt-ohmmeter is preferred over the older style analog multimeter for several reasons. The analog multimeter cannot display the volts-ohms or amps measurement in hundredths and thousandths increments. When working with electronic circuits which are often very low voltage, this accurate reading is most important. Another good reason for the digital multimeter is the high impedance circuit. The digital multimeter is equipped with a high resistance internal circuitry (10 million ohms). Because a voltmeter is hooked up in parallel with the circuit when testing, it is vital that none of the voltage being measured should be allowed to travel the parallel path set up by the meter itself. This dilemma does not show itself when measuring larger amounts of voltage (9 to 12 volt circuits) but if you are measuring a low voltage circuit such as the oxygen sensor signal voltage, a fraction of a volt may be a significant amount when diagnosing a problem. However, there are several exceptions where using an analog voltmeter may be necessary to test certain sensors.

2   Hand-held scanners are the most powerful and versatile tools for analyzing engine management systems used on later model vehicles **(see illustration)**. A scan tool has the ability to diagnose in-depth driveability

problems and it allows freeze frame data to be retrieved from the PCM stored memory. Freeze frame data is a feature that records all related sensor and actuator activity on the PCM data stream whenever an engine control or emissions fault is detected and a trouble code is set. This ability to look at the circuit conditions and values when the malfunction occurs provides a valuable tool when trying to diagnose intermittent driveability problems. If the tool is not available and intermittent driveability problems exist, have the vehicle checked at a dealer service department or other qualified repair shop.

General description

3   When they're functioning correctly, emission control systems significantly reduce the levels of certain pollutants (carbon monoxide, hydrocarbons and oxides of nitrogen) emitted by automotive engines. The purpose of On Board Diagnostics (OBD) is to ensure that these systems continue to function correctly. The OBD system is an integral part of the engine management system, which consists of information sensors, output actuators and the Powertrain Control Module (PCM) (referred to on older models as the Electronic Control Module, or ECM). When the OBD system detects a malfunction, it notifies the driver so that the vehicle can be serviced. It also stores enough information about the malfunction so that it can be identified and repaired.

4   In many ways, this system can be compared to the central nervous system in the human body. The sensors (nerve endings) constantly gather information and send this data to the PCM (brain), which processes the data and, if necessary, sends out a command to change the operation of the fuel, ignition and/or emission systems (limbs).

5   Here's a specific example of how one portion of this system operates: An oxygen sensor, mounted in the exhaust manifold and

protruding into the exhaust gas stream, constantly monitors the oxygen content of the exhaust gas as it travels through the exhaust pipe. If the percentage of oxygen in the exhaust gas is incorrect, an electrical signal is sent to the PCM. The PCM takes this information, processes it and then sends a command to the fuel injector(s) on the fuel injection system, telling it to change the air/fuel mixture. This happens in a fraction of a second, and it goes on continuously while the engine is running. The end result is an air/fuel mixture which is constantly kept at a predetermined ratio, regardless of driving conditions.

Testing

Refer to illustrations 2.7a and 2.7b

6   The OBD system indicates a problem by turning on a CHECK ENGINE light on the instrument panel when a fault has been detected. Perhaps more importantly, the PCM will recognize this fault, in a particular system monitored by one of the various information sensors, and store it in its memory in the form of a trouble code. Although the trouble code cannot reveal the exact cause of the malfunction, it greatly facilitates diagnosis as you or a dealer mechanic can ''tap into'' the PCMs memory and be directed to the problem area.

7   To retrieve this information from the PCM on 2.3L feedback carburetor systems, 2.6L and 3.2L MPFI and 3.1L TBI engines, you must link the diagnostic terminals. 2.3L and 2.6L engines use two long terminals, one male and the other female. 3.2L engines use a short jumper wire to ground a diagnostic terminal located under the left end of the instrument panel (sometimes behind the kick panel) **(see illustration)**. 3.1L engines are equipped with a specialized diagnostic terminal. This terminal is part of an electrical connector called the Assembly Line Data Link (ALDL) **(see illustration)**. The ALDL is located under the dashboard, just below the instrument

2.7b The 12 pin Assembly Line Data Link (ALDL) terminal is located under the dash on the driver's side on 3.1L engines. To retrieve the trouble codes, connect a jumper wire to terminals A and B

2.7a On 2.3L and 2.6L engines, connect the two diagnostic wires that are taped together near the PCM; on 1995 and earlier 3.2L engines, connect terminals 1 and 3 of the diagnostic connector

panel and to the left of the center console. To use the ALDL, remove the plastic cover (if equipped). With the electrical connector exposed to view, push one end of the paper clip into the diagnostic TEST terminal (A) and the other end into the GROUND terminal (B).

8   Turn the ignition to the On position. **Caution:** *Do not start the engine with the TEST terminal grounded.* The "CHECK ENGINE" light should flash Trouble Code 12, indicating that the diagnostic system is working. Code 12 will consist of one flash, followed by a short pause, then two more flashes in quick succession. After a longer pause, the code will repeat itself two more times. If no other codes have been stored, Code 12 will continue to repeat itself until the jumper wire is disconnected. If additional

Trouble Codes have been stored, they will follow Code 12. Again, each Trouble Code will flash three times before moving on.

9   Once the code(s) have been noted, use the Trouble Code Identification information which follows to locate the source of the fault. **Note:** *Whenever the battery cable is disconnected, all stored Trouble Codes in the PCM are erased. Be aware of this before you disconnect the battery.*

10   It should be noted that the self-diagnosis feature built into this system does not detect all possible faults. If you suspect a problem with the On Board Diagnostic (OBD) system, but the CHECK ENGINE light has not come on and no trouble codes have been stored, take the vehicle to a dealer service department or other qualified repair shop for diagnosis.

11   Furthermore, when diagnosing an engine performance, fuel economy or exhaust emissions problem (which is not accompanied by a CHECK ENGINE light) do

not automatically assume the fault lies in this system. Perform all standard troubleshooting procedures, as indicated elsewhere in this manual, before turning to the On Board Diagnostic (OBD) system.

12   Finally, since this is an electronic system, you should have a basic knowledge of automotive electronics before attempting any diagnosis. Damage to the PCM, Programmable Read Only Memory (PROM) calibration unit or related components can easily occur if care is not exercised.

Trouble Code Identification

13   Following is a list of the typical Trouble Codes which may be encountered while diagnosing the On Board Diagnostic (OBD) system. Also included are simplified troubleshooting procedures. If the problem persists after these checks have been made, the vehicle must be diagnosed by a professional mechanic who can use specialized diagnostic tools and advanced troubleshooting methods to check the system. Procedures marked with an asterisk (*) indicate component replacements which may not cure the problem in all cases. For this reason, you may want to seek professional advice before purchasing replacement parts

Erasing trouble codes

14   To clear the Trouble Code(s) from the PCM memory, momentarily remove the PCM MAIN fuse from the fuse box for 10 seconds. Clearing codes may also be accomplished by disconnecting the cable from the negative terminal of the battery. **Caution:** *To prevent damage to the PCM, the ignition switch must be OFF when disconnecting or connecting power to the PCM.*

OBD trouble codes for 2.3L, 2.6L, 3.1L and 3.2L models

Trouble codes	Circuit or system	Probable cause
Code 12	Diagnostic mode	This code will flash whenever the diagnostic terminal is grounded with the ignition turned on and the engine not running. If additional trouble codes are stored in the PCM, they will appear after this code has flashed three times.
Code 13	Oxygen sensor circuit	The wiring is broken or the oxygen sensor is deteriorated. Check the wiring, then replace the sensor.*
Code 14	Engine Coolant Temperature (ECT) Sensor (2.3L, 2.6L and 3.1L engines)	Sensor signal shorted (2.3L), grounded (2.6L) or out of range (3.1L) Repair the circuit or replace the sensor.*
Code 15	Engine Coolant Temperature (ECT) Sensor (2.3L and 2.6L engines)	Sensor signal open (2.3L), out of range (2.6L) or out of range (1989 through 1994 2.6L). Repair the circuit or replace the sensor.
Code 16	Engine Coolant Temperature (ECT) Sensor (2.6L engine)	Sensor signal open. Repair the circuit or replace the sensor.
Code 21	Idle switch (2.3L engine)	Open circuit. Repair the circuit or replace the idle switch.*
Code 21	Wide open throttle (WOT) switch (2.3L engine)	Short circuit. Repair the circuit or replace the WOT switch.*
Code 21	Throttle Valve Switch (TVS) (2.6L engine)	Idle contact and full contact made simultaneously. Replace the TVS.*
Code 21	Throttle Position Sensor (TPS) (3.1L engine)	Signal voltage high. Check for a sticking or incorrectly adjusted TPS plunger, check the TPS circuit and check the TPS adjustment. Repair the circuit or replace the TPS.*
Code 21	Throttle Position Sensor (TPS) (3.1L engine)	Signal voltage low. Check for a sticking or incorrectly adjusted TPS, check the TPS circuit and check the TPS adjustment. Repair the circuit or replace the TPS.*

OBD trouble codes for 2.3L, 2.6L, 3.1L and 3.2L models (continued)

Trouble codes	Circuit or system	Probable cause
Code 21	Throttle Position Sensor (TPS) (3.2L engines)	Signal voltage out of range. Check the TPS, the TPS circuit and the TPS adjustment. Repair the circuit or replace the TPS.*
Code 22	Fuel cut solenoid circuit (2.3L engine)	The fuel cut solenoid circuit is open or shorted.
Code 22	Starter signal system	No starter signal. The starter signal circuit is open or grounded.
Code 22	Throttle position sensor signal voltage low (3.1L engine)	Check the TPS adjustment. Check the PCM connector. Replace the TPS.*
Code 23	Ignition control module power transistor circuit	The ignition module for the ignition system is grounded or shorted at the output terminal. Check the circuit for a short or replace the ignition control module.*
Code 23	Intake Air Temperature (IAT) Sensor (3.1L engine)	Sensor signal falsely indicates a high temperature.
Code 23	Intake Air Temperature (IAT) Sensor (3.2L engine)	Sensor signal out of range.
Code 24	Vehicle Speed Sensor (VSS) (2.3L and 3.1L engines)	Circuit fault, should be indicated only when the vehicle is in motion. Disregard Code 24 if it is set when the drive wheels are not turning. Check the connections at the PCM.
Code 24	Vehicle Speed Sensor (VSS) (3.2L engine)	No input signal, which should be indicated only when the vehicle is in motion. Disregard Code 24 if it is set when the drive wheels are not turning. Check the connections at the PCM.
Code 24	Pressure regulator vacuum switching valve (1989 through 1994 2.3L engine)	Check and repair the pressure regulator vacuum switching valve circuit or connectors or replace the pressure regulator vacuum switching valve.*
Code 25	Air Injection Reactor (AIR) vacuum switching valve (VSV) (2.3L and 2.6L engines)	The VSV circuit is open, grounded or shorted. Check the circuit.
Code 25	Intake Air Temperature (IAT) Sensor	High temperature indicated. Check the IAT sensor circuit or replace the sensor.*
Code 26	Canister vacuum switching valve (VSV) system for canister purge (2.3L and 2.6L engines)	Circuit open or grounded. Check and repair the circuit or replace the VSV.*
Code 27	Canister vacuum switching valve (VSV) system for canister purge (2.3L engine)	Constant high voltage to PCM. Check and repair the circuit or replace the VSV.*
Code 27	Canister purge vacuum switching valve (VSV) (2.6L engine)	Faulty transistor or bad ground circuit. Check and repair the circuit or replace the VSV.*
Code 31	Ignition reference signal (2.3L engine)	No ignition reference signal circuit. Have the vehicle checked by a dealer service department.
Code 32	EGR sensor and circuit	The EGR sensor is faulty or the circuit is shorted. Check vacuum hoses and connections for leaks and restrictions. Check and repair the circuit or replace the EGR sensor, solenoid or valve as necessary.*
Code 33	Fuel injector circuit (2.6L engine)	The fuel injector circuit is open or grounded. Check and repair the circuit or replace the faulty injector.*
Code 33	MAP sensor (3.1L engine)	Signal voltage high. Check the MAP sensor and circuit. Replace the MAP sensor.*
Code 33	MAP sensor (3.2L engine)	Signal voltage out of range. Check for a loose or damaged air duct, incorrectly adjusted minimum idle speed and vacuum leaks. Inspect the MAP sensor, the circuit and the electrical connections. Replace the MAP sensor.*
Code 34	MAP sensor (3.1L engine)	Signal voltage low. Check for a loose or damaged air duct, incorrectly adjusted minimum idle speed and vacuum leaks. Check the MAP sensor, the circuit and the electrical connections. Replace the MAP sensor.*
Code 34	EGR temperature sensor circuit	The EGR temperature sensor circuit is shorted. Check the circuit for shorts.
Code 35	Ignition control module power transistor circuit (2.6L engine)	The power transistor circuit of the ignition control module is open. Check the wiring.
Code 41	Distributor signal (2.3L and 2.6L engines)	Check the distributor signal wiring. Replace the pick-up coil (2.3L engine) or the distributor (2.6L engine).
Code 41	Crankshaft Position Sensor (2.6L engine)	The crankshaft sensor wiring is open. Check the crankshaft sensor and harness.
Code 42	Fuel cut-off relay malfunction or circuit fault (2.3L engine)	Check the wiring or replace the fuel cut-off relay.*

Trouble codes	Circuit or system	Probable cause
Code 42	Electronic spark timing circuit (3.1L engine)	The electronic spark timing circuit of the ignition control module is open. Check the wiring and connections or replace the ignition control module as necessary.
Code 43	Electronic spark control	Knock circuit fault. Check and repair the knock circuit or replace the knock sensor.*
Code 43	Throttle valve switch (2.6L engine)	The throttle valve switch continuously makes contact at idle. Replace the switch.*
Code 44	Oxygen sensor (lean exhaust)	The oxygen sensor signal is low (lean condition). Check the wires between the PCM and the oxygen sensor for a short or ground or bad connection Check the vacuum hoses and intake manifold gasket for leaks.
Code 45	Oxygen sensor (rich exhaust)	The oxygen sensor signal is high (rich condition). Check the wires between the PCM and the oxygen sensor for a problem. Check the EVAP canister and system for the presence of fuel.
Code 51	Fuel cut-off solenoid	The fuel cut-off solenoid circuit is shorted or the PCM is faulty. Have the vehicle checked by a dealer service department.
Code 51	PCM (3.1L engine)	Faulty PROM or EEPROM, or incorrect PROM or EEPROM installation. Have the vehicle checked by a dealer service department.
Code 51	PCM (3.1L and 3.2L engines)	Faulty PCM. Have the vehicle checked by a dealer service department.
Code 52	PCM (2.3L and 2.6L engines)	Faulty PCM. Have the vehicle checked by a dealer service department.
Code 52	CALPAK (3.1L engine)	Faulty CALPAK or incorrect installation. Have the vehicle checked by a dealer service department.
Code 53	PCM, air switching solenoid (ASS) or air injection system (2.3L engine)	Faulty PCM, air switching solenoid or air injection system. Have the vehicle checked by a dealer service department.
Code 53	Vacuum switching valve (VSV) (2.6L engine)	VSV grounded, or faulty power transistor. Have the vehicle checked by a dealer service department.
Code 54	Fuel pump circuit (3.1L engine)	Low voltage in fuel pump circuit. Locate and repair circuit short or ground.
Code 54	Ignition control module power transistor (2.6L engine)	Grounded or defective power transistor. Repair bad ground or replace the ignition control module.*
Code 54	Mixture control solenoid (2.3L engine)	Shorted mixture control solenoid or defective PCM. Have the vehicle checked by a dealer service department.
Code 55	PCM (2.3L, 2.6L and 3.1L engines)	Have the vehicle checked by a dealer service department.
Code 61	Air flow sensor circuit (2.6L engine)	Check the air flow sensor circuit for a broken, grounded or shorted hot wire.
Code 62	Air flow sensor circuit (2.6L engine)	Check the air flow sensor circuit for a broken cold wire.
Code 63	Vehicle Speed Sensor (VSS) circuit (2.6L engine)	No VSS signal. The VSS is broken or the circuit is shorted. Repair the circuit or replace the VSS.*
Code 64	Fuel injector driver transistor	A fuel injector driver transistor in the PCM is faulty. Have the vehicle checked by a dealer service department.
Code 65	Throttle valve switch (2.6L engine)	The throttle valve switch continuously makes contact. Replace the switch.*

* Component replacement may not cure problem in all cases. For this reason, you may want to seek professional advice before purchasing replacement parts.

3   On-Board Diagnostics II (OBD-II) systems and trouble codes

General information

1   OBD-I systems are proprietary, i.e. each manufacturer designs its OBD-I system differently from other manufacturers. The OBD-I systems used on the various models of a particular manufacturer are often similar, and even compatible, but there is virtually no compatibility between the systems employed by different manufacturers. Each manufacturer uses its own diagnostic connector, trouble codes, engine management computer, "map" (program), etc. Because of increasingly stringent government regulation of automotive emissions, the latest OBD systems must monitor increasingly complex and inter-related emission control systems. They must not only provide enough data to identify a system or component malfunction but they must also be able to *predict* a malfunction that will occur in the future. As a result, the complexity, and dissimilarity between, various proprietary systems has increased markedly in recent years.

2   Because of this incompatibility problem, automobile manufacturers and the Society of Automotive Engineers (SAE) have been engaged in a collaborative process to create a standard diagnostic interface and a universal set of computer codes that could be applied to all on-board diagnostic systems regardless of the actual internal workings of the system. In 1996, manufacturers began standardizing the diagnostic connectors on all vehicles, and they began using a universal set of trouble codes, known as "P-Zero" codes (because they all begin with the letter P, followed by the number 0). This standardization of the diagnostic connector and the trouble codes is known as On Board Diagnostics II (OBD II). A number of generic scan tools have been developed by the aftermarket specifically for the do-it-yourselfer who wants to be able to access these standardized codes (see "Retrieving codes on OBD-II systems" below)

3   Besides the P-Zero codes, manufacturers still use many other new codes that are proprietary, i.e. *specific,* to their vehicles. These proprietary codes are used by dealer technicians for advanced diagnosis. They're not available to the home mechanic (they're not even available to professional mechanics working in independent service facilities). In most cases, this shouldn't be a problem for a do-it-yourselfer because the P-Zero codes

are quite specific, far more so than the old OBD-I codes.

4   OBD-II systems have self-diagnostic capabilities well beyond those of OBD-I systems. They are much better at detecting, and even predicting, system and component malfunctions. For example, where there might have been only one code for an oxygen sensor malfunction, there might now be six codes that help determine the specific nature of an oxygen sensor malfunction. OBD-II systems also perform many additional diagnostic and predictive functions in areas not required under earlier Federal laws, such as monitoring engine misfires and deterioration of the catalytic converter.

Diagnostic tool information

5   The diagnostic tool information in the previous Section also applies to OBD-II systems. Several generic OBD-II capable generic scanners (from manufacturers such as Actron and AutoXray, for example) are already available in the aftermarket for the home mechanic. Check with your local auto parts store for more information about these units.

OBD-II system general description

6   All 1996 and later models covered by this manual are equipped with OBD-II systems. The OBD-II system consists of an onboard computer, known as the Powertrain Control Module (PCM), information sensors and output actuators. The information sensors monitor various functions of the engine and send data to the PCM. Based on the data and the information programmed into the computer's memory, the PCM generates output signals to control various engine functions via control relays, solenoids and other output actuators. The PCM is programmed to optimize the emissions, fuel economy and driveability of the vehicle.

7   Because of a Federally mandated warranty which covers the emissions systems and all related components, and because any owner-induced damage to the PCM, the sensors or the control devices might void this warranty, it isn't a good idea to attempt diagnosis or replacement of the system components while the vehicle is still under warranty. Take the vehicle to a dealer service department if the PCM or a system component malfunctions.

Information sensors

8   **Camshaft Position (CMP) Sensor** - The camshaft position sensor provides information on camshaft position. The PCM uses this information, along with the crankshaft position sensor information, to control fuel injection synchronization.

9   **Crankshaft Position (CKP) Sensor** - The crankshaft position sensor senses crankshaft position (TDC) during each engine revolution. The PCM uses this information to control ignition timing and fuel injection synchronization.

10   **Engine Coolant Temperature (ECT) sensor** - The Engine Coolant Temperature (ECT) Sensor senses engine coolant temperature. The PCM uses this information to control fuel injection duration and ignition timing.

11   **Intake Air Temperature (IAT) Sensor** - The intake air temperature senses the temperature of the air entering the intake manifold. The PCM uses this information to control fuel injection duration.

12   **Knock Sensor (KS)** - The knock sensor is a piezoelectric element that detects the sound of engine detonation, or "pinging". The PCM uses the input signal from the knock sensor to recognize detonation and retard spark advance to avoid engine damage.

13   **Manifold Absolute Pressure (MAP) sensor** - The manifold absolute pressure monitors intake manifold pressure and ambient barometric pressure. The PCM uses this input signal to determine engine load and adjusts fuel injection duration accordingly.

14   **Mass Airflow (MAF) sensor** - The mass airflow sensor measures the amount of air passing through the sensor body and ultimately entering the engine. The PCM uses this information to control fuel delivery.

15   **Oxygen (O2) sensor** - The oxygen sensors generate a voltage signal that varies with the varying oxygen content of the exhaust gas. The PCM uses this information to determine if the fuel system is running rich or lean and make adjustments accordingly.

16   **Power Steering Pressure (PSP) switch** - The power steering pressure switch monitors the pressure of the power steering fluid. When the pressure exceeds a predetermined threshold, the switch signals the PCM, which raises the engine speed slightly to compensate for the additional load placed on the engine.

17   **Throttle Position (TP) sensor** - The throttle position sensor senses throttle movement and position. This signal enables the PCM to determine when the throttle is closed, in a cruise position, or wide open. The PCM uses this information to control fuel delivery and ignition timing.

18   **Vehicle Speed Sensor (VSS)** - The vehicle speed sensor provides information to the PCM to indicate vehicle speed.

19   **Miscellaneous PCM inputs** - In addition to the various sensors, the PCM monitors various switches and circuits to determine vehicle operating conditions. The switches and circuits include:

a) Air conditioning system
b) Battery voltage
c) Brake On/Off switch
d) Cruise control system
e) EGR valve position
f) Engine oil level and pressure
g) EVAP system
h) Fuel level and fuel tank pressure
i) Ignition switch
j) Park/neutral position switch
k) Sensor signal and ground circuits
l) Transmission controls

Output actuators

20   **Air conditioning clutch relay** - The PCM controls the operation of the air conditioning compressor clutch with the air conditioning clutch relay.

21   **Cruise control module** - The cruise control system operation is controlled by the PCM.

22   **Engine cooling fan relay** - The engine cooling fan is controlled by the PCM according to information received from the Engine Coolant Temperature (ECT) Sensor.

23   **EGR valve** - The electronic EGR valve is controlled by the PCM. Ideal EGR flow is determined by the PCM and the EGR valve pintle position is adjusted accordingly.

24   **EVAP canister purge and vent valve solenoids** - The evaporative emission canister purge and vent valve solenoids are operated by the PCM to purge the fuel vapor canister and route fuel vapor to the intake manifold for combustion.

25   **Fuel injectors** - The PCM opens the fuel injectors individually in firing order sequence. The PCM also controls the time the injector is held open (pulse width). The pulse width of the injector (measured in milliseconds) determines the amount of fuel delivered. For more information on the fuel delivery system and the fuel injectors, including injector replacement, refer to Chapter 4.

26   **Fuel pump relay** - The fuel pump relay is activated by the PCM with the ignition switch in the Start or Run position. When the ignition switch is turned on, the relay is activated to supply initial line pressure to the system. For more information on fuel pump check and replacement, refer to Chapter 4.

27   **Idle Air Control valve (IAC)** - The idle air control valve controls the amount of air allowed to bypass the throttle plate when the throttle valve is closed or at idle position. The more air allowed to bypass the throttle plate, the higher the idle speed. The idle air control valve opening and the resulting idle speed is controlled by the PCM.

28   **Ignition coils/control module** - The PCM controls ignition timing through the ignition coils/control module depending on engine operation conditions. Refer to Chapter 5 for more information on the ignition coil(s) or ignition control module.

Obtaining diagnostic trouble codes

Refer to illustration 3.30

Note: *The diagnostic trouble codes on OBD-II models can only be extracted from the Powertrain Control Module (PCM) with an OBD-II scanner. If you don't have a suitable scan tool, have the vehicle diagnosed by a dealer service department or other qualified automotive repair facility.*

29   The PCM illuminates the SERVICE ENGINE SOON light (also known as the Malfunction Indicator Lamp) on the dash if it recognizes a fault in the system. The light remains illuminated until the problem is repaired and the code is cleared or the PCM

does not detect any malfunction for several consecutive drive cycles.

30   The diagnostic codes for the OBD-II system can only be extracted from the PCM using a scan tool. The scan tool is programmed to interface with the system by plugging into the diagnostic connector **(see illustration)**. The OBD-II program mandates a standard 16-pin diagnostic link connector (DLC) for all vehicles. The DLC is also referred to as a J1962 connector (a designation taken from the physical and electrical specification number assigned by the SAE). Besides its standard pin configuration, the J1962 must also provide power and ground circuits for scan tool hook-up.

Clearing diagnostic trouble codes

31   After the system has been repaired, the codes must be cleared from the PCM memory. The preferred method is with a scan tool, but the codes can be cleared by disconnecting battery power from the PCM for a minimum of thirty seconds. Battery power can be disconnected from the PCM by removing the PCM fuse, disconnecting the PCM power connector near the positive battery terminal (if equipped) or by disconnecting the negative battery cable from the battery. **Caution:** *On models equipped with an anti-theft audio system, make sure the lockout feature is turned off before performing any procedure which requires disconnecting the battery.*

32   Always clear the codes from the PCM before starting the engine after a new elec-

3.30 On OBD-II vehicles, the Data Link Connector (DLC) (or J1962 connector) is located under the left end of the dash

tronic emission control component is installed onto the engine. The PCM stores the operating parameters of each sensor. The PCM may set a trouble code if a new sensor is allowed to operate before the parameters from the old sensor have been erased.

Diagnostic trouble code identification

33   The accompanying list of diagnostic trouble codes is a compilation of all the codes that may be encountered using a generic scan tool. Additional trouble codes may be obtainable with the use of the manu-

facturer specific scan tool. Not all codes pertain to all models and not all codes will illuminate the Service Engine Soon light when set. All models require a scan tool to access the diagnostic trouble codes.

Retrieving codes on OBD-II systems

34   Now that all new vehicles have a standardized connector and a universal set of diagnostic trouble codes, the same scan tool can be used on any vehicle, and any home mechanic can access these codes with a relatively affordable generic scan tool.

35   All aftermarket generic scan tools include good documentation, so refer to the manufacturer's hook-up instructions in your scan tool manual. Before plugging a scan tool into the DLC, inspect the condition of the DLC housing; make sure that all the wires are connected and that the contacts are fully seated in the housing. Inside the connector, make sure that there's no corrosion on the pins and that no pins are bent or damaged.

Clearing codes on OBD-II systems

36   After you've made the necessary repairs, you'll need to "clear" (erase) any stored trouble codes from the PCM,. Refer to the manufacturer's documentation that comes with the aftermarket generic scan tool you're using. It will guide you through the correct procedure for clearing any codes stored in the PCM.

OBD-II trouble codes for 2.2L and 3.2L models

Note: *The following list of OBD II trouble codes is applicable to all models equipped with an OBD II system, although not all codes apply to all models.*

Trouble code	Probable cause
P0101	Mass Air Flow (MAF) system performance problem
P0102	Mass Air Flow (MAF) sensor circuit, low frequency
P0103	Mass Air Flow (MAF) sensor circuit, high frequency
P0106	Manifold Absolute Pressure (MAP) sensor circuit range or performance problem
P0107	Manifold Absolute Pressure (MAP) sensor circuit, low frequency
P0108	Manifold Absolute Pressure (MAP) sensor circuit, high frequency
P0112	Intake Air Temperature (IAT) sensor circuit, low voltage
P0113	Intake Air Temperature (IAT) sensor circuit, high voltage
P0117	Engine Coolant Temperature (ECT) sensor circuit, low voltage
P0118	Engine Coolant Temperature (ECT) sensor circuit, high voltage
P0121	Throttle Position (TP) sensor circuit range or performance problem
P0122	Throttle Position (TP) sensor circuit, low voltage
P0123	Throttle Position (TP) sensor circuit, high voltage
P0125	Engine Coolant Temperature (ECT) sensor, insufficient coolant temperature for closed loop fuel control or excessive time to closed loop
P0128	Engine Coolant Temperature (ECT) sensor below thermostat regulating temperature
P0131	Oxygen sensor circuit, low voltage (cylinder bank no. 1, sensor no. 1)
P0132	Oxygen sensor circuit, high voltage (cylinder bank no. 1, sensor no. 1)
P0133	Oxygen sensor circuit, slow response (cylinder bank no. 1, sensor no. 1)
P0134	Oxygen sensor circuit - no activity detected (cylinder bank no. 1, sensor no. 1)

OBD-II trouble codes for 2.2L and 3.2L models (continued)

Trouble code	Probable cause
P0135	Oxygen sensor heater circuit malfunction (cylinder bank no. 1, sensor no. 1)
P0137	Oxygen sensor circuit, low voltage (cylinder bank no. 1, sensor no. 2)
P0138	Oxygen sensor circuit, high voltage (cylinder bank no. 1, sensor no. 2)
P0140	Oxygen sensor circuit - no activity detected (cylinder bank no. 1, sensor no. 2)
P0141	Oxygen sensor heater circuit malfunction (cylinder bank no. 1, sensor no. 2)
P0151	Oxygen sensor circuit, low voltage (cylinder bank no. 2, sensor no. 1)
P0152	Oxygen sensor circuit, high voltage (cylinder bank no. 2, sensor no. 1)
P0153	Oxygen sensor circuit, slow response (cylinder bank no. 2, sensor no. 1)
P0154	Oxygen sensor circuit - no activity detected (cylinder bank no. 2, sensor no. 1)
P0155	Oxygen sensor heater circuit malfunction (cylinder bank no. 2, sensor no. 1)
P0157	Oxygen sensor circuit, low voltage (cylinder bank no. 2, sensor no. 2)
P0158	Oxygen sensor circuit, high voltage (cylinder bank no. 2, sensor no. 2)
P0160	Oxygen sensor circuit - no activity detected (cylinder bank no. 2, sensor no. 2)
P0161	Oxygen sensor heater circuit malfunction (cylinder bank no. 2, sensor no. 2)
P0171	Fuel trim system too lean (cylinder bank no. 1)
P0172	Fuel trim system too rich (cylinder bank no. 1)
P0174	Fuel trim system too lean (cylinder bank no. 2)
P0175	Fuel trim system too rich (cylinder bank no. 2)
P0201	Injector circuit malfunction - cylinder no. 1
P0202	Injector circuit malfunction - cylinder no. 2
P0203	Injector circuit malfunction - cylinder no. 3
P0204	Injector circuit malfunction - cylinder no. 4
P0205	Injector circuit malfunction - cylinder no. 5
P0206	Injector circuit malfunction - cylinder no. 6
P0218	Automatic transmission overheating condition
P0300	Random/multiple cylinder misfire detected
P0301	Cylinder no. 1misfire detected
P0302	Cylinder no. 2 misfire detected
P0303	Cylinder no. 3 misfire detected
P0304	Cylinder no. 4 misfire detected
P0305	Cylinder no. 5 misfire detected
P0306	Cylinder no. 6 misfire detected
P0325	Knock sensor module or circuit malfunction
P0327	Knock sensor circuit, low input
P0336	Crankshaft Position (CKP) sensor circuit, range or performance problem
P0337	Crankshaft Position (CKP) sensor circuit, low input
P0341	Camshaft Position (CMP) sensor circuit, range or performance problem
P0342	Camshaft position (CMP) sensor circuit, low input
P0351	Ignition coil 1, primary or secondary circuit malfunction
P0352	Ignition coil 2, primary or secondary circuit malfunction
P0353	Ignition coil 3, primary or secondary circuit malfunction
P0354	Ignition coil 4, primary or secondary circuit malfunction
P0355	Ignition coil 5, primary or secondary circuit malfunction
P0356	Ignition coil 6, primary or secondary circuit malfunction
P0401	Exhaust Gas Recirculation (EGR) system, insufficient flow detected
P0402	Exhaust Gas Recirculation (EGR), excessive flow detected
P0404	Exhaust Gas Recirculation (EGR) circuit, range or performance problem
P0405	Exhaust Gas Recirculation (EGR) sensor circuit, low voltage
P0406	Exhaust Gas Recirculation (EGR) sensor circuit, high voltage

Trouble code	Probable cause
P0420	Catalyst system efficiency below threshold (cylinder bank no. 1)
P0430	Catalyst system efficiency below threshold (cylinder bank no. 2)
P0440	Evaporative emission (EVAP) control system malfunction
P0442	Evaporative emission (EVAP) control system, small leak detected
P0443	Evaporative emission (EVAP) control system, purge control valve circuit malfunction
P0444	Evaporative emission (EVAP) control system, purge control circuit open
P0445	Evaporative emission (EVAP) control system, purge control circuit shorted
P0446	Evaporative emission (EVAP) control system, vent control circuit malfunction
P0449	Evaporative emission (EVAP) control system, vent valve/solenoid circuit malfunction
P0452	Evaporative emission (EVAP) control system, fuel tank pressure sensor circuit, low voltage
P0453	Evaporative emission (EVAP) control system, fuel tank pressure sensor circuit, high voltage
P0456	Evaporative emission (EVAP) control system, very small leak detected
P0461	Fuel level sensor circuit, range or performance problem
P0462	Fuel level sensor circuit, low voltage
P0463	Fuel level sensor circuit, high voltage
P0464	Fuel level sensor circuit, noisy signal
P0480	Cooling fan no. 1, control circuit malfunction
P0481	Cooling fan no. 2, control circuit malfunction
P0502	Vehicle Speed Sensor (VSS) circuit, low voltage
P0506	Idle Air Control (IAC) system, rpm lower than expected
P0507	Idle Air Control (IAC) system, rpm higher than expected
P0532	Air conditioning refrigerant pressure sensor circuit, low voltage
P0533	Air conditioning refrigerant pressure sensor circuit, high voltage
P0560	System voltage low
P0562	System voltage low
P0563	System voltage high
P0565	Cruise control main switch circuit error
P0566	Cruise control cancel switch circuit error
P0567	Cruise control resume switch circuit error
P0568	Cruise control set switch circuit error
P0571	No brake switch signal
P0601	Powertrain Control Module (PCM), memory check sum error
P0602	Powertrain Control Module (PCM) programming error
P0604	Powertrain Control Module (PCM) Random Access Memory (RAM) error
P0606	Powertrain Control Module (PCM) internal performance
P0705	Transmission Range (TR) sensor, illegal position
P0706	Transmission Range (TR) sensor performance problem
P0712	Transmission Fluid Temperature (TFT), low voltage
P0713	Transmission Fluid Temperature (TFT), high voltage
P0719	Brake switch circuit low
P0722	Output speed sensor circuit, no signal
P0723	Output speed sensor circuit, intermittent signal
P0724	Brake switch circuit high
P0730	Incorrect gear ratio
P0742	Torque Converter Clutch (TCC) circuit stuck in "on" position
P0748	Transmission Pressure Control Solenoid, electrical circuit problem
P0751	Transmission shift solenoid A, performance problem or stuck in "off" position
P0753	Transmission shift solenoid A, electrical circuit problem
P0756	Transmission shift solenoid B, performance problem or stuck in "off" position
P0758	Transmission shift solenoid B, electrical circuit problem

4.2 On 1997 and earlier 3.2L models, the Camshaft Position (CMP) sensor is located on the front left upper corner of the timing belt cover (on 1998 and later models, the CMP sensor is located on the upper right rear corner of the timing belt cover, near the right cylinder bank); to remove the CMP sensor, remove the retaining bolt

4.3 On 2.2L models, the Camshaft Position (CMP) Sensor is located on the top of the front part of the cylinder head, between the camshaft sprockets (timing belt cover, timing belt and sprockets removed for clarity; it's not necessary to remove the timing belt or the sprockets to replace the CMP sensor)

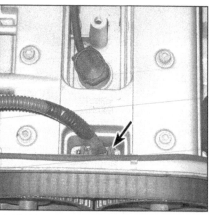

4.13 After removing the spark plug cable cover on 2.2L models, disconnect the electrical connector from the CMP sensor

13   Unplug the electrical connector from the CMP sensor **(see illustration)**.
14   Remove the serpentine drivebelt (see Section 25 in Chapter 1).
15   Remove the wiring harness cover from the top of the timing belt cover (see Section 4 in Chapter 2B).
16   Remove the crankshaft pulley (see Section 4 in Chapter 2B).
17   Remove the lower timing belt cover (see Section 4 in Chapter 2B).
18   Remove the CMP sensor retaining bolt **(see illustration)** and remove the CMP sensor.
19   Installation is the reverse of removal.

4   Camshaft Position (CMP) Sensor - check and replacement

General information

Refer to illustrations 4.2 and 4.3

1   The Camshaft Position (CMP) Sensor is a Hall Effect switching device that sends a "square-wave" signal to the PCM. The PCM uses this signal as a "sync pulse" to trigger the injectors in the correct sequence. The PCM also uses the signal from the CMP sensor to indicate the position of the No. 1 piston during its power stroke. If the PCM detects an incorrect CMP signal during engine operation, it will set diagnostic trouble code P0341.
2   On 1997 and earlier 3.2L engines, the CMP sensor is located in the upper front left corner of the timing belt cover, near the camshaft sprocket **(see illustration)**. On 1998 and later models, the CMP sensor is located at the front of the right cylinder head, on the inside surface of the head.
3   On 2.2L engines, the CMP sensor is located in the top of the front part of the cylinder head, between the camshaft sprockets **(see illustration)**.

Check

4   Disconnect the CMP sensor electrical connector and with the ignition key ON (engine not running), check for REFERENCE voltage to the CMP sensor. It should be approximately 5.0 volts. Refer to the wiring schematics at the end of Chapter 12 for additional information on the color of the harness wires to the CMP.
5   If reference voltage is reaching the crankshaft sensor, but the MIL is displaying a code indicating a problem with the CKP sensor, have the sensor diagnosed by a dealer service department.

Replacement

6   Disconnect the cable from the negative battery terminal.

3.2L engine

7   On 1998 and later models, remove the intake manifold assembly (see Chapter 2D).
8   Unplug the electrical connector from the CMP sensor.
9   Remove the CMP sensor retaining bolt and remove the sensor.
10   Inspect the sensor O-ring for cracks and leaks. If the O-ring is damaged, replace it. Be sure to lubricate the new O-ring with clean engine oil.
11   Installation is the reverse of removal.

2.2L engine

Refer to illustrations 4.13 and 4.18

12   Remove the spark plug wire cover from the valve cover (see Chapter 5).

4.18 To remove the CMP sensor on 2.2L models, unplug the electrical connector (1), remove the retaining bolt (2) and loosen the timing belt cover stud (3) enough to pull the CMP sensor straight up

5   Crankshaft Position (CKP) Sensor - check and replacement

General information

Refer to illustrations 5.1, 5.2 and 5.3

1   The Crankshaft Position (CKP) Sensor, which is adjacent to a pulse wheel on the crankshaft **(see illustration)**, is a Hall Effect type switching device that monitors the pulse

5.1 The Crankshaft Position (CKP) sensor wheel is attached to the crankshaft inside the engine block (3.2L V6 shown, 2.2L four-cylinder similar)

5.2 Location of the Crankshaft Position (CKP) sensor on the 3.2L engine

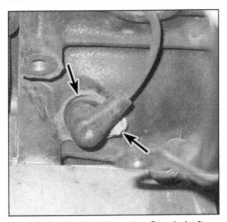

5.3 On 2.2L engines, the Crankshaft Position (CKP) sensor is located on the left side of the block (power steering pump removed for clarity)

5.4 On 2.2L engines, the electrical connector for the Crankshaft Position (CKP) sensor is located on the *right* side of the engine, between the alternator and the block

wheel as the teeth pass under the magnetic field created by the sensor. The pulse wheel has 58 teeth and a spot where one tooth is missing. Every time the missing tooth passes the CKP sensor, the sensor's voltage output is interrupted. This "square-wave" (ON-OFF) output is used by the PCM to determine piston speed and crankshaft position. If the PCM receives an incorrect number of pulses from the CKP sensor, it will set diagnostic trouble code P0336; if it receives no pulses from the CKP sensor, it will set code P0337. The engine won't start or run without the correct reference signal from the CKP sensor.

2   On 3.2L engines, the Crankshaft Position (CKP) Sensor **(see illustration)** is mounted on the right side of the block, behind the right engine mount. **Note:** *1993 through 1995 3.2L engines are equipped with a CKP sensor. 1996 and later 3.2L engines with OBD II systems are also equipped with a Camshaft Position (CMP) Sensor. Both the CKP and CMP sensors are Hall Effect type switching devices, but they perform different functions. Don't confuse the two sensors.*

3   On 2.2L engines, the Crankshaft Position (CKP) Sensor **(see illustration)** is mounted on the left side of the block, adjacent to the power steering pump.

Check

Refer to illustration 5.4

4   Disconnect the CKP sensor electrical connector **(see illustration)** and with the ignition key ON (engine not running), check for REFERENCE voltage to the CKP sensor. It should be approximately 5.0 volts. Refer to the wiring schematics at the end of Chapter 12 for additional information on the color of the harness wires to the CKP.

5   If reference voltage is reaching the CKP sensor, but the MIL is displaying a code indicating a problem with the CKP sensor or circuit, have the CKP sensor and circuit diagnosed by a dealer service department.

Replacement

6   Disconnect the cable from the negative battery terminal.

3.2L engine

7   Unplug the electrical connector from the CKP sensor.

8   Remove the CKP sensor retaining bolt and lift the crankshaft sensor from the engine block.

9   Inspect the sensor O-ring for cracks and leaks. If the O-ring is worn or damaged, replace it. Be sure to lubricate the new O-ring with clean engine oil before installing it.

10   Installation is the reverse of removal. Be sure to tighten the retaining bolt to the torque listed in this Chapter's Specifications.

2.2L engine

11   Remove the power steering pump drivebelt (see Chapter 1).

12   Remove the power steering pump (see Chapter 10).

13   Unplug the electrical connector from the CKP sensor **(see illustration 5.4)**

14   Remove the CKP sensor retaining bolt **(see illustration 5.3)** and remove the sensor from the engine block.

15   Installation is the reverse of removal. Tighten the mounting bolt securely.

16   After a new CKP sensor has been installed, the PCM must "re-learn" the new sensor, i.e. calibrate the sensor. If the MIL indicates a diagnostic trouble code, have the new CKP sensor calibrated by a dealer service department.

---

6   Engine Coolant Temperature (ECT) Sensor - check and replacement

General description

1   The Engine Coolant Temperature (ECT) Sensor is a thermistor (a resistor that varies the value of its voltage output in response to a change in temperature). The change in the resistance values will directly affect the voltage signal from the coolant sensor. As the sensor temperature DECREASES, the resistance values will INCREASE. As the sensor temperature INCREASES, the resistance values will DECREASE. A failure in the coolant sensor circuit should set a Code 13. This code indicates a failure in the coolant temperature circuit, so in most cases the appropriate solution to the problem will be either repair of a wire or replacement of the sensor.

Check

Refer to illustrations 6.2a and 6.2b

2   The Engine Coolant Temperature (ECT) Sensor on the 2.3L and 2.6L engines is located under the intake manifold. On 3.1L engines it is located up front near the thermostat housing. On 3.2L engines, it is located near the back of the left (driver's side) cylinder head **(see illustration)**. On 2.2L engines, it's located on the intake manifold, above the ignition coil **(see illustration)**. Check the resistance value of the Engine Coolant Temperature (ECT) Sensor while it is completely cold and compare the indicated resistance to the value specified in the accompanying chart. Then start the engine, warm it up until it reaches operating temperature, note the

6.2a On 3.2L V6 engines, the Engine Coolant Temperature (ECT) sensor is located on the coolant jacket on the rear side of the engine

6.2b On 2.2L four-cylinder engines, the Engine Coolant Temperature (ECT) sensor is located at the rear of the cylinder head, behind and below the valve cover

7.1 On 2.2L engines, the Intake Air Temperature (IAT) sensor is located on the air intake duct between the air cleaner assembly and the throttle body (on 3.2L V6 engines, the IAT sensor is located closer to the throttle body)

indicated resistance of the sensor as the coolant warms up and compare your measurements to the values specified in the accompanying chart.

Temperature (F)	Resistance (ohms)
212	177
194	241
176	332
158	467
140	667
122	973
113	1,188
104	1,459
95	1,802
86	2,238
77	2,796
68	3,520
59	4,450
50	5,670
41	7,280
32	9,420
23	12,300
14	16,180
5	21,450
-4	28,680
-22	52,700
-40	100,700

Note: *Access to the Engine Coolant Temperature (ECT) Sensor makes it difficult to position electrical probes on the terminals. If necessary, remove the sensor and perform the tests in a pan of heated water to simulate the conditions.*

3   If the resistance values on the sensor are correct, check the reference voltage to the sensor from the PCM. It should be approximately 5.0 volts.

Replacement

Warning: *The engine must be completely cool before starting this procedure.*

4   Before installing the new sensor, wrap the threads with Teflon sealing tape to prevent leakage and thread corrosion.
5   Disconnect the electrical connector and remove the sensor. **Caution:** *Handle the coolant sensor with care. Damage to this sen-*

sor will affect the operation of the entire fuel injection system. Install the sensor and tighten it securely. Check the coolant level and add some, if necessary (see Chapter 1).

---

7   Intake Air Temperature (IAT) sensor - check and replacement

General description

Refer to illustration 7.1

1   The Intake Air Temperature (IAT) Sensor **(see illustration)** is located on the air intake duct. On 3.2L engines, it's near the throttle body; on 2.2L engines, it's about midway between the air cleaner housing and the throttle body. The IAT sensor is a thermistor which changes its resistance in proportion to the temperature of the air entering the engine. Low temperatures produce a high resistance value (for example, at 68 degrees F the resistance is 2,700 ohms); high temperatures produce low resistance values (for instance, at 212-degrees F the resistance is 177 ohms). The PCM supplies a 5-volt reference voltage to the IAT sensor. The voltage signal returned to the PCM will change according to the temperature of the incoming air. The voltage will be high when the air temperature is cold and low when the air temperature is warm. By measuring the voltage, the PCM is able to calculate the temperature of incoming air. The PCM uses the IAT signal to adjust spark timing in accordance with air density. A failure of the IAT sensor will cause the PCM to set diagnostic trouble code P0112 or P0113.

Check

2   To check the IAT sensor, disconnect the two prong electrical connector and turn the ignition key ON but do not start the engine.
3   Measure the reference voltage. Refer to the wiring schematics at the end of Chapter 12 for harness wire colors and designations.

The voltmeter should indicate 5-volts. If the voltage signal is not correct, have the PCM diagnosed by a dealer service department.
4   Measure the resistance across the sensor terminals. The resistance should be high when the air temperature is low. Start the engine, wait awhile and let the engine reach operating temperature. Turn the ignition off, disconnect the IAT sensor and measure the resistance across the terminals. The resistance should be low when the air temperature is high. Compare your readings to the following table:

Temperature (F)	Resistance (ohms)
100	177
80	332
60	667
45	1,188
35	1,802
25	2,796
15	4,450
5	7,280
-5	12,300
-15	21,450
-30	52,700
-40	100,700

If the sensor does not change resistance as indicated, replace it.

Replacement

5   Disconnect the cable from the negative battery terminal.
6   Unplug the electrical connector from the IAT sensor.
7   The IAT sensor is installed in a rubber grommet. To disengage the sensor from the grommet, pull on it while rocking it from side to side.
8   While the IAT sensor is removed, inspect the condition of the grommet. If it's hard, cracked or deteriorated, replace it.
9   Installation is the reverse of removal.

---

8   Knock sensor - replacement

General description

Refer to illustrations 8.1a and 8.1b

1   The knock sensor detects abnormal vibration in the engine. The sensor produces an AC output voltage which increases with the severity of the knock. The signal is fed into the PCM and the timing is retarded up to 10 degrees to compensate for severe detonation. On 3.2L engines, the knock sensor **(see illustration)** is located in the top of the block, under the intake manifold. On 2.2L engines, the knock sensor **(see illustration)** is located on the right (passenger) side of the engine block, just behind the right engine mount.

Check

2   You need a special scan tool to check the knock sensor and the knock sensor circuit. Take the vehicle to a dealer service department to have the sensor checked.

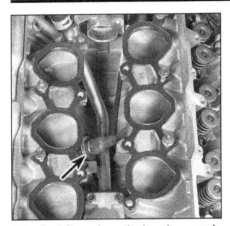

8.1a On 3.2L engines, the knock sensor is located in the valley between the cylinder heads (upper intake manifold removed for clarity)

8.1b On 2.2L engines, the knock sensor is located on the right side of the engine block, behind the right engine mount

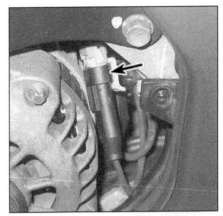

8.10 On 2.2L engines, the electrical connector for the knock sensor is located between the alternator and the cylinder head

Replacement

3   Disconnect the cable from the negative battery terminal.

3.2L engine

4   Drain the cooling system (see Chapter 1).
5   Remove the engine cover.
6   Remove the air intake plenum (upper intake manifold) (see Chapter 2D).
7   Unplug the electrical connector from the knock sensor **(see illustration 8.1a)**.
8   Unscrew the knock sensor from the engine block.
9   Installation is the reverse of removal. Tighten the knock sensor to the torque listed in this Chapter's Specifications.

2.2L engine

Refer to illustration 8.10

10   Trace the electrical lead from the knock sensor to the electrical connector **(see illus-**tration), which is located between the alternator and the engine, and unplug the connector.
11   Remove the knock sensor retaining bolt **(see illustration 8.1b)** and remove the knock sensor.
12   Installation is the reverse of removal. Tighten the knock sensor retaining bolt securely.

---

9   Manifold Absolute Pressure (MAP) sensor - check and replacement

---

General description

Refer to illustrations 9.1a and 9.1b

1   The Manifold Absolute Pressure (MAP) sensor **(see illustrations)** monitors the intake manifold pressure changes resulting from changes in engine load and speed and con-

verts the information into a voltage output. The PCM uses the MAP sensor to control fuel delivery and ignition timing. A failure in the MAP sensor circuit should set a trouble code.
2   The MAP sensor SIGNAL voltage to the PCM varies from below 2 volts at idle (high vacuum) to above 4 volts with the ignition key ON (engine not running) or wide open throttle (WOT) (low vacuum). These values correspond with the altitude and pressure changes the vehicle experiences while driving.

Check

Refer to illustrations 9.3a, 9.3b and 9.4

3   First, locate the MAP sensor connector by tracing the wires from the sensor back to the connector. Using a digital voltmeter and a pair of leads equipped with probes, check the reference voltage from the PCM to the MAP sensor as follows. Unplug the sensor, turn the ignition key to ON and probe the terminals for the reference and the ground wires (the two

9.1a On 1997 and earlier 3.2L V6 engines, the Manifold Absolute Pressure (MAP) sensor is located on the right side of the intake manifold (on 1998 and later 3.2L models, the MAP sensor, which is physically similar in appearance to the MAP sensor on 2.2L engines, is located at the right rear upper corner of the intake manifold)

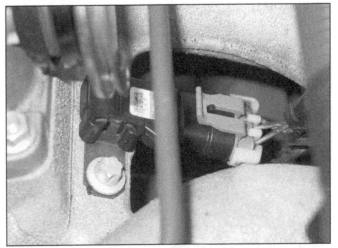

9.1b On 2.2L four-cylinder engines, the Manifold Absolute Pressure (MAP) sensor is located on the upper side of the intake manifold, between the throttle body and the cylinder head

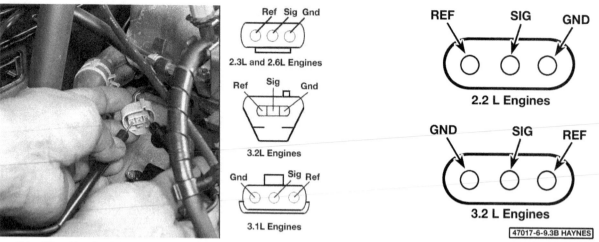

9.3a Checking for REFERENCE voltage on the MAP sensor harness (1997 and earlier models, 3.2L engine shown)

9.3b MAP sensor electrical connector terminal guide (1998 and later models)

outside terminals) **(see illustrations)**. The voltage should be approximately 5.0 volts.

4   Turn off the ignition, plug in the MAP sensor connector and, using the voltmeter with a pair of leads suitable for back-probing the connector, check the signal voltage as follows. Back-probe the signal terminal (the middle terminal) with the positive probe and the ground terminal with the negative probe **(1997 and earlier models, see illustration; 1998 and later models, see illustration 9.3b)**. With the ignition key turned to ON (engine not running), the voltage reading should be about 4.5 to 5 volts. **Caution:** *Be careful when back-probing the electrical connector. Do not damage the wiring harness or pull on any connectors to make clean contact.* **Note:** *If your voltmeter leads don't have probes small enough in diameter to back-probe the connector, use T-head pins, straight-pins, or straighten out a pair of paper clips, insert them into the backside of the connector and grasp them with alligator clips on the ends of the leads.*

5   Remove the vacuum line from the MAP sensor and install a hand-held vacuum pump

onto the MAP sensor port, apply vacuum and observe that the vacuum decreases. As vacuum increases voltage decreases.

Replacement

6   Disconnect the vacuum hose.
7   Disconnect the electrical connector from the MAP sensor.
8   Remove the MAP sensor mounting screws (3.2L engine) or unclip the sensor from its mounting bracket (2.2L engine).
9   Installation is the reverse of removal.

---

10   Mass Airflow (MAF) sensor - check and replacement

General Information

Note: *1991 through 1995 2.6L four-cylinder models and 1996 and later 3.2L models are equipped with a Mass Airflow (MAF) sensor.*
1   The MAF sensor is located on the air intake duct, at the air filter housing. The MAF sensor uses a hot wire sensing element to measure the amount of air entering the engine.

Think of the hot wire as a variable resistor: The PCM sends a reference voltage through the hot wire, which cools as air passes over it, which increases the resistance of the wire, which produces a variable voltage signal back to the PCM. The PCM monitors this signal to calculate the fuel injector pulse width. If the MAF sensor malfunctions, the PCM will set a trouble code (2.6L models, see Section 9; 3.2L models, see Section 10).

Check

2   If the problem is intermittent, look for a poor connection at both the MAF sensor and at the PCM. Verify that the wiring harness is correctly routed and undamaged. Inspect the harness for chafed insulation or a broken wire inside the insulation. If you have a scan tool, hook it up and watch the MAF values while you jiggle the wiring and connectors. Look for a leak in the duct work near the MAF sensor. Check for an engine or PCV system vacuum leak. Make sure that the correct PCV valve is installed. Make sure that the oil filler cap is tight and that the engine oil dipstick is fully seated.

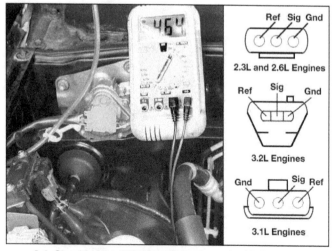

9.4 Checking for SIGNAL voltage from the MAP sensor (1997 and earlier models, 2.6L engine shown)

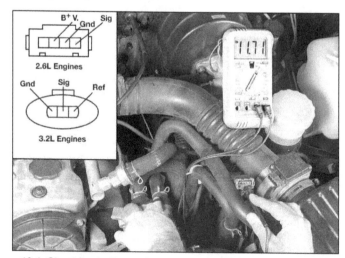

10.4 Checking battery voltage to the MAF sensor (2.6L engine)

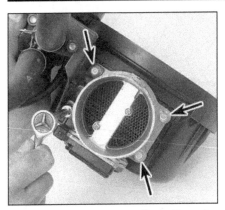

10.12 Remove the mounting bolts and separate the MAF sensor from the air cleaner (2.6L engine shown)

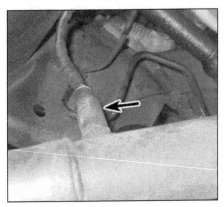

11.1 A typical upstream (pre-converter) oxygen sensor (2.2L engine shown)

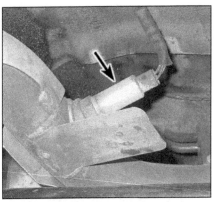

11.2 A typical downstream (post-converter) oxygen sensor (2.2L engine shown)

2.6L four-cylinder models

Refer to illustration 10.4

3   Pull back the rubber boot from the MAF sensor electrical connector and unplug it. There are three wires to the connector: red wire (the ignition feed wire to the MAF sensor, carries battery voltage); black wire (1991 and 1992 models) or violet wire(1993 through 1995 models), which is the signal wire from the MAF to the PCM; and the white wire (ground).

4   With the ignition switch turned to ON (engine not running), hook up the positive lead of a voltmeter to the connector terminal for the red wire **(see illustration)**, ground the other lead and verify that there's battery voltage.

5   Reconnect the MAF sensor connector and then start the engine. With the engine running, back-probe the connector terminal for the white wire, ground the other lead and verify that there's an indicated voltage of 0.5 to 2.0 volts. Turn off the engine.

6   With the ignition switch turned on (engine not running), back-probe the black wire (1991 and 1992 models) or the violet wire (1993 through 1995 models), ground the other lead (*to an engine ground*) and verify that there is less than 0.2 volt.

7   If the MAF sensor fails the tests in Steps 4 and/or 6, inspect the MAF sensor circuit for opens and shorts and make any necessary repairs. There should be continuity in each of the three wires between the MAF sensor connector terminals and their respective terminals at the PCM, but there should be no continuity between the white wire and (engine) ground, or between the red wire and (engine) ground.

8   If the MAF sensor passes the tests in Steps 4 and 6, but fails the test in the Step 5, replace the MAF sensor.

3.2L models

9   Checking the MAF sensor on 3.2L models requires a factory scan tool and is therefore beyond the scope of the home mechanic.

Replacement

Refer to illustration 10.12

10   Disconnect the negative battery cable.
11   Disconnect the electrical connector

from the MAF sensor.
12   On 2.6L models, remove the air cleaner assembly (see Chapter 4B), then remove the four MAF sensor mounting bolts **(see illustration)** and separate the MAF sensor from the air cleaner assembly.
13   On 3.2L models, loosen the clamps which secure the MAF sensor to the intake air duct and the air cleaner assembly, remove the intake air duct from the MAF sensor and remove the MAF sensor from the air cleaner assembly.
14   Installation is the reverse of removal.

11 Oxygen sensor - check and replacement

General information

Refer to illustrations 11.1 and 11.2
Note: *OBD-I models are equipped with one oxygen sensor per exhaust manifold: four-cylinder models have one sensor, V6 models have two (one in each manifold). OBD-II four-cylinder models have two sensors: a pre-converter sensor and a post-converter sensor. OBD-II V6 models have three sensors: a pre-converter sensor in each exhaust manifold and a post-converter sensor.*

1   The oxygen sensor **(see illustration)**, which is located in the exhaust manifold, measures the residual oxygen content of the exhaust gas stream. (On OBD-II models, an exhaust manifold sensor is referred to as a *pre-converter* sensor.) The leftover oxygen in the exhaust reacts with the oxygen sensor to produce a voltage output which varies from 0.1-volt (high oxygen, lean mixture) to 0.9-volts (low oxygen, rich mixture). The PCM interprets this variable voltage output to determine the ratio of oxygen to fuel in the mixture. The PCM alters the air/fuel mixture ratio by controlling the pulse width (open time) of the fuel injector(s). A mixture ratio of 14.7 parts air to 1 part fuel is the ideal mixture ratio for minimizing exhaust emissions, thus allowing the catalytic converter to operate at maximum efficiency. It is this ratio of 14.7 to 1 which the PCM attempts to maintain at all

times. **Note 1:** *Early models are equipped with a single-wire oxygen sensor but most models are equipped with a heated oxygen sensor, which has four wires. Refer to the wiring schematics at the end of Chapter 12 if necessary.* **Note 2:** *1996 and later OBD-II systems are equipped with upstream (pre-converter) and downstream (post-converter) oxygen sensors. On V6 models, there are two upstream oxygen sensors, one in the left exhaust manifold and one in the right manifold.*

2   The post-converter oxygen sensor **(see illustration)** on OBD-II models, which is mounted in the exhaust system after the catalytic converter, has no effect on PCM control of the air/fuel ratio. However, the post-converter sensor is identical to the pre-converter sensor and operates in the same way. The PCM uses the post-converter signal to monitor the efficiency of the catalytic converter. A post-converter oxygen sensor produces a more slowly fluctuating voltage signal than a pre-converter sensor; this reflects the lower oxygen content in the post-catalyst exhaust.

3   An oxygen sensor produces no voltage when it is below its normal operating temperature of about 600-degrees F. During this warm-up period, the PCM operates in an open-loop fuel control mode. It does not use the oxygen sensor signal as a feedback indication of residual oxygen in the exhaust. Instead, the PCM controls fuel metering based on the inputs of other sensors and its own programs. All oxygen sensors are equipped with a heating element, powered by fused ignition voltage, to heat the oxygen sensor to operating range as quickly as possible.

4   Correct operation of an oxygen sensor depends on four conditions:

   a) *Electrical - The low voltages generated by the sensor require good, clean connections which should be checked whenever a sensor problem is suspected or indicated.*

   b) *Outside air supply - The sensor needs air circulation to the internal portion of the sensor. Whenever the sensor is installed, make sure the air passages are not restricted.*

11.6 Back-probe the oxygen sensor and measure the voltage output the sensor produces as it goes from cold to fully warm

11.8 Checking battery voltage to the oxygen sensor heater

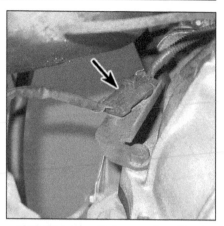

11.14 Because of intense exhaust heat, the oxygen sensor electrical connector is often somewhat removed from the sensor; to locate the connector, trace the electrical lead from the sensor to the connector

c) *Correct operating temperature - The PCM will not react to the sensor signal until the sensor reaches approximately 600-degrees F. This factor must be considered when evaluating the performance of the sensor.*

d) *Unleaded fuel - Unleaded fuel is essential for proper operation of the sensor.*

5   The PCM can detect several different oxygen sensor problems and set diagnostic trouble codes to indicate the specific fault (OBD vehicles, see Section 2; OBD-II vehicles, see Section 3). When an oxygen sensor fault occurs, the PCM will disregard the oxygen sensor signal voltage and revert to open-loop fuel control as described previously.

Check

Refer to illustrations 11.6 and 11.8

Caution: *The oxygen sensor is very sensitive to excessive circuit loads and circuit damage of any kind. For safest testing, disconnect the oxygen sensor connector, install jumper wires between the two connectors and connect your voltmeter to the jumper wires. If jumper wires aren't available, carefully back-probe the wires in the connector shell with suitable probes (such as T-pins). Do not puncture the oxygen sensor wires or try to back-probe the sensor itself. Use only a digital voltmeter to test an oxygen sensor.*

Note: *Performing the following tests will set a diagnostic trouble code. Be sure to erase the trouble code after performing the tests and making the necessary repairs (OBD vehicles, see Section 2; OBD-II vehicles, see Section 3).*

6   Locate the oxygen sensor electrical connector and, using a digital multimeter, back-probe the connector. **Caution:** *Be careful when back-probing the electrical connector. Do not damage the wiring harness or pull on any connectors to make clean contact.* **Note:** *If your meter leads don't have probes small enough in diameter to back-probe the connector, use T-head pins, straight-pins, or straighten out a pair of paper clips, insert*

them into the backside of the connector and grasp them with alligator clips on the ends of the leads. Attach the positive probe of the voltmeter to the pin with an alligator clip and attach the negative probe to ground. On models equipped with a heated oxygen sensor system, back-probe the signal wire **(see illustration)** (refer to the wiring diagrams at the end of Chapter 12 if necessary) and note the sensor signal voltage as the engine warms up.

7   In open loop (engine not warmed up), the oxygen sensor will produce a steady voltage signal of about 0.1 to 0.2 volts. In about two minutes, the engine will reach operating temperature, the oxygen sensor will go into closed loop and it will begin to fluctuate between 0.1 and 0.9 volts. If the oxygen sensor fails to go into closed loop mode or takes a very long time to reach closed loop, replace the oxygen sensor.

8   Also check the oxygen sensor heater. First, check for supply voltage to the heater **(see illustration)**. Disconnect the harness connector and measure the voltage on the harness side of the oxygen sensor electrical connector. There should be battery voltage with the ignition key turned to ON (engine not running). If there is no voltage, check the circuit between the EFI relay, the PCM and the sensor (refer to the wiring diagrams at the end of Chapter 12 if necessary).

9   Next, disconnect the oxygen sensor electrical connector, connect an ohmmeter between the heater terminals and verify that there's continuity (refer to the wiring diagrams at the end of Chapter 12 if necessary). Extremely high or low resistance readings indicate a defective sensor heater.

10   If the oxygen sensor fails any of these tests, replace it with a new part.

Replacement

Refer to illustrations 11.14 and 11.15

Note: *Because it is installed in the exhaust manifold or pipe, which contracts when cool, the oxygen sensor may be very difficult to loosen when the engine is cold. Rather than risk damage to the sensor (assuming you are*

planning to reuse it in another manifold or pipe), start and run the engine for a minute or two, then shut it off. Be careful not to burn yourself during the following procedure.

11   Take the following special precautions when servicing the oxygen sensor.

a) *The oxygen sensor has a permanently attached pigtail and electrical connector which should not be removed from the sensor. Damage or removal of the pigtail or electrical connector can adversely affect operation of the sensor.*

b) *Grease, dirt and other contaminants should be kept away from the electrical connector and the louvered end of the sensor.*

c) *Do not use cleaning solvents of any kind on the oxygen sensor.*

d) *Do not drop or roughly handle the sensor.*

e) *The silicone boot must be installed in the correct position to prevent the boot from being melted and to allow the sensor to operate correctly.*

12   Disconnect the cable from the negative terminal of the battery.

13   Raise the vehicle and place it securely on jackstands, if necessary on some models.

14   Carefully disconnect the electrical connector from the sensor **(see illustration)**.

15   Carefully unscrew the sensor from the exhaust manifold **(see illustration)**. **Caution:** *Excessive force may damage the threads.*

16   Installation is the reverse of removal. Be sure to use anti-seize compound on the sensor threads to facilitate future removal. The threads of new sensors will already be coated with this compound, but if an old sensor is removed and reinstalled, recoat the threads. Install the sensor and tighten it securely.

17   Reconnect the electrical connector of the pigtail lead to the main engine wiring harness.

18   Lower the vehicle and reconnect the cable to the negative terminal of the battery.

11.15 Remove the oxygen sensor using a special slotted oxygen sensor socket

12 Power Steering Pressure (PSP) switch - replacement

General description

Refer to illustration 12.1

1   Turning the steering wheel increases power steering fluid pressure, which increases the load on the engine, particularly at idle. The normally-open Power Steering Pressure (PSP) switch, which is located in the hydraulic pressure line on some models or on the power steering pump **(see illustration)** on other models, closes before the extra load can lower the engine idle speed. A pressure switch that will not close when the power steering system pressure increases will allow the engine rpm to drop at idle, and might even allow the engine to stall if the pressure is high enough.

Check

Refer to illustration 12.2

2   Disconnect the electrical connector to the PSP switch **(see illustration)**.
3   Connect an ohmmeter across the terminals of the PSP switch. Start the engine and

13.1 A typical Throttle Position (TP) sensor (TP sensor shown here is on a 2.2L engine; other TP sensor units are similar); to replace a TP sensor, simply unplug the electrical connector and remove the two mounting screws

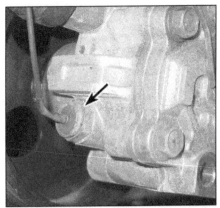

12.1 On some models, the Power Steering Pressure (PSP) switch is located right on the power steering pump (on other models, the PSP switch is located in the pressure line); to remove the switch, simply unscrew it

check for continuity as the steering wheel is turned. There should be no continuity when the wheels are pointing straight ahead (steering wheel not being turned), but there should be continuity when the wheels are turned.
4   If the PSP switch fails this test, replace the switch.

Replacement

5   Disconnect the PSP switch electrical connector **(see illustration 12.2)**.
6   Unscrew the PSP switch **(see illustration 12.1)**.
7   Before installing the PSP switch, coat the threads with Teflon tape or thread sealant to prevent leaks.
8   Installation is the reverse of removal.

13 Throttle Position (TP) sensor - check and replacement

General description

Refer to illustration 13.1

Note 1: *Refer to Chapter 4B, Section 9 for the adjustment and replacement procedure for the TPS used on TBI systems.*
Note 2: *Some 2.6L engines are equipped with a dual purpose TPS/throttle switch. This sensor has two harness connectors; one for the TPS and one for the throttle switch. Check this type of TPS using the test procedures for the TPS and throttle switch detailed in the next section.*
1   The Throttle Position (TP) sensor **(see illustration)** is located on the throttle body, on the end of the throttle valve shaft. The TP sensor is a variable potentiometer that varies its output voltage in proportion to the angle of the throttle valve. This output signal is monitored by the PCM, which uses this data to help determine the correct amount of fuel metered into the engine by the injectors. A broken or loose TP sensor can cause intermittent bursts of fuel from the injectors and

12.2 Power Steering Pressure (PSP) switch electrical connector on models with the PSP switch in the power steering pump (on models with the PSP switch in the pressure line, locate the switch first and then trace the lead back to the connector)

an unstable idle because the PCM thinks the throttle valve is moving. Any problems in the TP sensor or its circuit will set a diagnostic trouble code (see Sections 2 and 3).

Check

Note: *The following tests apply to 1997 and earlier models. Testing the TP sensor on 1998 and later models requires the use of a scan tool and is therefore beyond the scope of the home mechanic.*
2   The first test checks the signal voltage from the TP sensor. Backprobe the GROUND wire and the SIGNAL wire on the backside of the electrical connector with a digital voltmeter (see Wiring Diagrams at the end of Chapter 12 for the correct wire colors for the year and model of the vehicle that you're servicing). **Caution:** *Be careful when backprobing the electrical connector. Do not damage the wiring harness or pull on any connectors to make clean contact.*
3   Turn the ignition switch to ON (engine not running). The TP sensor should produce between 0.50 and 1.0 volt at closed throttle. Have an assistant depress the accelerator pedal to simulate full throttle and note the indicated voltage. The TP sensor should increase its voltage output to 4.0 to 5.0 volts. If the TP sensor output voltage is incorrect, replace the TP sensor.
4   The second test checks the TP sensor reference voltage. Backprobe the REFERENCE wire and the GROUND wire (see Wiring Diagrams at the end of Chapter 12 for the correct wire colors for the year and model of the vehicle that you're servicing). With the ignition key turned to ON (engine not running), there should be about 5.0 volts from the PCM to the TP sensor.

Replacement

5   Disconnect the negative battery cable.
6   Disconnect the electrical connector from the TP sensor.

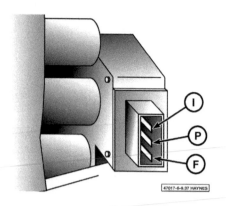

14.2 Throttle switch terminal designations (2.6L engine)

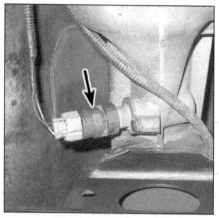

15.1 A typical Vehicle Speed Sensor (VSS) on a 2WD model; on 4WD models, the VSS is in the same general location, but is bolted onto the transfer unit instead of the transmission extension housing)

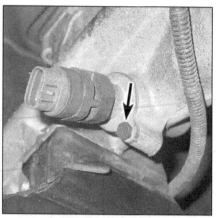

15.9 To remove the VSS from the transmission extension housing (2WD) or transfer unit (4WD), remove the hold-down bolt (2WD model shown)

7   Remove the two TP sensor retaining screws **(see illustration 13.1)** and separate the TP sensor from the throttle body.

8   Install the new TPS leaving the mounting screws loose. Connect a voltmeter to the signal and ground wires as in Step 2. Rotate the sensor until the voltmeter reading is about 0.25 volt. Tighten the screws securely.

14   TPS/throttle valve switch - check and replacement

Note: *Some 2.6L engines are equipped with a dual purpose TPS and throttle switch. This sensor has two harness connectors; one for the sensor portion and the other for the switch portion. Check this type of TPS using the test procedures for the TPS and throttle switch detailed in this section.*

General description

1   The throttle valve switch is located on the end of the throttle shaft on the throttle body. This switch operates simply as an ON - OFF switch mechanism for the different throttle ranges (idle, middle and wide open throttle). Any problems in the TPS or circuit will set a code 22.

Check

Refer to illustrations 14.2

2   To check the throttle valve switch, install the positive probe (+) of an ohmmeter onto terminal "I" and the negative probe (-) of the ohmmeter onto terminal "P" and check the resistance with the throttle completely closed (idle). It should be 0 ohms **(see illustration).**

3   Next, install the positive probe (+) of an ohmmeter onto terminal "P" and the negative probe (-) of the ohmmeter onto terminal "F" and check the resistance with the throttle completely closed (idle). It should be infinite resistance.

4   Next, check the mid-throttle position of the throttle switch. Install the positive probe (+) of an ohmmeter onto terminal "I" and the negative probe (-) of the ohmmeter onto ter-

minal "P" and check the resistance with the throttle at middle position. It should be infinite resistance.

5   Next, install the positive probe (+) of an ohmmeter onto terminal "P" and the negative probe (-) of the ohmmeter onto terminal "F" and check the resistance with the throttle at middle position. It should be infinite resistance.

6   Next, check the wide open throttle position on the throttle switch. Install the positive probe (+) of an ohmmeter onto terminal "I" and the negative probe (-) of the ohmmeter onto terminal "P" and check the resistance with at WOT. It should be infinite resistance.

7   Next, install the positive probe (+) of an ohmmeter onto terminal "P" and the negative probe (-) of the ohmmeter onto terminal "F" and check the resistance with at WOT. It should be 0 resistance.

8   If the resistance readings are incorrect, replace the throttle switch.

Replacement

9   Remove the two retaining screws and separate the throttle switch from the throttle body.

10   Install the new throttle switch leaving the mounting screws loose. Connect an ohmmeter to terminals I and P and rotate the switch until it reads 0 ohms. Tighten the screws securely.

15   Vehicle Speed Sensor (VSS) - check and replacement

General description

Refer to illustration 15.1

1   The Vehicle Speed Sensor (VSS) **(see illustration)** is located on the transfer case (4WD models) or on the transmission (2WD models). The VSS produces a pulsing voltage signal that is translated by the PCM into

miles-per-hour. The PCM uses this information to operate the cruise control system, the speedometer, and the torque converter clutch (TCC) and shift solenoid in the transmission (see Chapter 7B). A defective VSS can cause various driveability and transmission problems. A problem in the VSS circuit will set a trouble code.

2   On OBD-I vehicles (all 2.3L and 3.1L models and pre-1996 2.6L and 3.2L models), the VSS is a permanent magnet generator. The sensor is triggered by a toothed rotor on the transmission output shaft. As the output shaft rotates, the sensor produces an AC voltage, the frequency of which is proportional to vehicle speed.

3   On OBD-II vehicles (1996 and 1997 2.6L models, 1996 and later 3.2L models and all 2.2L models), the VSS is a hall-effect switch. A hall-effect circuit consists of a battery-powered hall-effect crystal, a permanent magnet and a shutter blade. The crystal and the magnet are stationary; the shutter blade is the only moving part. When the crystal is powered up by battery voltage, but is exposed to the permanent magnet, it puts out a weak (about 0.4 volt) signal. But when the shutter blade passes between the crystal and the magnet, the crystal emits a 12-volt signal.

Check

4   Raise the vehicle and support it securely on jackstands. Locate the vehicle speed sensor **(see illustration).** Unplug the electrical connector from the VSS and inspect the connector terminals and the VSS wire harness for any damage. Repair as necessary.

OBD-I vehicles

5   Using the a digital voltmeter, back-probe the two wire terminals of the VSS connector using suitable probes. **Caution:** *Be careful when back-probing the electrical connector. Do not damage the wiring harness or pull on any connectors to make clean contact.* **Note:** *If your voltmeter leads don't have probes small enough in diameter to back-*

15.10 Be sure to replace the O-rings on the VSS

16.1a On 2.3L engines with a feedback carburetor, and on 2.6L engine with MPFI (shown), the Powertrain Control Module (PCM) is located behind the left (driver's side) kick panel (on 3.1L V6 engines with TBI, the PCM is also located behind the kick panel, but it's higher up)

16.1b On 2.2L four-cylinder and on 3.2L V6 engines, the Powertrain Control Module (PCM) is located in the lower center of the dash

probe the connector, use T-head pins, straight-pins, or straighten out a pair of paper clips, insert them into the backside of the connector and grasp them with alligator clips on the ends of the leads. Connect a voltmeter to the probes and set the meter on the AC volts scale. Turn the ignition key On. Block one rear tire so it will not turn and rotate the other rear tire by hand while watching the voltmeter. **Note:** *If the vehicle is equipped with a limited-slip differential, it is not necessary to block one of the drive wheels.* The sensor should produce a minimum of 0.5 volts and the voltage should increase as the transmission output shaft rotates faster. If the VSS is defective, replace it.

OBD-II vehicles

6   Testing the VSS on these vehicles is beyond the scope of the home mechanic.

Replacement

Refer to illustrations 15.9 and 15.10

7   Raise the vehicle and support it securely on jackstands.
8   Disconnect the electrical connector from the VSS.
9   Remove the hold-down bolt **(see illustration)** and the hold-down clamp and pull the VSS out of the transfer case or transmission.
10   Whether you're installing the old VSS or a new unit, be sure to replace the O-rings **(see illustration)**.
11   Tighten the hold down bolt to the torque listed in this Chapter's Specifications. Installation is otherwise the reverse of removal.

---

16   Powertrain Control Module (PCM) - removal and installation

Refer to illustrations 16.1a and 16.1b
Caution: *To prevent damage to the PCM, the ignition switch must be turned to OFF when disconnecting or connecting the PCM connectors.*
Note: *3.1L TBI models are equipped with a replaceable calibration chip, known as a Pro-*

grammable Read-Only Memory (PROM), which is located in the PCM and which must be installed into a new PCM when replaced or exchanged.
1   The Powertrain Control Module (PCM) is the "brain" of the engine management system: it monitors voltage signals from various information sensors on the vehicle, processes this data and then alters the operation of the engine to keep it running smoothly under all conditions. The PCM is located in various places depending upon the year and model. On 2.3L models with a feedback carburetor and on 2.6L MPFI models, the computer is located behind the left (driver's side) kick panel **(see illustrations)**. On 3.1L TBI models, the computer is also located behind the left kick panel, but it's higher up, near the dash. On 2.2L and on 3.2L models, the PCM is located under the dash, below the sound system, just in front of the center console. In order to access the PCM, you'll have to remove the kick panel (earlier models) or the center console (later models) (see Chapter 11).
2   Using the tips of your fingers, tap vigorously on the side of the computer while the engine is running. If the computer is not functioning correctly, or if there's an intermittent open, ground or short in the PCM connector(s), the engine may stumble or stall. If you're using a SCAN tool or other diagnostic equipment, it may display glitches on the engine data stream.
3   If the PCM fails this test, check the electrical connectors. Each connector is color coded to fit the respective slot in the computer body. If there are no obvious signs of damage, have the PCM checked at a dealer service department.

2.3L and 2.6L four-cylinder and 3.1L V6 models

4   Disconnect the cable from the negative battery terminal.

5   Remove the kick panel (see Chapter 11).
6   Remove the PCM mounting fasteners..
7   Carefully pull out the PCM without damaging the electrical connectors and wiring harness.
8   Unplug the electrical connectors from the PCM. Each connector is color coded to fit its respective receptacle in the computer.
9   Pry the tang and lift the plastic protector cover (if equipped) from the PCM.
10   Installation is the reverse of removal.

PROM replacement (3.1L TBI engines)

Refer to illustrations 16.13 and 16.14
11   3.1L TBI engines are equipped with a memory calibration chip called the PROM. New PCMs are not supplied with a PROM, therefore it is necessary to remove the original and install it into the new PCM. Use care not to damage the PROM when working with these delicate electrical components.
12   Remove the PCM (see Step 6).
13   Remove the screws and detach the PROM cover from the PCM **(see illustration)**.
14   Using two fingers, carefully push both retaining clips back away from the PROM

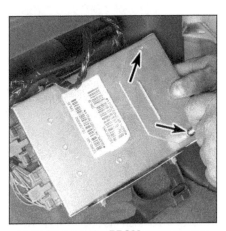

16.13 Remove the PROM cover screws from the PCM

16.14 Press the retaining tabs out to release the EEPROM

16.18a To detach the front of the PCM from the dash on a 2.2L model, remove the left mounting nut . . .

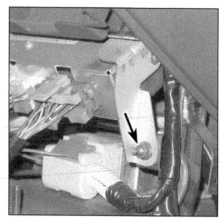

16.18b . . . and remove the right mounting nut (2.2L model shown; 3.2L model PCMs have screws instead of nuts)

and lift the unit straight out of the computer (PCM) **(see illustration)**. **Note:** *There are two types of clips used on the PROM sockets; hollow type tabs or solid type tabs.*

15   Installation is the reverse of removal. Be sure to use the alignment notches in the seat area of the computer when sliding the PROM back in place. Press only on the ends of the PROM assembly until it snaps back into place.

2.2L four-cylinder and 3.2L V6 models

Refer to illustrations 16.18a, 16.18b and 16.20

16   Disconnect the negative battery cable.
17   Remove the center console (see Chapter 11).
18   Remove the two PCM mounting nuts **(see illustrations)**.
19   Disconnect the PCM electrical connectors.
20   Remove the rear PCM mounting nuts or screws **(see illustration)**. (Some models have two rear PCM mounting screws or nuts and some models have only a left rear screw or nut.)
21   Remove the PCM.
22   Installation is the reverse of removal.

17 Air Injection Reactor (AIR) system

General information

Refer to illustrations 17.2 and 17.4

1   The Air Injection Reactor (AIR) system pump air into the exhaust manifold in order to reduce hydrocarbons (HC) and carbon monoxide (CO) in the exhaust gases. The extra oxygen in the exhaust stream helps the oxidation catalyst convert HC and CO into harmless oxygen (O), carbon dioxide (CO_2) and water vapor (H_2O).

2   On four-cylinder engines, the AIR system consists of a belt drive air pump, check valve, air switching valve (early) or air management valve (late), mixture control valve (2.3L feedback carburetor systems), air injection manifold and the associated hoses **(see illustration)**.

3   On 3.1L V6 engines, the AIR system consists of the air pump, the Electric Air Control (EAC) valve, an EAC solenoid, a check valve, the PCM and the associated hoses. Because the AIR system on 3.1L engines is

PCM-controlled, the checks that can be made to the system are limited.

4   Air is drawn through the air pump air cleaner, compressed, pumped through the check valve to the air injection manifold, and then through the injection nozzles into the exhaust manifold. When the pressure inside the exhaust manifold exceeds pump injection pressure, the check valve prevents exhaust gases from forcing their way into the air pump **(see illustration)**. During high-speed operation (when exhaust manifold pressure is relatively lower than pump pressure), excessive pump pressure is vented to the atmosphere through the relief valve (four-cylinder models) or to the air cleaner (3.1L models).

Check

5   Inspect all hoses, air gallery pipes and nozzles for security and condition.
6   Check and adjust the air pump drivebelt tension (see Chapter 1).
7   With the engine at normal operating temperature, disconnect the hose leading to the check valve.

Check valve

8   Run the engine at approximately 2,000

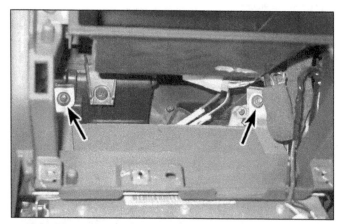

16.20 To detach the rear of the PCM from the dash on a 2.2L model, remove these two mounting nuts (3.2L model PCMs have only a left rear screw)

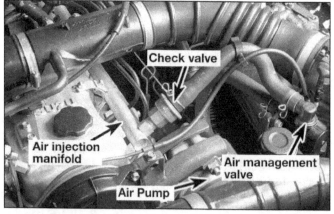

17.2 Typical Air Injection Reactor (AIR) system component locations (2.6L MPFI engine shown, others similar)

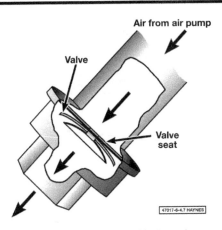

17.4 The AIR system check valve should allow air to pass through it in only one direction

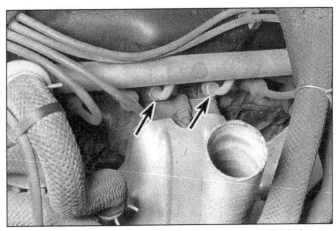

17.31 Apply penetrating oil to the air injection manifold pipes

rpm and then let it return to idling speed, while watching for exhaust gas leaking from the valve. If the valve is leaking, replace it.

Air pump relief valve

9   Run the engine at a steady 3000 rpm and place your hand on the air outlet of the emergency relief valve. A definite air pressure should be felt. If it is not, replace the valve.

Air switching valve/air management valve

10   The air switching valve is installed near the air pump. The early systems use the air management valve which is activated by vacuum from the vacuum switching valve (VSV) which is controlled by the PCM. Later systems use the air management valve which is activated by the Powertrain Control Module (PCM). If normal, the secondary air continues to blow out from the valve if the accelerator pedal is pressed quickly and then released. Air should blow for not more than 5 seconds. If the valve continues to blow air, replace the valve.

Mixture control valve (2.3L feedback carburetor)

11   Locate the mixture control valve near the air cleaner. The air cleaner may have to be removed for access.
12   Remove the air hose from the intake manifold to the mixture control valve and plug it. If the mixture control valve is functioning correctly, the secondary air continues to blow from the mixture control valve for only a few seconds when the accelerator pedal is pressed and released quickly. If the valve continues to blow air for more than 5 seconds, then replace the valve. Refer to Section 5 for additional information on the mixture control valve system.

Component replacement

Air pump

13   Remove the air hoses from the pump.
14   Remove the drivebelt from the pump (see Chapter 1).

15   Disconnect all hoses from the pump.
16   Remove the pump mounting and adjusting bolts and remove the pump.
17   If necessary, remove the pulley from the pump.
18   Installation is the reverse of the removal procedure.

Check valve

19   Remove the air cleaner.
20   Disconnect the air hose from the check valve.
21   Remove the check valve from the air gallery pipe, using two wrenches.
22   Installation is the reverse of the removal procedure.

Mixture control valve

23   Remove the valve from the clamp bracket and disconnect the hose.
24   Installation is the reverse of removal.

Air switching valve

25   Remove the mounting hardware and hoses from the valve and remove the valve.
26   Installation is the reverse of the removal procedure.

Air gallery pipe and injection nozzles

Refer to illustration 17.31

27   Because of the likelihood of bending or damaging the pipes during removal, the pipes should not be removed unless they are already damaged and need replacing.
28   Remove the air cleaner.
29   Disconnect the hose from the check valve.
30   Disconnect or remove all lines and hoses that will interfere with removal of the air gallery assembly.
31   Apply penetrating oil to the nuts that attach the air gallery to the exhaust manifold **(see illustration)**, then loosen them until the gallery can be lifted out.
32   Lift out the air gallery pipes from the threaded openings in the exhaust manifold.
33   Installation is the reverse of the removal procedure.

18   Air regulator (2.6L engine) - check and replacement

General information

1   The air regulator valve (located under the air intake plenum) provides air bypass when the engine is cold for fast idle. It consists of a contained wax element, a piston and a spring built into the regulator assembly. When the temperature of the wax is low, the shutter opens and allows air to bypass from the air intake duct into the intake manifold. As the engine warms, the wax expands and does not allow any air to circulate into the intake manifold. This condition continues until the engine is stopped and the temperature of the engine block cools down.

Check

2   Disconnect the electrical connector from the air regulator. The air regulator is mounted under the intake manifold.
3   Working on the air regulator side, check the resistance. It should be between 45 to 50 ohms resistance. Replace the air regulator if the specifications are incorrect.

Replacement

4   Raise the vehicle and secure it with jackstands.
5   Remove the air regulator mounting bolts.
6   Installation is the reverse of removal.

19   Catalytic converter - general information

General description

1   The catalytic converter is an emission control device in the exhaust system that reduces hydrocarbons (HC), carbon monoxide (CO) and oxides of nitrogen (NOx) in the exhaust gas stream.

20.2a Location of the vacuum switching valve on early models (2.6L engine shown)

20.2b The EVAP system assembly (canister, purge control solenoid, hoses, etc.) is located under the vehicle attached to the frame near the fuel tank

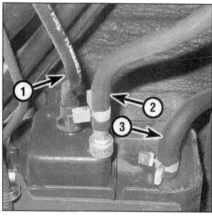

20.16 Clearly label and then disconnect the hoses from the EVAP canister

1   *Hose to EVAP canister purge valve solenoid (in engine compartment, on ignition coil bracket)*
2   *Hose to fuel filler neck/fuel tank*
3   *Hose to EVAP canister vent solenoid (right side of EVAP canister)*

Check

2   The test equipment for a catalytic converter is expensive. If you suspect that the converter on your vehicle is malfunctioning, take it to a dealer or authorized emissions inspection facility for diagnosis and repair.

3   Whenever the vehicle is raised for servicing of underbody components, check the converter for leaks, corrosion and other damage. If damage is discovered, the converter should be replaced.

Replacement

4   Because the converter is part of the exhaust system, converter replacement requires removal of the exhaust pipe assembly (see Chapter 4). Take the vehicle to a dealer service department or to a muffler shop.

20   Evaporative Emissions Control System (EVAP) - check and component replacement

General description

Refer to illustrations 20.2a and 20.2b

1   The Evaporative Emissions Control (EVAP) system collects and stores fuel vapors that evaporate from the fuel tank and from the intake manifold.

2   The EVAP system consists of a charcoal-filled canister, canister purge valve and the lines connecting the canister to the fuel tank, ported vacuum and intake manifold vacuum. Some models are equipped with a vacuum switching valve that regulates vacuum to several components - the EGR valve, purge valve, distributor, etc. **(see illustration)**. 1996 and later OBD II systems are equipped with a computer-controlled purge control solenoid, a PCM relay and an evaporative canister. This system is mounted below the trunk area on the frame **(see illustration)**.

3   Fuel vapors are transferred from the fuel tank, throttle body and intake manifold to a canister where they are stored when the

engine is not operating. When the engine is running, the fuel vapors are purged from the canister by a purge control solenoid and consumed in the normal combustion process.

Check

4   Poor idle, stalling and poor driveability can be caused by an inoperative purge control solenoid, a damaged canister, split or cracked hoses or hoses connected to the wrong tubes.

5   Evidence of fuel loss or fuel odor can be caused by fuel leaking from fuel lines or the throttle body, a cracked or damaged canister, an inoperative bowl vent valve, an inoperative purge valve, disconnected, misrouted, kinked, deteriorated or damaged vapor or control hoses or an incorrectly seated air cleaner or air cleaner gasket.

6   Inspect each hose attached to the canister for kinks, leaks and breaks along its entire length. Repair or replace as necessary.

7   Inspect the canister. If it is cracked or damaged, replace it.

8   Look for fuel leaking from the bottom of the canister. If fuel is leaking, replace the canister and check the hoses and hose routing.

9   Apply a short length of hose to the lower tube of the purge control valve and attempt to blow through it. Little or no air should pass into the canister (a small amount of air will pass because the canister has a constant purge hole).

10   With a hand-held vacuum pump, apply vacuum through the control vacuum signal tube near the throttle body to the EGR and canister purge control valve.

11   If the purge control valve does not hold vacuum for at least 20 seconds, the purge control valve is leaking and must be replaced.

Component replacement

Earlier models

12   Clearly label, then detach, all vacuum lines from the canister.

13   Loosen the canister mounting clamp

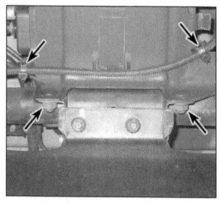

20.17 Before removing the EVAP canister mounting bracket, unplug the electrical connector from the EVAP canister vent solenoid and detach the harness clips from the underside of the canister mounting bracket; to detach the canister mounting bracket, remove these two nuts

bolt(s) and pull the canister out.

14   Installation is the reverse of removal.

1996 and later OBD-II models

EVAP canister

Refer to illustrations 20.16 and 20.17

15   Raise the rear of the vehicle and place it securely on jackstands.

16   Clearly label and then disconnect all hoses from the EVAP canister **(see illustration)**.

17   Disconnect the electrical connector from the EVAP canister vent solenoid **(see illustration 20.20)** and detach the vent solenoid electrical harness from the underside of the EVAP canister mounting bracket **(see illustration)**.

18   Remove the EVAP canister mounting

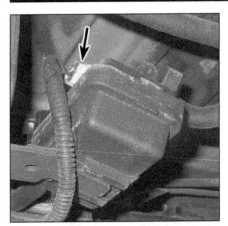

20.20 Disconnect the electrical connector from the EVAP canister vent solenoid

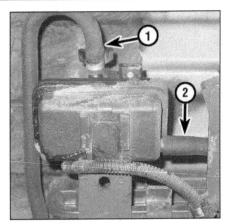

20.21 Clearly label and then disconnect the hoses from the EVAP canister vent solenoid

1   *Hose to EVAP canister*
2   *Hose to air separator*

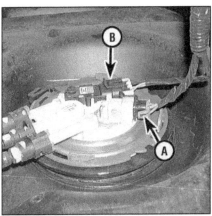

20.25 Lower the fuel tank and disconnect the electrical connector from the fuel tank pressure sensor (it's not necessary to unplug the electrical connector from the fuel pump/fuel level sending unit unless you're going to actually remove the tank)

A   *Fuel pump/fuel level sending unit electrical connector*
B   *Fuel tank pressure sensor electrical connector*

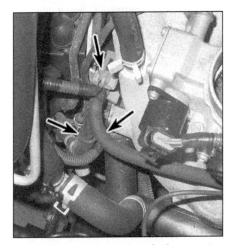

20.28 Disconnect the electrical connector and the two vapor hoses from the EVAP canister purge valve solenoid

bracket nuts **(see illustration 20.17)** and remove the EVAP canister and mounting bracket as a single assembly. Separate the EVAP canister from the canister mounting bracket.

EVAP canister vent solenoid

Refer to illustrations 20.20 and 20.21
19   Raise the rear of the vehicle and place it securely on jackstands.
20   Disconnect the electrical connector from the EVAP canister vent solenoid **(see illustration)**.
21   Clearly label and then disconnect the hoses from the EVAP canister vent solenoid **(see illustration)**.
22   To disengage the EVAP canister vent solenoid from its mounting bracket, slide it to the right.
23   Installation is the reverse of removal.

Fuel tank pressure sensor

Refer to illustration 20.25
24   Raise the rear of the vehicle and place it securely on jackstands.
25   Lower the fuel tank (see Chapter 4) and disconnect the electrical connector from the

fuel tank pressure sensor **(see illustration)**. (As long as the tank is adequately supported, it's not absolutely necessary to actually *remove* the fuel tank, i.e. disconnect all the electrical connectors, fuel lines and vapor hoses.)
26   Remove the fuel tank pressure sensor.
27   Installation is the reverse of removal.

EVAP canister purge valve solenoid (2.2L engines)

Refer to illustrations 20.28, 20.31a, 20.31b, 20.34a and 20.34b
Note: *The EVAP canister purge valve solenoid, which is located on the front right side of the ignition coil bracket, below the rear part of the intake manifold, is extremely difficult to access. There are two methods for removing and installing it: The first method involves less work, but requires small hands. The second method is easier, but requires more disassembly.*
28   Disconnect the electrical connector

from the EVAP canister purge valve solenoid **(see illustration)**.
29   Disconnect the two hoses from the EVAP canister purge valve solenoid **(see illustration 20.28)**.
30   If you want to try the first method, raise the front of the vehicle and place it securely on jackstands.
31   Working from underneath the engine, use a small screwdriver to pry the locking tang down **(see illustrations)** and then slide off the EVAP canister purge valve solenoid.
32   If you're unable to extricate the EVAP canister purge valve solenoid from its mounting bracket with the ignition coil bracket in place, the alternative is to remove the ignition coil bracket and then try again:
33   Disconnect the spark plug wires from the ignition coil (see Chapter 1).

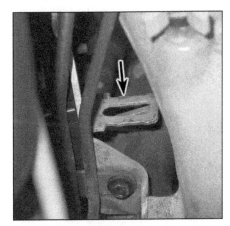

20.31a The EVAP canister purge valve solenoid is mounted on this bracket (which is attached to the front side of the ignition coil bracket) (solenoid removed for clarity)

20.31b To disengage the EVAP canister purge valve solenoid from its mounting bracket, pry the locking tang down and slide of the solenoid (solenoid and bracket removed for clarity)

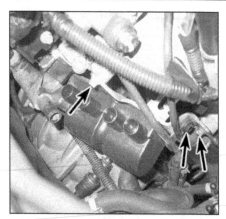

20.34a To detach the ignition coil bracket from the engine, remove these four bolts (bolt on far left not visible in this photo)

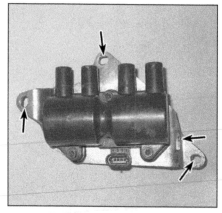

20.34b The ignition coil bracket, removed from the engine (far left hole is for bolt not visible in previous photo)

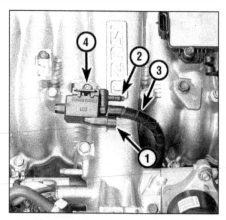

20.36 On 1997 and earlier 3.2L models, the EVAP canister purge valve solenoid is located on top of the upper intake manifold; to remove it, unplug the electrical connector (1), disconnect the vacuum hoses (2) and (3), and remove the retaining bolt (4)

34   Remove the ignition coil bracket (see illustrations).
35   Detach the EVAP canister purge valve solenoid from the ignition coil bracket (see illustrations 20.31a and 20.31b).

EVAP canister purge valve solenoid (3.2L engines)

Refer to illustration 20.36
Note: *On 1997 and earlier 3.2L models, the EVAP canister purge valve solenoid is bolted to the top of the upper intake manifold. On 1998 and later models, it's bolted to the coolant pipe, adjacent to the left front corner of the throttle body.*
36   On 1997 and earlier models, locate the EVAP canister purge valve solenoid on top of the upper intake manifold. Disconnect the electrical connector and the vacuum hoses from the solenoid and then remove the solenoid retaining bolt (see illustration).
37   On 1998 and later models, locate the EVAP canister purge valve solenoid on the coolant pipe adjacent to the left front corner of the throttle body. Disconnect the electrical connector and the vacuum hoses from the solenoid and then remove the solenoid retaining bolt.
38   Installation is the reverse of removal.

---

21   Exhaust Gas Recirculation (EGR) system - check and component replacement

---

General description

1   The EGR system is used to lower NOx (oxides of nitrogen) emission levels caused by high combustion temperatures. The EGR recirculates a small amount of exhaust gases into the intake manifold. The additional mixture lowers the temperature of combustion thereby reducing the formation of NOx compounds.
2   Early EGR systems are equipped with either a Thermal Vacuum Valve (2.3L and 2.6L engines) that receives a ported vacuum signal or an Electronic Vacuum Regulator Valve (EVRV) (3.1L engine) which is controlled

by the Powertrain Control Module (PCM). On early mechanical systems, the venturi vacuum system utilizes a vacuum tap at the throat of the carburetor venturi to provide the control signal. The ported vacuum control system uses a slot in the carburetor body which is exposed to an increasing percentage of manifold vacuum as the throttle valve is opened up during acceleration. On the early systems, the backpressure transducer monitors exhaust pressure and venturi vacuum in order to activate the EGR diaphragm. The TVV controls the throttle vacuum that is in turn applied to the EGR control valve. The thermal vacuum valve detects coolant temperature and opens or closes the air passage according to temperature.
3   1996 and later EGR systems are fully controlled by the PCM. The principle component of these later systems is the EGR valve, referred to by Isuzu as a "linear" EGR valve. On these systems, the PCM monitors and controls the actual position of the pintle valve that controls the flow of exhaust gases through the EGR valve. Aside from making sure that the electrical connectors are secure, checking linear type EGR valves is beyond the scope of the home mechanic. These EGR systems can only be diagnosed with a special scan tool.

Check

Refer to illustrations 21.5 and 21.6
4   Check all hoses for cracks, kinks, broken sections and correct connection. Inspect all system connections for damage, cracks and leaks.
5   To check the EGR system operation, bring the engine up to operating temperature and, with the transmission in Neutral (parking brake set and tires blocked to prevent movement), allow it to idle for 70 seconds. Open the throttle abruptly so the engine speed is between 2,000 and 3,000 rpm and then allow it to close. The EGR valve stem should move if the control system is working correctly. The test should be repeated several times. Movement of the stem indicates the control system is functioning correctly (see illustration).
6   If the EGR valve stem does not move, check all of the hose connections to make sure they are not leaking or clogged. Disconnect the vacuum hose and apply ten inches of vacuum with a hand pump (see illustration). If the stem still does not move, replace the EGR valve with a new one. If the valve does

21.5 Check to make sure the EGR pintle moves freely (if the valve is hot, wear a glove to prevent burns)

21.6 Apply vacuum to the EGR valve and make sure the EGR holds vacuum and does not leak down

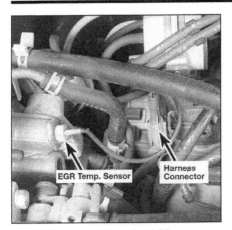

21.27 Location of the EGR gas temperature sensor on the 2.6L engine

21.31a Typical linear EGR valve (1997 and earlier 3.2L V6 engines)

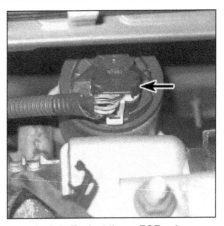

21.31b Typical linear EGR valve (2.2L engine shown, 1998 and later 3.2L engines similar)

open, measure the valve travel to make sure it is approximately 1/8-inch. Also, the engine should run roughly when the valve is open. If it doesn't, the passages are probably clogged.

7 Apply vacuum with the pump and then clamp the hose shut. The valve should stay open for 30 seconds or longer. If it does not, the diaphragm is leaking and the valve should be replaced with a new one.

8 If the EGR valve is not receiving vacuum, the carburetor or throttle body unit must be removed to check and clean the slotted port in the throttle bore and the vacuum passages and orifices in the throttle body. Use solvent to remove deposits and check for flow with light air pressure.

9 If the engine idles roughly and it is suspected the EGR valve is not closing, remove the EGR valve and inspect the poppet and seat area for deposits.

10 If the deposits are more than a thin film of carbon, the valve should be cleaned. To clean the valve, apply solvent and allow it to penetrate and soften the deposits, making sure that none gets on the valve diaphragm, as it could be damaged.

11 Use a vacuum pump to hold the valve open and carefully scrape the deposits from the seat and poppet area with a tool. Inspect the poppet and stem for wear and replace the valve with a new one if wear is found.

12 Locate the TVV or BPT valve and plug one of the ports with a finger. Use a hand-held vacuum pump and apply vacuum to the valve. Check for any sign of leaks.

13 With the engine running, check for a vacuum signal from the intake manifold. This test will determine if manifold vacuum is reaching the various valves and components of the EGR system.

14 Also apply vacuum to the port (vacuum hose) that is routed to the EGR valve. The gauge should hold vacuum. If not, check for broken hoses or ruptured diaphragms in the EGR valve.

15 Check the EVRV (3.1L engines). Check for battery voltage to the valve with the ignition key ON. If battery voltage is reaching the EVRV, have the PCM checked by a dealer service department.

16 Also remove the vacuum hose(s) from the EVRV and check for vacuum to the EGR valve.

Component replacement

EGR valve

17 When buying a new EGR valve, make sure that you have the right EGR valve. Use the stamped code located on the top of the EGR valve.

18 Detach the cable from the negative terminal of the battery.

19 Remove the air cleaner housing assembly (see Chapter 4A or 4B).

20 Detach the vacuum line from the EGR valve.

21 Raise the vehicle and support it securely on jackstands. Remove the EGR pipe from the exhaust manifold. Lower the vehicle.

22 Remove the EGR valve mounting bolts.

23 Remove the EGR valve and gasket from the manifold. Discard the gasket.

24 With a wire wheel, buff the exhaust deposits from the EGR valve mounting surface on the manifold and, if you plan to use the same valve, the mounting surface of the valve itself. Look for exhaust deposits in the valve outlet. Remove deposit build-up with a screwdriver. **Caution:** *Never wash the valve in solvents or degreaser - both agents will permanently damage the diaphragm. Sand blasting is also not recommended because it will affect the operation of the valve.*

25 If the EGR passage contains an excessive build-up of deposits, clean it out with a wire wheel. Make sure that all loose particles are completely removed to prevent them from clogging the EGR valve or from being ingested into the engine.

26 Installation is the reverse of removal.

EGR gas temperature sensor

General Description

Refer to illustration 21.27

27 Some models are equipped with an EGR gas temperature sensor mounted near the EGR valve **(see illustration)**, installed into the EGR tube. This sensor detects the temperature of the exhaust as it moves through the EGR valve. The information is sent to the

PCM and in turn the EGR on/off time is regulated precisely and more efficiently. Any malfunction with the EGR gas temperature sensor will set a code 34 on MPFI systems.

Check

28 Disconnect the harness connector for the EGR gas temperature sensor and measure the resistance of the sensor. Resistance should decrease as temperature increases and at 212-degrees F should measure 77 to 93 K-ohms.

Removal and installation

29 Disconnect the harness connector for the EGR gas temperature sensor and using and open-end wrench, remove the sensor from the intake manifold.

30 Installation is the reverse of removal.

Linear EGR valve system (some later 2.6L engines, all 2.2L and 3.2L engines)

General description

Refer to illustrations 21.31a and 21.31b

31 The linear EGR valve **(see illustrations)** feeds small amounts of exhaust gas back into the intake manifold and then into the combustion chamber, independently of intake manifold vacuum.

32 The valve controls exhaust gas flow from the exhaust manifold into the intake manifold through a single orifice with a PCM-controlled pintle. This type of EGR valve functions in a fashion similar to the stepper motor used in IAC valves that control idle quality.

33 If the EGR system malfunctions, the PCM will set a diagnostic trouble code P0401, P0402, P0404, P0405 or P0406. Have the EGR system checked by a dealer service department in the event of EGR system failure. Special electronic diagnostic equipment is needed to check this valve.

Replacement

Refer to illustrations 21.36, 21.37 and 21.38

34 Disconnect the electrical connector

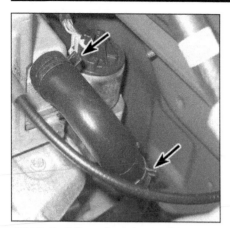

21.36 On 2.2L engines, loosen these two hose clamps and remove the crankshaft breather hose

21.37 On 1997 and earlier 3.2L engines, unscrew the EGR exhaust tube nut from the linear EGR valve and disconnect the tube from the valve

21.38 To detach the EGR valve from the engine, remove these two bolts (2.2L engine shown, 3.2L engine similar)

from the EGR valve **(3.2L engines, see illustration 21.31a; 2.2L engines, see illustration 21.31b)**.
35   On 2.2L engines, disconnect the electrical connector from the Intake Air Temperature (IAT) sensor (see Section 7) and remove the air intake duct (see Chapter 4B).
36   On 2.2L engines, remove the crankshaft breather hose **(see illustration)**.
37   On 1997 and earlier 3.2L engines, remove the EGR tube from the EGR valve **(see illustration)**.
38   Remove the two mounting bolts **(see illustration)** and remove the EGR valve from the air intake plenum.
39   Remove the EGR valve gasket.
40   Clean the mounting surface of the EGR valve. Remove all traces of gasket material from the intake manifold and from the valve if it is to be reinstalled. Clean both mating surfaces with a cloth dipped in lacquer thinner or acetone.
41   Install a new gasket and the EGR valve and tighten the bolts securely.
42   Connect the electrical connector onto the EGR valve.

---

22   Fuel cut-off system (2.3L engine) - general information and check

---

General information

1   The fuel cut-off system trims the fuel supply during deceleration to prevent backfiring, overheating and after-burning in the exhaust system. The carbureted version is mechanically actuated; the fuel-injected version is operated by the Powertrain Control Module (PCM).
2   The fuel cut-off system consists principally of a slow cut solenoid valve incorporated in the carburetor, an engine speed sensor, a transmission switch, an accelerator switch and a clutch switch.
3   This system acts as an auxiliary fuel system operated by the slow cut solenoid. When the solenoid is de-energized, it pulls the plunger OUT to close the fuel flow passage.

4   The engine speed sensor monitors the engine rpm by sensing the electric pulses from the ignition coil. When the engine speed exceeds 2,000 RPM, the engine speed sensor turns OFF and the coasting fuel cut solenoid becomes de-energized.
5   When the transmission is shifted into any gear, the neutral switch is activated, allowing the coasting richer circuit to energize.
6   The accelerator switch is connected to the accelerator linkage and is turned ON when the pedal is not pressed. When the accelerator is pressed, the switch is turned OFF and the circuit is open.
7   The clutch switch is installed near the clutch pedal and turns OFF when the pedal is pressed, de-energizing the coasting richer circuit.

Check

8   Check the engine speed sensor by connecting the color coded wiring terminals BY and LgW with a suitable jumper wire. Start the engine and check for voltage between Lg and B coded terminals. The engine speed sensor is functioning correctly if the voltage is 0 when the engine runs over 2,000 rpm.
9   Check the setting of the clutch switch and the accelerator switch and adjust them if necessary (see Chapter 8).
10   Check each switch with an ohmmeter. The switches are functioning correctly if they turn OFF (open circuit) when the pedal is pressed.

---

23   High altitude emission control system - general information

---

General information

1   Some 2.3L engines are equipped with a high altitude emission control system to compensate for the density of air at higher elevations. This system will allow the engine to perform better and the emission system to control the release of excess emissions at higher altitudes. This system consists of the MAP sensor, the vacuum switching valve, the Pow-

ertrain Control Module (PCM), the main relay and all the vacuum lines attaching the components. **Note:** *If the vehicle has been in use in high altitudes and is equipped with this system, and is returning to low altitudes (permanently), consult a dealer service department to alter the system for low altitude operation.*

Installation

2   Obtain the correct high altitude emission control system kit from an Isuzu dealer parts department. Dealers located in cities situated at high altitudes will have the necessary information.
3   Adjust the ignition timing to the Specification on the label that came with the kit (usually 4 degrees more advance than the original timing specification). Check the idle speed and adjust it if necessary.
4   Attach the label in a prominent location under the hood.

---

24   Idle Air Control (IAC) valve - check and replacement

General Information

Refer to illustrations 24.1a and 24.1b
1   The IAC valve **(see illustrations)** is attached to the throttle body. The Powertrain Control Module (PCM) controls the IAC valve with an on/off pulse signal. The longer the on-duty signal is left on, the larger the amount of air that's allowed to flow through the valve. Before diagnosing the IAC system, make sure there are no obvious intake leaks (broken hoses, leaking gaskets etc.) where they are mounted.

Check

2   With the engine running, disconnect the IAC valve electrical connector and confirm that the idle speed drops immediately. If it does not, continue checking the idle control system.
3   First, remove the IAC valve and check the pintle for excessive carbon deposits. If necessary, clean it with carburetor cleaner spray. Also clean the valve housing to

24.1a Typical Idle Air Control (IAC) valve used on 3.2L V6 engines (1997 and earlier unit shown, later units similar)

24.1b Idle Air Control (IAC) valve (2.2L engine)

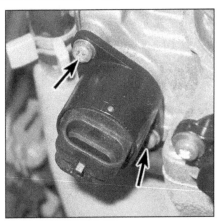

24.6 To detach the Idle Air Control (IAC) valve from the throttle body, remove these two screws (2.2L engine shown, 3.2L V6 engine similar)

remove any deposits.

4   With the ignition key ON (engine not running), check for battery voltage to the IAC valve. Battery voltage should exist. If not, check the harness and PCM (refer to the wiring diagrams at the end of Chapter 12).

Replacement

Refer to illustration 24.6

5   Disconnect the electrical connector from the IAC valve **(see illustration 24.1a or 24.1b)**.
6   Remove the valve attaching screws **(see illustration)** and separate the assembly from the air intake plenum.
7   Check the condition of the O-ring. Replace it with a new one if it's cracked or otherwise deteriorated.
8   Clean the sealing surface and the bore of the throttle body assembly to ensure a good seal. **Caution:** *The IAC valve is an electrical component and must not be soaked in any liquid cleaner, as damage may result.*
9   Install a new gasket on the IAC valve assembly.
10   Install the IAC valve and tighten the screws securely.
11   Plug in the electrical connector to the IAC valve assembly.

25   Mixture control valve (2.3L engine) - check and replacement

General information

1   During sudden deceleration on a carbureted engine, the throttle valve closes, cutting off the supply of intake air and increasing intake manifold vacuum, which produces a momentarily rich air/fuel mixture ratio. The mixture control valve system allows additional air to enter the intake manifold during sudden deceleration, which reduces hydrocarbon (HC) and carbon monoxide (CO) emissions caused by the portion of the air/fuel mixture that would otherwise remain unburned because of the lack of oxygen.
2   Here's how it works: When the vehicle suddenly decelerates, intake manifold vac-

uum increases, sending a vacuum signal to the mixture control valve. A chamber inside the valve opens, allowing air to enter the intake manifold. After about five seconds, the valve closes, cutting off this additional supply of air. If the mixture control valve continues to blow air for more than five seconds, replace it.

Check

3   Disconnect the large rubber hose from the mixture control valve and plug the intake manifold side to prevent the engine from stalling. Raise the engine speed to wide-open throttle and feel the air with your finger as the air is pushed through the valve for about five seconds.
4   If there is no response, or if the valve continues to supply air for more than five seconds, replace the valve.

Replacement

5   Remove the air cleaner.
6   Remove the mixture control valve mounting brackets.
7   Lift the valve from the engine.
8   Installation is the reverse of removal.

26   Positive Crankcase Ventilation (PCV) system - general information

Refer to illustration 26.1

1   The Positive Crankcase Ventilation (PCV) system **(see illustration)** reduces hydrocarbon emissions by scavenging crankcase vapors. It does this by circulating fresh air from the air cleaner through the crankcase, where it mixes with blow-by gases and is then rerouted through a PCV valve to the intake manifold.
2   The main components of the PCV system are the PCV valve (not used on 1998 and later models), a fresh air filtered inlet and the hoses connecting the valve cover to the air intake system.
3   To maintain idle quality, the PCV valve restricts the flow when the intake manifold vacuum is high. If abnormal operating conditions arise, the system is designed to allow

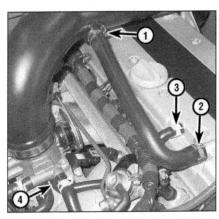

26.1 Typical Positive Crankcase Ventilation (PCV) system (2.2L engine shown, other PCV systems similar in function)

1   Fresh air from the air intake duct enters the PCV system
2   After being drawn through the larger PCV hose, fresh air form the air intake duct enters the crankcase here
3   Intake vacuum pulls crankcase vapors from engine into smaller PCV hose
4   Crankcase vapors from engine are drawn through smaller PCV hose into throttle body and then into intake manifold

excessive amounts of blow-by gases to flow back through the crankcase vent tube into the air cleaner to be consumed by normal combustion.
4   Checking and replacement of the PCV valve and filter is covered in Chapter 1.

27   Thermostatically controlled air cleaner (THERMAC) system - check and component replacement

Note: *The THERMAC system is installed on all 2.3L engines with feedback carburetor systems and on 3.1L engines.*

General information

Refer to illustration 27.2

1   The Thermostatically Controlled Air Cleaner (THERMAC) system improves engine efficiency and reduces hydrocarbon emissions during the initial warm-up period of the engine by maintaining a controlled air temperature into the carburetor or throttle body. Temperature control of the incoming air allows leaner fuel mixtures and helps prevent carburetor icing in cold weather.

2   The system **(see illustration)** uses a hot air control valve located in the snorkel of the air cleaner housing to control the ratio of cold and warm air into the carburetor or TBI unit. This valve is controlled by a vacuum motor which is, in turn, modulated by a thermo sensor or temperature sensor in the air cleaner. This air bleed valve opens when the intake air temperature is cold, thus allowing warm air from the heat stove to reach the carburetor. When the air is hot, the valve closes, thus closing off the warm air.

3   It is during the first few miles of driving (depending on outside temperature) that this system has its greatest effect on engine performance and emissions output. When the engine is cold, the air control valve blocks off the air cleaner inlet snorkel, allowing only warm air from the exhaust manifold to enter the carburetor. Gradually, as the engine warms up, the valve opens the snorkel passage, increasing the amount of cold air allowed in. Once the engine reaches normal operating temperature, the valve completely opens, allowing only cold, fresh air to enter.

4   Because of this cold-engine-only function, it is important to periodically check this system to prevent poor engine performance when cold, or overheating of the fuel mixture once the engine has reached operating temperatures. If the air cleaner valve sticks in the "no heat" position, the engine will run poorly, stall and waste gas until it has warmed up on its own. A valve sticking in the "heat" position causes the engine to run as if it is out of tune due to the constant flow of hot air to the carburetor.

Check

Thermo sensor

Refer to illustration 27.8

5   With the engine off, note the position of the hot air control valve inside the air cleaner snorkel. If the vehicle is equipped with an air duct on the end of the snorkel, it will have to be removed prior to this check. If visual access to the valve is difficult, use a mirror. The valve should be down, meaning that all air would flow through the snorkel (fresh air) and none through the exhaust manifold hot-air duct at the underside of the air cleaner housing (hot air).

6   Now have an assistant start the engine and continue to watch the valve inside the snorkel. With the engine cold and at idle, the valve should close off all air from the snorkel (fresh air), allowing heated air from the exhaust manifold to enter the air cleaner

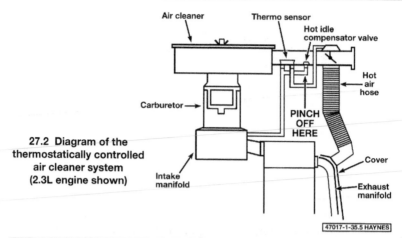

27.2 Diagram of the thermostatically controlled air cleaner system (2.3L engine shown)

27.8 Location of the thermo sensor and idle compensator on carbureted vehicles

1   *Thermo sensor*
2   *Hot idle compensator*

intake. As the engine warms to operating temperature the valve should move, allowing outside air through the snorkel to be included in the mixture. Eventually, the valve should move down to the point where most of the incoming air is through the snorkel (fresh air) and not the exhaust manifold duct (hot air).

7   If the valve did not close off snorkel air (fresh air) when the cold engine was first started, disconnect the vacuum hose at the snorkel vacuum motor and place your thumb over the hose end, checking for vacuum. If there is vacuum going to the motor, check that the valve and link are not seized or binding in the air cleaner snorkel. Replace the vacuum motor if the hose routing is correct and the valve moves freely.

8   If there was no vacuum going to the motor in the above test, check the hoses to make sure they are not cracked, crimped or disconnected. If the hoses are clear and in good condition, replace the thermo sensor inside the air cleaner housing **(see illustration)**.

Vacuum motor

Refer to illustration 27.9

9   Detach the vacuum hose from the vacuum motor and connect a hand-held vacuum pump in its place **(see illustration)**.

10   Apply vacuum to the motor. The hot air control valve should move, closing off fresh

27.9 The vacuum motor should activate the hot air control valve when vacuum is applied

air from the snorkel inlet.

11   If the vacuum motor does not perform as described it should be replaced.

Component replacement

Thermo sensor

12   Using pliers, flatten the clip that retains the vacuum hose to the sensor and remove the clips.

13   Disconnect the hoses from the sensor.

14   Carefully note the position of the sensor. The new sensor must be installed in exactly the same position.

15   Lift the sensor from the air cleaner.

Note: *The gasket between the sensor and air cleaner is bonded to the air cleaner and should not be removed.*

16   Install the new sensor with new gaskets in the same position as the old one.

17   Press the retaining clip on the sensor. Do not damage the control mechanism in the center of the sensor.

18   Connect the vacuum hoses and install the air cleaner on the engine.

Vacuum motor

19   Remove the screws or drill out the rivets that attach the vacuum motor retainer to the air cleaner.

20   Turn the vacuum motor to disengage it from the air control valve and lift it off.

21   Installation is the reverse of the removal procedure.

Chapter 7 Part A
Manual transmission

Contents

Specifications

General

Lubricant type .... See Chapter 1

Torque specifications   **Ft-lbs** (unless otherwise indicated)

Transmission-to-engine bolts
  Borg Warner
    M10 30
    M6 .... 52 in-lbs
  MSG .... 22 to 33
  MUA
    25 mm 30
    40 mm (M6) .... 52 in-lbs
    45 mm 56
    50 mm 56
    85 mm 56
Shift lever retaining bolts 15

1   General information

All vehicles covered in this manual come equipped with either a five-speed manual transmission or an automatic transmission. All information on the manual transmission is included in this Part of Chapter 7. Information on the automatic transmission can be found in Part B of this Chapter. Information on the transfer case used on 4WD models is in Part C.

The manual transmission used in these models is a five-speed unit with the fifth gear being an overdrive.

Due to the complexity, unavailability of replacement parts and the special tools necessary, internal repair by the home mechanic is not recommended. The information in this Chapter is limited to general information and removal and installation of the transmission.

Depending on the expense involved in having a faulty transmission overhauled, it may be a good idea to replace the unit with either a new or rebuilt one. Your local dealer or transmission shop should be able to supply you with information concerning cost, availability and exchange policy. Regardless of how you decide to remedy a transmission problem, you can still save a lot of money by removing and installing the unit yourself.

2.3 To access the shift lever retaining bolts, pull up this boot (there may also be a smaller rubber cover over the retainer and bolts; if so, peel that back as well)

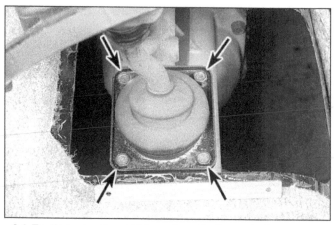

2.4 To disengage the shift lever from the transmission, remove these four bolts (arrows)

2   Shift lever - removal and installation

Refer to illustrations 2.3 and 2.4

1   Remove the shift lever knob and the center console (see Chapter 11).
2   Remove the carpet and sound deadener mat from the area around the shift lever.
3   Pull up the shift boot **(see illustration)**.
4   Pry out the rubber cover, if necessary, and remove the shift lever retaining bolts **(see illustration)**.
5   Lift the shift lever up and out; it's spring loaded, so ease it out carefully, or the spring will disengage and fall out.
6   Installation is the reverse of removal.

3   Manual transmission - removal and installation

Removal

Refer to illustrations 3.5a, 3.5b, 3.13, 3.14, 3.15, 3.16a and 3.16b

1   Disconnect the negative cable from the battery.

2   Remove the shift lever from the transmission (see Section 2) and, on 4WD models, from the transfer case (see Chapter 7C).
3   Raise the vehicle and support it securely on jackstands.
4   Drain the transmission and, on 4WD models, the transfer case (see Chapter 1).
5   Disconnect the speedometer cable **(see illustration)** or vehicle speed sensor, unplug the electrical connector from the back-up light switch **(see illustration)** and any other electrical connectors, then detach all wire harness clips and brackets from the transmission/transfer case and tuck them out of the way where they'll be safe during transmission removal.
6   Remove the rear and, on 4WD models, the front driveshafts (see Chapter 8). Use a plastic bag to cover the end of the transmission/transfer case to prevent fluid loss and contamination.
7   Unbolt the exhaust pipe(s) from the exhaust manifold(s). Remove all exhaust system components that would interfere with transmission removal (see Chapter 4).
8   Remove the starter motor (see Chapter 5).
9   Disconnect the clutch cable (2.3L Amigo models) or remove the clutch release cylinder

(all other models) (see Chapter 8).
10   Support the engine, either with an engine hoist from above, or with a jack under the engine oil pan (put a block of wood between the jack head and the pan). **Caution:** *The engine should remain supported at all times while the transmission/transfer case is out of the vehicle.*
11   Support the transmission/transfer case with a jack - preferably a special jack made for this purpose. Safety chains will help steady the transmission/transfer case on the jack.
12   Remove the transfer case skid plate bolts **(see illustration 7.2** in Chapter 7B) and remove the skid plate.
13   Remove the transmission mount-to-crossmember nuts **(see illustration)**.
14   Remove the crossmember retaining bolts **(see illustration)**, raise the transmission/transfer case slightly and remove the crossmember.
15   Remove the nuts and bolts securing the transmission mount to the transmission **(see illustration)**.
16   Remove the bolts and nuts securing the transmission bellhousing to the engine **(see illustrations)**. **Note:** *As the bolts and nuts are*

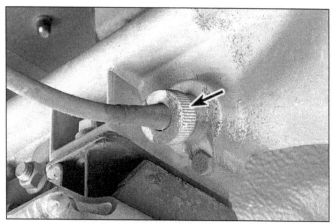

3.5a Unscrew the speedometer cable (arrow) from the transmission (some models use a vehicle speed sensor instead; on those vehicles, unplug the electrical connector from the sensor)

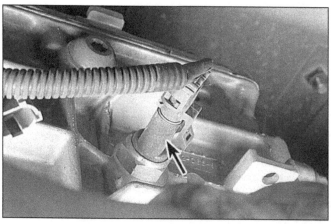

3.5b Unplug the electrical connector from the back-up light switch (arrow)

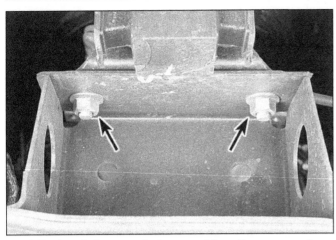

3.13 To detach the transmission mount from the crossmember, remove these insulator nuts (arrows)

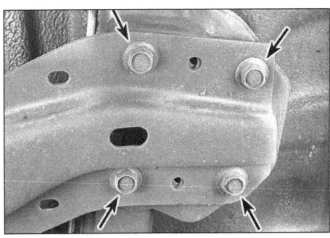

3.14 To detach the rear crossmember from the vehicle, remove the four bolts (arrows) from each end of the crossmember

removed, make detailed notes indicating the size, type and location of each fastener. This is very important, as the various fasteners have different torque values.

17   Make a final check that all wires and hoses have been disconnected from the transmission/transfer case, then move the transmission/transfer case and jack toward the rear of the vehicle until the transmission input shaft is clear of the clutch housing. Keep the transmission/transfer case level as this is done.

18   Once the input shaft is clear, lower the transmission/transfer case and remove it from under the vehicle. **Caution:** *Do not depress the clutch pedal while the transmission/transfer case is out of the vehicle.*

19   The clutch components can be inspected by removing the clutch housing from the engine (see Chapter 8). In most cases, new clutch components should be routinely installed if the transmission is removed.

3.15 Raise the transmission slightly with a floor jack and remove these mount-to-transmission nuts and bolts (arrows) (right side shown; after removing the nuts from the two long through-bolts, knock out the through-bolts and remove the other transmission-to-mount bolt from the left side)

Installation

20   If removed, install the clutch components (see Chapter 8).

21   With the transmission/transfer case secured to the jack as on removal, raise the transmission/transfer case into position behind the clutch housing and then carefully slide it forward, engaging the input shaft with the clutch plate hub. Do not use excessive force to install the transmission - if the input shaft does not slide into place, readjust the

angle of the transmission/transfer case so it is level and/or turn the input shaft so the splines engage properly with the clutch.

22   Install the transmission-to-engine fasteners, using your notes made in Step 16 to ensure that they are returned to their original positions. Tighten the bolts to the torque values listed in this Chapter's Specifications.

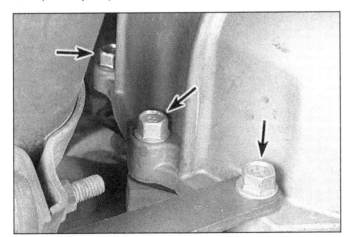

3.16a Remove the left bellhousing-to-engine bolts (arrows) (bolts on top, not visible in this photo, cannot be accessed until the engine and transmission are lowered slightly)

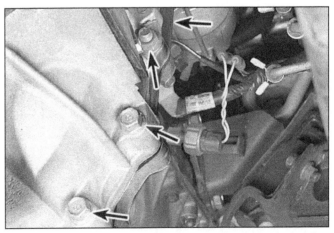

3.16b Remove the right bellhousing-to-engine bolts (arrows) (bolts on top, not visible in this photo, cannot be accessed until the engine and transmission are lowered slightly)

23 Install the crossmember and transmission/transfer case support. Tighten all nuts and bolts securely.

24 Remove the jacks supporting the transmission/transfer case and the engine.

25 Install the various items removed previously, referring to Chapter 8 for the installation of the driveshaft and Chapter 4 for information regarding the exhaust system components.

26 Make a final check that all wires, hoses and the speedometer cable or vehicle speed sensor have been connected and that the transmission and transfer case have been filled with lubricant to the proper level (see Chapter 1). Lower the vehicle.

27 Working inside the vehicle, install the shift lever (see Section 2). On 4WD models, install the shift lever for the transfer case (see Chapter 7C).

28 Adjust the clutch cable, if equipped (see Chapter 8).

29 Connect the negative battery cable. Road test the vehicle for proper operation and check for leakage.

4 Manual transmission overhaul - general information

Overhauling a manual transmission is a difficult job for the do-it-yourselfer. It involves the disassembly and reassembly of many small parts. Numerous clearances must be precisely measured and, if necessary, changed with select fit spacers and snap-rings. As a result, if transmission problems arise, it can be removed and installed by a competent do-it-yourselfer, but overhaul should be left to a transmission repair shop. Rebuilt transmissions may be available - check with your dealer parts department and auto parts stores. At any rate, the time and money involved in an overhaul is almost sure to exceed the cost of a rebuilt unit.

Nevertheless, it's not impossible for an inexperienced mechanic to rebuild a transmission if the special tools are available and the job is done in a deliberate step-by-step manner so nothing is overlooked.

The tools necessary for an overhaul include internal and external snap-ring pliers, a bearing puller, a slide hammer, a set of pin punches, a dial indicator and possibly a hydraulic press. In addition, a large, sturdy workbench and a vise or transmission stand will be required.

During disassembly of the transmission, make careful notes of how each piece comes off, where it fits in relation to other pieces and what holds it in place. Noting how they are installed when you remove the parts will make it much easier to get the transmission back together. Before taking the transmission apart for repair, it will help if you have some idea what area of the transmission is malfunctioning. Certain problems can be closely tied to specific areas in the transmission, which can make component examination and replacement easier. Refer to the *Troubleshooting* section at the front of this manual for information regarding possible sources of trouble.

Chapter 7 Part B
Automatic transmission

Contents

Specifications

Fluid type	See Chapter 1
Throttle valve (TV) cable adjustments*	
Boot-to-stopper distance	1/32 to 1/16 inch
Stroke	1-5/16 to 1-23/64 inch
Torque converter-to-bellhousing clearance (Amigo models)	1-13/64 inches

*Note: These adjustments apply to both carbureted and fuel-injected models.

Torque specifications

	Ft-lbs (unless otherwise indicated)
Transmission bellhousing-to-engine bolts	
Amigo models	47
Rodeo/Passport models	
16 mm	52 in-lbs
35 mm (lower three bolts)	
Bellhousing-to-stiffener (2)	35
Other bolt	20
40 mm (upper left)	56
45 mm (upper right and top two)	56
50 mm (middle left)	56
68 mm (middle right)	56
Driveplate-to-torque converter bolts	
Amigo models	22
Rodeo/Passport	41

1 General information

All vehicles covered in this manual are equipped with either a five-speed manual transmission or a four-speed automatic transmission. All information on the automatic transmission is included in this Part of Chapter 7. Information for the manual transmission can be found in Part A. Information on the transfer case used on 4WD models can be found in Part C.

Amigo models equipped with an automatic transmission use an AW03-72L four-speed transmission (available only on 2.6L models); Rodeo models with an automatic use a 4L30-E electronic four-speed.

Specialized techniques and equipment are required when working on automatic transmissions, due to their complexity. Consequently, this Chapter addresses only those procedures concerned with routine maintenance, general diagnosis and removal and installation.

If the transmission requires major repair work, it should be left to a dealer service department or an automotive transmission repair shop. You can, however, remove and install the transmission yourself and save the expense, even if the repair work is done by a transmission specialist.

2 Diagnosis - general

Note: Automatic transmission malfunctions may be caused by five general conditions: poor engine performance, improper adjustments, hydraulic malfunctions, mechanical malfunctions or computer malfunctions. Diagnosis of these problems should always begin with a check of the easily repaired items: fluid level and condition (Chapter 1), and shift linkage adjustment. Next, perform a road test to determine if the problem has been corrected or if more diagnosis is necessary. If the problem persists after the preliminary tests and cor-

3.5 Remove the cotter and washer and detach the shift linkage trunnion from the shift lever

rections are completed, additional diagnosis should be done by a dealer service department or transmission repair shop. Refer to the Troubleshooting *Section at the front of this manual for information on symptoms of transmission problems.*

Preliminary checks

1   Drive the vehicle to warm the transmission to normal operating temperature.
2   Check the fluid level as described in Chapter 1:

a) *If the fluid level is unusually low, add enough fluid to bring the level within the designated area of the dipstick, then check for external leaks (see below).*

b) *If the fluid level is abnormally high, drain off the excess, then check the drained fluid for contamination by coolant. The presence of engine coolant in the automatic transmission fluid indicates that a failure has occurred in the internal radiator walls that separate the coolant from the transmission fluid (see Chapter 3).*

c) *If the fluid is foaming, drain it and refill the transmission, then check for coolant in the fluid or a high fluid level.*

3   Check the engine idle speed. **Note:** *If the engine is malfunctioning, do not proceed with the preliminary checks until it has been repaired and runs normally.*
4   Inspect the shift linkage (Section 3). Make sure that it's properly adjusted and that the linkage operates smoothly.

Fluid leak diagnosis

5   Most fluid leaks are easy to locate visually. Repair usually consists of replacing a seal or gasket. If a leak is difficult to find, the following procedure may help.
6   Identify the fluid. Make sure it's transmission fluid and not engine oil or brake fluid (automatic transmission fluid is a deep red color).
7   Try to pinpoint the source of the leak. Drive the vehicle several miles, then park it over a large sheet of cardboard. After a

minute or two, you should be able to locate the leak by determining the source of the fluid dripping onto the cardboard.
8   Make a careful visual inspection of the suspected component and the area immediately around it. Pay particular attention to gasket mating surfaces. A mirror is often helpful for finding leaks in areas that are hard to see.
9   If the leak still cannot be found, clean the suspected area thoroughly with a degreaser or solvent, then dry it.
10   Drive the vehicle for several miles at normal operating temperature and varying speeds. After driving the vehicle, visually inspect the suspected component again.
11   Once the leak has been located, the cause must be determined before it can be properly repaired. If a gasket is replaced but the sealing flange is bent, the new gasket will not stop the leak. The bent flange must be straightened.
12   Before attempting to repair a leak, check to make sure that the following conditions are corrected or they may cause another leak. **Note:** *Some of the following conditions cannot be fixed without highly specialized tools and expertise. Such problems must be referred to a transmission repair shop or a dealer service department.*

Gasket leaks

13   Check the pan periodically. Make sure the bolts are tight, no bolts are missing, the gasket is in good condition and the pan is flat (dents in the pan may indicate damage to the valve body inside).
14   If the pan gasket is leaking, the fluid level or the fluid pressure may be too high, the vent may be plugged, the pan bolts may be too tight, the pan sealing flange may be warped, the sealing surface of the transmission housing may be damaged, the gasket may be damaged or the transmission casting may be cracked or porous. If sealant instead of gasket material has been used to form a seal between the pan and the transmission housing, it may be the wrong sealant.

Seal leaks

15   If a transmission seal is leaking, the fluid level or pressure may be too high, the vent may be plugged, the seal bore may be damaged, the seal itself may be damaged or improperly installed, the surface of the shaft protruding through the seal may be damaged or a loose bearing may be causing excessive shaft movement.
16   Make sure the dipstick tube seal is in good condition and the tube is properly seated. Periodically check the area around the speedometer gear or sensor for leakage. If transmission fluid is evident, check the O-ring for damage.

Case leaks

17   If the case itself appears to be leaking, the casting is porous and will have to be repaired or replaced.

18   Make sure the oil cooler hose fittings are tight and in good condition.

Fluid comes out vent pipe or fill tube

19   If this condition occurs, the transmission is overfilled, there is coolant in the fluid, the case is porous, the dipstick is incorrect, the vent is plugged or the drain back holes are plugged.

---

3   Shift linkage - check and adjustment

Check

1   With the engine running and your foot placed firmly on the brake, place the shift lever in each detent and make sure that the transmission shifts into the corresponding gear as the indicator is moved. If adjustment is required, turn the engine off and proceed as follows.

Adjustment

2   Raise the vehicle and place it securely on jackstands.

Amigo

3   Loosen the nut on the shift linkage **(see illustration 3.5)**, push the shift lever on the transmission all the way to the rear, then return it two notches to the Neutral position.
4   Holding the shift lever lightly toward the Reverse position, tighten the shift linkage nut.

Rodeo/Passport

Refer to illustrations 3.5, 3.8a, 3.8b and 3.8c
5   Working under the vehicle, disconnect the shift linkage from the shift lever **(see illustration)**.
6   Inside the vehicle, put the shift lever in the Neutral position. Hold the lever against the Neutral detent with a rubber band (the shift lever must be held snugly against the forward edge of the Neutral detent to eliminate freeplay during the following adjustment).
7   Back under the vehicle, put the manual lever in the Neutral position by clicking it all the way forward, then bringing it back two detent positions.
8   Verify that the pin on the shift linkage trunnion block is centered on the hole in the bottom of the shift lever **(see illustration)**.

a) *If it is, the shift linkage is adjusted; reconnect the trunnion to the shift lever.*

b) *If it isn't, remove the nut and retainer clip, then turn the adjusting nuts to center the trunnion pin on the hole in the shift lever.*

c) *When the trunnion pin is centered on the hole in the shift lever, reconnect the trunnion to the shift lever, tighten the adjusting nuts* **(see illustration)** *and install the retainer* **(see illustration)**.

9   Install the nut and retainer clip.

3.8a With the shift lever (A) and the manual lever in the Neutral position, verify that the pin (B) on the shift linkage trunnion block is centered on the hole in the shift lever; if (like the one shown here) it isn't, remove the retainer nut and retainer, then turn the adjusting nuts to move the trunnion pin into alignment with the hole in the lever

3.8b When the trunnion pin is aligned with the hole in the shift lever, reattach the trunnion to the shift lever, install the washer and cotter, and snug up the two trunnion adjusting nuts

All models

10   Check shift lever operation and make sure that the gear position indicator points to the correct gear in each position. If it doesn't, readjust the shift linkage; you're probably off by one detent position.

---

4   Throttle Valve (TV) cable (Amigo) - adjustment

Carbureted models

Refer to illustrations 4.5 and 4.6

1   Loosen the TV cable adjusting nuts.
2   Make sure the throttle adjust screw contacts the stopper. If the screw is not in contact with the stopper, the fast idle mechanism is activated and must be de-activated as follows.
3   Disconnect the negative battery cable and remove the air cleaner cover.
4   Have an assistant depress the accelera-

tor pedal, open the choke plate completely, then have the assistant release the pedal. Don't depress the pedal again or the fast idle will engage. Connect the battery cable and install the air cleaner cover.
5   With the fast idle mechanism de-activated, the TV cable may now be adjusted. Pull back the boot on the outer TV cable, then use the adjusting nuts to adjust the inner and outer cable to the distance listed in this Chapter's Specifications **(see illustration)**. Tighten the adjusting nuts.
6   After adjustment, measure the stroke between the open and closed position of the cable and compare this measurement to the distance listed in this Chapter's Specifications **(see illustration)**. Adjust the stroke as necessary, using the adjusting nuts.
7   Move the boot back into position.

Fuel-injected models

Refer to illustration 4.9

8   Depress the accelerator pedal fully and

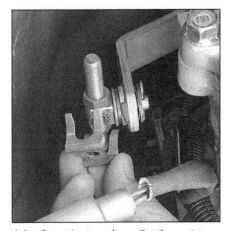

3.8c Once the trunnion adjusting nuts are snug, reattach the retainer to make sure they stay tight

make sure the throttle valve opens all the way, adjusting the accelerator linkage as necessary (see Chapter 4).

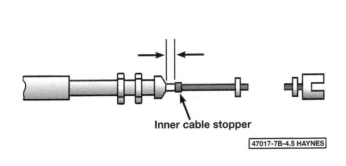

4.5 Throttle Valve (TV) cable adjustment details (carbureted models)

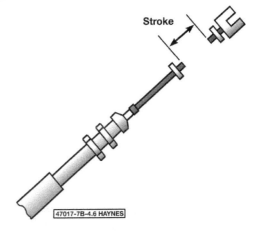

4.6 Measure the TV cable stroke (the distance between the closed and wide-open positions) (carbureted models)

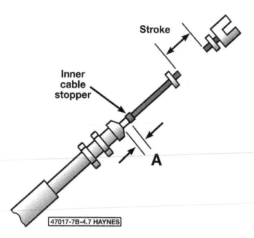

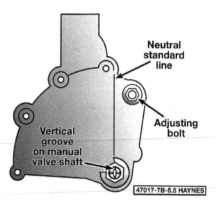

4.9 Throttle valve (TV) cable adjustment details - dimension A is the boot-to-stopper distance (fuel-injected models)

5.5 To adjust the Park/Neutral switch on an Amigo model, loosen the adjusting bolt and rotate the switch to line up the Neutral standard line on the switch with the vertical groove on the manual valve shaft

9   Hold the pedal down, loosen the cable adjusting nuts and adjust the cable boot-to-stopper clearance to the dimension listed in this Chapter's Specifications **(see illustration)**.
10   After adjusting the distance, check the cable stroke (distance the cable travels between the open and closed position), adjusting as necessary, using the adjusting nuts. Tighten the adjusting nuts.

5   Park/Neutral switch - check, adjustment and replacement

1   The Park/Neutral switch, located on the right-hand side of the transmission on Amigo models, and on the left side of the transmission on Rodeo and Passport models, prevents the engine from starting with the transmission in gear.

Check

2   Make sure the engine will start only with the selector lever in Park and Neutral. With the key in the On position, make sure the backup lights function when the lever is put into Reverse.
3   If a malfunction is noted, try adjusting the switch.

Adjustment

4   Place the shift lever in Neutral. Raise the vehicle and support it on jackstands. Set the parking brake securely and block the wheels.

Amigo

Refer to illustration 5.5

5   Loosen the adjusting bolt and rotate the switch to line up the Neutral standard line on the switch with the vertical groove on the manual valve shaft **(see illustration)**.
6   Tighten the adjusting bolt, then recheck the switch operation as described above.

Rodeo/Passport

Refer to illustration 5.7, 5.8, 5.9 and 5.10

7   Remove the manual lever from the Park/Neutral switch **(see illustration)**.
8   Remove the switch cover **(see illustration)**.

9   Loosen the two switch retaining bolts **(see illustration)**.
10   Rotate the switch until the slot in the switch housing aligns with the manual lever shaft bushing, then insert a 3/32-inch drill bit or punch into the slot **(see illustration)**.
11   Tighten the bolts securely.
12   Snap the switch cover into place.
13   Install the manual lever.

Replacement

Refer to illustrations 5.17a, 5.17b and 5.17c

14   Disconnect the negative battery cable.
15   Remove the air cleaner housing assembly (see Chapter 4).
16   Raise the vehicle and support it on jackstands. Set the parking brake securely and block the wheels. Shift the transmission into Neutral.
17   From below, remove the electrical harness heat shield, then from above, trace the harness along the left side of the engine compartment and detach the harness from all retainer clips. Unplug the electrical connector **(see illustrations)**.

5.7 To remove the manual lever from the Park/Neutral switch, remove this retaining nut (arrow) (Rodeo/Passport models)

5.8 To remove the Park/Neutral switch cover, simply pop it off (Rodeo/Passport models)

5.9 Loosen the two switch retaining bolts (socket and arrow) (Rodeo/Passport models)

5.10 Rotate the switch until the slot in the switch housing aligns with the manual lever shaft bushing, then insert a 3/32-inch drill bit or punch into the slot (switch removed for clarity) (Rodeo/Passport models)

18   Remove the manual lever nut **(see illustration 5.7)** and remove the manual lever.
19   Remove the switch cover **(see illustration 5.8)**.
20   Remove the switch retaining bolts **(see illustration 5.9)** and remove the switch.
21   Installation is the reverse of removal. Do not tighten the switch retaining screws fully until the switch has been adjusted (see above).
22   Connect the battery negative cable.
23   Start the engine in both Park and Neutral to verify that the switch is properly adjusted.

6   Oil seal replacement

Refer to illustrations 6.4, 6.8a, 6.8b, 6.9a, 6.9b and 6.9c
1   Oil leaks frequently occur due to wear of the extension housing oil seal and/or the speedometer driven gear O-ring. Replacement of these seals is relatively easy, since the repairs can usually be performed without

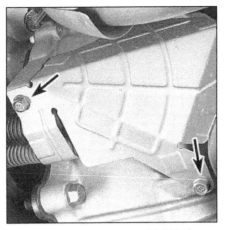

5.17a Remove the heat shield bolts (arrows), remove the heat shield, trace the harness forward . . .

removing the transmission from the vehicle.
2   The extension housing oil seal is located at the extreme rear of the transmission, where the driveshaft is attached. If leakage at the seal is suspected, raise the vehicle and support it securely on jackstands. If the seal is leaking, transmission lubricant will be built up on the front of the driveshaft and may be dripping from the rear of the transmission.
3   Remove the driveshaft (see Chapter 8).
4   Using a screwdriver or oil seal replacement tool, carefully pry the oil seal out of the rear of the transmission **(see illustration)**. Do not damage the splines on the transmission output shaft.
5   Using a large section of pipe or a very large deep socket as a drift, install the new oil seal. Drive it into the bore squarely and make sure it's completely seated.
6   Lubricate the splines of the transmission output shaft and the outside of the driveshaft sleeve yoke with multi-purpose grease, then install the driveshaft. Be careful not to damage the lip of the new seal.
7   The speedometer cable and driven gear housing is located on the side of the exten-

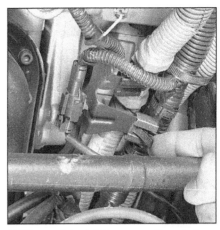

5.17b . . . detach it from any clips such as this one (on a late-model Rodeo) . . .

5.17c . . . and unplug the black electrical connector (arrow) (Rodeo/Passport models)

sion housing. Look for transmission oil around the cable housing to determine if the O-ring is leaking.
8   Disconnect the speedometer cable or vehicle speed sensor **(see illustrations)**.
9   Remove the driven gear, install a new O-

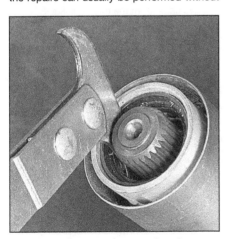

6.4 Using a seal removal tool or screwdriver, carefully pry the oil seal out of the rear of the transmission case; don't damage the splines on the output shaft

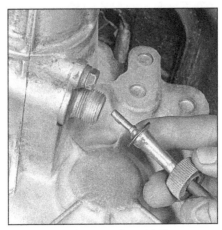

6.8a Unscrew the collar and disconnect the speedometer cable (Amigo) . . .

6.8b . . . or unplug the electrical connector for the vehicle speed sensor (Rodeo/Passport)

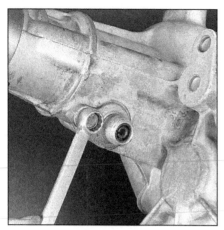

6.9a To replace the speedometer/speed sensor driven gear, remove the hold-down bolt . . .

6.9b . . . pull out the driven gear assembly . . .

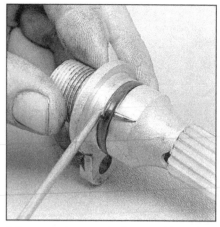

6.9c . . . and use a small screwdriver or an awl to remove the O-ring from its groove in the driven gear assembly

7.2 To check the transmission mount for wear, insert a large screwdriver or pry bar into the space between the insulator and the crossmember and try to pry the transmission up slightly

7.4 Remove the skidplate bolts (arrows) and remove the skidplate

7.5 Remove the nuts (arrows) attaching the mount to the crossmember

ring on the driven gear housing and reinstall the driven gear housing and cable assembly on the extension housing **(see illustrations)**.

7   Transmission mount - check and replacement

Check

Refer to illustration 7.2

1   Raise the vehicle and place it securely on jackstands.
2   Insert a large screwdriver or pry bar into the space between the extension housing and the crossmember and try to pry the transmission up slightly **(see illustration)**.
3   The transmission should not move away from the insulator much at all.

Replacement

Refer to illustrations 7.4, 7.5 and 7.6

4   Remove the skidplate **(see illustration)**.

5   Remove the nuts attaching the mount to the crossmember **(see illustration)**.
6   Raise the transmission slightly with a jack and remove the bolts from the insulator, noting which holes are used in the cross-member for proper alignment during installation **(see illustration)**.
7   Installation is the reverse of the removal procedure. Be sure to tighten the nuts/bolts securely.

8   Automatic transmission - removal and installation

Removal

Refer to illustrations 8.9a, 8.9b, 8.9c, 8.10, 8.15, 8.18, 8.19 and 8.24
Note: *On 4WD models, the transmission and transfer case are removed and installed as a unit.*

1   Disconnect the negative cable from the battery. Unsnap the upper half of the air cleaner housing from the air cleaner assembly (see Chapter 4). (Unsnapping the upper

half of the air cleaner housing will protect the air induction system from damage when the rear of the engine is tilted down slightly to disengage the transmission.)
2   On Amigo models, disconnect the throttle valve cable from the throttle linkage.

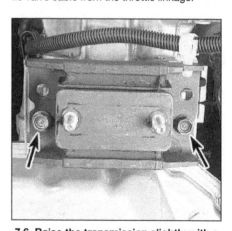

7.6 Raise the transmission slightly with a jack, remove these bolts (arrows) and remove the transmission mount (crossmember removed for clarity)

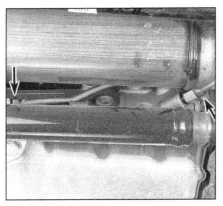

8.9a To disconnect the oil cooler lines from the transmission, unscrew these fittings (arrows)

8.9b To detach the oil cooler line bracket from the engine stiffener, remove this nut (arrow)

8.9c To detach the oil cooler line bracket from the block, remove this nut (arrow)

8.10 Remove the front crossmember nuts and bolts (arrows) and remove the crossmember

8.15 To detach the engine stiffener from the transmission and the engine, remove these bolts (arrows)

3   On 4WD models, unscrew the shift lever knobs and remove the center console (see Chapter 11), then remove the transfer case shift lever (see Chapter 7C).
4   Raise the vehicle and support it securely on jackstands.
5   Remove the skid plate (see illustration 7.2).
6   Remove any exhaust components which will interfere with transmission/transfer case removal (see Chapter 4).
7   Unplug all electrical connectors and detach all wire harnesses from the transmission and, on 4WD models, the transfer case. (On some models, the harnesses are protected by a heat protector - see Step 16).
8   Drain the transmission fluid and, on 4WD models, the transfer case lubricant (see Chapter 1).
9   Disconnect the oil cooler lines from the transmission (see illustration). Remove any oil cooler line brackets (see illustrations), detach the forward ends of the lines from the oil cooler hoses and remove the lines.
10   On Rodeo/Passport models, remove the front crossmember (see illustration).
11   Remove the driveshaft (all models); remove the front driveshaft (4WD models) (see Chapter 8).

12   Disconnect the shift lever from the shift linkage (see Section 3).
13   On Amigo models, disconnect the speedometer cable (see illustration 6.8a).
14   Remove the dipstick tube, if equipped.
15   On Rodeo/Passport models, remove the engine stiffener (see illustration).
16   On Rodeo/Passport models, remove the heat protector (see illustration 5.17a), unplug the electrical connectors, detach the harness

clip (see illustration 5.17b), unplug the harness connectors (see illustration 5.17c) and set the harness aside.
17   Remove the starter motor (see Chapter 5).
18   On Rodeo/Passport models, remove the torque converter access cover (see illustration). On Amigo models, the torque converter can be accessed through the starter motor opening.

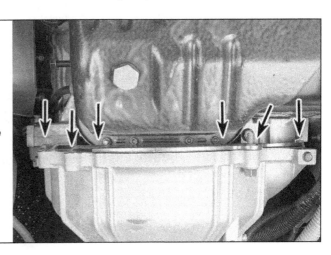

8.18 To remove the torque converter access cover, remove these bolts (arrows)

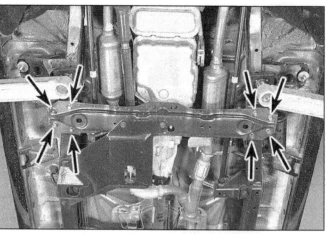

8.19 Rotate the crankshaft to remove each driveplate-to-torque converter bolt (arrow); use white paint on the driveplate and inside a bolt hole to mark the relationship of the driveplate to the torque converter

8.24 To detach the rear crossmember, remove these eight bolts (arrows)

19   Remove the torque converter-to-driveplate bolts **(see illustration)**. Turn the crankshaft for access to each bolt.

20   Using one of the bolt holes in the driveplate, mark the relationship of the torque converter and driveplate with white paint to ensure that the dynamic balance of the driveplate and torque converter is re-established during reassembly.

21   Support the engine with a jack. Use a block of wood under the oil pan to spread the load.

22   Support the transmission/transfer case with a jack - preferably a jack made for this purpose. Place a block of wood between the jack head and the transmission to protect the transmission oil pan. Safety chains will help steady the transmission/transfer case on the jack.

23   Remove the rear mount-to-rear crossmember nuts (see Section 7).

24   Remove the skidplate, if equipped **(see illustration 7.4)**, and the rear crossmember-to-frame bolts **(see illustration)**.

25   Raise the transmission/transfer case enough to allow removal of the crossmember.

26   Remove the bolts securing the transmission to the engine. (You won't be able to reach the bolts at the top of the bellhousing until the transmission is lowered in the next step.) **Note:** *As the bolts and nuts are removed, make detailed notes indicating the size, length, type and location of each fastener. This is very important, as the various fasteners have different torque values.*

27   Lower the transmission/transfer case slightly and remove the upper bolts securing the transmission to the engine. Make a final inspection of the top of the transmission and

transfer case to ensure that everything is disconnected.

28   Move the transmission/transfer case to the rear to disengage it from the engine block dowel pins. Make sure the torque converter is detached from the driveplate. Secure the torque converter to the transmission so it won't fall out during removal.

Installation

29   Prior to installation, make sure the torque converter is securely engaged with the pump and the input shaft. To do this, rotate the torque converter while pushing it in. If it isn't already seated properly, it will clunk into place (it may even do this a couple of times).

30   With the transmission/transfer case secured to the jack, raise it into position. Be sure to keep it level so the torque converter does not slide forward. Connect the transmission fluid cooler lines.

31   Turn the torque converter to line up the holes with the holes in the driveplate. The white paint mark on the torque converter and the driveplate must line up.

32   Move the transmission forward carefully until the dowel pins and the torque converter are engaged.

33   Install the transmission-to-engine bolts/nuts, in their proper positions, and tighten the bolts to the torque values listed in this Chapter's Specifications.

34   Install the torque converter-to-driveplate bolts. Tighten the bolts to the torque listed in this Chapter's Specifications.

35   On Rodeo/Passport models, install the torque converter cover and tighten the bolts securely.

36   Install the transmission mount, if removed (see Section 7).

37   Install the rear crossmember and tighten the bolts.

38   Lower the transmission/transfer case, guiding the mount studs through their respective holes in the crossmember, install the nuts and tighten them securely.

39   Remove the jacks supporting the engine and the transmission/transfer case.

40   Install the dipstick tube, if equipped.

41   Install the starter motor (see Chapter 5).

42   Install the engine stiffener and tighten the bolts securely.

43   Connect the shift linkage to the manual lever and adjust the linkage (see Section 3).

44   Reroute all transmission/transfer case wire harnesses and plug in the electrical connectors.

45   Install the driveshaft(s) (see Chapter 8).

46   On Amigo models, connect the speedometer cable.

47   Install any exhaust system components that were removed or disconnected. Install any heat shields that were removed.

48   Refill the transfer case, if equipped, with the specified fluid (see Chapter 1).

49   On models without a dipstick tube, refill the transmission with the specified fluid (see Chapter 1).

50   Remove all jackstands and lower the vehicle.

51   Fill the transmission with the specified fluid (Chapter 1), run the engine and check for fluid leaks.

52   On Amigo models, adjust the throttle valve cable (see Section 4).

53   On 4WD models, install the transfer case shift lever (see Chapter 7C).

54   Snap the upper half of the air cleaner housing into place (see Chapter 4).

Chapter 7 Part C
Transfer case

Contents

Specifications

Torque specifications

Transfer case nuts/bolts   **Ft-lbs**

Adapter-to-transfer bolts (4)	30
Adapter-to-transfer nuts (2)	50
Transfer-to-adapter nut (1)	50
Companion flange retaining nut	123

1 General information

Four-wheel drive models are equipped with a transfer case mounted on the rear of the transmission housing. Drive is passed through the transmission and transfer case to the front and rear wheels by the driveshafts.

On these models the transfer case is combined with the transmission to form one unit. The transfer case cannot be removed separately; the transmission/transfer case must be disassembled as a unit. Because of the special tools and techniques required, disassembly and overhaul of the transfer case should be left to a dealer or properly equipped shop. You can, however, remove and install the transfer case/transmission yourself and save the expense, even if the repair work is done by a specialist.

2 Shift lever - removal and installation

Refer to illustrations 2.5 and 2.7

1   Remove the carpet and sound deadener mat.

2   Unscrew the transmission and transfer case shift lever knobs and remove the center console (see Chapter 11).

3   Raise the vehicle and place it securely on jackstands.

4   Disconnect the shift linkage from the shift lever (see Chapter 7B).

5   Remove the bolts retaining the base plate for the transmission shift lever assembly and the transfer case shift lever boot **(see illustration)**. Lift up the entire assembly, unplug the two electrical connectors and set the assembly aside.

2.5 Remove the shift lever boot mounting plate bolts (arrows), lift off the plate and boot assembly and unplug the two electrical connectors (Rodeo shown)

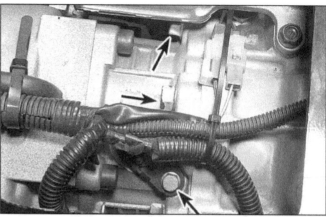

2.7 Remove the shift lever retaining bolts (arrows) (Rodeo shown)

3.9 To detach the wiring harness from the transfer case, remove this bolt (lower arrow); note the two upper adapter-to-transfer bolts (upper arrows)

6   On some early models, there may be a rubber cover over the shift lever mounting bolts. If so, remove it.

7   Remove the shift lever mounting bolts **(see illustration)**.

8   Lift the shift lever up and out of the transfer case. On some models the lever is spring loaded, so ease it out of the case or the spring will be disengaged and will fall out.

9   Installation is the reverse of removal.

---

3   Transfer case - removal and installation

Refer to illustrations 3.9, 3.13a and 3.13b

Note: *The following procedure applies only to the transfer case used with automatic transmissions; the transfer case used with manual transmissions is an integral part of the transmission (see Chapter 7A).*

1   Disconnect the negative cable from the battery. Unsnap the upper half of the air cleaner housing from the air cleaner assembly (see Chapter 4). (Unsnapping the upper half of the air cleaner housing will protect the air induction system from damage when the rear of the engine is tilted down slightly to disengage the transfer case.)

2   Unscrew the transmission and transfer case shift lever knobs and remove the center console (see Chapter 11).

3   Remove the shift lever base retaining bolts **(see illustration 2.5)**, but don't try to lift off the shift lever assembly until you have disconnected the shift lever from the shift linkage, in step 5.

4   Raise the vehicle and support it securely on jackstands.

5   Disconnect the transmission shift lever from the shift linkage (see Chapter 7B). Lift up the shift lever assembly, unplug the electrical connectors for the harnesses on the left side of the shift lever assembly and set it aside. Don't disconnect anything more than what's absolutely necessary; the only reason for removing the shift lever assembly is to gain access to the upper adapter-to-transfer

case bolt.

6   Remove the skid plate (see Chapter 7B).

7   Remove any exhaust components which will interfere with transfer case removal (see Chapter 4).

8   On Amigo models, disconnect the speedometer cable (see Chapter 7B).

9   Unplug the vehicle speed sensor (see Chapter 7B) and any other electrical connectors, and detach all wire harnesses from the transfer case **(see illustration)**.

10   Drain the transfer case lubricant (see Chapter 1).

11   Disconnect the front and rear driveshafts from the transfer case (see Chapter 8).

12   Support the transfer case with a jack - preferably a jack made for this purpose. Place a block of wood between the jack head and the transfer case. Safety chains will help steady the transfer case on the jack.

13   Remove the bolts securing the transfer case to the transmission **(see illustrations)**. There are three nuts and four bolts securing the transfer case to the adapter. The two lower nuts face forward and the third nut, on the left side, faces to the rear; all four bolts

(two on top and two on the lower right) face forward. **Note:** *As the bolts and nuts are removed, make detailed notes indicating the size, length, type and location of each fastener.*

14   Make a final inspection of the top of the transmission/transfer case to ensure that everything is disconnected.

15   Move the transfer case to the rear to disengage it from the transmission dowel pins.

Installation

16   With the transfer case secured to the jack, raise it into position. Apply a thin coat of moly grease to the input shaft spline.

17   Move the transfer case forward carefully until the dowel pins and the torque converter are engaged.

18   Install the adapter-to-transfer bolts and nuts and the transfer-to-adapter nut and tighten them to the torque listed in this Chapter's Specifications.

19   Remove the jack supporting the transfer case.

20   Place the shift lever assembly back in

3.13a The two lower adapter-to-transfer case nuts face forward; left lower nut (arrow) shown, right nut in same position on other side of transfer case

3.13b The left transfer case-to-adapter nut (arrow) faces to the rear (it's the only one that does so); two of the four adapter-to-transfer case bolts are shown in illustration 3.9

5.5 Unstake the companion flange retaining nut (it's staked in two places)

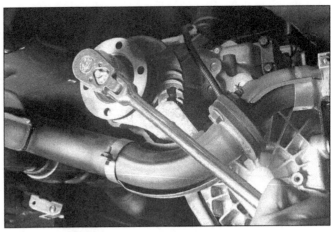

5.6 Holding the companion flange with a suitable tool, loosen the flange nut

position, install the retaining bolts and plug in any electrical connectors that were unplugged. Connect the shift linkage to the shift lever and adjust the linkage (Chapter 7B).

21 Reroute all transfer case wire harnesses and plug in the electrical connectors.

22 Install the driveshafts (see Chapter 8).

23 On Amigo models, connect the speedometer cable.

24 Install any exhaust system components that were removed or disconnected. Install any heat shields that were removed.

25 Refill the transfer case with the specified fluid (see Chapter 1).

26 Remove all jackstands and lower the vehicle.

27 Install the transfer case shift lever.

28 Install the center console (see Chapter 11).

4 Transfer case overhaul - general information

On these models the transmission/transfer case assembly is designed to be overhauled as a unit. Consequently, over-

haul should be left to a repair shop specializing in both transmissions and transfer cases. Rebuilt units may be available - check with your dealer parts department and auto parts stores. At any rate, some cost savings can be realized by removing the transmission/transfer case unit and taking it to the shop (see Part A or B of this Chapter).

5 Oil seal - replacement

Refer to illustrations 5.5, 5.6, 5.7, 5.8, 5.9 and 5.11

1 Disconnect the negative battery cable.

2 Raise the vehicle and place it securely on jackstands.

3 If you're planning to replace the seal for the front driveshaft companion flange, remove the skidplate (see Chapter 7B).

4 Remove the front or rear driveshaft (see Chapter 8).

5 Unstake the companion flange retaining nut **(see illustration)**. It's staked in two places; make sure you unstake both.

6 Holding the companion flange with a suitable tool, remove the flange nut **(see illustration)**.

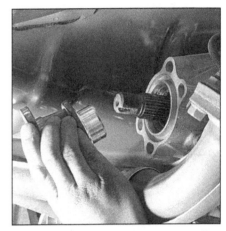

5.7 Remove the companion flange; if necessary, use a small puller

7 Remove the companion flange **(see illustration)**. Use a small puller if necessary.

8 Using a seal removal tool or a large screwdriver, pry out the old seal **(see illustration)**.

9 Using a seal installation tool or a large socket with an outside diameter slightly smaller than the outside diameter of the new

5.8 Using a seal removal tool, pry out the old seal

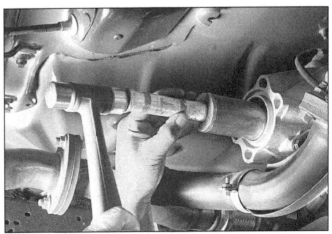

5.9 Using a seal installer or a big socket, install the new seal

seal, install the seal **(see illustration)**.

10   Install the companion flange and a new companion flange nut and tighten the nut to the torque listed in this Chapter's Specifications.

11   Stake the nut in two places **(see illustration)**.

12   Install the driveshaft (see Chapter 8).

13   Install the skid plate, if removed.

14   Remove the jackstands and lower the vehicle.

15   Connect the negative battery cable.

5.11 Stake the flange nut at two points

Chapter 8   Clutch and driveline

Contents

Specifications

Clutch

Hydraulic system fluid type	See Chapter 1
Clutch disc minimum lining thickness	1/64 inch
Pedal freeplay	See Chapter 1
Pedal height	See Chapter 1

Driveaxle CV joint length

1995 and earlier models	5-51/64 inches
1996 and later models	6-1/2 inches
Front hub bearing preload (4WD)	
New bearing and seal	4.4 to 5.5 lbs
Used bearing and seal	2.6 to 4.0 lbs

Torque specifications

Ft-lbs (unless otherwise indicated)

Clutch

Pressure plate-to-flywheel bolts	
Four-cylinder engines	156 in-lbs
V6 engines	180 in-lbs

Driveline

Center bearing	
Flange nut	112
Bracket-to-crossmember bolts	45
Universal joints	
U-joint-to-flange bolts/nuts (all)	46
U-joint strap bolts	180 in-lbs

Front axle (4WD)

Front axle-to-frame bolts	100 to 120
Hub nut	
Initial torque	22, then loosen fully
Final torque	
New bearing and new oil seal	53 to 66 in-lbs
Used bearing and new oil seal	31 to 48 in-lbs
Hub bearing cap bolts	44
Rotor-to-hub mounting bolts	76

Rear axle

Rear axle bearing retainer nuts (Amigo)	
Four cylinder	54
V6	76
Differential pinion shaft lock bolt (Rodeo and Passport)	25
Differential cover bolts	20

1 General information

The information in this Chapter deals with the components from the rear of the engine to the rear wheels and to the front wheels on 4WD models, except for the transmission and transfer case (4WD models), which are in the previous Chapter. For the purposes of this Chapter, these components are grouped into four categories: clutch, driveshaft, rear axle and front axle. Separate Sections within this Chapter offer general descriptions and checking procedures for each of these groups.

Since nearly all the procedures covered in this Chapter involve working under the vehicle, make sure it's securely supported on sturdy jackstands or on a hoist where the vehicle can be easily raised and lowered.

2 Clutch - description and check

1   All models equipped with a manual transmission feature a single dry plate, diaphragm spring-type clutch. The actuation is through a cable (2.3L models) or hydraulic system (all other models).

2   When the clutch pedal is depressed on 2.3L models, the clutch cable pulls the release lever, which moves the release bearing into contact with the pressure plate release fingers, disengaging the clutch disc. When the pedal is depressed on models with a hydraulically actuated clutch, the clutch master cylinder pushes against hydraulic fluid in the cylinder, the lines and the release cylinder, forcing a pushrod in the release cylinder to push against the clutch release lever, which moves the release bearing into contact with the pressure plate release fingers, disengaging the clutch disc.

3   Terminology can be a problem regarding clutch components because the names of the parts used in this book may differ from those used by the manufacturer. For example, the release cylinder is also called the slave cylinder, the release lever is also referred to as the release fork, the clutch release bearing is also called a throwout bearing, the pressure plate is also referred to as a clutch cover, the driven plate is also called the clutch plate or disc, and so on.

4   Due to the slow wearing qualities of the clutch, it is not easy to decide when to go to the trouble of removing the transmission in order to check the wear on the friction lining. The only positive indication that something should be done is when it starts to slip or when squealing noises during engagement indicate that the friction lining has worn down to the rivets. In such instances it can only be hoped that the friction surfaces on the flywheel and pressure plate have not been badly worn or scored.

5   A clutch will wear according to the way in which it is used. Intentionally slipping the clutch, i.e. using the transmission to help slow down the vehicle during deceleration, will accelerate wear. The disc may need to be replaced in as little as 40,000 miles, or it may last 100,000. It all depends on how it's used.

6   Because of the clutch's location between the engine and transmission, it cannot be worked on without removing either the engine or transmission. If repairs requiring engine removal are not needed, the quickest way to gain access to the clutch is by removing the transmission (see Chapter 7).

7   Unless you're replacing obviously damaged components, the easiest way to diagnose a clutch system failure is to perform the following preliminary checks.

   a) *On 2.3L models, check the clutch cable adjustment (see Section 3).*
   b) *Check the fluid level in the clutch master cylinder (models with a hydraulic release system). If the fluid level is low, add fluid as necessary and re-test. If the master cylinder runs dry, or if any of the hydraulic components are serviced, bleed the hydraulic system as described in Section 6.*
   c) *Check "clutch spin down time": Run the engine at normal idle speed with the transmission in Neutral (clutch pedal up - engaged). Disengage the clutch (pedal down), wait nine seconds and shift the transmission into Reverse. No grinding noise should be heard. A grinding noise would indicate component failure in the pressure plate assembly or the clutch disc.*
   d) *Verify that the clutch is releasing completely: Run the engine (with the brake on to prevent movement) and hold the clutch pedal approximately 1/2-inch from the floor mat. Shift the transmission between 1st gear and Reverse several times. If the shift is not smooth, component failure is indicated. Measure the hydraulic release cylinder pushrod travel. With the clutch pedal completely depressed the release cylinder pushrod should extend substantially. If the pushrod will not extend very far or not at all, check the fluid level in the clutch master cylinder.*
   e) *Visually inspect the clutch pedal bushing at the top of the clutch pedal to make sure there is no sticking or excessive wear.*
   f) *Under the vehicle, verify that the release lever is solidly mounted on the ball stud.*

Note: *Because the clutch components are difficult to access, any time the engine or the transmission is removed, the clutch disc, pressure plate assembly and release bearing should be carefully inspected and, if necessary, replaced with new parts. Because the clutch disc is normally a high-wear item, it should be replaced if there's any question about its condition.*

3 Clutch cable - removal, installation and adjustment

1   Locate the clutch cable rubber grommet where it goes through the firewall. Loosen the cable adjusting nut until there's no tension on the cable.

2   Working under the dash with a flashlight, disengage the cable from the top of the clutch pedal.

3   Raise the vehicle and support it securely on jackstands.

4   Unhook the return spring from the release lever.

5   Disconnect the cable from the release lever, pull it forward through the bracket, detach any retaining clips and remove it from the vehicle.

6   Installation is the reverse of removal, taking care there are no kinks in the cable which will cause it to bind.

7   After installation, adjust the clutch pedal freeplay (see Chapter 1).

4 Clutch master cylinder - removal, overhaul and installation

Removal

Refer to illustrations 4.1 and 4.2

Caution: *Do not allow brake fluid to contact any painted surfaces of the vehicle, as damage to the finish may result.*

1   Disconnect the hydraulic line from the master cylinder **(see illustration)** and drain the fluid into a suitable container. Use a flare-nut wrench, if available, to prevent damaging the corners of the fitting.

2   Working under the dash, inside the vehicle, remove the clevis pin that attaches the clutch master cylinder pushrod to the clutch pedal, then remove the master cylinder flange mounting nuts **(see illustration)** and remove the master cylinder.

Overhaul

Refer to illustrations 4.4 and 4.5

3   Remove the retaining clamp and pull off the reservoir.

4   Secure the master cylinder in a bench

4.1 To disconnect the hydraulic line from the clutch master cylinder, unscrew this tube nut (arrow)

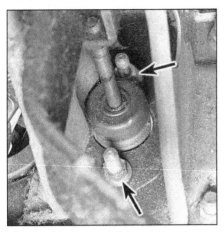

4.2 To detach the clutch master cylinder from the firewall, remove these two nuts (arrows)

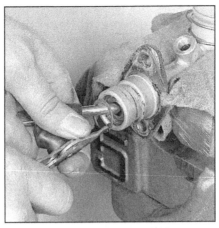

4.4 Push in on the pushrod and remove the snap-ring with a pair of snap-ring pliers

4.5 Remove the piston and spring from the bore of the master cylinder; note that the seal lip faces forward, toward the front of the master cylinder

vise, push down the piston and remove the snap-ring with a pair of snap-ring pliers or a small screwdriver **(see illustration)**.

5   Pull out the piston and spring assembly **(see illustration)**.

6   Examine the inner surface of the cylinder bore. If it is scored or exhibits bright wear areas, the entire master cylinder should be replaced.

7   If the cylinder bore is in good condition, obtain a clutch master cylinder rebuild kit, which will contain all of the necessary replacement parts.

8   Prior to installing any parts, first dip them in brake fluid to lubricate them.

9   Installation of the parts in the cylinder is the reverse of removal. Make sure that the piston seal lip faces forward, toward the front of the master cylinder.

Installation

10   Position the clutch master cylinder against the firewall, inserting the pedal pushrod into the piston. Install the nuts, tightening them securely.

11   Bleed the clutch hydraulic system (see Section 6), then check the pedal height and freeplay (see Chapter 1).

---

5   Clutch release cylinder - removal, overhaul and installation

Removal

Refer to illustrations 5.2, 5.3a and 5.3b

1   Raise the vehicle and support it securely on jackstands.

2   Locate the clutch release cylinder on the side of the transmission bellhousing. Disconnect the hydraulic line from the release cylinder **(see illustration)**. On some models (such as the one shown here), remove the bolt from the banjo fitting on the cylinder body. If there is no banjo fitting at the release cylinder, follow the hose up to the fitting where it meets the metal line, loosen the tube nut (using a flare-nut wrench, if available), then unscrew the hose fitting from the cylinder body.

3   If the hydraulic hose itself is damaged,

replace it **(see illustrations)**.

4   Remove the two retaining nuts and pull off the release cylinder **(see illustration 5.2)**.

Overhaul

5   Pull off the dust boot and pushrod and then tap the cylinder gently on a block of wood to extract the piston and spring.

6   Unscrew and remove the bleeder screw.

7   Examine the surfaces of the piston and cylinder bore for scoring or bright wear areas. If any are found, discard the cylinder and purchase a new one.

8   If the components are in good condition, wash them in clean brake fluid. Remove the seal and discard it, noting carefully which way the seal lips face.

9   Obtain a repair kit which will contain all the necessary new items.

10   Install the new seal using your fingers only to manipulate it into position. Be sure the lips face in the proper direction.

11   Dip the piston assembly in clean brake

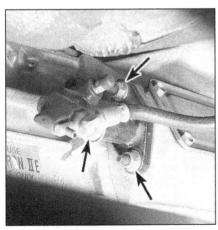

5.2 Before detaching the clutch release cylinder from the transmission, remove this banjo bolt (arrow), then remove these two retaining nuts (arrows)

5.3a If the clutch hydraulic hose must be replaced, unscrew the tube nut (on the metal line side of the bracket); use a back-up wrench on the hose fitting so the bracket doesn't get bent . . .

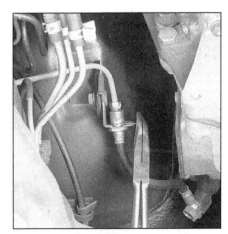

5.3b . . . then remove the retainer clip from the bracket and pull the hose and line apart

fluid before installing it and the spring into the cylinder.

12   Reinstall the bleeder screw.

13   Complete the reassembly by installing the pushrod and the dust cover. Be sure the dust cover is secure on the cylinder housing.

Installation

14   Installation is the reverse of the removal procedure. After the cylinder has been installed, bleed the clutch hydraulic system (see Section 6).

6   Clutch hydraulic system - bleeding

Caution: *Do not allow the brake fluid to contact any painted surface of the vehicle, as damage to the finish will result.*

1   Bleeding will be required whenever the hydraulic system has been dismantled and reassembled and air has entered the system.

2   First fill the fluid reservoir with clean brake fluid which has been stored in an airtight container. Never use fluid which has drained from the system or has bled out previously, as it may contain grit and moisture.

3   Attach a rubber or plastic bleed tube to the bleeder screw on the release cylinder and immerse the open end of the tube in a glass jar containing an inch or two of clean brake fluid.

4   Open the bleeder screw about half a turn and have an assistant depress the clutch pedal completely. Tighten the screw and then have the clutch pedal slowly released completely. Repeat this sequence of operations until air bubbles are no longer ejected from the open end of the tube beneath the fluid in the jar.

5   After two or three strokes of the pedal, make sure the fluid level in the reservoir has not fallen too low. Keep it full of fresh fluid, otherwise air will be drawn into the system.

6   Tighten the bleeder screw on a pedal down stroke (do not overtighten it), remove the bleed tube and jar, top-up the reservoir and install the cap.

7   If an assistant is not available, alternative 'one-man' bleeding operations can be carried out using a bleed tube equipped with a one-way valve or a pressure bleed kit, both of which should be used in accordance with the manufacturer's instructions.

7   Clutch release bearing and lever - removal and installation

Removal

1   Disconnect the negative cable from the battery.

2   Remove the transmission (see Chapter 7).

3   On MSG and MUA transmissions, detach the retaining spring holding the release bearing to the release lever, then slide the release bearing off the transmission input shaft. On Borg-Warner transmissions, the release lever uses a flat retaining clip welded to the backside of the lever. On MUA transmissions bolted to a V6 engine, the release lever pivots on a pin supported by a fulcrum bridge, allowing the lever to see-saw; on these models, remove the retainer, washer and pivot pin from the fulcrum bridge, then disengage the retaining clips that attach the release bearing to the release lever fingers.

4   Remove the clutch release fork from the ball stud by pulling it straight off.

5   Hold the bearing and turn the inner portion. If the bearing doesn't turn smoothly or if it's noisy, replace it with a new one. Wipe the bearing with a clean rag and inspect it for damage, wear and cracks. Don't immerse the bearing in solvent - it's sealed for life and to do so would ruin it.

Installation

6   Lightly lubricate the release lever fingers, where they contact the bearing, with moly-based grease. Apply a thin coat of the same grease to the inner diameter of the bearing and also to the transmission input shaft bearing retainer.

7   Engage the release bearing and the release lever so that both of the lever fingers fit into the bearing groove. Make sure the spring or retainer, if equipped, is properly engaged.

8   Lubricate the clutch release lever ball socket, if applicable, with molybdenum disulfide grease and push the lever onto the ball stud until it's firmly seated. Verify that the bearing slides back and forth smoothly on the input shaft.

9   On V6 models, lubricate the pivot pin, then install the pin, washer and retainer.

10   The remainder of the installation is the reverse of the removal procedure, tightening all bolts to the specified torque.

8   Clutch components - removal, inspection and installation

Removal

Refer to illustration 8.5

Warning: *Dust produced by clutch wear and deposited on clutch components may contain asbestos, which is hazardous to your health. DO NOT blow it out with compressed air and DO NOT inhale it. An approved filtering mask should be worn when working on the clutch components. DO NOT use gasoline or petroleum-based solvents to remove the dust. Brake system cleaner should be used to flush the dust into a drain pan. After the clutch components are wiped clean with a rag, dispose of the contaminated rags and cleaner in a covered container.*

1   Access to the clutch components is normally accomplished by removing the transmission, leaving the engine in the vehicle. If, of course, the engine is being removed for

8.5 If you're going to re-use the same pressure plate, mark its relationship to the flywheel (arrow)

major overhaul, then the opportunity should always be taken to check the clutch for wear and replace worn components as necessary. The following procedures assume that the engine will stay in place.

2   Remove the release cylinder (if equipped) (see Section 5).

3   Referring to Chapter 7 Part A, remove the transmission from the vehicle. Support the engine while the transmission is out. Preferably, an engine hoist should be used to support it from above. However, if a jack is used underneath the engine, make sure a piece of wood is used between the jack and oil pan to spread the load. **Caution:** *The pickup for the oil pump is very close to the bottom of the oil pan. If the pan is bent or distorted in any way, engine oil starvation could occur.*

4   To support the clutch disc during removal, install a clutch alignment tool through the clutch disc hub.

5   Carefully inspect the flywheel and pressure plate for indexing marks. The marks are usually an X, an O or a white letter. If they cannot be found, apply marks yourself so the pressure plate and the flywheel will be in the same alignment during installation **(see illustration)**.

6   Turning each bolt only 1/2-turn at a time, slowly loosen the pressure plate-to-flywheel bolts. Work in a diagonal pattern and loosen each bolt a little at a time until all spring pressure is relieved. Then hold the pressure plate securely and completely remove the bolts, followed by the pressure plate and clutch disc.

Inspection

Refer to illustrations 8.12a and 8.12b

7   Ordinarily, when a problem occurs in the clutch, it can be attributed to wear of the clutch disc assembly. However, all components should be inspected at this time.

8   Inspect the flywheel for cracks, heat checking, grooves or other signs of obvious defects. If the imperfections are slight, a machine shop can machine the surface flat

NORMAL FINGER WEAR

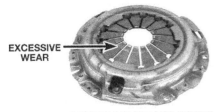

EXCESSIVE FINGER WEAR

BROKEN OR BENT FINGERS

8.12a Replace the pressure plate if excessive wear is noted

8.12b Examine the pressure plate friction surface for score marks, cracks and evidence of overheating

8.14a The flywheel side of the clutch disc is usually marked

8.14b Use a clutch alignment tool to hold the clutch plate to center it before tightening the pressure plate bolts

and smooth, which is highly recommended regardless of the surface appearance. Refer to Chapter 2 for the flywheel removal and installation procedure.

9   Inspect the pilot bearing (see Section 9).

10   Inspect the lining on the clutch disc. There should be at least 1 mm (3/64-inch) of lining above the rivet heads. Check for loose rivets, warpage, cracks, distorted springs or damper bushings and other obvious damage. As mentioned above, ordinarily the clutch disc is replaced as a matter of course, so if in doubt about the condition, replace it with a new one.

11   Ordinarily, the release bearing is also replaced along with the clutch disc (see Section 7).

12   Check the machined surfaces of the pressure plate **(see illustrations)**. If the surface is grooved or otherwise damaged, replace the pressure plate. Also check for obvious damage, distortion, cracking, etc. Light glazing can be removed with emery cloth or sandpaper. If a new pressure plate is indicated, new or factory-rebuilt units are available.

Installation

Refer to illustrations 8.14a and 8.14b

13   Before installation, carefully wipe the flywheel and pressure plate machined surfaces clean with brake system cleaner or a rubbing-alcohol dampened rag. It's important that no

oil or grease is on these surfaces or the lining of the clutch disc. Handle these parts only with clean hands.

14   Position the clutch disc and pressure plate with the clutch held in place with an alignment tool **(see illustration)**. Make sure it's installed properly; most replacement clutch discs will be marked "flywheel side" or something similar **(see illustration)**.

15   Tighten the pressure plate-to-flywheel bolts only finger tight, working around the pressure plate.

16   Center the clutch disc by ensuring the alignment tool is through the splined hub and into the pilot bearing in the crankshaft. Wiggle the tool up, down or side-to-side, as needed, to bottom the tool in the pilot bearing. Tighten the pressure plate-to-flywheel bolts a little at a time, working in a criss-cross pattern to prevent distorting the cover. After all of the bolts are snug, tighten them to the specified torque. Remove the alignment tool.

17   Using high temperature grease, lubricate the inner groove of the release bearing (see Section 7). Also place grease on the fork fingers.

18   Install the clutch release bearing (see Section 7).

19   Install the transmission, release cylinder and all components removed previously, tightening all fasteners to the proper torque specifications.

9   Pilot bearing - inspection, removal and installation

Refer to illustrations 9.10 and 9.11

1   The clutch pilot bearing is a ball-type bearing which is pressed into the rear of the crankshaft. Its primary purpose is to support the front of the transmission input shaft. The pilot bearing should be inspected and, if necessary, replaced whenever the clutch components are removed from the engine. **Note:** *If the engine has been removed from the vehicle, disregard any of the following steps which don't apply.*

2   Remove the transmission (see Chapter 7A).

3   Remove the clutch components (see Section 8).

4   Using a clean rag, wipe the bearing clean and inspect for any excessive wear, scoring or obvious damage. A flashlight will be helpful to direct light into the recess.

5   Check to make sure the pilot bearing turns smoothly and quietly. If the transmission input shaft contact surface is worn or damaged, replace the bearing with a new one.

6   Removal can be accomplished with a slide hammer and internal puller attachment, but an alternative method also works sometimes.

7   Find a solid steel bar slightly smaller in diameter than the bearing. Alternatives to a solid bar would be a wood dowel or a socket with a bolt fixed in place to make it solid.

8   Check the bar for fit - it should just slip into the bearing with very little clearance.

9   Pack the bearing and the area behind it (in the crankshaft recess) with heavy grease. Pack it tightly to eliminate as much air as possible.

9.10 You can remove the old pilot bearing by packing the recess behind the bearing with heavy grease and forcing it out hydraulically with a steel rod slightly smaller than the bore in the bearing - when the hammer strikes the rod, the bearing will pop out of the crankshaft

10   Insert the bar into the bearing bore and lightly hammer on the bar **(see illustration)**; this will force the grease to the backside of the bearing and push it out. **Warning:** *Cover the bar and bearing with a rag to catch any grease that may squirt out.* Remove the bearing and clean all grease from the crankshaft recess.

11   To install the new bearing, lubricate the outside surface with oil then drive it into the recess with a hammer and a socket with an outside diameter that matches the bearing outer race **(see illustration)**.

12   Pack the bearing with lithium base grease (NLGI No. 2). Wipe off all excess grease so the clutch lining will not become contaminated.

13   Install the clutch components, transmission and all other components removed to gain access to the pilot bearing.

10   Driveshafts, differentials and axles - general information

Driveshafts

A variety of different driveshaft assemblies are used on the vehicles covered in this manual. All Rodeo models use a driveshaft with a center bearing; Amigo models use a one-piece driveshaft with no center bearing. The driveshafts used with 2WD manual transmissions use a slip yoke at the extension housing and are bolted to the pinion flange at the differential end. These units have two universal joints: one at the slip yoke and the other at the rear end of the driveshaft, where it bolts to the pinion flange. The driveshafts used with automatics and 4WD models are attached to flanges at both the extension housing and the differential (the slip yoke is located in the rear portion of the driveshaft, behind the center bearing). Two-piece driveshafts with a center bearing use three U-

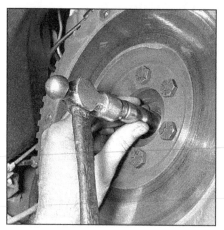

9.11 Tap the bearing into place with a bushing driver or a socket slightly smaller than the outside diameter of the bearing

joints: one at the extension housing companion flange, one behind the center bearing and one at the differential pinion flange. 4WD models also have a front driveshaft, which is bolted to a flange on the front of the transfer case and a pinion flange at the front axle differential. Front driveshafts have two U-joints, one at each end. All U-joints can be replaced separately from the driveshaft.

The driveshafts are finely balanced during production and whenever they are removed or disassembled, they must be reassembled and reinstalled in the exact manner and positions they were originally in, to avoid excessive vibration.

Axles

The rear axle is "semi-floating," i.e. the axleshafts support the weight of the vehicle. The axle housing is held in proper alignment with the body by the rear suspension. In the center of the rear axle is the differential, which transfers the turning force of the driveshaft to the rear axleshafts. The axleshafts are splined at their inner ends to fit into the splines in the differential gears; outer support for the shafts is provided by the rear wheel bearings, which are pressed onto the axle shafts.

The front axle on 4WD vehicles consists of a frame-mounted differential assembly and two driveaxles. The driveaxles incorporate constant velocity (CV) joints at each end, enabling them to transmit power at various suspension angles independent from each other.

Because of the complexity and critical nature of the differential adjustments, as well as the special equipment needed, we recommend any disassembly of the differential be done by a dealer service department or other repair shop.

11   Driveline inspection

1   Raise the rear of the vehicle and support it securely on jackstands.

2   Slide under the vehicle and visually inspect the condition of the driveshaft. Look for any dents or cracks in the tubing. If any are found, the driveshaft must be replaced.

3   Check for any oil leakage at the front and rear of the driveshaft. Leakage where the driveshaft enters the transmission indicates a defective rear transmission seal. Leakage where the driveshaft enters the differential indicates a defective pinion seal.

4   While still under the vehicle, have an assistant turn the rear wheel so the driveshaft will rotate. As it does, make sure that the universal joints are operating properly without binding, noise or looseness. On models so equipped, listen for any noise from the center bearing, indicating it is worn or damaged. Also check the rubber portion of the center bearing for cracking or separation, which will necessitate replacement.

5   The universal joint can also be checked with the driveshaft motionless, by gripping your hands on either side of the joint and attempting to twist the joint. Any movement at all in the joint is a sign of considerable wear. Lifting up on the shaft will also indicate movement in the universal joints.

6   Finally, check the driveshaft mounting bolts at the ends to make sure they are tight.

7   On 4WD models, the above driveshaft checks should be repeated on all driveshafts. In addition, check for grease leakage around the sleeve yoke, indicating failure of the yoke seal.

8   Check for leakage at each connection of the driveshafts to the transfer case and front differential. Leakage indicates worn oil seals.

9   At the same time, check for looseness in the joints of the front driveaxles.

12   Driveshafts - removal and installation

Front driveshaft (4WD models)

Refer to illustrations 12.1 and 12.3

1   Raise the front of the vehicle and place it on jackstands. Mark the relationship of the front driveshaft flange to the front differential companion flange so they can be realigned upon installation. Also mark the direction the driveshaft faces **(see illustration)**.

2   Mark the relationship of the rear driveshaft flange to the transfer case companion flange.

3   Lock the driveshaft from turning with a large screwdriver or prybar, then remove the four nuts and bolts from the front flange **(see illustration)**.

4   Again lock the driveshaft from turning and remove the four nuts and bolts from the rear flange. On some models, it may be necessary to support the transmission with a jack and remove the crossmember for access to the bolts and nuts.

5   Slide the front portion of the driveshaft toward the rear, so it telescopes along the splined portion of the shaft. Lower the driveshaft from the vehicle.

12.1 If you're removing the front driveshaft on a 4WD model, mark the relationship of the U-joint at each end of the driveshaft to its corresponding companion flange

12.3 To detach a front driveshaft from the companion flanges, remove these four nuts and bolts (arrows)

12.8 To detach the center bearing from the crossmember, remove these two bolts (arrows) (the nuts are welded to the center bearing bracket)

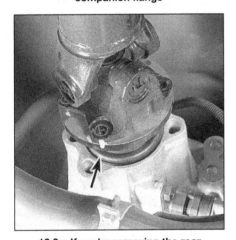

12.9a If you're removing the rear driveshaft on an automatic or a 4WD model, mark the relationship of the U-joint to the corresponding companion flange (arrow) (on 2WD models with a manual transmission, the front U-joint has a slip yoke instead)

6   Installation is the reverse of removal. Be sure to align all marks and tighten the flange bolts to the torque listed in this Chapter's Specifications.

Rear driveshaft

Refer to illustrations 12.8, 12.9a, 12.9b and 12.10

7   Raise the rear of the vehicle and support it on jackstands.
8   Remove the bolts holding the center support bearing bracket to the frame, if equipped **(see illustration)**.
9   Mark the relationship of the U-joints to the driveshaft and to the front and rear flanges so they can be realigned upon installation **(see illustrations)**.
10   Remove the four nuts and bolts **(see illustration)**.
11   On 2WD models, support the driveshaft, pull the yoke from the transmission extension housing and remove the driveshaft. On auto-

12.9b If you're removing the rear driveshaft, mark the relationship of the U-joint to the pinion flange (arrow)

matics and 4WD models, move the driveshaft slightly forward or backward to disengage the U-joints from the flanges and remove the driveshaft and center bearing assembly as a unit.
12   While the driveshaft is removed, insert a plug in the transmission to prevent lubricant leakage (2WD models).
13   Installation is the reverse of the removal procedure. During installation, make sure all flange marks line up. When connecting the center bearing support to the frame, first finger-tighten the two mounting bolts, then make sure that the bearing bracket is at right angles to the driveshaft. Tighten all nuts and bolts to the torque listed in this Chapter's Specifications.

13 Center bearing - replacement

Refer to illustrations 13.3 and 13.4

1   Remove the rear driveshaft assembly (see Section 12).
2   Support the front driveshaft in a vise with padded jaws, with the center bearing

12.10 To detach a rear driveshaft from the rear axle, the extension housing on an automatic transmission or the transfer case on a 4WD model, unbolt the U-joint from the companion flange by removing all four nuts and bolts

assembly facing out.
3   Mark the relationship of the U-joint to the flange **(see illustration)**.

13.3 Mark the relationship of the U-joint to the flange

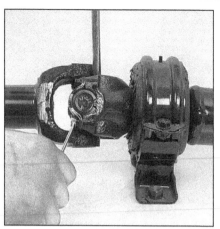

13.4 To separate the front part of the driveshaft from the center bearing, remove the four strap bolts that secure the U-joint to the flange

14.2 Use needle-nose pliers to remove the U-joint snap-rings

14.4 To press the U-joint out of the driveshaft yoke, set it up in a vise with the small socket pushing the joint and bearing cap into the large socket

14.8 If a snap-ring won't seat in the groove, strike the yoke with a brass hammer - this will slightly spring the yoke ears and relieve tension in the yoke (this is also a good method for relieving tension if the joint feels tight when assembled)

15.3a Mark the relationship of the flange nut to the pinion shaft

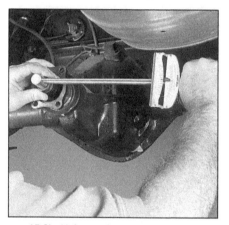

15.3b Using an inch-pound torque wrench, measure the torque required to start the pinion flange turning; jot down this figure and save it for reassembly

4   Remove the four bolts retaining the U-joint straps **(see illustration)** and remove the front part of the driveshaft.

5   Take the rear part of the driveshaft and the center bearing to an automotive machine shop, have the old bearing removed and a new unit installed.

6   Installation is the reverse of removal. Be sure to tighten the strap bolts to the torque listed in this Chapter's Specifications.

14 Universal joints - replacement

Refer to illustrations 14.2, 14.4 and 14.8

1   Remove the driveshaft (see Section 12).
2   Clean away all dirt from the ends of the bearings on the yokes so the snap-rings can be removed with a pair of needle-nose pliers or a screwdriver **(see illustration)**.
3   Supporting the driveshaft, place it in a bench vise.

4   Place a piece of pipe or a large socket with the same inside diameter over one end of the bearing caps. Position a socket of slightly smaller diameter than the cap on the opposite bearing cap **(see illustration)** and use the vise or press to force the cap out (inside the pipe or large socket, stopping just before it comes completely out of the yoke. Use the vise or large pliers to work the cap the rest of the way out.

5   Transfer the sockets to the other side and press the opposite bearing cap out in the same manner.

6   Pack the new U-joint bearings with grease. Ordinarily, specific instructions for lubrication will be included with the U-joint servicing kit and should be followed carefully.

7   Position the spider into the bearing cap, then partially install the other cap. Align the spider and press the bearing caps into position, being careful not to damage the dust seals.

8   Install the snap-rings. If the snap-rings are difficult to seat, strike the driveshaft yoke sharply with a hammer **(see illustration)**. This will spring the yoke ears slightly, allowing the snap-rings to seat in their grooves.

9   Install the grease fitting and fill the joint with grease. Be careful not to overfill the joint, as this could blow out the grease seals.

10   Install the driveshaft, tightening the companion flange bolts to the torque listed in this Chapter's Specifications.

15 Differential pinion seal - replacement

Refer to illustrations 15.3a, 15.3b, 15.4, 15.5, 15.6 and 15.7

1   Raise the vehicle and place it securely on jackstands.
2   Remove the driveshaft (see Section 12).
3   Mark the relationship of the flange nut to the pinion shaft **(see illustration)**. Using an inch-pound torque wrench, measure the torque required to start the pinion flange turning **(see illustration)**. Jot down this figure and save it for reassembly.
4   Holding the flange with a suitable tool, remove the retaining nut **(see illustration)**.
5   Remove the pinion flange; use a puller if

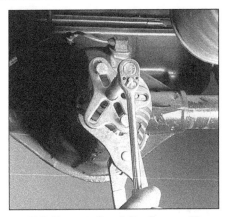

15.4 Holding the flange with a suitable tool (such as a pair of big water pump pliers), remove the retaining nut

15.5 Remove the pinion flange with a puller, if necessary

15.6 Remove the old seal with a seal removal tool or a large screwdriver

15.7 Install a new seal with a seal installer tool or a big socket

16.3 To detach the axleshaft bearing retainer from the axle tube, remove these four nuts (arrows)

16.4a Install a slide-hammer and flange puller attachment to the axleshaft flange as shown . . .

necessary **(see illustration)**.

6   Remove the old seal **(see illustration)**.

7   Install a new seal **(see illustration)**.

8   Apply sealant to the splines of the pinion flange, then install it onto the pinion, aligning the marks you made in Step 3. Apply a non-hardening thread locking compound to the threads of pinion, then install and tighten the pinion flange retaining nut until the previously made matchmarks are in alignment. When the marks are aligned, the turning torque of the pinion flange should be the same as it was before disassembly. When this figure is attained, tighten the nut to obtain another five inch-pounds of turning torque.

9   Install the driveshaft (see Section 12).

10   Lower the vehicle.

---

16   Rear axleshaft, bearing and seal - removal and installation

Refer to illustrations 16.3, 16.4a, 16.4b, 16.5 and 16.6

Note: *Models covered in this manual are equipped with either a Dana or Saginaw rear axle. You can identify which axle your vehicle has by the shape of the rear cover and the number of bolts used to attach it. Dana differentials have an odd-shaped rear cover with a 12-bolt*

pattern; it has an angular look to it because the ring-gear side is longer than the other side when facing the rear cover. Saginaw differentials have a uniformly oval-shaped cover with an 11-bolt pattern that has a round look to it.

Dana rear axles (Amigo and Rodeo models)

1   Loosen the rear wheel lugs nuts, raise the vehicle and place it securely on jackstands.

16.4b . . . then remove the axleshaft

2   Remove the rear drum brake shoes, or the rear caliper and disc, and disconnect the parking brake cable (see Chapter 9).

3   Remove the four nuts that attach the axleshaft bearing retainer to the axle tube **(see illustration)**.

4   Install a slide hammer and puller attachment to the axleshaft flange and remove the axleshaft **(see illustrations)**.

5   Remove the brake backing plate **(see illustration)**.

16.5 Remove the brake backing plate from the bearing retainer and axleshaft

16.6 The collar and bearing are pressed onto the axleshaft and retained by a large snap-ring; the seal is located below the bearing

6   Take the axleshaft and bearing retainer assembly to an automotive machine shop to have the collar, bearing and seal **(see illustration)** replaced.
7   Installation is the reverse of removal. Tighten the bearing retainer nuts to the torque listed in this Chapter's Specifications. Tighten the wheel lug nuts to the torque listed in the Chapter 1 Specifications.

Saginaw rear axles (Rodeo and Passport models)

Axleshaft - removal and installation

Refer to illustrations 16.10 and 16.11

8   Raise the rear of the vehicle, support it securely and remove the wheel and brake drum (see Chapter 9).
9   Unscrew and remove the pressed steel cover from the differential housing and allow the lubricant to drain into a container.
10   Remove the lock bolt from the differential pinion shaft **(see illustration)**. Remove the pinion shaft.
11   Push the outer (flanged) end of the axleshaft in and remove the C-lock from the inner end of the shaft **(see illustration)**.

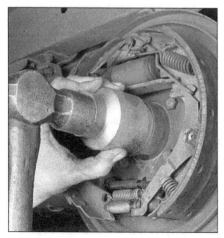

16.18 Installing a rear axleshaft oil seal with a large socket

16.10 Remove the lock screw and pull the pinion shaft out of the differential far enough to clear the inner end of the axleshaft

12   Withdraw the axleshaft, taking care not to damage the oil seal in the end of the axle housing as the splined end of the axleshaft passes through it.
13   Installation is the reverse of removal. Tighten the lock bolt to the torque listed in this Chapter's Specifications. **Note:** *The manufacturer recommends using a new lock bolt.*
14   Always use a new cover gasket and tighten the cover bolts to the torque listed in this Chapter's Specifications.
15   Refill the differential with the correct quantity and grade of lubricant (see Chapter 1).

Axleshaft oil seal - replacement

Refer to illustration 16.18

16   Remove the axleshaft (see Step 8).
17   Pry out the old oil seal from the end of the axle housing, using a large screwdriver or the inner end of the axleshaft itself as a lever.
18   Apply high melting point grease to the oil seal recess and tap the seal into position **(see illustration)** so that the lips are facing in and the metal face is visible from the end of the axle housing. When correctly installed,

16.21 Bearing puller and slide hammer being used to remove a rear axleshaft bearing

16.11 Push the axleshaft in and remove the C-lock

the face of the oil seal should be flush with the end of the axle housing.
19   Install the axleshaft.

Axleshaft bearing - replacement

Refer to illustration 16.21

20   Remove the axleshaft (see Step 8) and the oil seal (see Step 17).
21   A bearing puller will be required or a tool which will engage behind the bearing will have to be fabricated **(see illustration)**.
22   Attach a slide hammer and pull the bearing from the axle housing.
23   Lubricate the new bearing with gear lubricant. Clean out the bearing recess and drive in the new bearing using a bearing driver. If you don't have a bearing driver, use a piece of pipe or a socket of the proper diameter, applied against the outer race of the bearing. Make sure that the bearing is tapped into the full depth of its recess and that the numbers on the bearing are visible from the outer end of the housing.
24   Discard the old oil seal and install a new one, then install the axleshaft.

17   Rear axle assembly - removal and installation

Refer to illustration 17.6

1   Loosen the rear wheel lug nuts, raise the vehicle and support it securely on jackstands placed underneath the frame. Remove the wheels.
2   Support the rear axle assembly with a floor jack placed underneath the differential.
3   Disconnect the lower ends of the shock absorbers from the axle (see Chapter 10).
4   Disconnect the driveshaft from the rear axle (see Section 12) and hang it with a piece of wire from the underbody.
5   Remove the rear drum brake assemblies or the rear brake calipers and discs (see Chapter 9). Disconnect the parking brake cables from the parking brake levers (rear drums) or the parking brake shoes (rear discs).
6   Detach the brake junction block and all brake line clips from the axle tube **(see illustration)**. Also disconnect the breather hose

17.6 Before lowering the rear axle, be sure to detach this brake line junction block (arrow) and the brake line clips (arrow points to clip on the right axle tube; there's another clip on the left tube); there's also a breather hose (not visible in this photo) on top of the differential housing that must be detached

18.3 Mark the relationship of the hub cap with white paint

from the top of the differential housing.

7   Support the rear axle assembly with a jack.

8   Remove the U-bolt nuts from the leaf spring seats (see Chapter 10) and raise the axle assembly slightly with the jack.

9   Remove the rear spring shackle bolts and lower the rear of each leaf spring to the floor. Carefully lower the axle assembly to the floor with the jack, then remove it from under the vehicle. It would be a good idea to have an assistant on hand - the axle assembly is very heavy.

10   Installation is the reverse of removal. Be sure to tighten all fasteners to the torque listed in this and in the Chapter 10 Specifications. Tighten the wheel lug nuts to the torque listed in the Chapter 1 Specifications.

11   Bleed the brakes (see Chapter 9).

---

18   Front hub assembly (4WD) - removal, bearing repack, installation and adjustment

---

Older style

Removal

Refer to illustrations 18.3, 18.4a, 18.4b, 18.5a, 18.5b, 18.6, 18.7, 18.8, 18.9a, 18.9b, 18.10, 18.11 and 18.12

1   Set the hub control lever to the 2H posi-

tion and the free-wheeling hub to the Free position. Loosen the wheel lug nuts, raise the vehicle and place it securely on jackstands. Remove the wheel.

2   Remove the brake caliper and hang it out of the way on a piece of wire (see Chapter 9).

3   Mark the position of the hub cap so it can be reinstalled in the same position **(see illustration)**.

4   Remove the six bolts and lift the cap off **(see illustrations)**.

5   Mark the position of the hub housing, then remove it **(see illustrations)**.

6   On free-wheeling hubs, use snap-ring pliers to remove the snap-ring which retains the inner assembly. Use a screwdriver and needle-nose pliers to remove the snap-ring that retains the drive clutch assembly **(see illustration)**. Be sure to keep track of the number and locations of any shims so they can be reinstalled in the same positions.

18.4a Lock the hub from turning, remove the bolts . . .

18.4b . . . then lift the cap off

18.5a Mark the relationship of the housing . . .

18.5b . . . then remove it

18.6 A small screwdriver and needle-nose pliers can be used to remove the snap-ring from the hub

18.7 Lift the drive clutch assembly
out of the hub

18.8 Lift the inner cam off

18.9a Remove the retaining
screws (arrows) . . .

18.9b . . . then lift the lock washer off

18.10 Use a punch or Phillips screwdriver
to unscrew the locknut

18.11 Dislodge the outer bearing,
then lift it out of the hub

7   Remove the drive clutch assembly (**see illustration**).
8   Remove the inner cam (**see illustration**).
9   Remove the screws and lift the lock washer off (**see illustrations**).
10   Unscrew the hub nut, using a punch (**see illustration**).
11   Pull the hub out slightly, then push it back into its original position. This should force the outer wheel bearing off the spindle enough so it can be removed (**see illustration**).
12   Pull the hub off the spindle (**see illustration**).
13   Use a screwdriver to pry the grease seal out of the rear of the hub. As this is done, note how the seal is installed.
14   Remove the inner wheel bearing from the hub.

Bearing repack

15   Use solvent to remove all traces of the old grease from the bearings, hub and spindle. A small brush may prove helpful; however make sure no bristles from the brush embed themselves inside the bearing rollers. Allow the parts to air dry.
16   Carefully inspect the bearings for cracks, heat discoloration, worn rollers, etc. Check the bearing races inside the hub for wear and damage. If the bearing races are

defective, the hubs should be taken to a machine shop with the facilities to remove the old races and press new ones in. Note that the bearings and races come as matched sets and old bearings should never be installed on new races.
17   Use high-temperature moly-based front wheel bearing grease to pack the bearings. Work the grease completely into the bearings, forcing it between the rollers, cone and cage from the back side (**see illustra-**

18.12 Grasp the hub and disc assembly
securely and lift if off the spindle

tion **37.15** in Chapter 1).
18   Apply a thin coat of grease to the spindle at the outer bearing seat, inner bearing seat, shoulder and seal seat.

Installation and adjustment
Refer to illustration 18.22
19   Place the grease-packed inner bearing into the rear of the hub and put a little more grease outside of the bearing.
20   Place a new seal over the inner bearing

18.22 Use needle-nose pliers to tighten
the hub nut

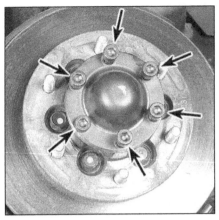

18.36 Remove these bolts (arrows) and remove the grease cover

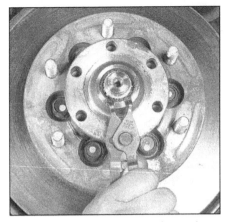

18.37 Remove this snap-ring and the shim behind it

18.38 Remove the hub flange

18.39a Remove these lock screws (arrows) . . .

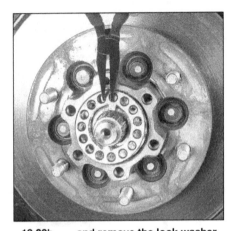

18.39b . . . and remove the lock washer

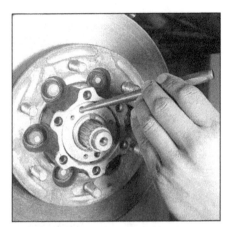

18.40 Loosen the hub nut with a hammer and punch, then unscrew and remove it

and tap the seal evenly into place with a hammer and block of wood until it's flush with the hub.

21   Carefully place the hub assembly onto the spindle and push the grease-packed outer bearing into position.

22   Install the hub nut and tighten it only slightly **(see illustration)**.

23   Spin the hub in a forward direction to seat the bearings and remove any grease or burrs which could cause excessive bearing play later.

24   Attach a spring scale to one of the wheel studs at right angles to the hub and measure the amount of pull necessary to rotate the hub. This is the preload. Compare the preload reading to the Specifications at the front of this Chapter.

25   Tighten or loosen the hub nut to achieve the specified preload.

26   Install the lock washer and see if the bolt holes in the washer are aligned with the holes in the hub nut. If they are not, turn the lock washer over. The holes should now line up. If they don't, turn the hub nut just enough to align them. Check the preload to make sure it is still within specification.

27   Install the retaining screws.

28   Clean the contact surfaces of the lock washer and the inner cam and then slide the inner cam into place. If the inner cam is difficult to install, tap it lightly with a plastic hammer to seat it in the hub.

29   Install the drive clutch assembly and secure it with the snap-ring.

30   The remainder of installation is the reverse of removal. When installing the snap-ring on the drive clutch assembly, thread a bolt into the hub and pull the hub out.

31   Install the caliper (Chapter 9).

32   Install the wheel and lug nuts.

33   Lower the vehicle and tighten the lug nuts to the torque listed in the Chapter 1 Specifications.

Later style

Refer to illustrations 18.36, 18.37, 18.38, 18.39a, 18.39b, 18.40, 18.41, 18.43 and 18.49

34   Loosen the wheel lug nuts, raise the vehicle and place it securely on jackstands. Remove the wheel.

35   Remove the brake caliper and hang it out of the way, then remove the caliper anchor bracket (see Chapter 9).

36   Remove the grease cover **(see illustration)**.

37   Remove the snap-ring and shim **(see illustration)**.

38   Remove the hub flange **(see illustration)**.

39   Remove the lock screws and lift off the lock washer **(see illustrations)**.

40   Remove the hub nut **(see illustration)**.

41   Remove the outer bearing **(see illustration)**.

42   Remove the hub and disc assembly and place it on a workbench.

18.41 Remove the outer bearing, then remove the hub and disc assembly

18.43 Remove the seal from the backside of the disc with a seal removal tool or a screwdriver, then remove the inner bearing assembly

18.49 Install a new seal with a seal installer tool (shown) or a large socket

19.2 To remove the rock shield that protects the vacuum switching valves and the vacuum shift motor, remove these two bolts (arrows)

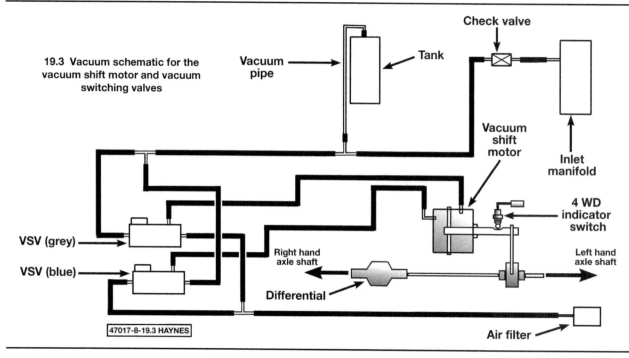

19.3 Vacuum schematic for the vacuum shift motor and vacuum switching valves

43 Remove the seal from the backside of the disc **(see illustration)**, then remove the inner bearing assembly.

44 Thoroughly wash the disc and the inner and outer wheel bearings in clean solvent and blow them dry with compressed air.

45 Inspect the bearings for excessive wear. If either bearing is worn, replace both. **Note:** *If the bearings are replaced, be sure to replace the races, too.*

46 Inspect the bearing races (pressed into the hub and disc assembly). If either is excessively worn, take the hub and disc to an automotive machine shop and have new ones installed. **Note:** *If the races are replaced, replace the bearings also.*

47 Pack the bearings with high-temperature wheel bearing grease (NLGI No. 2, or equivalent) **(see illustration 37.15 in Chapter 1)**.

48 Install the inner bearing.

49 Install a new seal **(see illustration)**.

50 Installation is the reverse of removal. Make sure that all parts are clean and be sure to tighten the hub nut to the torque listed in this Chapter's Specifications. Tighten the wheel lug nuts to the torque listed in the Chapter 1 Specifications.

19 Vacuum shift motor and solenoids - check and component replacement

Check

Refer to illustration 19.2

1 Raise the front of the vehicle and support it securely on jackstands. Remove the wheels.

2 Remove the rock shield **(see illustration)** that protects the vacuum switching valves and the vacuum shift motor.

Vacuum hoses

Refer to illustration 19.3

3 Inspect the system vacuum hoses **(see illustration)** for leaks. Make sure all hoses are in good condition and tightly attached. Repair as necessary.

Check valve

Refer to illustration 19.4

4 The check valve is located near the intake manifold. Detach the valve from both vacuum hoses, apply 15 or 16 in-Hg of vacuum to the orange end of the valve with a hand-operated vacuum pump **(see illustration)** and verify that the valve holds a vac-

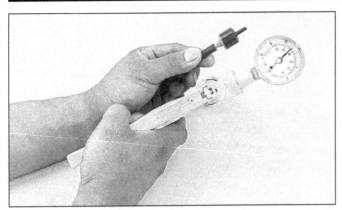

19.4 To test the check valve for the vacuum shift motor, apply 10 or 15 in-Hg of vacuum to the orange end of the valve and verify that the valve holds vacuum; then try to apply vacuum to the other end of the valve and verify that it doesn't hold vacuum

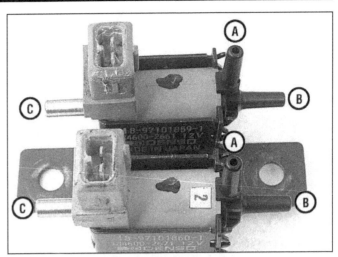

19.5 To check either switching valve, verify that you can blow through pipe A to pipe C (manifold vacuum is vented to atmosphere); then apply battery voltage to each valve and verify that you can blow through pipe A to pipe B (vacuum goes to vacuum shift motor and engages [blue] or disengages [gray] axle) but not from A to C (valves removed for clarity; this test can be performed on the vehicle)

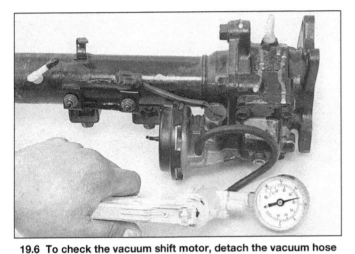

19.6 To check the vacuum shift motor, detach the vacuum hose from the upper diaphragm chamber, hook up a vacuum pump, apply 15 or 16 in-Hg of vacuum to the upper chamber and verify that the shift motor engages the front axle; then perform the same check on the lower diaphragm chamber: the motor should now disengage the front axle (vacuum shift motor and axle tube removed for clarity)

19.7a To replace either vacuum switching valve, unplug the electrical connector, label and detach the vacuum hoses, remove the bracket retaining nuts (arrows) . . .

uum. Try to apply vacuum to the black end and verify that the valve doesn't hold vacuum. If the check valve doesn't operate as described, replace it.

Vacuum switching valves

Refer to illustration 19.5

5   The vacuum switching valves are located on the axle tube, next to the vacuum shift motor. Unplug the electrical connectors and the vacuum hoses from the valves (mark them first to prevent confusion on reassembly) from the valves, and connect a clean vacuum hose to pipe A **(see illustration)** of the blue switching valve. Apply battery voltage to the blue switching valve and blow into the hose; air should pass through and come out port B. With no voltage applied, air should pass through to port C. Repeat this check on the gray switching valve. If either switching valve fails to perform as described, replace it.

Vacuum shift motor

Refer to illustration 19.6

6   Detach the vacuum hose from the upper chamber of the diaphragm chamber on the vacuum shift motor and hook up a hand-operated vacuum pump to the upper chamber. Apply 15 or 16 in-Hg of vacuum to the upper chamber **(see illustration)** and verify that the shift motor engages the front axle. Detach the vacuum pump from the upper chamber and reattach the hose to that chamber. Detach the hose from the lower chamber, attach the vacuum pump to the lower chamber and apply 15 or 16 in-Hg of vacuum to the lower chamber and verify that the shift motor disengages the front axle. If the vacuum motor fails to operate as described, replace it.

Component replacement

Refer to illustrations 19.7a, 19.7b and 19.8

Vacuum switching valves

7   To replace either vacuum switching

19.7b . . . turn the bracket and valves over, and remove the vacuum switching valve retaining screw(s) (arrows)

19.8 To detach the vacuum shift motor from the axle tube, unplug the electrical connector, detach the vacuum hoses and remove these four bolts (arrows)

20.8 Detach the vent hose from the front axle

20.10a To detach the front axle assembly from the frame, remove these four big nuts (arrows)

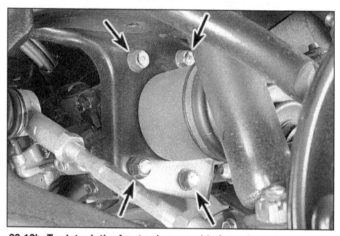

20.10b To detach the front axle assembly from the axle mounting brackets, remove these four bolts (arrows) from each bracket (left bracket shown, right bracket similar on later style axles, different on earlier style axles)

valve, unplug the electrical connector from the valve, label and detach the vacuum hoses, remove the valve bracket nuts **(see illustration)**, remove the bracket and valves, turn the assembly over and remove the valve retaining screw **(see illustration)**. Installation is the reverse of removal.

Vacuum shift motor

8   To replace the vacuum shift motor, unplug the electrical connector, label and detach the vacuum lines and remove the four mounting bolts **(see illustration)**. Installation is the reverse of removal. Be sure to apply moly-based grease to the shift fork.

---

20 Front axle assembly (4WD) - removal and installation

Refer to illustrations 20.8, 20.10a, 20.10b, 20.11a and 20.11b
Note: *The illustrations accompanying this section show a later style front axle with an*

automatic "shift-on-the-fly" system. Earlier units have different axle mounting brackets and they don't have a vacuum shift motor and vacuum switching valves. However, the removal procedure for these models is generally similar.

1   Loosen the wheel lug nuts, raise the front of the vehicle and support it securely on jackstands. Remove the wheels.
2   Remove the steering linkage (see Chapter 10).
3   Detach the front driveshaft from the front axle (see Section 12), then hang it out of the way with a piece of wire.
4   Remove the left brake caliper and hang it out of the way (see Chapter 9).
5   Remove the skid plate (if equipped).
6   Disconnect and remove the steering linkage (Pitman arm, idler arm, center link and both tie rods) as a single assembly (see Chapter 10).
7   Separate the left steering knuckle from the left upper control arm balljoint (see Chapter 10).
8   Detach the vent hose **(see illustration)**

from the front axle.
9   Unplug the electrical connectors from the vacuum switching valves and from the vacuum shift motor, label and detach the vacuum hoses from the valves and the motor, and detach the switching valve mounting bracket and remove the motor from the axle (see Section 19).
10   Support the front axle assembly with a floor jack. Remove the four nuts - two at each end - attaching the axle mounting brackets to the frame **(see illustration)** and remove the eight bolts - four on each end - attaching these same brackets to the ends of the axle assembly **(see illustration)**.
11   With the axle mounting brackets detached from the frame and the axle assembly detached from the brackets, disengage the splined inner end of the left driveaxle from the axle assembly **(see illustration)**, carefully lower the axle assembly slightly, disengage it from the inner end of the right driveaxle assembly **(see illustration)**, and remove the axle.
12   Installation is the reverse of the removal

20.11a Disengage the inner end of the left driveaxle from the front axle assembly

20.11b Disengage the inner end of the right driveaxle from the front axle assembly

21.4 Disengage the outer end of the driveaxle from the steering knuckle and hub (left driveaxle shown, right driveaxle is removed same way)

22.3 Cut off the boot clamps and discard them

procedure. Be sure to tighten all brake, steering and suspension fasteners to the torque listed in the Chapter 9 and 10 Specifications. Tighten the wheel lug nuts to the torque listed in the Chapter 1 Specifications.

21 Front driveaxle (4WD) - removal and installation

Refer to illustration 21.4

1 Loosen the wheel lug nuts, raise the vehicle and place it securely on jackstands. Remove the wheel.
2 Remove the front axle assembly (see Section 20).
3 Remove the driveaxle-to-hub snap-ring from the hub assembly (see Section 18).
4 Disengage the outer end of the driveaxle from the steering knuckle and hub and remove the driveaxle assembly towards the center of the vehicle **(see illustration)**.
5 Take the driveaxle and axle mounting bracket to an automotive machine shop and have the bearing pressed off the inner end of the driveaxle assembly so that the driveaxle assembly can be separated from the bracket.
6 Inspect the bearing for excessive wear.

If it's worn, replace it. Discard the old seal.
7 After you have replaced a CV joint and/or boot (see Section 22), take the driveaxle assembly back to the automotive machine shop and have the axle mounting bracket and the new seal installed and the bearing pressed on. Even if you're re-using the old bearing, be sure to use a new seal.
8 Installation is the reverse of removal.

22.4 Pry the wire retainer ring from the CV joint housing with a small screwdriver

22 Front driveaxle - boot replacement and constant velocity (CV) joint inspection (4WD)

Note: *If the CV joints exhibit signs of wear indicating need for an overhaul (usually due to torn boots), explore all options before beginning the job. Complete rebuilt driveaxles are available on an exchange basis, which eliminates much time and work. Whichever route you choose to take, check on the cost and availability of parts before disassembling the vehicle.*

Inner CV joint

Disassembly

Refer to illustrations 22.3, 22.4, 22.5, 22.7, 22.9, 22.10 and 22.11

1 Remove the driveaxle from the vehicle (see Section 21).
2 Mount the driveaxle in a vise. The jaws of the vise should be lined with wood or rags to prevent damage to the axleshaft.
3 Cut the boot clamps from the boot and discard them **(see illustration)**.
4 Slide the boot back on the axleshaft and

22.5 With the retainer removed, the outer race can be pulled off the bearing assembly

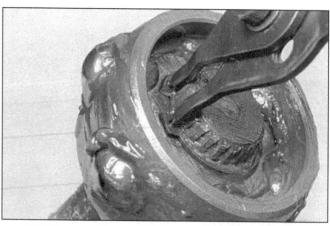

22.7 Remove the snap-ring from the end of the axleshaft

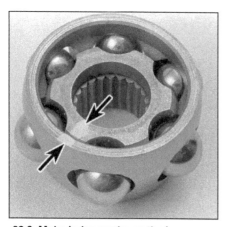

22.9 Make index marks on the inner race and cage so they'll both be facing the same direction when reassembled

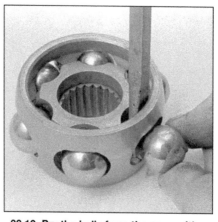

22.10 Pry the balls from the cage with a screwdriver (be careful not to nick or scratch them)

22.11 Tilt the inner race 90-degrees and rotate it out of the cage

pry the wire ring ball retainer from the outer race **(see illustration)**.
5   Pull the outer race off the inner bearing assembly **(see illustration)**.
6   Wipe as much grease off the inner bearing as possible.
7   Remove the snap-ring from the end of the axleshaft **(see illustration)**.

8   Slide the inner bearing assembly off the axleshaft.
9   Mark the inner race and cage to ensure that they are reassembled with the correct sides facing out **(see illustration)**.
10   Using a screwdriver or piece of wood, pry the balls from the cage **(see illustration)**. Be careful not to scratch the inner race, the balls or the cage.

11   Rotate the inner race 90-degrees, align the inner race lands with the cage windows and rotate the race out of the cage **(see illustration)**.

Inspection

Refer to illustrations 22.12a and 22.12b
12   Clean the components with solvent to remove all traces of grease. Inspect the cage and races for pitting, score marks, cracks

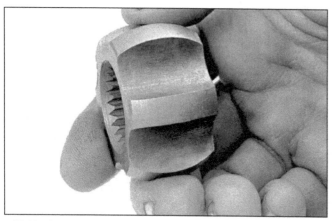

22.12a Inspect the inner race lands and grooves for pitting and score marks

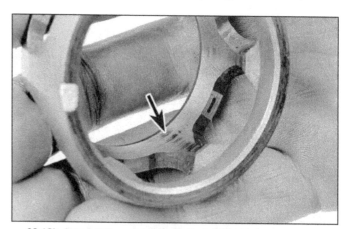

22.12b Inspect the cage for cracks, pitting and score marks (shiny spots are normal and don't affect operation)

22.14 Press the balls into the cage through the windows

22.16 Wrap the splined area of the axle with tape to prevent damage to the boot

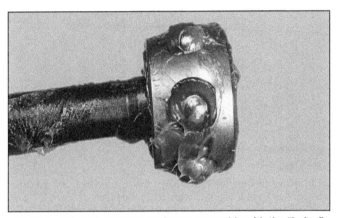

22.17 Install the inner race and cage assembly with the "bulge" facing the end of the axleshaft

22.20 Pack grease into the bearing until it's completely full

and other signs of wear and damage. Shiny, polished spots are normal and will not adversely affect CV joint performance **(see illustrations)**.

Reassembly

Refer to illustrations 22.14, 22.16, 22.17, 22.20, 22.23, 22.24a and 22.24b

13   Insert the inner race into the cage. Verify that the matchmarks are on the same side. However, it's not necessary for them to be in direct alignment with each other.

14   Press the balls into the cage windows with your thumbs **(see illustration)**.

15   Wrap the axleshaft splines with tape to avoid damaging the boot

16   Slide the small boot clamp and boot onto the axleshaft, then remove the tape **(see illustration)**.

17   Install the inner race and cage assembly on the axleshaft with the larger diameter side or "bulge" of the cage facing the axleshaft end **(see illustration)**.

18   Install the snap-ring **(see illustration 22.7)**.

19   Fill the outer race with CV joint grease (normally included with the new boot kit).

20   Pack the inner race and cage assembly with grease, by hand, until grease is worked completely into the assembly **(see illustration)**.

22.23 Equalize the pressure inside the boot by inserting a small screwdriver between the boot and the outer race

21   Slide the outer race down onto the inner race and install the wire ring retainer.

22   Wipe any excess grease from the axle boot groove on the outer race. Seat the small diameter of the boot in the recessed area on the axleshaft and install the clamp. Push the other end of the boot onto the outer race and move the race in-or-out to adjust the joint to the proper length, as listed in this Chapter's

22.24a To install the new clamps, bend the tang down . . .

Specifications. **Note:** *The joint length is measured from the bottom (inner end) of the outer race to the edge of the small diameter of the boot.*

23   With the joint set to the proper length, equalize the pressure in the boot by inserting a dull screwdriver between the boot and the outer race **(see illustration)**. Don't damage the boot with the tool.

24   Install the boot clamps, bend the tangs

over and secure them in place **(see illustrations)**.

25   Install the driveaxle (see Section 21).

Outer CV joint and boot

Refer to illustration 22.30

Note: *The outer CV joint is a non-serviceable item and is permanently retained to the driveaxle. If any damage or excessive wear occurs to the axle or the outer CV joint, the entire driveaxle assembly must be replaced (excluding the inner CV joint). Service to the outer CV joints is limited to boot replacement and grease repacking only.*

26   Remove the driveaxle from the vehicle (see Section 21).

27   Mount the driveaxle in a vise. The jaws of the vise should be lined with wood or rags to prevent damage to the axleshaft.

28   Cut the boot clamps from both inner and outer boots and discard them **(see illustration 22.3)**.

29   Remove the inner CV joint and boot (see Steps 4 through 11).

30   Remove the outer CV joint boot. Wash the outer CV joint assembly in solvent and inspect it as described in Step 12 **(see illustration)**. Replace the axle assembly if any CV joint components are excessively worn. Install the new, outer boot and clamps onto the axleshaft **(see illustration 22.16)**.

31   Repack the outer CV joint with CV joint grease and spread grease inside the new boot as well.

32   Position the outer boot on the CV joint and install new boot clamps **(see illustrations 22.24a and 22.24b)**.

33   Reassemble the inner CV joint and boot (see Steps 13 through 24).

34   Install the driveaxle (see Section 21).

22.24b . . . and tap the tabs down to hold it in place

22.30 After the old grease has been rinsed away and the cleaning solvent has been blown out with compressed air, rotate the outer joint housing through its full range of motion and inspect the bearing surfaces for wear or damage - if any of the balls, the race or cage look damaged, replace the driveaxle and outer joint

Chapter 9 Brakes

Contents

Specifications

General

Brake fluid type	See Chapter 1
Brake booster flange face-to-pushrod end clearance	0.717 inch
Brake light switch plunger-to-brake pedal clearance	0.020 to 0.040 inch

Disc brakes

Minimum brake pad thickness	See Chapter 1
Disc minimum thickness	Refer to the dimension marked on the disc
Lateral runout	0.005 inch
Parallelism	0.0004 inch

Drum brakes

Minimum brake shoe lining thickness	See Chapter 1
Brake drum maximum diameter	Refer to the dimension marked on the drum

Torque specifications

Ft-lbs (unless otherwise indicated)

Brake caliper pins	
Front	
1995 V6 models, all 1996 and later models	54
All others	24
Rear	32
Caliper support bracket bolts	
Front	115
Rear	77
Caliper banjo bolt	26
Master cylinder mounting nuts	120 in-lbs
Power brake booster mounting nuts	16
Brake rotor-to-hub bolts	76
Wheel cylinder mounting bolts	72 to 104 in-lbs

2.3 Before removing the caliper, use a large C-clamp to depress the piston back into the caliper (so there will be room for the new pads)

2.4a Wash down the brake assembly with brake cleaner - do NOT use compressed air!

1   General information

The vehicles covered by this manual are equipped with hydraulically operated front and rear brake systems. The front brakes are disc-type. The rear brakes are either drums or discs.

These models are equipped with a dual master cylinder which allows the operation of half of the system if the other half fails. This system also incorporates a proportioning bypass valve which limits pressure to the rear brakes under heavy braking to prevent rear wheel lock-up.

All models are equipped with a power brake booster which utilizes engine vacuum to assist in application of the brakes.

The parking brake system uses a parking brake lever to apply the rear brakes through cables to each rear brake. On models with rear drum brakes, the cables are attached to a lever that actuates the rear brake shoes; on models with rear disc brakes, they're attached to special parking brake shoe assemblies housed inside the discs.

There are some notes and cautions involving the brake system on this vehicle:

a)   Use only DOT 3 brake fluid in this system.

b)   The brake pads and linings may contain asbestos fibers which are hazardous to your health if inhaled. Whenever you work on the brake system components, wear an approved filtering mask and carefully clean all parts with brake cleaner. Do not allow the fine asbestos dust to become airborne.

c)   Safety should be paramount whenever any servicing of the brake components is performed. Do not use parts or fasteners which are not in perfect condition, and be sure that all clearances and torque specifications are adhered to. If you are at all unsure about a certain procedure, seek professional advice. Upon

2.4b To detach the caliper from the support bracket, remove these two bolts (wrench on lower bolt, arrow on upper bolt)

completion of any brake system work, test the brakes carefully in a controlled area before putting the vehicle into normal service. If a problem is suspected in the brake system, do not drive the vehicle until the fault is corrected.

d)   Tires, load and front end alignment are factors which also affect braking performance.

2   Disc brake pads - replacement

Refer to illustration 2.3

Warning: Disc brake pads must be replaced on both wheels at the same time - never replace the pads on only one wheel. Also, brake system dust may contain asbestos, which is hazardous to your health. DO NOT blow it out with compressed air and DO NOT inhale it. An approved filtering mask should be worn when working on the brakes. DO NOT use gasoline or solvents to remove the dust. Use brake system cleaner only!

1   Loosen the front wheel lug nuts, raise

2.4c Remove the caliper . . .

the front of the vehicle and support it securely on jackstands. Apply the parking brake. Remove the wheels.

2   Remove about two-thirds of the fluid from the master cylinder reservoir and discard it; as the pistons are pushed in for clearance to allow the pads to be removed, the fluid will be forced back into the reservoir. Position a drain pan under the brake assembly and clean the caliper and surrounding area with brake system cleaner.

3   Push the piston back into the bore with a C-clamp to provide room for the new brake pads **(see illustration)**. As the piston is depressed to the bottom of the caliper bore, the fluid in the master cylinder will rise. Make sure it doesn't overflow. If necessary, siphon off some of the fluid.

Front brake

Refer to illustrations 2.4a through 2.4t

4   To replace the brake pads, follow the accompanying photos, beginning with **illustration 2.4a**. Be sure to stay in order and read the caption under each illustration. Work on one brake assembly at a time so that you'll have something to refer to if you get in trouble.

2.4d . . . and hang it out of the way with a piece of wire; do **NOT** hang the caliper by the brake hose

2.4e Remove the shims from the outer brake pad

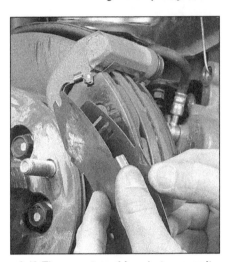

2.4f There are two shims, but you can't separate them until they're removed from the pad

2.4g Remove the outer brake pad

2.4h Remove the two inner shims

2.4i Remove the inner brake pad

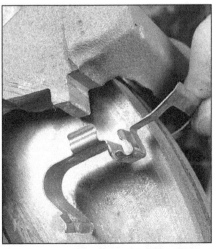

2.4j Remove the upper anti-rattle clip and inspect it for signs of wear. If it's in good shape, it's okay to reuse it; if not, replace it. Make sure the clip is fully seated into the support bracket

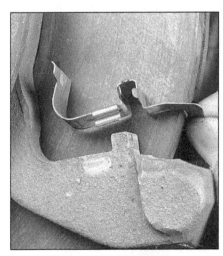

2.4k Do the same for the lower anti-rattle clip

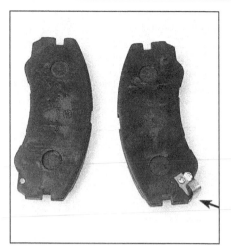

2.4l The inner pad has a wear indicator (arrow) riveted to it

2.4m With the anti-rattle clips installed, place the inner brake pad in position . . .

2.4n . . . and install the inner pad shims

2.4o Install the outer pad . . .

2.4p . . . and the outer pad shims

2.4q Remove each caliper pin dust boot and inspect it for cracks and tears; if either boot is damaged, replace it

5   While the pads are removed, inspect the caliper for brake fluid leaks and ruptures in the piston boot. Overhaul or replace the caliper as necessary (see Section 3). Also inspect the brake disc carefully (see Sec-

tion 4). If machining is necessary, follow the information in that Section to remove the disc.

6   Before installing the caliper mounting bolts, clean them and check them for corro-

sion and damage. If they're significantly corroded or damaged, replace them. Be sure to tighten the caliper mounting bolts to the

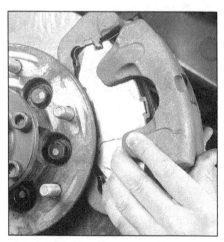

2.4r Install the caliper assembly over the pads and onto the support bracket; make sure the caliper is fully seated

2.4s Lubricate the caliper pins with high-temperature grease . . .

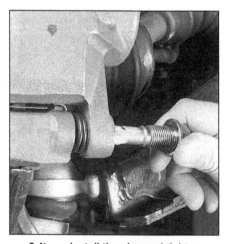

2.4t . . . install the pins and tighten them to the torque listed in this Chapter's Specifications

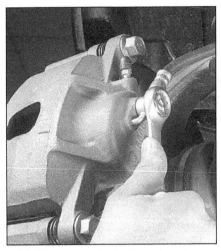

3.3 Remove the brake hose-to-caliper banjo bolt

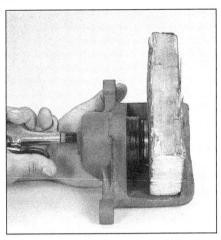

3.6 With a block of wood to cushion the piston, use compressed air to ease the piston out of the caliper (a very small burst of air is sufficient)

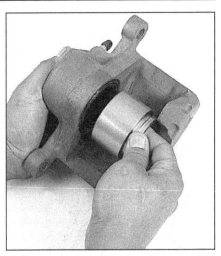

3.7 Remove the piston from the caliper

3.8 Remove the dust boot retainer ring

3.9 Remove the piston dust boot

torque listed in this Chapter's Specifications.

7   Install the brake pads on the opposite wheel, then install the wheels and lower the vehicle. Tighten the lug nuts to the torque listed in the Chapter 1 Specifications.

8   Add brake fluid to the reservoir until it's full (see Chapter 1). Pump the brakes several times to seat the pads against the disc, then check the fluid level again.

9   Check the operation of the brakes before driving the vehicle in traffic. Try to avoid heavy brake applications until the brakes have been applied lightly several times to seat the pads.

Rear brake

10   The rear brake pad procedure is virtually identical to the front pad procedure.

All models

11   Firmly depress the brake pedal a few times to bring the pads into contact with the disc. Check the fluid level in the master cylinder, topping it up if necessary.

12   Road test the vehicle carefully before placing it into normal use.

---

3   Disc brake caliper - removal, overhaul and installation

Warning: *Dust created by the brake system may contain asbestos, which is harmful to your health. Never blow it out with compressed air and don't inhale any of it. An approved filtering mask should be worn when working on the brakes. Do not, under any circumstances, use petroleum-based solvents to clean brake parts. Use brake system cleaner only!*

Note 1: *If an overhaul is indicated (usually because of fluid leakage) explore all options before beginning the job. New and factory rebuilt calipers are available on an exchange basis, which makes this job quite easy. If it's*

decided to rebuild the calipers, make sure that a rebuild kit is available before proceeding. Always rebuild the calipers in pairs - never rebuild just one of them.

Note 2: *This procedure applies to the front and rear calipers.*

Removal

Refer to illustration 3.3

1   Remove the cap from the brake fluid reservoir, siphon off two-thirds of the fluid into a container and discard it.

2   Loosen the wheel lug nuts, raise the front of the vehicle and support it securely on jackstands. Remove the wheels.

3   Remove the brake hose-to-caliper banjo bolt **(see illustration)** and detach the hose. Have a rag handy to catch spilled fluid and wrap a plastic bag tightly around the end of the hose to prevent fluid loss and contamination. **Note:** *Don't detach the hose if you are removing the caliper just for access to other components.*

4   Remove the caliper (see Section 2; caliper removal is part of the brake pad replacement procedure).

Overhaul

Refer to illustrations 3.6 through 3.16

Note: *The illustrations accompanying the following procedure depict a typical front caliper overhaul. However, they apply to the rear caliper as well, which is simply a smaller version of the front caliper.*

5   Clean the exterior of the caliper with brake system cleaner or denatured alcohol. Never use gasoline, kerosene or petroleum-based cleaning solvents. Place the caliper on a clean workbench.

6   Position a wooden block or several shop rags in the caliper as a cushion, then use compressed air to remove the piston from the caliper **(see illustration)**. Use only enough air pressure to ease the piston out of the bore. If the piston is blown out, even with the cushion in place, it may be damaged.

Warning: *Never place your fingers in front of the piston in an attempt to catch or protect it when applying compressed air, as serious injury could occur.*

7   Remove the piston from the caliper **(see illustration)**.

8   Remove the dust boot retainer ring **(see illustration)**.

9   Remove the dust boot **(see illustration)**. Discard the boot; it must be replaced.

3.10 To remove the seal from the caliper bore, use a plastic or wooden tool, such as a pencil

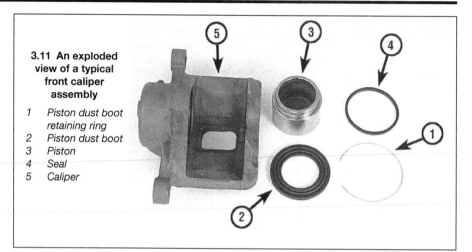

3.11 An exploded view of a typical front caliper assembly

1   *Piston dust boot retaining ring*
2   *Piston dust boot*
3   *Piston*
4   *Seal*
5   *Caliper*

10   Using a wood or plastic tool (metal tools may cause bore damage), remove the piston seal from the groove in the caliper bore **(see illustration)**. Discard the seal.

11   Clean all components in brake cleaner or clean brake fluid, blow them dry with compressed air and lay them out for inspection **(see illustration)**. Inspect the cylinder bore and piston for signs of wear, corrosion or surface defects such as scoring. If either part is damaged, replace the caliper assembly.

12   Lubricate the piston seal with brake lube or brake fluid, install it in the groove in the caliper bore, then lubricate the seal and the bore **(see illustration)**.

13   Install the new dust boot on the end of the piston **(see illustration)**.

14   Lubricate the piston with brake lube or clean brake fluid, then carefully insert the piston into the caliper using only finger pressure **(see illustration)**.

15   Seat the dust seal in the caliper **(see illustration)**.

16   Install the dust boot retainer ring **(see illustration)**.

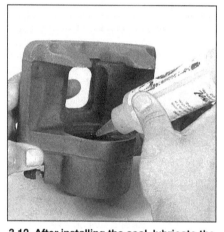

3.12 After installing the seal, lubricate the walls of the piston bore with brake assembly fluid or clean brake fluid

Installation

17   Install the caliper (see Section 3) and tighten the caliper pins to the torque listed in this Chapter's Specifications.

18   Install the brake hose-to-caliper banjo

3.13 Install the new dust boot onto the outer end of the piston (the inner lip seats into the groove in the piston)

bolt and tighten it to the torque listed in this Chapter's Specifications. Make sure the hose does not interfere with any suspension or steering components.

19   Install the wheel and lug nuts.

20   Bleed the brake system (see Section 9).

3.14 Position the piston square to the caliper bore, then press it all the way in with your fingers

3.15 Seat the outer bead of the new dust boot into its groove in the caliper housing

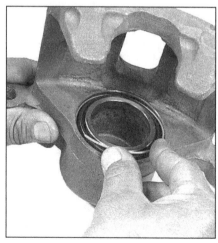

3.16 Install the dust boot retainer ring

4.4a Check disc runout with a dial indicator positioned about 1/2-inch from the edge of the disc - if the reading exceeds the maximum allowable runout, the disc will have to be resurfaced or replaced

4.4b Using a swirling motion, remove the glaze from the disc with sandpaper or emery cloth

21 Lower the vehicle to the ground and tighten the lug nuts to the torque listed in this Chapter's Specifications. Test the brakes carefully before placing the vehicle into normal operation.

4 Brake disc - inspection, removal and installation

Inspection

Refer to illustrations 4.4a, 4.4b and 4.5

1 Loosen the wheel lug nuts, raise the vehicle and support it securely on jackstands. Remove the wheel.

2 Remove the brake caliper as outlined in Sections 2 and 3. It's not necessary to disconnect the brake hose. After removing the caliper bolt, suspend the caliper out of the way with a piece of wire **(see illustration 2.4d)**. Don't let the caliper hang by the hose and don't

stretch or twist the hose.

3 Visually check the disc surface for score marks and other damage. Light scratches and shallow grooves are normal after use and may not always be detrimental to brake operation, but deep score marks - over 0.015-inch (0.38 mm) - require disc removal and refinishing by an automotive machine shop. Be sure to check both sides of the disc. If pulsating has been noticed during application of the brakes, suspect excessive disc runout.

4 To check disc runout, place a dial indicator at a point about 1/2-inch from the outer edge of the disc **(see illustration)**. Set the indicator to zero and turn the disc. The indicator reading should not exceed the specified allowable runout limit. If it does, the disc should be refinished by an automotive machine shop. **Note:** *Professionals recommend resurfacing of brake discs regardless of the dial indicator reading (to produce a smooth, flat surface, that will eliminate brake*

pedal pulsations and other undesirable symptoms related to questionable discs). At the very least, if you elect not to have the discs resurfaced, deglaze the brake pad surface with emery cloth or sandpaper (use a swirling motion to ensure a non-directional finish) **(see illustration)**.

5 The disc must not be machined to a thickness less than the specified minimum refinish thickness. The minimum wear thickness is cast into the disc. The disc thickness can be checked with a micrometer **(see illustration)**.

Removal

Refer to illustrations 4.6a, 4.6b and 4.8

6 Remove the two caliper support bracket bolts **(see illustrations)** and lift off the bracket.

7 On front disc brakes, remove the front hub/disc assembly (see Chapter 1 for 2WD models or Chapter 8 for 4WD models). The

4.5 Use a micrometer to measure the thickness of the disc

4.6a To remove the front caliper support bracket, remove these bolts (arrows)

4.6b To remove the rear caliper support bracket, remove these bolts (arrows)

4.8 Removing the rear disc; note the minimum thickness dimension cast into the disc

5.4a To unlock the front brake shoe hold-down spring, position the tool as shown, push in, turn the tool 90-degrees in either direction, then release pressure

5.4b Remove the rear brake shoe hold-down spring, washers and pin

5.4c Unhook the lower return spring from the front brake shoe

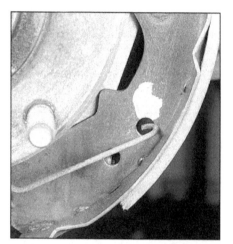

5.4d Unhook the lower return spring from the rear brake shoe

disc can then be unbolted from the hub.

8   On rear disc brakes, remove the disc **(see illustration)**. If the disc is stuck to the flange, screw a bolt into the threaded hole in the hub to break the disc loose.

Installation

9   Install the front disc to the hub and tighten the bolts in a criss-cross pattern to the torque listed in the Chapter 1 (2WD) or Chapter 8 (4WD) Specifications. To install the rear disc, simply slide it over the wheel studs onto the axleshaft flange and secure it with the small retaining screw.

10   Install the front disc and hub assembly and adjust the wheel bearing (2WD, see Chapter 1; 4WD, see Chapter 8).

11   Install the caliper support bracket and tighten the bracket bolts to the torque listed in this Chapter's Specifications. Install the pads and caliper (see Section 2) and tighten the caliper pins to the torque listed in this Chapter's Specifications.

12   Install the wheel, then lower the vehicle

to the ground. Depress the brake pedal a few times to bring the brake pads into contact with the disc. Bleeding of the system will not be necessary unless the brake hose was disconnected from the caliper. Check the operation of the brakes carefully before placing the vehicle into normal service.

5   Drum brake shoes - replacement

Refer to illustrations 5.4a through 5.4ff

Warning: *Brake shoes must be replaced on both wheels at the same time - never replace the shoes on only one wheel. Also, brake system dust often contains asbestos, which is hazardous to your health. DO NOT blow it out with compressed air and DO NOT inhale it. An approved filtering mask should be worn when working on the brakes. DO NOT use gasoline or solvent to remove the dust. Use brake system cleaner only.*

Caution: *Whenever the brake shoes are replaced, the retracting and hold-down*

springs should also be replaced. Due to the continuous heating/cooling cycle the springs are subjected to, they lose their tension over a period of time and may allow the shoes to drag on the drum and wear at a much faster rate than normal.

1   Loosen the wheel lug nuts, raise the rear of the vehicle and support it securely on jackstands. Block the front wheels to keep the vehicle from rolling off the stands. Remove the rear wheels.

2   Remove the brake drum. If the drum is stuck because of corrosion between the axle flange and the wheel studs or brake drum, spray penetrating oil around the flange and studs and allow it to soak in. Tap around the studs with a hammer and flange to break the drum loose, then tap around the back edge of the drum to remove it.

3   If the drum is locked onto the shoes because of excessive drum wear, knock out the plug in the access hole in the backing plate **(see illustration 6.2)**, insert a screwdriver through the hole, push the actuator lever off the star adjuster and, using another

5.4e Unhook the adjuster spring from the front brake shoe

5.4f Unhook the adjuster spring from the adjuster lever

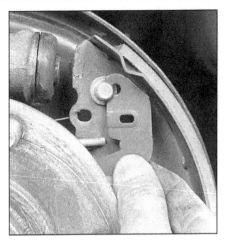

5.4g Remove the adjuster lever

5.4h Unhook the upper return spring from the rear brake shoe; the front end of the upper return spring is hooked through a hole in the front shoe

5.4i Detach the parking brake cable from the parking brake lever

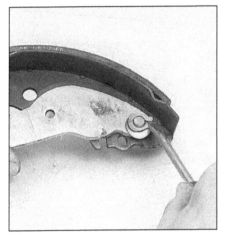

5.4j Pry off the C-clip from the pin on the backside of the rear parking brake shoe . . .

screwdriver or a brake adjuster tool, back off the adjuster wheel to retract the shoes. **Note:** *If the plug is already knocked out, make sure it's not lodged somewhere inside the brake*

assembly. Be sure to install a plug (available at most auto parts stores) in the hole when you're done to prevent water from entering the brake assembly.

4   Clean the brake assembly with brake system cleaner before beginning work. Fol-

low **illustrations 5.4a through 5.4ff** for the inspection and replacement of the brake shoes. Be sure to stay in order and read the caption under each illustration.

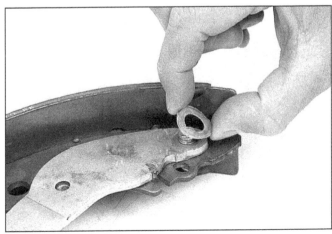

5.4k . . . remove the washer . . .

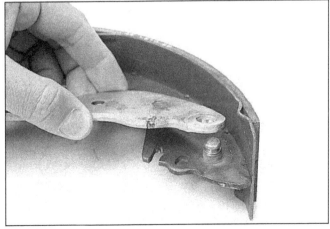

5.4l . . . and the parking brake lever . . .

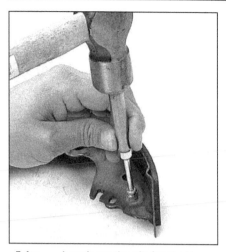

5.4m . . . knock out the pin for the lever, then install the pin, the parking brake lever, the washer and the C-clip on the new rear shoe

5.4n Clean the adjusting screw with solvent. Dry it off and lubricate the threads and end with multi-purpose grease

5.4o Lubricate the friction surfaces of the brake backing plate with high-temperature grease

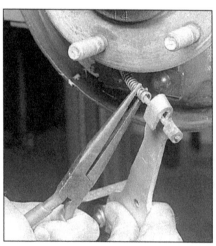

5.4p Reattach the parking brake cable to the parking brake lever

5.4q Place the rear shoe and parking brake lever in position and install the adjuster

5.4r Place the front shoe in position and engage it with the adjuster

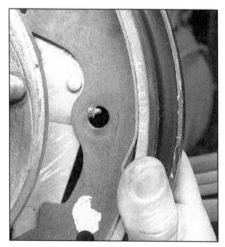

5.4s Insert the pin through the backing plate and through the hole in the rear parking brake shoe . . .

5.4t . . . install the lower washer over the pin . . .

5.4u . . . and install the spring and the upper washer, and lock them into place with a brake spring installer/remover tool (see illustration 5.4b)

5.4v Make sure the adjuster is engaged with the rear shoe as shown

5.4w Hook the upper return spring through its hole in the rear shoe as shown . . .

5.4x . . . then hook the other end of the spring through its hole in the front shoe; note that both ends of the spring enter their respective holes from the back side of the shoes and that the coil end of the spring is oriented toward the front shoe

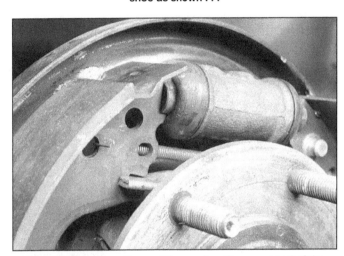

5.4y Before proceeding, pull forward on the upper end of the front shoe and engage the upper end of the shoe with the notch in the end of the front pushrod protruding from the wheel cylinder; also make sure that the front shoe is correctly engaged with the front end of the adjuster mechanism as shown

5.4z Install the pin, washers and hold-down spring for the front shoe and lock them into place with the brake spring installer tool

5.4aa Install the adjuster lever

5.4bb Make sure that the adjuster lever is properly engaged with the rear end of the adjuster as shown, then hook the long end of the adjuster spring through the hole in the lever as shown

5.4cc Hook the other end of the adjuster spring through its hole in the front shoe

5.4dd Hook the lower return spring through its hole in the rear brake shoe

5.4ee Hook the other end of the lower return spring through its hole in the front brake shoe

5   Clean the brake drum and check it for score marks, deep grooves, hard spots (which will appear as small discolored areas) and cracks. If the drum is worn, scored or out-of-round, it can be resurfaced by an automotive machine shop. **Note:** *Professionals recommend resurfacing the drums whenever a brake job is done. Resurfacing will eliminate the possibility of out-of-round drums. If the drums are worn so much they can't be resurfaced without exceeding the maximum allowable diameter (stamped into the drum), new ones will be required. At the very least, if you elect not to have the drums resurfaced, remove the glazing from the surface with emery cloth or sandpaper using a swirling motion.*
6   Repeat this procedure for the other rear brake assembly.
7   Install the brake drums. Pump the brake several times, then turn the adjuster star wheels using a screwdriver inserted through the hole in the backing plate until the shoes slightly drag on the drums as the drums are turned. Now, back off the adjuster until the shoes don't drag on the drums.

8   Install the rear wheels, install the lug nuts, lower the vehicle and tighten the wheel lug nuts to the torque listed in the Chapter 1 Specifications.
9   Check the brake pedal position. If the brake pedal goes too close to the floor, further adjustment of the brakes is required. Back the vehicle up, making repeated stops, to actuate the self-adjusters, which work only when the vehicle is in reverse. Test the brakes for proper operation before driving in traffic.

6   Wheel cylinder - removal, overhaul and installation

Note: *If an overhaul is indicated (usually because of fluid leakage or sticky operation) explore all options before beginning the job. New wheel cylinders are available, which makes this job quite easy. If you decide to rebuild the wheel cylinder, make sure a rebuild kit is available before proceeding. Never overhaul only one wheel cylinder. Always rebuild both of them at the same time.*

Removal

Refer to illustration 6.2
1   Remove the brake drum and brake shoes (see Section 6).
2   Unscrew the brake line fitting from the rear of the wheel cylinder **(see illustration)**. If available, use a flare-nut wrench to avoid rounding off the corners on the fitting. Don't pull the metal line out of the wheel cylinder - it could bend, making installation difficult.
3   Remove the two bolts securing the wheel cylinder to the brake backing plate.
4   Remove the wheel cylinder.
5   Plug the end of the brake line to prevent the loss of brake fluid and the entry of dirt.

Overhaul

Refer to illustration 6.6
6   To disassemble the wheel cylinder, remove the rubber boot from each end of the cylinder, push out the two pistons and cups, and remove the cup return spring **(see illustration)**. Discard the rubber parts and use new ones from the rebuild kit when reassem-

5.4ff And that's it! Here's how your completed rear brake assembly should look. Repeat the procedure on the other brake

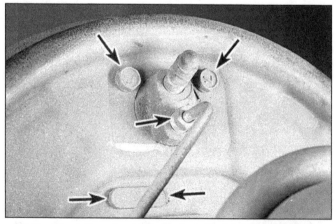

6.2 To detach the wheel cylinder from the brake backing plate, unscrew the tube nut (center arrow) with a flare-nut wrench, then remove the two wheel cylinder retaining bolts (upper arrows); to access the brake shoe adjuster wheel, remove the oblong plug (lower arrows) below the wheel cylinder

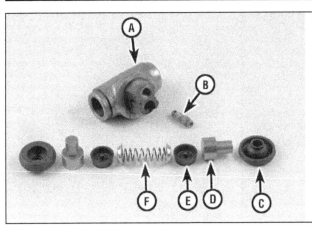

6.6 An exploded view of a typical wheel cylinder assembly

A   Wheel cylinder body
B   Bleeder screw
C   Boot
D   Piston
E   Cup
F   Spring

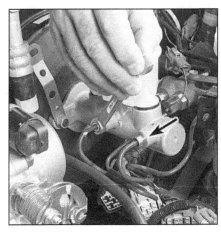

7.4 To disconnect the brake lines, unscrew the tube nut fittings with a flare nut wrench

bling the wheel cylinder.

7   Inspect the pistons for scoring and scuff marks. If present, the pistons should be replaced with new ones.

8   Examine the inside of the cylinder bore for score marks and corrosion. If these conditions exist, the cylinder can be honed slightly to restore it, but replacement is recommended.

9   If the cylinder is in good condition, clean it with brake system cleaner or denatured alcohol. **Warning:** *DO NOT, under any circumstances, use gasoline or petroleum-based solvents to clean brake parts!*

10   Remove the bleeder screw and make sure the hole is clean.

11   Lubricate the cylinder bore with clean brake fluid, then insert one of the new rubber cups into the bore. Make sure the lip on the rubber cup faces in.

12   Place the spring in the opposite end of the bore and push it in until it contacts the rear of the rubber cup.

13   Install the remaining cup in the cylinder bore.

14   Attach the rubber boots to the pistons, then install the pistons and boots.

15   The wheel cylinder is now ready for installation.

Installation

16   Installation is the reverse of removal.

Attach the brake line to the wheel cylinder *before* installing the wheel cylinder mounting bolts, but don't *tighten* the tube nut fitting until *after* the wheel cylinder mounting bolts have been tightened. If available, use a flare-nut wrench to tighten the tube nut fitting. Be sure to tighten the wheel cylinder mounting bolts to the torque listed in this Chapter's Specifications.

17   Bleed the brakes (see Section 9). Don't drive the vehicle in traffic until the operation of the brakes has been thoroughly tested.

7   Master cylinder - removal, overhaul and installation

Removal

Refer to illustrations 7.4 and 7.6

Note: *Before deciding to overhaul the master cylinder, check on the availability and cost of a new or factory rebuilt unit and also the availability of a rebuild kit.*

1   The master cylinder is mounted on the power brake booster.

2   Remove as much fluid as you can from the reservoir with a syringe.

3   Place rags under the fluid fittings and prepare caps or plastic bags to cover the ends of the lines once they are disconnected. **Caution:** *Brake fluid will damage paint. Cover*

all body parts and be careful not to spill fluid during this procedure.

4   Loosen the tube nuts at the ends of the brake lines where they enter the master cylinder **(see illustration)**. To prevent rounding off the flats on these nuts, the use of a flare nut wrench, which wraps around the nut, is preferred.

5   Pull the brake lines slightly away from the master cylinder and plug the ends to prevent contamination.

6   Remove the master cylinder mounting nuts **(see illustration)** and remove the master cylinder.

Overhaul

Refer to illustrations 7.8a, 7.8b, 7.8c, 7.9a, 7.9b, 7.10a, 7.10b, 7.11, 7.13a, 7.13b, 7.14a, 7.14b, 7.15a, 7.15b, 7.15c, 7.16a, 7.16b and 7.16c

7   Before attempting the overhaul of the master cylinder, obtain the proper rebuild kit, which will contain the necessary replacement parts and also any instructions which may be specific to your model.

8   Inspect the reservoir or inlet grommet(s) for indications of leakage near the base of the reservoir. Remove the reservoir and the

7.6 To detach the master cylinder from the power brake booster, remove these two nuts (arrows)

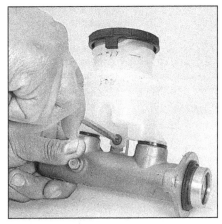

7.8a Remove the reservoir retaining screw . . .

7.8b . . . then pull off the reservoir . . .

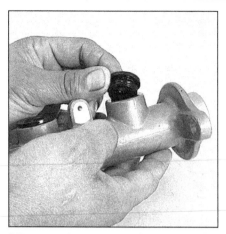

7.8c ... and remove the rubber grommets

7.9a Push in the primary piston with a large Phillips screwdriver and remove the snap-ring with snap-ring pliers ...

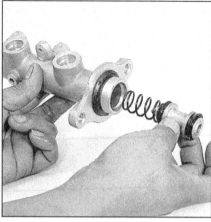

7.9b ... and remove the primary piston

grommets **(see illustrations)**.

9   Place the cylinder in a vise, use a punch or Phillips screwdriver to fully depress the primary piston and remove the snap-ring **(see illustration)**. Remove the primary piston **(see illustration)**.

10   Remove the stop bolt on the side of the master cylinder **(see illustration)**, then remove the secondary piston **(see illustration)**.

11   Wash all parts in brake system cleaner and lay them out for inspection **(see illustration)**. **Caution:** *Do not use any petroleum-based cleaners.*

12   Carefully inspect the bore of the master cylinder. Any deep scoring or other damage will mean a new master cylinder is required.

13   Replace all parts included in the rebuild kit, following any instructions in the kit **(see illustrations)**. During assembly, lubricate all parts liberally with clean brake fluid. Be sure to tighten all fittings and connections to the specified torque.

14   Push the assembled components into the bore **(see illustrations)**.

15   Depress the pistons and install the new snap-ring **(see illustration)**. Make sure it's

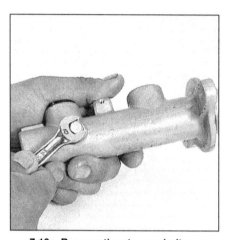

7.10a Remove the stopper bolt ...

properly seated in its groove. Place a new washer on the stop bolt, then depress the pistons again and install the stop bolt, tightening it securely **(see illustrations)**.

16   Install the new grommets and the reservoir **(see illustrations)**.

7.10b ... then remove the secondary piston; if the piston is stuck, tap the master cylinder body against a piece of wood to dislodge it

17   Before installing the new master cylinder it should be bench bled. Because it will be necessary to apply pressure to the master cylinder piston and, at the same time, control flow from the brake line outlets, it is recom-

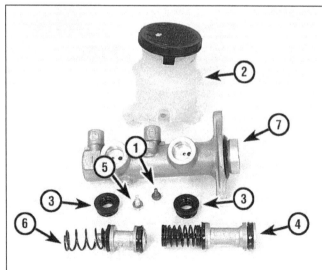

7.11 After washing all the parts in clean solvent and blowing them dry with compressed air, lay them out for inspection:

1   *Reservoir retaining screw*
2   *Reservoir and cap*
3   *Rubber grommets*
4   *Primary piston*
5   *Stopper bolt*
6   *Secondary piston*
7   *Master cylinder body*

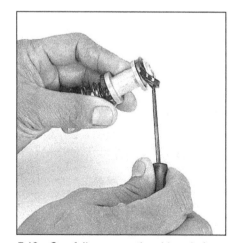

7.13a Carefully remove the old seals from the pistons, then coat the new seals with clean brake fluid and install them

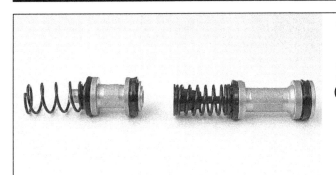

7.13b Make sure the piston seals are installed as shown, with the two seals on the secondary piston (left) facing away from each other, and the two on the primary piston (right) facing forward (toward the front of the master cylinder)

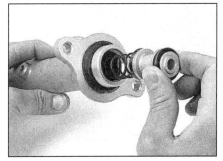

7.14a Coat the secondary piston with clean brake fluid and insert it into the master cylinder . . .

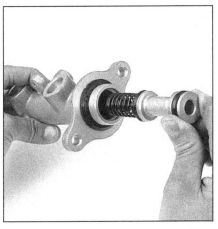

7.14b . . . coat the primary piston with clean brake fluid and install the primary piston

7.15a Depress the primary piston and install the snap-ring

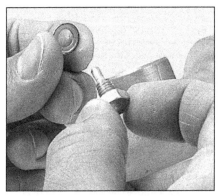

7.15b Install a new washer on the stopper bolt

mended that the master cylinder be mounted in a vise, with the jaws of the vise clamping on the mounting flange.

18   Insert threaded plugs into the brake line outlet holes and snug them down so that there will be no air leakage past them, but not so tight that they cannot be easily loosened.

19   Fill the reservoir with brake fluid of the recommended type (see Chapter 1).

20   Remove one plug and push the piston assembly into the master cylinder bore to

expel the air from the master cylinder. A large Phillips screwdriver can be used to push on the piston assembly.

21   To prevent air from being drawn back into the master cylinder the plug must be replaced and snugged down before releasing the pressure on the piston assembly.

22   Repeat the procedure until only brake fluid is expelled from the brake line outlet hole. When only brake fluid is expelled, repeat the procedure with the other outlet

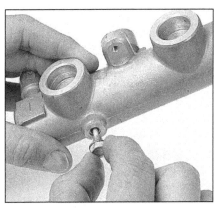

7.15c Depress the pistons again and install the stopper bolt

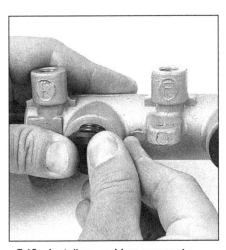

7.16a Install new rubber grommets, . . .

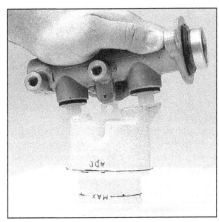

7.16b . . . lay the reservoir upside down, lubricate the grommets with clean brake fluid and push the master cylinder onto the reservoir

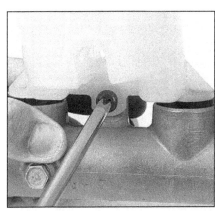

7.16c Install the reservoir retaining screw

8.4a To disconnect the brake flex hose from the metal brake line, unscrew this tube nut with a flare nut wrench . . .

8.4b . . . then pull out the large retainer clip and pull the hose through the bracket (front hose shown)

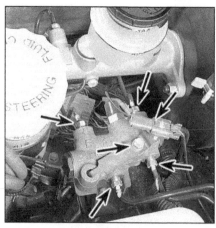

8.9 To remove the combination valve, unplug the electrical connector for the low-fluid-level warning switch, unscrew the four tube nuts for the brake lines and remove the valve retaining bolt (arrows)

hole and plug. Be sure to keep the master cylinder reservoir filled with brake fluid to prevent the introduction of air into the system.

23   Since high pressure is not involved in the bench bleeding procedure, an alternative to the removal and replacement of the plugs with each stroke of the piston assembly is available. Before pushing in on the piston assembly, remove the plug as described in Step 20. Before releasing the piston, however, instead of replacing the plug, simply put your finger tightly over the hole to keep air from being drawn back into the master cylinder. Wait several seconds for brake fluid to be drawn from the reservoir into the piston bore, then depress the piston again, removing your finger as brake fluid is expelled. Be sure to put your finger back over the hole each time before releasing the piston, and when the bleeding procedure is complete for that outlet, replace the plug and snug it before going on to the other port.

Installation

24   Install the master cylinder over the studs on the power brake booster and tighten the attaching nuts only finger tight at this time.
25   Thread the brake line fittings into the master cylinder. Since the master cylinder is still a bit loose, it can be moved slightly in order for the fittings to thread in easily. Do not strip the threads as the fittings are tightened.
26   Fully tighten the mounting nuts and the brake fittings.
27   Fill the master cylinder reservoir with fluid, then bleed the master cylinder (only if the cylinder has not been bench bled) and the brake system as described in Section 9. To bleed the cylinder on the vehicle, have an assistant pump the brake pedal several times and then hold the pedal to the floor. Loosen the fitting nut to allow air and fluid to escape. Repeat this procedure on both fittings until the fluid is clear of air bubbles. Test the operation of the brake system carefully before placing the vehicle in normal service.

8   Brake lines, hoses and combination valve - inspection and replacement

Refer to illustrations 8.4a, 8.4b and 8.6
1   About every six months the flexible hoses which connect the steel brake lines with the front and rear brakes should be inspected for cracks, chafing of the outer cover, leaks, blisters, and other damage (Chapter 1).
2   Replacement steel and flexible brake lines are commonly available from dealer parts departments and auto parts stores. Do not, under any circumstances, use anything other than genuine steel lines or approved flexible brake hoses as replacement items.
3   When installing the brake line, leave at least 3/4-inch (19 mm) clearance between the line and any moving or vibrating parts.
4   First, disconnect the metal brake line from the brake flex hose by unscrewing the tube nut fitting with a flare nut wrench **(see illustration)**. Then remove the spring clip with a pair of pliers **(see illustration)**.
5   On some models, there may be another clip and bracket. Simply remove the clip and pull the hose through the bracket.
6   The bracket and clip for the rear hose are located at a bracket on the axle tube. They're disconnected the same way as the front.
7   Steel brake lines are usually retained along their span with clips. Always remove these clips completely before removing a fixed brake line. Always reinstall these clips, or new ones if the old ones are damaged, when replacing a brake line, as they provide support and keep the lines from vibrating, which can eventually break them. When replacing brake lines be sure to use the correct parts. Don't use copper tubing for any brake system components. Purchase steel brake lines from a dealer or auto parts store. Prefabricated brake line, with the tube ends already flared and fittings installed, is avail-

able at auto parts stores and dealers. These lines are also sometimes bent to the proper shapes.
8   Remember to bleed the hydraulic system after replacing a hose or line.

Combination valve

Refer to illustration 8.9
9   The combination valve **(see illustration)**, which is located in the engine compartment, below the master cylinder, performs two functions: a proportioning valve limits hydraulic pressure to the rear brakes after a pressure threshold has been reached, and a bypass valve assures full pressure to the rear brakes in the event of a malfunction in the rear brake circuit.
10   To replace the combination valve, unplug the electrical connector for the low-fluid warning switch, unscrew the four tube nut fittings for the brake lines and remove the valve retaining bolt **(see illustration 8.9)**.

9   Brake system - bleeding

Refer to illustration 9.8
Warning: *Wear eye protection when bleeding the brake system. If the fluid comes in contact with your eyes, immediately rinse them with water and seek medical attention.*
Note: *Bleeding the hydraulic system is necessary to remove any air that manages to find its way into the system when it's been opened during removal and installation of a hose, line, caliper or master cylinder.*
1   It will probably be necessary to bleed the system at all four brakes if air has entered the system due to low fluid level, or if the brake lines have been disconnected at the master cylinder. If a brake line was disconnected only at a wheel, then only that caliper or wheel cylinder must be bled.
2   If a brake line is disconnected at a fitting

9.8 When bleeding the brakes, a hose is connected to the bleed screw at the caliper or wheel cylinder and then submerged in brake fluid - air will be seen as bubbles in the tube and container (all air must be expelled before moving to the next wheel)

10.11 To detach the power brake booster from the firewall, remove these four nuts

located between the master cylinder and any of the brakes, that part of the system served by the disconnected line must be bled.

3   Remove any residual vacuum from the brake power booster by applying the brake several times with the engine off.

4   Remove the master cylinder reservoir cover and fill the reservoir with brake fluid. Reinstall the cover. **Note:** *Check the fluid level often during the bleeding operation and add fluid as necessary to prevent the fluid level from falling low enough to allow air bubbles into the master cylinder.*

5   Have an assistant on hand, as well as a supply of new brake fluid, a clear container partially filled with clean brake fluid, a length of 3/16-inch plastic, rubber or vinyl tubing to fit over the bleeder screw and a wrench to open and close the bleeder screw.

6   If the vehicle is equipped with an Anti-lock Braking System (ABS), prime the hydraulic modulator. This is done by attaching the bleed hose to the rearward bleeder screw on the modulator unit and placing the other end of the hose into the jar partially filled with clean brake fluid. Open the bleeder screw slowly and have an assistant push down on the brake pedal until fluid flows from the hose, then close the bleeder screw. Do this until no bubbles are present in the fluid, then move the hose to the front bleeder screw and repeat the procedure.

7   Working at the right rear wheel, loosen the bleeder screw slightly, then tighten it to a point where it is snug but can still be loosened quickly and easily.

8   Place one end of the tubing over the bleeder screw and submerge the other end in brake fluid in the container **(see illustration)**.

9   Have the assistant pump the brakes slowly a few times to get pressure in the system, then hold the pedal firmly depressed.

10   While the pedal is held depressed, open the bleeder screw just enough to allow a flow of fluid to leave the valve. Watch for air bubbles to exit the submerged end of the tube. When the fluid flow slows after a couple of

seconds, close the valve and have your assistant release the pedal.

11   Repeat Steps 9 and 10 until no more air is seen leaving the tube, then tighten the bleeder screw and proceed to the left rear wheel, the right front wheel and the left front wheel, in that order, and perform the same procedure. Be sure to check the fluid in the master cylinder reservoir frequently.

12   If the vehicle is equipped with ABS, perform Step 6 once again.

13   Never use old brake fluid. It contains moisture which will lower the boiling point of the brake fluid and also deteriorate the brake system rubber components.

14   Refill the master cylinder with fluid at the end of the operation. **Warning:** *Do not operate the vehicle if you are in doubt of the effectiveness of the brake system. It is possible for air to become trapped in the hydraulic modulator of the anti-lock brake system, so, if the pedal continues to feel spongy after repeated bleedings or any warning lights on the instrument panel stay on, have the vehicle towed to a dealer service department or other qualified shop to be bled with the aid of a scan tool.*

10   Power brake booster - check, removal and installation

Operating check

1   Depress the brake pedal several times with the engine off and make sure that there is no change in the pedal reserve distance.

2   Depress the pedal and start the engine. If the pedal goes down slightly, operation is normal.

Airtightness check

3   Start the engine and turn it off after one or two minutes. Depress the brake pedal several times slowly. If the pedal goes down farther the first time but gradually rises after the second or third depression, the booster is airtight.

4   Depress the brake pedal while the engine is running, then stop the engine with the pedal depressed. If there is no change in the pedal reserve travel after holding the pedal for 30 seconds, the booster is airtight.

Removal

Refer to illustration 10.11

5   Power brake booster units should not be disassembled. They require special tools not normally found in most service stations or shops. They are fairly complex and because of their critical relationship to brake performance it is best to replace a defective booster unit with a new or rebuilt one.

6   To remove the booster, first remove the brake master cylinder (see Section 7). On some vehicles it's not necessary to disconnect the brake lines from the master cylinder; there's enough room to reposition the cylinder to allow booster removal.

7   Using a flashlight, locate the pushrod clevis connecting the booster to the top of the brake.

8   Remove the clevis pin retaining clip with pliers and pull out the pin.

9   Holding the clevis with a pair of pliers, disconnect the clevis locknut with a wrench. The clevis is now loose.

10   Disconnect the hose leading from the engine to the booster. Be careful not to damage the hose when removing it from the booster fitting.

11   Remove the four nuts and washers holding the brake booster to the firewall **(see illustration)**. Again, you will need a flashlight to find these nuts.

12   Slide the booster straight out from the firewall until the studs clear the holes and pull the booster, brackets and gaskets from the engine compartment area.

Installation

Refer to illustrations 10.14a and 10.14b

13   Installation procedures are basically the reverse of those for removal. Tighten the cle-

10.14a To measure the distance that the pushrod protrudes from the booster, apply about 20 in-Hg of vacuum with a hand-operated vacuum pump, measure the distance from the flange face of the booster body to the tip of the pushrod, then compare your measurement to the specified dimension (see Specifications)

10.14b To adjust the pushrod length, hold the serrated portion of the rod with a pair of pliers and turn the adjusting screw in or out, as necessary, to achieve the specified dimension

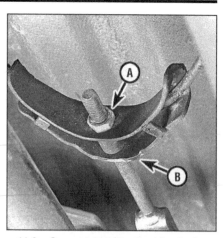

11.2a On models with a parking brake handle, you'll find the adjusting nut (A) at the equalizer under the vehicle; before trying to turn the adjusting nut, be sure to loosen the locknut (B) first

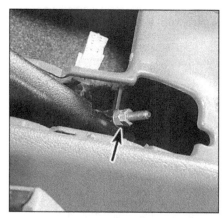

11.2b On models with a parking brake lever in the center console, you'll find the adjusting nut (arrow) on the lever itself (a plastic trim piece must be removed from the console first); to adjust the cable, loosen the locknut (the upper nut), then turn the adjusting nut (lower nut) in or out as necessary

vis locknut securely and the booster mounting nuts to the torque listed in this Chapter's Specifications.

14   If the power booster unit is being replaced, the pushrod protrusion must be checked and, if necessary, adjusted. Connect a hand-held vacuum pump to the fitting on the booster and apply 20 in-Hg of vacuum. Measure the distance from the master cylinder flange face on the power booster to the end of the pushrod **(see illustration)** and compare your measurement with the value listed in this Chapter's Specifications. Turn the end of the pushrod in or out to achieve the desired setting **(see illustration)**.

15   Install the master cylinder.

16   After the final installation of the master cylinder and brake hoses and lines, the brake pedal height and freeplay must be adjusted (see Chapter 1) and the system must be bled (see Section 9).

11 Parking brake cable - adjustment

Refer to illustrations 11.2a and 11.2b

1   The parking brake system should prevent the vehicle from rolling when the parking brake handle is pulled back 9 to 11 clicks (models with a parking brake handle under the dash) or pulled up 6 clicks (models with a parking brake lever in the center console). If it doesn't, adjust the cable.

2   Locate the cable adjusting nut for your particular model. On models with a parking brake handle under the dash, the adjusting nut is located under the vehicle, at the cable equalizer (where the front and rear cables meet) **(see illustration)**. On models with a parking brake lever in the console, the adjusting nut is located on the lever itself **(see illustration)**. To get at it, remove the plastic trim

piece from the console (see *Center console - removal and installation* in Chapter 11).

3   Turn the adjusting nut to remove all slack from the parking brake cables. After adjustment apply the parking brake several times and verify that handle or lever travel is as specified. Make sure the rear brakes don't drag when the parking brake is released and verify that the parking brake light on the dash glows when the parking brake handle or lever is applied.

4   Lower the vehicle and verify that the parking brake will hold the vehicle on a moderate incline. If it still won't keep the vehicle from rolling, inspect the rear brakes as described in Chapter 1 (models with drum brakes) or check the parking brake shoes as described in Section 13.

12 Parking brake cable(s) - replacement

1   On models with the parking brake lever in the center console, loosen the adjusting nut (see Section 11).

2   Raise the vehicle and place it securely on jackstands.

Models with a parking brake handle under the dash

Front lower cable

3   On models with the parking brake handle under the dash, loosen the locknut, then completely unscrew the adjusting nut and disconnect the equalizer from the rear relay lever assembly.

4   Remove the cotter pin, plain washer, waved washer and pin that secure the front lower cable to the rear relay lever assembly.

5   Disconnect the front lower cable from the front relay lever assembly and remove the cable.

6   Installation is the reverse of the removal procedure.

7   Following installation, adjust the parking brake (see Section 11).

Intermediate cable and rear cable assembly

8   If necessary for clearance, raise the vehicle at the front and support it on jackstands.

9   Remove the cotter pins, plain washers, waved washers and pins that secure the rear cables to the rear wheel.

10   Loosen the locknut, then completely unscrew the adjusting nut securing the equalizer bracket to the rear relay lever assembly.

11   Remove the intermediate cable and cable guides from the frame brackets.

12   Remove the clips and bolts attaching the intermediate cable and rear cable assembly to the frame brackets and remove the cables from the brackets and lift out the cable.

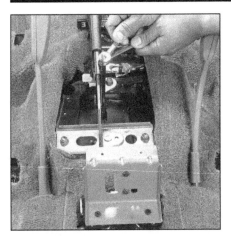

12.16 After removing the center console, disconnect the front cable from the parking brake lever

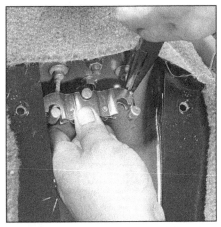

12.17 Disengage the front cable and equalizer from the rear cables

12.18 To detach the rear cable bracket from the pan, remove these two nuts (arrows)

12.19 After removing the bracket-to-floorpan nuts, go underneath the vehicle, pull off the cable bracket (arrow) and disengage the cable(s) from the bracket

12.20 To detach the rear cable from the leaf spring, remove these two bolts

12.21a On models with rear drum brakes, disengage the parking brake cable from the parking brake lever . . .

13   Installation is the reverse of the removal procedure.
14   Following installation, adjust the parking brake (see Section 11).

Models with a parking brake lever in the center console

Refer to illustrations 12.16, 12.17, 12.18, 12.19, 12.20, 12.21a, 12.21b, 12.22a and 12.22b

15   Remove the center console (see Chapter 11).
16   Disconnect the front cable from the parking brake lever **(see illustration)**.
17   Disengage the equalizer from the two rear parking brake cables **(see illustration)**.
18   Remove the two rear cable bracket retaining nuts **(see illustration)**.
19   Underneath the vehicle, remove the rear cable bracket **(see illustration)** from the pan, then disengage the cables from the bracket.
20   Detach any cable clamps from the pan, frame and/or springs **(see illustration)**.
21   On models with rear drum brakes, remove the drum and rear brake assembly

(see Section 5), disengage the parking brake cable from the parking brake lever **(see illustration)**, then detach the cable from the brake backing plate **(see illustration)** and pull the cable through the backing plate.
22   On models with rear disc brakes,

remove the caliper (see Section 3), disc/hub (see Section 4) and parking brake shoes (see Section 13), disengage the parking brake cable from the parking brake lever **(see illustration)** and detach the cable from the brake

12.21b . . . then squeeze the cable housing tangs together and pull the cable through the backing plate

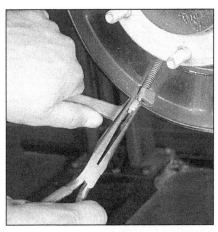

12.22a On models with rear disc brakes, disengage the parking brake cable from the parking brake lever . . .

12.22b ... then remove these two screws (arrows) and detach the cable housing from the brake backing plate

13.5a Unhook the rear shoe return spring from the anchor pin

13.5b Unhook the front shoe return spring from the anchor pin

backing plate **(see illustration)**.

23   Installation is the reverse of removal.

24   Be sure to adjust the cable when you're done (see Section 11).

---

13   Parking brake shoes - check, removal and installation

Warning: *Parking brake shoes must be replaced on both wheels at the same time - never replace the shoes on only one wheel. Also, brake system dust may contain asbestos, which is hazardous to your health. DO NOT blow it out with compressed air and DO NOT inhale it. An approved filtering mask should be worn when working on the brakes. DO NOT use gasoline or solvent to remove the dust. Use brake system cleaner only.*

Check

1   The parking brake system should be checked as a normal part of driving. With the vehicle parked on a hill, apply the brake, place the transmission in Neutral and check that the parking brake alone will hold the

13.5c Unhook the lower spring from the rear shoe

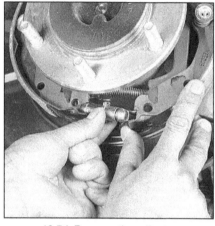

13.5d Remove the adjuster

vehicle (be sure to stay in the vehicle during this check). However, every 24 months (or whenever a fault is suspected), the assembly itself should be visually inspected.

2   With the vehicle raised and supported on jackstands, remove the rear wheels.

3   Remove the rear brake calipers and discs (see Sections 2, 3 and 4). Support the

caliper assemblies with a coat hanger or heavy wire and do not disconnect the brake line from the caliper.

4   With the disc removed, the parking brake components are visible and can be inspected for wear and damage. The linings should last the life of the vehicle. However, they can wear down if the parking brake sys-

13.5e Remove the strut

13.5f Remove the rear shoe hold-down spring, washer and pin

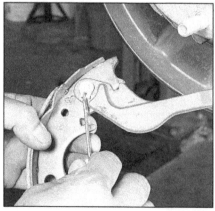

13.5g Pry off the C-clip that secures the parking brake lever to the rear shoe; don't lose the washer

13.5h Remove the front shoe hold-down spring, washer and pin

13.5i Lubricate the shoe contact points on the backing plate with high temperature grease

13.5j Attach the parking brake lever to the new rear shoe . . .

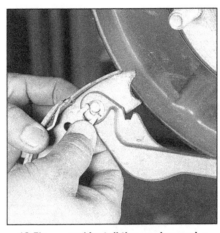

13.5k . . . and install the washer and a *new* C-clip

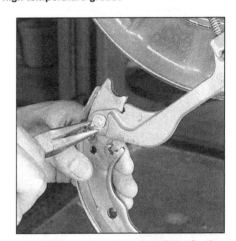

13.5l Firmly pinch the ends of the C-clip together to lock it onto the pin

tem has been improperly adjusted. There is no minimum thickness specification for the parking brake shoes, but as a rule of thumb, if the shoe material is less 1/32-inch thick, you should replace them. Also check the springs and adjuster mechanism and inspect

the drum for deep scratches and other damage.

Removal and installation

Refer to illustrations 13.5a through 13.5w

5   Loosen the wheel lug bolts, raise the

rear of the vehicle and place it securely on jackstands. Remove the rear wheels. Remove the brake discs (see Section 4) if you haven't already done so. Follow the accompanying photo sequence beginning with **illustration 13.5a**. Work on only one side at a time, so

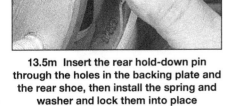

13.5m Insert the rear hold-down pin through the holes in the backing plate and the rear shoe, then install the spring and washer and lock them into place

13.5n Install the shoe guide plate on the anchor pin

13.5o Hook the rear shoe return spring over the anchor pin

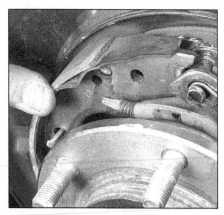

13.5p Make sure the front shoe and the strut are properly engaged as shown, then insert the pin for the front shoe hold-down spring through the holes in the backing plate and the front shoe

13.5q Install the front shoe pin, hold-down spring and washer and lock them into place

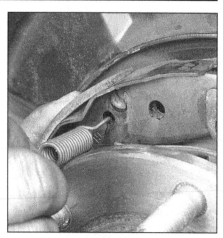

13.5r Hook the return spring into the front shoe

13.5s Hook the other end of the front shoe return spring onto the anchor pin

13.5t Hook the lower return spring onto the front shoe (note that the spring enters the hole from the backside of the shoe)

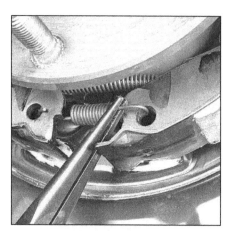

13.5u Hook the other end of the return spring into the rear shoe (note that the spring is hooked into the hole from the front side of the shoe)

you can use the other side as a reference during reassembly.

6   Installation is the reverse of removal.

7   After installing the brake disc, adjust the parking brake shoes. Temporarily install two lug bolts, turn the adjuster and expand the shoes until the disc locks, then back off the adjuster until the shoes don't drag.

8   Adjust the parking brake cable (see Section 11).

9   Remove the jackstands and lower the vehicle.

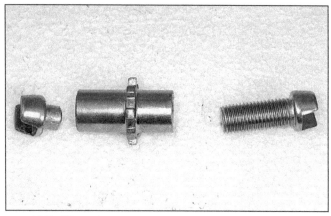

13.5v Disassemble the adjuster mechanism and clean it thoroughly, then lubricate the threads and the pivoting end with high temperature grease and reassemble it

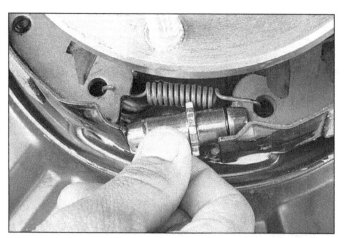

13.5w Install the adjuster mechanism between the two shoes, with the longer end pointing forward

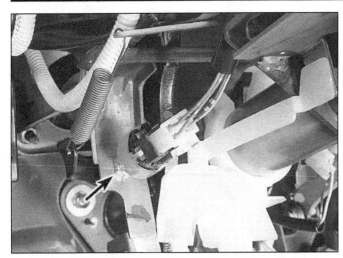

14.1 The brake light switch is located at the top of the brake pedal; to remove the switch, unplug the electrical connector, back off the locknut (arrow) and unscrew the switch

14.2 To check the brake light switch, hook up an ohmmeter to terminals 1 and 2 (two lower terminals) (switches without cruise control have only two terminals): with the pedal released, there should be infinite resistance; with the pedal applied, there should be continuity

14 Brake light switch - check and replacement

Check

Refer to illustrations 14.1 and 14.2

1   The brake light switch **(see illustration)** is mounted on a bracket at the upper end of the brake pedal. The brake light switch is normally open; when the pedal is applied, it releases a spring-loaded plunger on the front of the switch, closing the brake light circuit and turning on the brake lights. On models without cruise control, there are just two terminals; on models with cruise control, there two additional terminals for the cruise control circuit. The switch activates both circuits simultaneously on cruise-equipped models.

2   If the brake light circuit is not working properly, the following simple checks should help you identify the problem:

a)   *If both brake lights fail to come on when the brake pedal is applied, check the fuse (see Chapter 12).*

b)   *If only one light fails to come on, check that bulb (see Chapter 12).*

c)   *If the fuse is okay and both brake light bulbs are okay, but the brake lights still fail to come on when the pedal is applied, verify that the switch is getting battery voltage (it should be hot all the time).*

d)   *To check the switch, disconnect the negative battery cable, unplug the electrical connector from the switch, then hook up a an ohmmeter to the two lower terminals* **(see illustration)**. *There should be infinite resistance when the pedal is released, and continuity when the pedal is applied. It can also be checked with a test light; one side should be hot all the time, and the other side should receive voltage when the*

pedal is depressed. If the switch doesn't operate as described, replace it.

e)   *If the switch is working properly, verify that there's voltage between the switch and the brake lights when the pedal is applied (refer to the Wiring Diagrams at the end of Chapter 12).*

Replacement

3   Unplug the negative battery cable.

4   Unplug the electrical connector from the switch.

5   Loosen and back off the locknut **(see illustration 14.1)**, then unscrew and remove the switch.

6   Installation is the reverse of removal. Before tightening the locknut or plugging in the electrical connector, adjust the switch. To adjust the switch, insert a feeler gauge of the specified thickness between the switch plunger and the brake pedal, then screw the switch in until the plunger touches the feeler gauge. Tighten the locknut and plug in the electrical connector.

15 Anti-lock Brake Systems (ABS) - general information

Rear Wheel Anti-Lock (RWAL)

Some earlier models are equipped with a rear-wheel ABS system, sometimes referred to as RWAL (Rear Wheel Anti-Lock). The system is designed to maintain vehicle maneuverability, directional stability and optimum deceleration under severe braking conditions on most road surfaces. It does so by monitoring the rotational speed of the wheels and controlling the brake line pressure to the rear wheels during braking. This prevents the rear wheels from locking up prematurely during hard braking.

Anti-lock valve assembly

The anti-lock valve assembly consists of a dump valve and an isolation valve. The valve assembly operates by changing the rear brake fluid pressure in response to signals from the control module. The valve is mounted to the right frame rail, just behind the cab area.

Control module

The control module is mounted in the cab, under the seat and is the "brain" of the system. It accepts and processes information received from the speed sensor and brake light switch to control the hydraulic line pressure, avoiding wheel lock up. The control module also constantly monitors the system, even under normal driving conditions, to find faults within the system. If a problem develops within the system, the RWAL light will glow on the dashboard. A diagnostic code will also be stored, which, when retrieved, will indicate the problem area or component.

Speed sensor

A speed sensor is located on top of the rear axle housing. The speed sensor sends a signal to the control module, indicating rear wheel rotational speed.

Brake light switch

The brake light switch signals the control module when the driver steps on the brake pedal. Without this signal the anti-lock system won't activate.

Diagnosis and repair

If the RWAL warning light on the dashboard comes on and stays on, make sure the parking brake is not applied and there's no problem with the brake hydraulic system. If neither of these is the cause, the RWAL system is probably malfunctioning. The home mechanic can perform a few preliminary

checks before taking the vehicle to a dealer service department.

a) *Make sure the brakes and wheel cylinders are in good condition.*

b) *Check the electrical connectors at the control module assembly.*

c) *Check the fuses.*

d) *Retrieve the diagnostic code and follow the wiring harness to the indicated component (speed sensor, anti-lock valve, brake light switch, etc.) and make sure all connections are secure and the wiring or component isn't damaged.*

If the above preliminary checks don't rectify the problem, the vehicle should be diagnosed by a dealer service department.

Self-diagnostic codes - retrieving

To retrieve self-diagnostic information from the ECM memory, you must turn the ignition switch to on and ground the diagnostic terminal. The diagnostic terminal is located in the corner of the floor on the drivers side under the ECM. Grounding the diagnostic terminal with the rear anti-lock light on will cause the light to flash a code if a failure has been detected and is in memory. Count the number of flashes, starting with the long flash, and including the long flash as a count. The light will continue to flash, repeating the code, as long as the terminal is grounded. If there is more than one failure, only the first recognized code will be retained in memory and flashed. Removing battery voltage for 30 seconds will clear stored trouble codes. Trouble codes should be cleared after repairs have been completed. **Caution:** *To prevent damage to the ECM, the ignition must be off before disconnecting power to the ECM.*

The following is a list of the typical codes which may be encountered while diagnosing the computerized system. If the problem persists after checks have been made, more detailed service procedures will have to be done by a dealer service department.

Diagnosis codes

Code 2   Open isolation valve circuit or malfunctioning ECM

Code 3   Open dump valve circuit or malfunctioning ECM

Code 4   Closed rear wheel anti-lock valve switch

Code 5   System dumps too many times, condition occurs while making normal or hard stops, rear brakes may lock

Code 6   Erratic speed sensor signal while driving

Code 7   Isolation valve output signal missing or anti-lock valve wiring shorted to ground

Code 8   Dump valve output signal missing or anti-lock valve wiring shorted to ground

Code 9   High speed sensor resistance

Code 10 Low speed sensor resistance

Code 11 Stop light switch circuit defective, condition indicated only when driving over 35 mph

Code 13 Speed processor failure (ECM)

Code 14 Program check failure (ECM)

Code 15 Memory failure (ECM)

Note: *Component replacement may not cure the problem in all cases. For this reason, you may want to seek professional advice before purchasing replacement parts.*

Anti-lock Brake system (ABS)

The Anti-lock Brake System is designed to maintain vehicle steerability, directional stability and optimum deceleration under severe braking conditions on most road surfaces. It does so by monitoring the rotational speed of each wheel and controlling the brake line pressure to each wheel during braking. This prevents the wheels from locking up.

The ABS system has three main components - the wheel speed sensors, the electronic control unit and the modulator (hydraulic control unit). The sensors - one at each wheel (4WD models) or one at each front wheel and one at the differential (2WD models) - send a variable voltage signal to the electronic control unit, which monitors these signals, compares them to its program and determines whether a wheel is about to lock up. When a wheel is about to lock up, the control unit signals the hydraulic unit to

reduce hydraulic pressure (or not increase it further) at that wheel's brake caliper. Pressure modulation is handled by three electrically-operated solenoid valves - one for each front wheel and one for the rear wheels - inside the modulator.

If a problem develops within the system, an "ABS" warning light will glow on the dashboard. Sometimes, a visual inspection of the ABS system can help you locate the problem. Carefully inspect the ABS wiring harness. Pay particularly close attention to the harness and connections near each wheel. Look for signs of chafing and other damage caused by incorrectly routed wires. If a wheel sensor harness is damaged, the sensor should be replaced (the harness and sensor are integral). **Warning:** *Do NOT try to repair an ABS wiring harness. The ABS system is sensitive to even the smallest changes in resistance. Repairing the harness could alter resistance values and cause the system to malfunction. If the ABS wiring harness is damaged in any way, it must be replaced.* **Caution:** *Make sure the ignition is turned off before unplugging or reattaching any electrical connections.*

Diagnosis and repair

If a dashboard warning light comes on and stays on while the vehicle is in operation, the ABS system requires attention. Although special electronic ABS diagnostic testing tools are necessary to properly diagnose the system, you can perform a few preliminary checks before taking the vehicle to a dealer service department.

a) *Check the brake fluid level in the master cylinder reservoir.*

b) *Verify that all ABS system electrical connectors in the engine compartment are plugged in.*

c) *Check the fuses.*

d) *Follow the wiring harness to each front wheel and to the differential sensor and verify that all connections are secure and that the wiring is undamaged.*

If the above preliminary checks do not rectify the problem, the vehicle should be diagnosed by a dealer service department. Due to the complex nature of this system, all actual repair work must be done by a dealer service department.

Chapter 10
Suspension and steering systems

Contents

Specifications

Torque specifications
Ft-lbs (unless otherwise indicated)

Front suspension

Balljoints	
Balljoint-to-lower control arm bolts/nuts	76
Balljoint-to-upper control arm bolts/nuts	
1989 through 1995	24
1996 on	42
Balljoint-to-steering knuckle nut	
Upper balljoint	72
Lower balljoint	
1989 through 1995	94
1996 on	108
Lower control arm-to-frame pivot bolts/nuts	
Front bolt/nut	116
Rear bolt/nut	145
Shock absorber	
Upper nut	15
Lower nut/bolt	61
Stabilizer bar	
Bushing bracket nuts	
1989 through 1995	21
1996 on	16
Stabilizer bar-to-link nuts	
1989 through 1995	96 in-lbs
1996 on	37
Link-to-lower control arm nuts	
1989 through 1995	96 in-lbs
1996 on	37
Torsion bar-to-control arm torque arm nuts	85
Upper control arm	
Pivot shaft-to-frame bolts	112
Control arm-to-pivot shaft nuts	80

Torque specifications

Ft-lbs (unless otherwise indicated)

Rear suspension

1989 through 1997

Leaf spring

Shackle pin nuts	73
Front through-bolt	113

Shock absorber

Lower mounting nut	29

Upper mounting nut

Amigo	29
Rodeo/Passport	15
U-bolt nuts	49

1998 on

Lateral rod

Upper bolt and nut	101
Lower nut	58

Shock absorbers

Upper mounting nut	15
Lower mounting nut	61

Stabilizer bar

Bushing bracket bolts	19
Link nuts and bolts	27
Trailing arms (all fasteners)	101

Steering

Airbag module retaining bolts	69 in-lbs
Wheel lug nuts	See Chapter 1

1989 through 1997

Idler arm-to-center link nut	43
Inner tie-rod end-to-center link nuts	72
Outer tie-rod end-to-steering knuckle nut	72
Pitman arm-to-center link nut	72
Pitman arm-to-steering gear nut	159
Steering gear-to-frame bolts	33
Steering shaft-to-steering gear coupling pinch bolt	18
Steering wheel nut	25

1998 on

Steering gear mounting nuts	85
Tie-rod end-to-steering knuckle nut	87
U-joint connector bolts	23
Crossmember mounting bolts	140

1.1a Typical front steering and suspension components (4WD models)

1   Stabilizer brackets
2   Stabilizer bar
3   Pitman arm
4   Idler arm
5   Center link
6   Tie-rod ends
7   Torsion bars
8   Lower control arms
9   Lower control arm balljoints

1.1b Typical left front steering and suspension components

1   Stabilizer bar bracket
2   Stabilizer bar
3   Shock absorber
4   Upper control arm
5   Upper control arm balljoint
6   Lower control arm balljoint
7   Stabilizer-to-lower control arm link
8   Lower control arm
9   Tie-rod end
10 Center link
11 Steering knuckle

1   General information

Refer to illustrations 1.1a, 1.1b, 1.1c, 1.1d, 1.2a and 1.2b

The fully independent front suspension **(see illustrations)** consists of upper and lower control arms, torsion bars, balljoint mounted steering knuckles and shock absorbers. Some models are equipped with a stabilizer bar to limit body roll during corner-ing. Four wheel drive models use the same basic arrangement, but with the necessary modifications to accommodate the driveaxles.

The rear suspension system on 1989 through 1997 models **(see illustration)** con-sists of the rear axle housing, leaf springs and shock absorbers. The rear suspension sys-tem on 1998 and later models **(see illustra-tion)** consists of the rear axle housing, coil springs, shock absorbers, trailing arms, a lat-eral rod and a stabilizer bar. Information regarding axles and the rear axle housing can be found in Chapter 8.

The steering system on 1989 through 1997 models consists of the steering column, the steering gear, and the steering linkage: the Pitman arm, the idler arm, the center link, the steering damper (on some models) and

1.1c Typical front steering and suspension components (2WD models)

1   Stabilizer bar
2   Stabilizer bar bracket bolts
3   Pitman arm
4   Idler arm
5   Center link
6   Tie-rod ends
7   Torsion bars
8   Lower control arms
9   Lower control arm balljoints
10 Steering damper

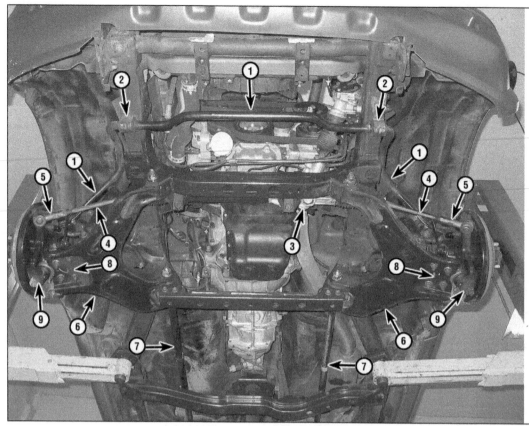

1.1d Typical front steering and suspension components (1998 and later 2WD models)

1   Stabilizer bar
2   Stabilizer bar brackets and bushings
3   Rack-and-pinion steering gear
4   Tie-rods
5   Tie-rod ends
6   Lower control arms
7   Torsion bar
8   Balljoints (balljoints, not visible in this photo, are bolted to the upper side of the lower control arms)
9   Steering knuckles

two tie-rod assemblies. Some models feature power assisted steering, which includes a belt-driven pump and associated hoses to provide hydraulic pressure to the steering gear.

All 1998 and later models are equipped with power-assisted rack-and-pinion steer-

ing. The steering system consists of a transfer gear unit (which connects the steering shaft to the steering gear input shaft), the steering gear assembly and a pair of tie-rods that connect the steering gear to the steering knuckles.

Frequently, when working on the sus-

pension or steering system components, you may come across fasteners which seem impossible to loosen. These fasteners on the underside of the vehicle are continually subjected to water, road grime, mud, etc., and can become rusted or "frozen," making them extremely difficult to remove. In order to

1.2a Typical rear suspension components (1989 through 1997 models)

1   Rear axle assembly
2   Leaf spring-to-axle plates
3   Leaf springs
4   Shackles
5   Shock absorbers

1.2b Typical rear suspension components (1998 and later models)

1 Shock absorbers
2 Coil springs
3 Lateral rod
4 Lower trailing arms (upper trailing arms not visible in this photo)
5 Stabilizer bar
6 Stabilizer bar brackets and rubber bushings

unscrew these stubborn fasteners without damaging them (or other components), be sure to use lots of penetrating oil and allow it to soak in for a while. Using a wire brush to clean exposed threads will also ease removal of the nut or bolt and prevent damage to the threads. Sometimes a sharp blow with a hammer and punch is effective in breaking the bond between a nut and bolt threads, but care must be taken to prevent the punch from slipping off the fastener and ruining the threads. Heating the stuck fastener and surrounding area with a torch sometimes helps too, but isn't recommended because of the obvious dangers associated with fire. Long

breaker bars and extension, or "cheater," pipes will increase leverage, but never use an extension pipe on a ratchet - the ratcheting mechanism could be damaged. Sometimes, turning the nut or bolt in the tightening (clockwise) direction first will help to break it loose. Fasteners that require drastic measures to unscrew should always be replaced with new ones.

Since most of the procedures that are dealt with in this Chapter involve jacking up the vehicle and working underneath it, a good pair of jackstands will be needed. A hydraulic floor jack is the preferred type of jack to lift the vehicle, and it can also be used to support certain components during various operations. **Warning:** *Never, under any circumstances, rely on a jack to support the vehicle while you're working under the vehicle. Whenever any of the suspension or steering*

fasteners are loosened or removed they must be inspected and, if necessary, replaced with new ones of the same part number or of original equipment quality and design. Torque specifications must be followed for proper reassembly and component retention. Never attempt to heat or straighten any suspension or steering component. Always replace them with new ones.

2   Stabilizer bar (front) - removal and installation

Removal

Refer to illustrations 2.2 and 2.3
Warning: *Whenever any of the suspension or steering fasteners are loosened or removed, they must be inspected and, if necessary, replaced with new ones of the same part number or of original equipment quality and design. Torque specifications must be followed for proper reassembly and component retention.*
1   Raise the vehicle and support it securely on jackstands. Apply the parking brake.
2   Remove the stabilizer bar-to-link nut **(see illustration)**. Note how the sequence of spacers, washers and bushings as they're removed. If you're planning to remove the lower control arm, remove the link as well.
3   Remove the stabilizer bar bracket nuts **(see illustration)** and detach the bar from the vehicle.
4   Pull the brackets off the stabilizer bar and inspect the bushings for cracks, hardening and other signs of deterioration. If the bushings are damaged, replace them.

2.2 To disconnect the stabilizer bar from the link, remove the upper nut (pay attention to the disassembly sequence of any washers, spacers and/or bushings used); to disconnect the link itself from the lower control arm, remove the lower nut (arrow)

2.3 To detach the stabilizer bar from the frame, remove these nuts (arrows)

Installation

5   Position the stabilizer bar bushings on the bar. Push the brackets over the bushings and raise the bar up to the frame. Install the bracket nuts but don't tighten them completely at this time.

6   Install the stabilizer bar-to-link nuts, washers, spacers and rubber bushings and tighten the nuts to the torque listed in this Chapter's Specifications.

7   Tighten the bracket nuts to the torque listed in this Chapter's Specifications.

3   Shock absorbers (front) - removal and installation

Refer to illustrations 3.3 and 3.5

Warning: *Whenever any of the suspension or steering fasteners are loosened or removed, they must be inspected and, if necessary, replaced with new ones of the same part number or of original equipment quality and design. Torque specifications must be followed for proper reassembly and component retention.*

1   Jack up the front of the vehicle and place it securely on jackstands.

2   Remove the front wheels.

3   Remove the nut holding the shock absorber to the mount **(see illustration)**. Clamp a pair of locking pliers to the flats at the top of the shock rod to prevent it from turning.

4   Remove the upper washer and rubber bushing from the shaft of the shock absorber.

5   Disconnect the shock from the lower arm by removing the single through bolt **(see illustration)**.

6   Fully compress the shock absorber.

7   Remove the shock absorber. Remove the lower rubber bushing and washer from the shock.

8   Installation is the reverse of the removal procedure. Make sure all washers and rubber bushings are assembled in the proper order.

3.3 To detach the shock absorber from its upper mounting bracket, remove this nut (you may have to clamp a pair of locking pliers on the "flats" of the shaft to prevent it from turning while removing the nut); note the washers and rubber bushings used above and below the bracket - be sure to put them back in the same order and replace either bushing if it's damaged or worn

Tighten the upper nut and lower nut and bolt to the torque listed in this Chapter's Specifications.

4   Torsion bars - removal and installation

Removal

Refer to illustrations 4.2 and 4.3

Warning: *Whenever any of the suspension or steering fasteners are loosened or removed, they must be inspected and, if necessary, replaced with new ones of the same part number or of original equipment quality and design. Torque specifications must be followed for proper reassembly and component retention.*

1   Loosen the wheel lug nuts, raise the

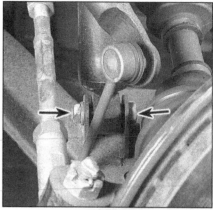

3.5 To detach the lower end of the shock absorber from the lower control arm, remove this nut and bolt (arrows); don't forget the washer when installing the new shock!

vehicle and support it securely on jackstands. Remove the wheel.

2   Use white paint to mark the relationship of the torsion bar to the torque arm **(see illustration)**.

3   Use white paint to mark the alignment of the torsion bar and the height control arm **(see illustration)**.

4   Count the number of threads showing on the height adjustment bolt **(see illustration 4.3)**, then remove the bolt and nuts.

5   Remove the torque arm nuts and bolts **(see illustration 4.2)**.

6   Lift up the height control arm and the rear end of the torsion bar, then remove the torque arm, torsion bar and height control arm as a single assembly.

7   Remove the torque arm and height control arm from the torsion bar.

8   Inspect the torsion bar, torque arm and height control arm splines for damage and wear. Inspect the bar for distortion, cracks or nicks which might cause the bar to fail prematurely. Replace any seriously damaged parts.

4.2 Mark the relationship of the torsion bar to the lower control arm as shown

4.3 Mark the relationship of the torsion bar to the height control arm as shown, then - before removing the height control arm adjustment bolt - count the number of threads showing on the upper end of the bolt and record this figure for reassembly

5.5 To detach the upper control arm assembly from the frame, remove the pivot shaft-to-frame bolts (arrows), as well as the washer, shims and nut plate; note the location of the shims and make sure you install the shims in their original locations when reattaching the upper arm

Installation

9   Apply a coat of multi-purpose grease to the splines of the torsion bar, then slide the torque arm and height control arm onto the torsion bar, aligning the marks you made before disassembly.

10   Raise the torsion bar assembly into position and reattach the torque arm to the lower control arm. Don't tighten the nuts yet.

11   Lower the height control arm into the frame bracket. Install the height control arm adjustment bolt and nuts, then tighten the bolt until the same number of threads are showing as before.

12   Tighten the torque arm retaining nuts to the torque listed in this Chapter's Specifications.

13   Install the wheel. Remove the jackstands and lower the vehicle. Tighten the wheel lug nuts to the torque listed in the Chapter 1 Specifications.

14   Drive the vehicle to an alignment shop to have the front end alignment checked and, if necessary, adjusted.

5   Upper control arm - removal and installation

Removal

Refer to illustration 5.5

Warning: *Whenever any of the suspension or steering fasteners are loosened or removed, they must be inspected and, if necessary, replaced with new ones of the same part number or of original equipment quality and design. Torque specifications must be followed for proper reassembly and component retention.*

1   Loosen the wheel lug nuts, raise the front of the vehicle and support it securely on jackstands. Apply the parking brake. Remove the wheel.

2   Support the lower arm with a jack or

jackstand. The support point must be as close to the balljoint as possible to give maximum leverage on the lower arm.

3   Disconnect the brake hose from the caliper, the upper control arm and the metal line (see "Brake lines and hoses - inspection and replacement" in Chapter 9).

4   Remove the cotter pin and nut and separate the upper balljoint from the steering knuckle (see Section 7). Be careful not to let the steering knuckle/hub/brake assembly fall outward, which will damage the brake hose.

5   Remove the upper arm pivot shaft-to-frame bolts, washer, shims and nut plate **(see illustration)**. Place the washer, shims and plate in a marked container. The original shims must be reinstalled in their original location to maintain wheel alignment.

6   Detach the upper arm from the vehicle. Remove the pivot shaft nuts and washers, if necessary. Inspect the bushings for cracks and deterioration. If they're damaged, have them replaced at an automotive machine shop.

Installation

7   Install the washers and nuts on the ends of the pivot shaft (if removed), but don't tighten them yet. Place the upper control arm assembly in position on the frame, install the pivot shaft bolts and the nut plate. Install the alignment shims in the original location from which they were removed. Tighten the pivot shaft-to-frame bolts to the torque listed in this Chapter's Specifications. Tighten the pivot shaft nuts to the torque listed in this Chapter's Specifications.

8   Connect the balljoint to the steering knuckle and tighten the nut to the torque listed in this Chapter's Specifications.

9   Reattach the brake hose to the metal line, the upper arm and the caliper, and bleed the brakes (see Chapter 9).

10   Install the wheel and lug nuts and lower the vehicle. Tighten the lug nuts to the torque specified in Chapter 1.

11   Drive the vehicle to an alignment shop to have the front end alignment checked and, if necessary, adjusted.

6   Lower control arm - removal and installation

Removal

Refer to Illustration 6.8

Warning: *Whenever any of the suspension or steering fasteners are loosened or removed, they must be inspected and, if necessary, replaced with new ones of the same part number or of original equipment quality and design. Torque specifications must be followed for proper reassembly and component retention.*

1   Loosen the wheel lug nuts, raise the front of the vehicle and support it securely on jackstands. Remove the wheel.

2   Support the lower control arm with a

6.8 To detach the lower control arm from the frame brackets, remove the front and rear pivot bolts and nuts (front nut and bolt shown, rear similar)

floor jack.

3   Disconnect the stabilizer bar from the lower arm (Section 2).

4   Disconnect the shock absorber from the lower control arm (see Section 3).

5   Remove the torsion bar (see Section 4).

6   Remove the floor jack.

7   Disconnect the balljoint from the steering knuckle (see Section 7).

8   Remove the lower arm pivot bolt nuts and washers **(see illustration)**, then pull out the bolts; drive them out with a hammer and drift punch if necessary.

9   Work the lower arm free from the frame brackets and remove it.

Installation

10   Inspect the lower arm bushings for deterioration, cracking and other damage. If either bushing is in need of replacement, take the lower arm to an automotive machine shop to have the old bushings pressed out and new ones pressed in.

11   Place the lower arm in position and install the pivot bolts, washers and nuts, but don't tighten them yet.

12   Attach the steering knuckle the lower balljoint and tighten the nut to the torque listed in this Chapter's Specifications. Install a new cotter pin. If necessary, tighten the nut a little bit more to line up the slots in the nut with the hole in the balljoint stud (never *loosen* the nut to align the hole).

13   Tighten the pivot bolt nuts to the torque listed in this Chapter's Specifications.

14   Support the lower control arm with a floor jack.

15   Install the torsion bar (see Section 4).

16   Reattach the shock absorber to the lower arm (see Section 3).

17   Reattach the stabilizer bar to the lower arm (Section 2).

18   Install the wheel and lug nuts. Lower the vehicle and tighten the lug nuts to the torque listed in this Chapter's Specifications.

19   Drive the vehicle to an alignment shop to have the front end alignment checked and, if necessary, adjusted.

7.8a With the lower balljoint stud nut loosened (but not removed), install a small puller on the steering knuckle boss as shown, pop the ballstud loose from the knuckle, then remove the nut and separate the balljoint from the knuckle

7.8b On 4WD models, use a pickle fork tool to separate the upper balljoint from the knuckle; unless you're very careful, the pickle fork will probably damage the balljoint seal, but there's no other way to split the upper balljoint unless you want to remove the driveaxle!

7   Balljoints - check and replacement

Check

Warning: *Whenever any of the suspension or steering fasteners are loosened or removed, they must be inspected and, if necessary, replaced with new ones of the same part number or of original equipment quality and design. Torque specifications must be followed for proper reassembly and component retention.*

1   Raise the vehicle and support it securely on jackstands.

2   Visually inspect the rubber seal for cuts, tears or leaking grease. If any of these conditions are noticed, the balljoint should be replaced.

3   Place a large pry bar under the balljoint and attempt to push the balljoint upwards. Next, position the pry bar between the steering knuckle and the arm and apply downward pressure. If any movement is seen or felt during either of these checks, a worn out

balljoint is indicated.

4   Have an assistant grasp the tire at the top and bottom and shake the top of the tire in an in-and-out motion. Touch the balljoint stud castellated nut. If any looseness is felt, suspect a worn out balljoint stud or a widened hole in the steering knuckle boss. If the latter problem exists, the steering knuckle should be replaced as well as the balljoint.

Replacement

Refer to illustration 7.8a, 7.8b, 7.10a and 7.10b

5   With the vehicle still raised and supported, position a floor jack under the lower control arm - it must stay there throughout the entire operation. Remove the wheel.

6   Separate the driveaxle from the steering knuckle (see Chapter 8).

7   Remove the cotter pin and loosen the castellated nut a couple of turns, but don't remove it (it will prevent the balljoint and steering knuckle from separating violently).

8   Using a balljoint separating tool, separate the balljoint from the steering knuckle **(see illustrations)**. There are several types of balljoint tools available, but the kind that

pushes the balljoint stud out of the knuckle boss works best. A wedge, or "pickle fork" tool works well, but it tends to damage the balljoint seal. Some two-jaw pullers will also do the job. However, on 4WD models, you won't be able to use a puller on the upper balljoint because the outer CV joint is in the way; on these models, you'll have to use a pickle fork to separate the balljoint from the steering knuckle.

9   Remove the castle nut and disconnect the balljoint from the steering knuckle. If the upper arm balljoint is being removed, be careful not to let the steering knuckle/hub/brake assembly fall outward, because the brake hose might be damaged. If necessary, wire the assembly to the frame to prevent this from happening.

10   The balljoint is bolted to the control arm. Remove the nuts and bolts **(see illustrations)**, then detach the balljoint from the arm.

11   Position the new balljoint on the arm, install the bolts and nuts and tighten them to the torque listed in this Chapter's Specifications.

7.10a To remove the upper balljoint from the upper control arm, remove these bolts and nuts

7.10b To detach the lower control arm from the balljoint, remove these four nuts (arrows) and bolts

8.4 To detach the splash shield from the steering knuckle, remove these bolts (arrows)

9.3 To detach the lower end of the rear shock from the spring plate, remove this nut (arrow)

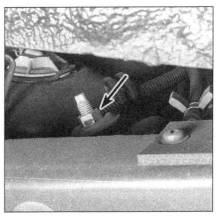

9.4 You'll find the upper ends of the rear shocks above the frame; to detach the upper end, remove this nut (arrow)

12   Insert the balljoint stud into the steering knuckle boss, install the castellated nut and tighten it to the torque listed in this Chapter's Specifications. Install a new cotter pin. If necessary, tighten the nut a little bit more to line up the slots in the nut with the hole in the balljoint stud (never *loosen* the nut to align the hole).
13   The remainder of installation is the reverse of removal.
14   Install the wheel and lug nuts. Lower the vehicle and tighten the lug nuts to the specified torque.

8   Steering knuckle - removal and installation

Warning: *Whenever any of the suspension or steering fasteners are loosened or removed, they must be inspected and, if necessary, replaced with new ones of the same part number or of original equipment quality and design. Torque specifications must be followed for proper reassembly and component retention.*

Removal

Refer to illustration 8.4

1   Loosen the front wheel lug nuts, raise the front of the vehicle and support it

securely on jackstands placed under the lower control arms. Apply the parking brake. Remove the front wheel.
2   Remove the brake caliper and place it on top of the upper arm or wire it up out of the way. Remove the caliper anchor bracket from the steering knuckle (see Chapter 9).
3   Remove the front hub assembly and brake disc (2WD models, see Chapter 1; 4WD models, see Chapter 8).
4   Remove the splash shield from the steering knuckle **(see illustration)**.
5   Separate the tie-rod end from the knuckle arm (see Section 16).
6   Support the lower control arm with a floor jack.
7   Separate both balljoints from the steering knuckle (see Section 7).
8   Remove the steering knuckle.

Installation

9   Place the knuckle between the upper and lower suspension arms and insert the balljoint studs into the knuckle, beginning with the lower balljoint. Install the nuts and tighten them to the torque listed in this Chapter's Specifications. Install new cotter pins, tightening the nuts slightly to align the slots in the nuts with the holes in the balljoint studs, if

necessary. Remove the jack from underneath the lower arm.
10   Connect the tie-rod end to the knuckle arm (see Section 16).
11   Install the splash shield.
12   Install the hub and brake disc assembly (2WD models, see Chapter 1; 4WD models, see Chapter 8).
13   Install the brake caliper mounting bracket and caliper (see Chapter 9).
14   Install the wheel and lug nuts. Lower the vehicle to the ground and tighten the nuts to the torque listed in the Chapter 1 Specifications.

9   Shock absorbers (rear) - removal and installation

Refer to illustrations 9.3 and 9.4
Warning: *Whenever any of the suspension or steering fasteners are loosened or removed, they must be inspected and, if necessary, replaced with new ones of the same part number or of original equipment quality and design. Torque specifications must be followed for proper reassembly and component retention.*
1   If the shock absorber is to be replaced

10.6 Remove the nuts (arrows) retaining the U-bolts and the shock absorber to the spring seat

10.8a To detach the rear end of the leaf spring assembly, remove these two shackle pin nuts (arrows) and the shackles

10.8b To detach the front end of the leaf spring assembly, remove this through-bolt and nut (arrows)

with a new one, it is recommended that both shocks on the rear of the vehicle be replaced at the same time.

2   Raise and support the rear of the vehicle according to the jacking and towing procedures in this manual. Use a jack to raise the differential until the tires clear the ground, then place jackstands under the axle housing. Do not attempt to remove the shock absorbers with the vehicle raised and the axle unsupported.

3   Unscrew and remove the lower shock absorber mounting nut **(see illustration)**.

4   Unscrew and remove the upper mounting nut **(see illustration)** and remove the shock absorber.

5   If the unit is defective, it must be replaced with a new one. Replace worn rubber bushings with new ones.

6   Install the shocks in the reverse order of removal, but lower the vehicle to the ground before tightening the mounting bolts and nuts.

7   Bounce the rear of the vehicle a couple of times to settle the bushings into place, then tighten the nuts to the torque listed in this Chapter's Specifications.

10   Leaf springs (1989 through 1997 models) - removal and installation

Refer to illustrations 10.6, 10.8a and 10.8b
Warning: *Whenever any of the suspension or steering fasteners are loosened or removed, they must be inspected and, if necessary, replaced with new ones of the same part number or of original equipment quality and design. Torque specifications must be followed for proper reassembly and component retention.*

1   Jack up the vehicle and support the frame securely on jackstands.

2   Remove the rear wheels and tires.

3   Place a jack under the rear differential housing.

4   Lower the axle housing until the leaf spring tension is relieved, and lock the jack in this position.

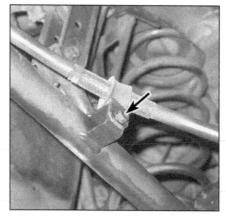

11.2 To detach the parking brake cable from the lower trailing arm, remove this bolt

5   Disconnect the shock absorber from the spring seat (see Section 9).

6   Remove the U-bolt mounting nuts **(see illustration)**.

7   Remove the spring seat.

8   Remove the shackle pin assembly and the front through-bolt **(see illustrations)** and remove the spring.

9   If the bushings in the spring or frame are deteriorated, have an automotive machine shop press out the old bushings and press in new ones.

10   Installation is the reverse of the removal procedure. Lubricate the shackle pins and bushings with lithium base grease. Tighten the shackle pin nuts, through-bolt nut and U-bolt nuts to the torque listed in this Chapter's Specifications.

11   Trailing arms (1998 and later models) - removal and installation

1   Loosen the rear wheel lug nuts. Raise the rear of the vehicle and place it securely on jackstands. Remove the rear wheels. Support the rear axle with a floor jack.

11.3 To detach the rear end of the lower trailing arm from the axle bracket, remove this nut and bolt

Lower trailing arms

Refer to illustrations 11.2, 11.3 and 11.4

2   Detach the parking brake cable from the trailing arm **(see illustration)**.

3   Remove the nut and bolt from the rear end of the lower trailing arm, at the axle bracket **(see illustration)**.

4   Remove the nut and bolt from the forward end of the lower trailing arm, at the frame bracket **(see illustration)**. If you're removing the *left* lower trailing arm, remove the protector.

5   Remove the lower trailing arm assembly.

6   Inspect the rubber bushings inside each end of the upper trailing arm. If they're worn or damaged, have them replaced by an automotive machine shop.

7   Installation is the reverse of removal. Be sure to tighten the lower trailing arm fasteners to the torque listed in this Chapter's Specifications, after the vehicle has been lowered.

Upper trailing arms

Refer to illustrations 11.8, 11.10 and 11.11

8   If you're removing the *left* upper trailing arm, detach the Anti-Lock Brake System (ABS) speed sensor wire harness from the

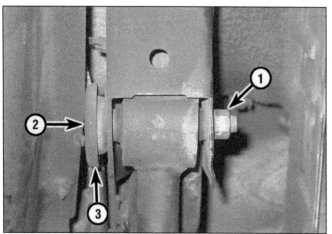

11.4 To detach the forward end of the lower trailing arm from the frame bracket, remove the nut (1), bolt (2) and protector (3)

11.8 To detach the ABS harness from the left upper trailing arm, detach these two cable clips

11.10 To detach the rear end of the upper trailing arm from the axle bracket, remove this nut and bolt

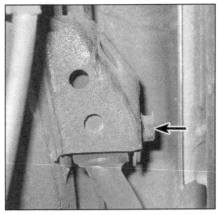

11.11 To detach the forward end of the upper trailing arm from the frame bracket, remove this nut and bolt

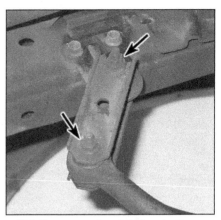

12.2 To remove the stabilizer-to-frame bracket links, remove the two nuts and bolts

arm **(see illustration)**.

9   If you're removing the *left* upper trailing arm, remove the fuel tank (see Chapter 4).

10   Remove the nut and bolt from the rear end of the upper trailing arm, at the axle bracket **(see illustration)**. Remove and discard the two old rubber washers.

11   Remove the nut and bolt from the forward end of the upper trailing arm, at the frame bracket **(see illustration)**. Remove and discard the two old rubber washers.

12   If you're removing the *left* upper trailing arm, remove the protector (similar to the protector shown in **illustration 11.4**).

13   Remove the upper trailing arm assembly.

14   Inspect the rubber bushings inside each end of the upper trailing arm. If they're worn or damaged, have them replaced by an automotive machine shop.

15   Installation is the reverse of removal. Be sure to use a pair of new rubber washers and tighten the upper trailing arm fasteners to the torque listed in this Chapter's Specifications, after the vehicle has been lowered.

---

12   Stabilizer bar (rear) (1998 and later models) - removal and installation

---

Refer to illustrations 12.2 and 12.4

1   Loosen the rear wheel lug nuts. Raise the rear of the vehicle and place it securely on jackstands. Remove the rear wheels.

2   Remove the nuts and bolts from each end of the links that connect the stabilizer to the frame brackets **(see illustration)**.

3   Remove the links.

4   Remove the two bolts from each of the brackets for the rubber bushings **(see illustration)**.

5   Remove the bushing brackets and the rubber bushings.

6   Remove the stabilizer bar.

7   Inspect the condition of the rubber bushings. If they're cracked or torn, replace them.

8   Installation is the reverse of removal. Be

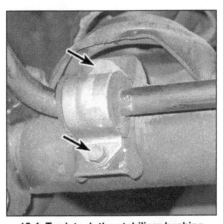

12.4 To detach the stabilizer bushing brackets, remove these two bolts

sure to tighten the bushing bracket bolts and link nuts and bolts to the torque listed in this Chapter's Specifications, after the vehicle has been lowered.

---

13   Coil springs (1998 and later models) - removal and installation

---

Refer to illustrations 13.3 and 13.4

1   Loosen the rear wheel lug nuts. Raise the rear of the vehicle and place it securely on jackstands. Remove the rear wheels.

2   Support the rear axle with a floor jack.

3   Disconnect the upper end of the brake hose from the junction at the crossmember **(see illustration)**.

4   Detach the upper end of the breather hose from its clip on the crossmember **(see illustration)**.

5   Disconnect the rear nut and bolt that attach the rear end of the left upper trailing arm to the axle bracket (see Section 11).

6   Disconnect the ends of the stabilizer bar from the lower ends of the links (see Section 12).

7   Disconnect the shock absorbers from the axle (see Section 9).

13.3 Before lowering the axle, trace the rear brake hose from the junction on the axle up to this bracket, pull off the retainer clip with a pair of pliers and then disconnect the hose from the metal line by unscrewing this fitting

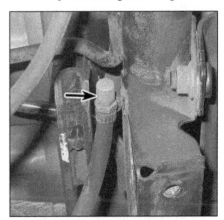

13.4 Also unclip the upper end of the breather hose from the crossmember before dropping the axle

8   Carefully and slowly lower the axle until all spring tension is removed from the coil springs.

9   Remove the coil springs and the spring insulators (the piece at the top of each spring).

14.2 To disconnect the upper end of the lateral rod from the frame bracket, remove this nut and bolt

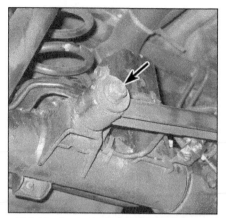

14.3 To disconnect the lower end of the lateral rod from the axle bracket, remove this nut

15.2 On models without an airbag, simply pull off the horn pad and unplug the electrical connector

10   Installation is the reverse of removal. Be sure to tighten all suspension fasteners to the torque listed in this Chapter's Specifications, after the vehicle has been lowered. **Caution:** *The coil springs must be installed with the correct end facing up. Verify that each spring is installed with its paint mark below the midpoint of the spring. If the paint mark is near the top of the spring, the spring is upside down.*

14   Lateral rod (1998 and later models) - removal and installation

Refer to illustrations 14.2 and 14.3
1   Raise the rear of the vehicle and place it securely on jackstands.
2   Remove the upper bolt and nut **(see illustration)**.
3   Remove the lower nut and washer **(see illustration)**.
4   Remove the lateral rod.
5   Inspect the rubber bushings in each end of the lateral rod. If they're cracked or torn or badly worn, have them replaced by an automotive machine shop.
6   Installation is the reverse of removal. Be

sure to tighten the lateral rod fasteners to the torque listed in this Chapter's Specifications, after the vehicle has been lowered.

15   Steering wheel - removal and installation

Removal

Refer to illustrations 15.2, 15.4a, 15.4b, 15.5, 15.6 and 15.7
Warning: *Some models covered by this manual are equipped with airbags. Always disable the airbag system (see Chapter 12) before working in the vicinity of the impact sensors, steering column or instrument panel. Failure to follow these procedures may cause accidental deployment of the airbag, which could cause personal injury.*
1   Disconnect the cable from the negative battery terminal.
2   On models without an airbag, pry off the horn button or center pad and unplug the horn switch connector **(see illustration)**.
3   On models with airbags, disable the airbag system (Chapter 12).
4   On models with airbags, remove the

airbag module retaining bolts, lift off the module and unplug the electrical connector **(see illustrations)**. **Warning:** *Use care when handling the airbag module. Accidental deployment of the airbag could cause personal injury. Carry the module with the trim side facing away from your body and store the airbag module in a safe location with the trim side facing up.*
5   Unplug the electrical connector for the cruise control system, if equipped **(see illustration)**.
6   Unscrew the steering wheel nut and mark the steering wheel hub and the column shaft with paint to ensure correct repositioning during reassembly **(see illustration)**.
7   Remove the wheel using a puller **(see illustration)**. **Caution:** *Do not hammer on the wheel or the shaft to separate them.*

Installation

8   Realign the steering wheel and the column shaft using the match marks. Install and tighten the retaining nut to the torque listed in this Chapter's Specifications.
9   Plug in the electrical connectors for the horn and, if equipped, the cruise control system.

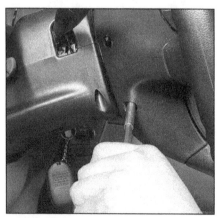

15.4a To detach the airbag module from the steering wheel, remove these two retaining bolts (left bolt shown) . . .

15.4b . . . lift off the module, turn it over and unplug the yellow electrical connector

15.5 On models equipped with cruise control, unplug the electrical connector

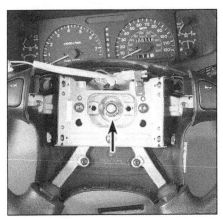

15.6 Remove the steering wheel to steering shaft nut (arrow), then mark the relationship of the steering wheel to the steering shaft

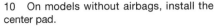

15.7 Use a steering wheel puller to remove the steering wheel from the shaft

16.4 Loosen the jam nut and mark the relationship of the outer tie-rod end to the tie-rod by painting an alignment mark on the threads adjacent to the tie-rod end

10 On models without airbags, install the center pad.

11 On models with airbags, plug in the connector to the airbag module, place the airbag module in position, install the module retaining bolts and tighten them to the torque listed in this Chapter's Specifications. Then refer to Chapter 12 and enable the airbag system.

12 Reconnect the negative battery cable.

16 Steering linkage - removal and installation

Warning: *Whenever any of the suspension or steering fasteners are loosened or removed, they must be inspected and, if necessary, replaced with new ones of the same part number or of original equipment quality and design. Torque specifications must be followed for proper reassembly and component retention.*

1 All steering linkage removal and installation procedures should be performed with the front end of the vehicle raised and placed securely on jackstands.

2 Before removing any steering linkage components, obtain a balljoint separator. It may be a screw-type puller or a wedge-type (pickle fork) tool, although the wedge-type tool tends to damage the balljoint seals. It is possible to jar a balljoint taper pin free from its eye by striking opposite sides of the eye simultaneously with two large hammers, but the space available to do so is usually very limited.

3 After installing any of the steering linkage components, the front wheel alignment should be checked by a reputable front end alignment shop.

1989 through 1997 models

Tie-rod ends

Refer to illustrations 16.4, 16.5 and 16.6

4 Loosen the jam nut and mark the relationship of the outer tie-rod end to the tie-rod with white paint **(see illustration)**.

5 Remove the cotter pin and loosen the castle nut on the balljoint stud, then install a small puller and separate the outer tie-rod

end from the steering knuckle arm **(see illustration)**.

6 If replacing the inner tie-rod end balljoint, detach the inner tie-rod end from the center link the same way: loosen the inner tie-rod end jam nut, mark the threads on the tie rod, remove the cotter pin, loosen the castle nut securing the inner tie-rod end to the center link, install a small puller on the balljoint stud **(see illustration)** and separate the inner tie-rod end from the center link. You'll have to turn the steering wheel all the way to the left or right to get at the left or right inner tie-rod ballstud.

7 When installing a new tie-rod rod end, thread it onto the tie-rod until it reaches the mark, then turn the jam nut until it contacts the rod end. Don't tighten it fully at this time.

8 Turn the tie-rod ends so they are at a 90-degree angle to each other, then tighten the jam nuts to lock the ends in position.

9 The remainder of installation is the reverse of removal. Be sure to tighten the castle nuts to the torque listed in this Chapter's Specifications.

16.5 Remove the cotter pin, back off the castle nut (don't remove it), install a small puller as shown and separate the outer tie-rod end from the steering knuckle

16.6 To replace an inner tie-rod end, back off the jam nut, mark the relationship of the tie-rod end to the tie rod, remove the cotter pin, loosen the castle nut, install a small puller as shown and separate the inner tie-rod end from the center link

16.10 Remove the Pitman arm-to-steering gear shaft nut (right arrow), remove the cotter pin and loosen the castle nut (left arrow) on the other end of the Pitman arm

16.12 To separate the Pitman arm from the steering gear, use a Pitman arm puller

16.14 To separate the Pitman arm from the center link, remove the cotter pin, back off the castle nut and install a small puller as shown

Pitman arm

Refer to illustrations 16.10, 16.12 and 16.14

10   Remove the nut securing the Pitman arm to the steering gear sector shaft **(see illustration)**.
11   Scribe or paint match marks on the arm and shaft.
12   Using a puller, disconnect the Pitman arm from the shaft splines **(see illustration)**.
13   Remove the cotter pin and castle nut securing the Pitman arm to the center link.
14   Using a puller, disconnect the Pitman arm from the center link **(see illustration)**.
15   Installation is the reverse of the removal procedure. Be sure to tighten the nuts to the specified torque and install a new cotter pin.

Idler arm

Refer to illustration 16.17

16   Remove the cotter pin, loosen the castle nut and install a puller on the balljoint stud **(see illustration 16.14)**.
17   Unbolt the idler arm from the frame **(see illustration)**.

Center link

18   Separate the tie-rod ends from the center link **(see illustration 16.6)**.
19   Separate the Pitman arm from the center link **(see illustration 16.14)**.
20   Separate the idler arm from the center link **(see illustration 16.14)**.
21   Installation is the reverse of removal. Be sure to tighten all nuts to the torque listed in this Chapter's Specifications.

Steering damper

22   Remove the cotter pin, nut and washer from the bolt that attaches the damper to the center link.
23   Remove the cotter pin, nut and washer attaching the damper to the crossmember (2WD) or idler arm bracket.
24   Remove the steering damper from the vehicle.
25   Installation is the reverse of removal.

16.17 To remove the idler arm, remove the cotter pin, loosen the castle nut (arrow), remove the big nut (arrow) and separate the arm from the idler shaft and from the center link the same way you would remove a Pitman arm (see illustrations 12.12 and 12.14); to detach the idler shaft from the frame, remove these four nuts (arrows)

1998 and later models

Tie-rod ends

Refer to illustrations 16.26a, 16.26b and 16.27

26   Loosen the jam nut and mark the relationship of the outer tie-rod end to the tie-rod with white paint **(see illustrations)**.
27   Loosen the castle nut on the balljoint stud, then install a small puller and separate

16.26a Using a back-up wrench to hold the tie-rod end, loosen the jam nut . . .

the outer tie-rod end from the steering knuckle arm **(see illustration)**.
28   When installing a new tie-rod rod end, thread it onto the tie-rod until it reaches the mark, then turn the jam nut until it contacts the rod end. Don't tighten it fully at this time.
29   Insert the tie-rod end balljoint stud into the hole in the steering knuckle arm, then

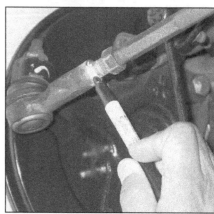

16.26b . . . and then mark the location of the tie-rod end to ensure that the new tie-rod end is screwed on the same number of turns as the old unit

16.27 Using a small puller, separate the tie-rod end balljoint stud from the steering knuckle arm

17.2 Remove this splash shield to get to the steering shaft-to-steering gear coupling (1989 through 1997 models)

tighten the castle nut to the torque listed in this Chapter's Specifications.
30   Tighten the castle nut to the torque listed in this Chapter's Specifications.

17   Steering gear - removal and installation

Warning 1: *Some models covered by this manual are equipped with airbags. Always disable the airbag system (see Chapter 12) before working in the vicinity of the impact sensors, steering column or instrument panel. Failure to follow these procedures may cause accidental deployment of the airbag, which could cause personal injury. Also DO NOT allow the steering column to rotate with the steering gear removed or damage to the airbag system could occur. Position the front wheels pointing straight ahead, lock the steering column and remove the key before beginning this procedure.*
Warning 2: *Whenever any of the suspension or steering fasteners are loosened or*

removed, they must be inspected and, if necessary, replaced with new ones of the same part number or of original equipment quality and design. Torque specifications must be followed for proper reassembly and component retention.
Note: *If you find that the steering gear is defective, it is not recommended that you overhaul it. Because of the special tools needed to do the job, it is best to let your dealer service department overhaul it for you (or replace it with a factory rebuilt unit). However, you can remove it yourself using the procedure which follows.*

1989 through 1997 models

Refer to illustrations 17.2, 17.3 and 17.6
1   Apply the parking brake. Loosen the left front wheel lug nuts. Raise the front of the vehicle and place it securely on jackstands. Remove the left front wheel.
2   Remove the splash shield **(see illustration)**.
3   Mark the relationship of the steering shaft-to-sector shaft coupling **(see illustra-**

tion), then remove the coupling bolt.
4   Disconnect the power steering lines (if applicable), and plug them immediately to prevent loss or contamination of power steering fluid.
5   Remove the Pitman arm (see Section 16).
6   Remove the bolts securing the steering gear to the frame **(see illustration)** and remove the steering gear.
7   Installation is the reverse of the removal procedure except that the Pitman arm nut should be tightened before the Pitman arm is connected to the center link. Be sure to tighten all nuts and bolts to the torque listed in this Chapter's Specifications.

1998 and later models

Refer to illustrations 17.9, 17.11, 17.12a, 17.12b, 17.13 and 17.17
8   Apply the parking brake. Loosen the front wheel lug nuts. Raise the front of the vehicle and place it securely on jackstands. Remove the front wheels.
9   Remove the engine under-cover **(see illustration)**.

17.3 Mark the relationship of the steering shaft-to-steering gear coupling, then remove the coupling pinch bolt (arrow); then disconnect both power steering lines (arrows) (1989 through 1997 models)

17.6 To detach the steering gear from the frame, remove these nuts (1989 through 1997 models)

17.9 To remove the engine under-cover, remove these fasteners (1998 and later models)

17.11 To disconnect the power steering fluid lines from the steering gear, unscrew these two fittings with a wrench (use a flare-nut wrench, if available, to protect the corners of the nuts from being rounded off) (1998 and later models)

10   Disconnect the tie-rod ends from the steering knuckle arms (see Section 16).

11   Disconnect the power steering line fittings **(see illustration)**.

12   Mark the relationship of the U-joint connector between the transfer gear output shaft and the steering gear input shaft **(see illustrations)**.

13   Loosen the two pinch bolts on the U-joint connector between the transfer gear and the steering gear **(see illustration)**.

14   Remove the torsion bars (see Section 4).

15   Detach the lower control arms from the frame (see Section 6).

16   Mark the relationship of the crossmember to the frame, then unscrew the cross-member bolts and detach the crossmember and stearing gear as a unit.

17   Remove the nuts and clamps that attach the steering gear assembly to the crossmember **(see illustration)**.

18   Installation is the reverse of removal. Be sure to tighten the crossmember bolts, the steering gear nuts and the U-joint connector bolts to the torque listed in this Chapter's Specifications.

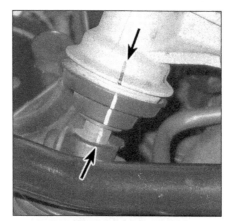

17.12a Mark the relationship of the U-joint connector to the transfer gear output shaft . . .

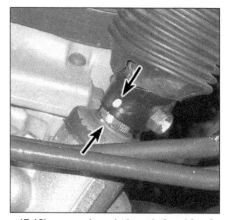

17.12b . . . and mark the relationship of the U-joint connector to the steering gear input shaft (1998 and later models)

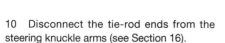

18   Steering freeplay - adjustment

Refer to illustration 18.6

Note: *This procedure applies to 1997 and earlier models only.*

1   Raise the vehicle with a jack so that the front wheels are off the ground and place the vehicle securely on jackstands.

2   Point the wheels straight ahead.

3   Using a wrench, loosen the locknut on the steering gear.

17.13 Loosen the two pinch bolts at the upper and lower ends of the connector (1998 and later models)

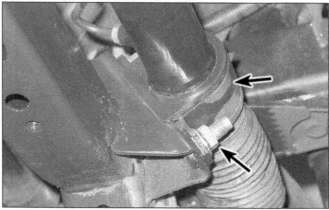

17.17 To detach the steering gear from the crossmember, remove the two big nuts and clamps (right nut and clamp shown) (1998 and later models)

18.6 When tightening the adjusting screw locknut, be sure to hold the screw to keep it from turning

4   Turn the adjusting screw clockwise to decrease wheel freeplay and counterclockwise to increase it. **Note:** *Turn the adjusting screw in small increments, checking the steering wheel freeplay between them.*

5   Turn the steering wheel halfway around in both directions, checking that the freeplay is correct and that the steering is smooth.

6   Hold the adjusting screw so that it will not turn and tighten the locknut **(see illustration)**.

7   Remove the jackstands and lower the vehicle.

---

19 Power steering pump - removal and installation

Warning: *Whenever any of the suspension or steering fasteners are loosened or removed, they must be inspected and, if necessary, replaced with new ones of the same part number or of original equipment quality and design. Torque specifications must be followed for proper reassembly and component retention.*

Note: *If you find that the steering pump is defective, it is not recommended that you overhaul it. Because of the special tools needed to do the job, it is best to let your dealer service department overhaul it for you (or replace it with a factory rebuilt unit). However, you can remove it yourself using the procedure which follows.*

Removal

Refer to illustrations 19.1 and 19.2

1   Position a container to catch the power steering fluid. Disconnect the hoses from the pump body **(see illustration)**. As each is disconnected, cap or tape over the hose opening and then secure the end in a raised position to prevent leakage and contamination. Cover or plug the pump openings to so dirt won't enter the unit.

2   Loosen the pump pulley bolts **(see illustration)** (the pump can't be unbolted from its mounting bracket until the pulley is removed).

3   Remove the pump drivebelt (see Chapter 1).

4   Remove the pump pulley bolts and remove the pulley.

5   Remove the pump-to-bracket bolts and remove the pump.

Installation

6   Installation is the reverse of the removal procedure. When installing the pressure line, make sure there is sufficient clearance between the line and the exhaust manifold.

7   Be sure to adjust the drivebelt tension (see Chapter 1).

8   Fill the power steering fluid reservoir with the specified fluid and bleed the power steering system (see Section 20).

9   Check for fluid leaks.

---

20 Power steering system - bleeding

1   Check the fluid in the reservoir and add fluid of the specified type if it is low.

2   Jack up the front of the vehicle and place it securely on jackstands.

3   With the engine off, turn the steering

wheel fully in both directions two or three times.

4   Recheck the fluid in the reservoir and add more fluid if necessary.

5   Start the engine and turn the steering wheel fully in both directions two or three times. The engine should be running at 1000 rpm or less.

6   Remove the jackstands and lower the vehicle completely.

7   With the engine running at 1000 rpm or less, turn the steering wheel fully in both directions two or three times.

8   Return the steering wheel to the center position.

9   Check that the fluid is not foamy or cloudy.

10   Measure the fluid level with the engine running.

11   Turn the engine and again measure the fluid level. It should rise no more than 0.20 in (5 mm) when the engine is turned off.

12   If a problem is encountered, repeat Steps 7 through 11.

13   If the problem persists, remove the pump (see Section 19), and have it repaired by a dealer service department.

---

21 Front end alignment - general information

Refer to illustration 21.1

A front end alignment refers to the adjustments made to the front wheels so they are in proper angular relationship to the suspension and the ground. Front wheels that are out of proper alignment not only affect steering control, but also increase tire wear. The front end adjustments normally required are camber, caster and toe-in **(see illustration)**.

Getting the proper front wheel alignment is a very exacting process, one in which complicated and expensive machines are necessary to perform the job properly. Because of

19.1 Detach the power steering lower pressure return hose and high pressure line (arrows) from the power steering pump

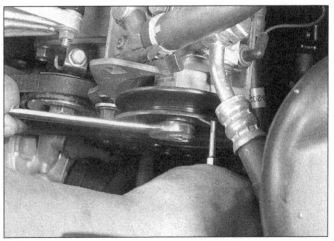

19.2 Insert a screwdriver through a pulley hole and lodge it against a boss on the power steering pump to hold the pulley while breaking loose the pulley bolts

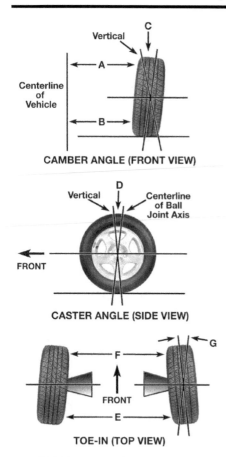

CAMBER ANGLE (FRONT VIEW)

CASTER ANGLE (SIDE VIEW)

TOE-IN (TOP VIEW)

21.1 Front end alignment details

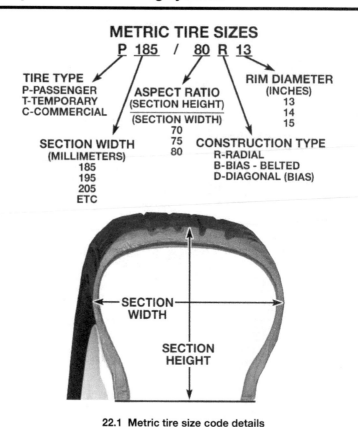

22.1 Metric tire size code details

this, you should have a technician with the proper equipment perform these tasks. We will, however, use this space to give you a basic idea of what is involved with front end alignment so you can better understand the process and deal intelligently with the shop that does the work.

Toe-in is the turning in of the front wheels. The purpose of a toe specification is to ensure parallel rolling of the front wheels. In a vehicle with zero toe-in, the distance between the front edges of the wheels will be the same as the distance between the rear edges of the wheels. The actual amount of toe-in is normally only a fraction of an inch. Toe-in adjustment is controlled by the tie-rod end position on the inner tie-rod. Incorrect toe-in will cause the tires to wear improperly by making them scrub against the road surface.

Camber is the tilting of the front wheels from the vertical when viewed from the front of the vehicle. When the wheels tilt out at the top, the camber is said to be positive (+). When the wheels tilt in at the top the camber is negative (-). The amount of tilt is measured in degrees from the vertical and this mea-

surement is called the camber angle. This angle affects the amount of tire tread which contacts the road and compensates for changes in the suspension geometry when the vehicle is cornering or traveling over an undulating surface. Camber is adjusted by adding or subtracting shims at the upper control arm mount.

Caster is the tilting of the front steering axis from the vertical. A tilt toward the rear is positive caster and a tilt toward the front is negative caster. Caster angle affects the self-centering action of the steering, which governs straight-line stability. Caster is adjusted by moving shims from one end of the upper arm mount to the other.

22   Wheels and tires - general information

Refer to illustration 22.1

All models covered by this manual are equipped with metric-sized radial tires **(see illustration)**. Use of other size or type of tires may affect the ride and handling of the vehicle. Don't mix different types of tires, such as radials and bias belted, on the same vehicle as handling may be seriously affected. It's recommended that tires be replaced in pairs

on the same axle, but if only one tire is being replaced, be sure it's the same size, structure and tread design as the other.

Because tire pressure has a substantial effect on handling and wear, the pressure on all tires should be checked at least once a month or before any extended trips (see Chapter 1).

Wheels must be replaced if they are bent, dented, leak air, have elongated bolt holes, are heavily rusted, out of vertical symmetry or if the lug nuts won't stay tight. Wheel repairs that use welding or peening are not recommended.

Tire and wheel balance is important to the overall handling, braking and performance of the vehicle. Unbalanced wheels can adversely affect handling and ride characteristics as well as tire life. Whenever a tire is installed on a wheel, the tire and wheel should be balanced by a shop with the proper equipment.

Wheels can be damaged by an impact with a curb or other solid object. If the wheels are bent, the result is a hazardous condition which must be corrected. To check the wheels, raise the vehicle and set it on jackstands. Visually inspect the wheels for obvious signs of damage such as cracks and deformation.

Chapter 11
Body

Contents

1 General information

The vehicles covered by this manual are built with a body-on-frame construction. The frame is a ladder-type, consisting of two box steel side rails joined by crossmembers. These crossmembers are welded to the side rails, with the exception of the transmission crossmember which is bolted into place for easy removal. The vehicle bodies are secured to the chassis by rubber insulated mounts and can be completely removed from the chassis.

Certain components are particularly vulnerable to accident damage and can be unbolted and repaired or replaced. Among these parts are the body moldings, front fenders, doors, bumpers, the hood and tailgate and all glass. Only general body maintenance practices and body panel repair procedures within the scope of the do-it-yourselfer are included in this Chapter.

2 Body - maintenance

1   The condition of your vehicle's body is very important, because the resale value depends a great deal on it. It's much more difficult to repair a neglected or damaged body than it is to repair mechanical components. The hidden areas of the body, such as the wheel wells, the frame and the engine compartment, are equally important, although they don't require as frequent attention as the rest of the body.
2   Once a year, or every 12,000 miles, it's a good idea to have the underside of the body steam cleaned. All traces of dirt and oil will be removed and the area can then be inspected carefully for rust, damaged brake lines, frayed electrical wires, damaged cables and other problems. The front suspension components should be greased after completion of this job.
3   At the same time, clean the engine and the engine compartment with a steam cleaner or water-soluble degreaser.
4   The wheel wells should be given close attention, since undercoating can peel away and stones and dirt thrown up by the tires can cause the paint to chip and flake, allowing rust to set in. If rust is found, clean down to the bare metal and apply an anti-rust paint.
5   The body should be washed about once a week. Wet the vehicle thoroughly to soften the dirt, then wash it down with a soft sponge and plenty of clean soapy water. If the surplus dirt is not washed off very carefully, it can wear down the paint.
6   Spots of tar or asphalt thrown up from the road should be removed with a cloth soaked in solvent.
7   Once every six months, wax the body and chrome trim. If a chrome cleaner is used to remove rust from any of the vehicle's plated parts, remember that the cleaner also removes part of the chrome, so use it sparingly.

3   Vinyl trim - maintenance

Don't clean vinyl trim with detergents, caustic soap or petroleum-based cleaners. Plain soap and water works just fine, with a soft brush to clean dirt that may be ingrained. Wash the vinyl as frequently as the rest of the vehicle. After cleaning, application of a high-quality rubber and vinyl protectant will help prevent oxidation and cracks. The protectant can also be applied to weatherstripping, vacuum lines and rubber hoses, which often fail as a result of chemical degradation, and to the tires.

4   Upholstery and carpets - maintenance

1   Every three months, remove the floor-mats and clean the interior of the vehicle (more frequently if necessary). Use a stiff whisk broom to brush the carpeting and loosen dirt and dust, then vacuum the upholstery and carpets thoroughly, especially along seams and crevices.
2   Dirt and stains can be removed from carpeting with basic household or automotive carpet shampoos available in spray cans. Follow the directions and vacuum again, then use a stiff brush to bring back the "nap" of the carpet.
3   Most interiors have cloth or vinyl upholstery, either of which can be cleaned and maintained with a number of material-specific cleaners or shampoos available in auto supply stores. Follow the directions on the product for usage, and always spot-test any upholstery cleaner on an inconspicuous area (bottom edge of a back seat cushion) to ensure that it doesn't cause a color shift in the material.
4   After cleaning, vinyl upholstery should be treated with a protectant. **Note:** *Make sure the protectant container indicates the product can be used on seats - some products can make a seat slippery.* **Caution:** *Do not use protectant on vinyl-covered steering wheels.*
5   Leather upholstery requires special care. It should be cleaned regularly with saddlesoap or leather cleaner. Never use alcohol, gasoline, nail polish remover or thinner to clean leather upholstery.
6   After cleaning, regularly treat leather upholstery with a leather conditioner, rubbed in with a soft cotton cloth. Never use car wax on leather upholstery.
7   In areas where the interior of the vehicle is subject to bright sunlight, cover leather seating areas of the seats with a sheet if the vehicle is to be left out for any length of time.

5   Body repair - minor damage

Repair of scratches

1   If the scratch is superficial and does not penetrate to the metal of the body, repair is very simple. Lightly rub the scratched area with a fine rubbing compound to remove loose paint and built-up wax. Rinse the area with clean water.
2   Apply touch-up paint to the scratch, using a small brush. Continue to apply thin layers of paint until the surface of the paint in the scratch is level with the surrounding paint. Allow the new paint at least two weeks to harden, then blend it into the surrounding paint by rubbing with a very fine rubbing compound. Finally, apply a coat of wax to the scratch area.
3   If the scratch has penetrated the paint and exposed the metal of the body, causing the metal to rust, a different repair technique is required. Remove all loose rust from the bottom of the scratch with a pocket knife, then apply rust inhibiting paint to prevent the formation of rust in the future. Using a rubber or nylon applicator, coat the scratched area with glaze-type filler. If required, the filler can be mixed with thinner to provide a very thin paste, which is ideal for filling narrow scratches. Before the glaze filler in the scratch hardens, wrap a piece of smooth cotton cloth around the tip of a finger. Dip the cloth in thinner and then quickly wipe it along the surface of the scratch. This will ensure that the surface of the filler is slightly hollow. The scratch can now be painted over as described earlier in this Section.

Repair of dents

See photo sequence
4   When repairing dents, the first job is to pull the dent out until the affected area is as close as possible to its original shape. There is no point in trying to restore the original shape completely as the metal in the damaged area will have stretched on impact and cannot be restored to its original contours. It is better to bring the level of the dent up to a point which is about 1/8-inch below the level of the surrounding metal. In cases where the dent is very shallow, it is not worth trying to pull it out at all.
5   If the back side of the dent is accessible, it can be hammered out gently from behind using a soft-face hammer. While doing this, hold a block of wood firmly against the opposite side of the metal to absorb the hammer blows and prevent the metal from being stretched.
6   If the dent is in a section of the body which has double layers, or some other factor makes it inaccessible from behind, a different technique is required. Drill several small holes through the metal inside the damaged area, particularly in the deeper sections. Screw long, self tapping screws into the holes just enough for them to get a good grip in the metal. Now the dent can be pulled out by pulling on the protruding heads of the screws with locking pliers.
7   The next stage of repair is the removal of paint from the damaged area and from an inch or so of the surrounding metal. This is easily done with a wire brush or sanding disk in a drill motor, although it can be done just as effectively by hand with sandpaper. To complete the preparation for filling, score the surface of the bare metal with a screwdriver or the tang of a file or drill small holes in the affected area. This will provide a good grip for the filler material. To complete the repair, see the subsection on *filling and painting*.

Repair of rust holes or gashes

8   Remove all paint from the affected area and from an inch or so of the surrounding metal using a sanding disk or wire brush mounted in a drill motor. If these are not available, a few sheets of sandpaper will do the job just as effectively.
9   With the paint removed, you will be able to determine the severity of the corrosion and decide whether to replace the whole panel, if possible, or repair the affected area. New body panels are not as expensive as most people think and it is often quicker to install a new panel than to repair large areas of rust.
10   Remove all trim pieces from the affected area except those which will act as a guide to the original shape of the damaged body, such as headlight shells, etc. Using metal snips or a hacksaw blade, remove all loose metal and any other metal that is badly affected by rust. Hammer the edges of the hole on the inside to create a slight depression for the filler material.
11   Wire brush the affected area to remove the powdery rust from the surface of the metal. If the back of the rusted area is accessible, treat it with rust inhibiting paint.
12   Before filling is done, block the hole in some way. This can be done with sheet metal riveted or screwed into place, or by stuffing the hole with wire mesh.
13   Once the hole is blocked off, the affected area can be filled and painted. See the following subsection on *filling and painting*.

Filling and painting

14   Many types of body fillers are available, but generally speaking, body repair kits which contain filler paste and a tube of resin hardener are best for this type of repair work. A wide, flexible plastic or nylon applicator will be necessary for imparting a smooth and contoured finish to the surface of the filler material. Mix up a small amount of filler on a clean piece of wood or cardboard (use the hardener sparingly). Follow the manufacturer's instructions on the package, otherwise the filler will set incorrectly.
15   Using the applicator, apply the filler paste to the prepared area. Draw the applicator across the surface of the filler to achieve the desired contour and to level the filler surface. As soon as a contour that approximates the original one is achieved, stop working the paste. If you continue, the paste will begin to stick to the applicator. Continue to add thin layers of paste at 20-minute intervals until the level of the filler is just above the surrounding metal.
16   Once the filler has hardened, the excess can be removed with a body file. From then

on, progressively finer grades of sandpaper should be used, starting with a 180-grit paper and finishing with 600-grit wet-or-dry paper. Always wrap the sandpaper around a flat rubber or wooden block, otherwise the surface of the filler will not be completely flat. During the sanding of the filler surface, the wet-or-dry paper should be periodically rinsed in water. This will ensure that a very smooth finish is produced in the final stage.

17   At this point, the repair area should be surrounded by a ring of bare metal, which in turn should be encircled by the finely feathered edge of good paint. Rinse the repair area with clean water until all of the dust produced by the sanding operation is gone.

18   Spray the entire area with a light coat of primer. This will reveal any imperfections in the surface of the filler. Repair the imperfections with fresh filler paste or glaze filler and once more smooth the surface with sandpaper. Repeat this spray-and-repair procedure until you are satisfied that the surface of the filler and the feathered edge of the paint are perfect. Rinse the area with clean water and allow it to dry completely.

19   The repair area is now ready for painting. Spray painting must be carried out in a warm, dry, windless and dust-free atmosphere. These conditions can be created if you have access to a large indoor work area, but if you are forced to work in the open, you will have to pick the day very carefully. If you are working indoors, dousing the floor in the work area with water will help settle the dust which would otherwise be in the air. If the repair area is confined to one body panel, mask off the surrounding panels. This will help minimize the effects of a slight mismatch in paint color. Trim pieces such as chrome strips, door handles, etc., will also need to be masked off or removed. Use masking tape and several thickness of newspaper for the masking operations.

20   Before spraying, shake the paint can thoroughly, then spray a test area until the spray painting technique is mastered. Cover the repair area with a thick coat of primer. The thickness should be built up using several thin layers of primer rather than one thick one. Using 600-grit wet-or-dry sandpaper, rub down the surface of the primer until it is very smooth. While doing this, the work area should be thoroughly rinsed with water and the wet-or-dry sandpaper periodically rinsed as well. Allow the primer to dry before spraying additional coats.

21   Spray on the top coat, again building up the thickness by using several thin layers of paint. Begin spraying in the center of the repair area and then, using a circular motion, work out until the whole repair area and about two inches of the surrounding original paint is covered. Remove all masking material 10 to 15 minutes after spraying on the final coat of paint. Allow the new paint at least two weeks to harden, then use a very fine rubbing compound to blend the edges of the new paint into the existing paint. Finally, apply a coat of wax.

6   Body repair - major damage

1   Major damage must be repaired by an auto body shop specifically equipped to perform body and frame repairs. These shops have the specialized equipment required to do the job properly.

2   If the damage is extensive, the body must be checked for proper alignment or the vehicle's handling characteristics may be adversely affected and other components may wear at an accelerated rate.

3   Due to the fact that all of the major body components (hood, fenders, etc.) are separate and replaceable units, any seriously damaged components should be replaced rather than repaired. Sometimes the components can be found in a wrecking yard that specializes in used vehicle components, often at considerable savings over the cost of new parts.

7   Hinges and locks - maintenance

Once every 3000 miles, or every three months, the hinges and latch assemblies on the doors, hood and trunk should be given a few drops of light oil or lock lubricant. The door latch strikers should also be lubricated with a thin coat of grease to reduce wear and ensure free movement. Lubricate the door and trunk locks with spray-on graphite lubricant.

8   Windshield and fixed glass - replacement

Replacement of the windshield and fixed glass requires the use of special fast-setting adhesive/caulk materials and some specialized tools and techniques. These operations should be left to a dealer service department or a shop specializing in glass work.

9   Hood - removal, installation and adjustment

Removal and installation

Refer to illustration 9.4

Caution: *The hood is heavy and somewhat awkward to remove and install - at least two people should perform this procedure.*

1   Use blankets or pads to cover the cowl area of the body and fenders. This will protect the body and paint as the hood is lifted off.

2   Make marks or scribe a line around the hood hinge to ensure proper alignment during installation.

3   Disconnect any cables or wires that will interfere with removal.

4   Have an assistant support the hood. Remove the hinge-to-hood nuts or bolts **(see illustration)**.

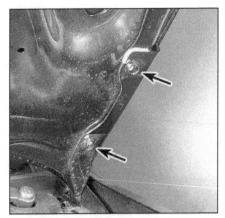

9.4 Detach the hinge-to-hood bolts at each end and lift off the hood with the help of an assistant

5   Lift off the hood.

6   Installation is the reverse of removal.

Adjustment

Refer to illustrations 9.10, 9.11a and 9.11b

7   Fore-and-aft and side-to-side adjustment of the hood is done by moving the hinge plate slot after loosening the bolts or nuts.

8   Scribe a line around the entire hinge plate so you can determine the amount of movement.

9   Loosen the bolts or nuts and move the hood into correct alignment. Move it only a little at a time. Tighten the hinge bolts and carefully lower the hood to check the position.

10   If necessary after installation, the entire hood latch assembly can be adjusted up-and-down as well as from side-to-side on the radiator support so the hood closes securely and flush with the fenders. To make the adjustment, scribe a line or mark around the hood latch mounting bolts to provide a reference point, then loosen them and reposition the latch assembly, as necessary **(see illustration)**. Following adjustment, retighten the mounting bolts.

9.10 To adjust the hood latch, loosen the retaining bolts, move the latch and retighten bolts, then close the hood to check the fit

These photos illustrate a method of repairing simple dents. They are intended to supplement *Body repair - minor damage* in this Chapter and should not be used as the sole instructions for body repair on these vehicles.

1 If you can't access the backside of the body panel to hammer out the dent, pull it out with a slide-hammer-type dent puller. In the deepest portion of the dent or along the crease line, drill or punch hole(s) at least one inch apart . . .

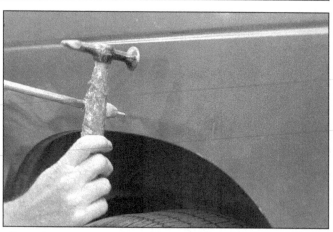

2 . . . then screw the slide-hammer into the hole and operate it. Tap with a hammer near the edge of the dent to help 'pop' the metal back to its original shape. When you're finished, the dent area should be close to its original contour and about 1/8-inch below the surface of the surrounding metal

3 Using coarse-grit sandpaper, remove the paint down to the bare metal. Hand sanding works fine, but the disc sander shown here makes the job faster. Use finer (about 320-grit) sandpaper to feather-edge the paint at least one inch around the dent area

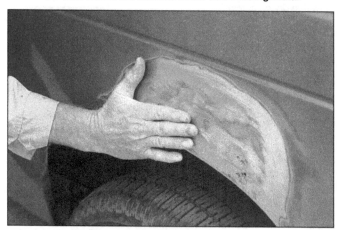

4 When the paint is removed, touch will probably be more helpful than sight for telling if the metal is straight. Hammer down the high spots or raise the low spots as necessary. Clean the repair area with wax/silicone remover

5 Following label instructions, mix up a batch of plastic filler and hardener. The ratio of filler to hardener is critical, and, if you mix it incorrectly, it will either not cure properly or cure too quickly (you won't have time to file and sand it into shape)

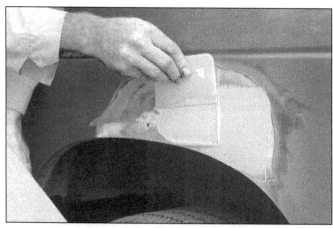

6 Working quickly so the filler doesn't harden, use a plastic applicator to press the body filler firmly into the metal, assuring it bonds completely. Work the filler until it matches the original contour and is slightly above the surrounding metal

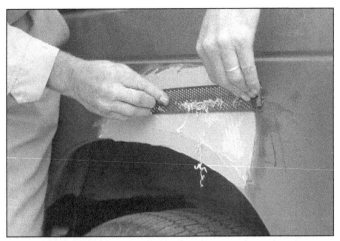

7 Let the filler harden until you can just dent it with your fingernail. Use a body file or Surform tool (shown here) to rough-shape the filler

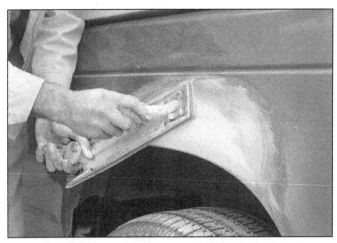

8 Use coarse-grit sandpaper and a sanding board or block to work the filler down until it's smooth and even. Work down to finer grits of sandpaper - always using a board or block - ending up with 360 or 400 grit

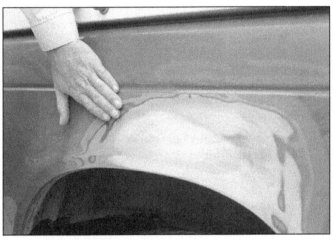

9 You shouldn't be able to feel any ridge at the transition from the filler to the bare metal or from the bare metal to the old paint. As soon as the repair is flat and uniform, remove the dust and mask off the adjacent panels or trim pieces

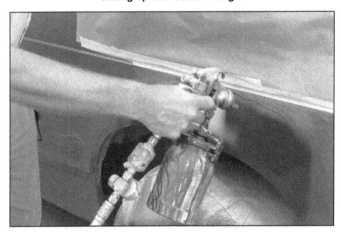

10 Apply several layers of primer to the area. Don't spray the primer on too heavy, so it sags or runs, and make sure each coat is dry before you spray on the next one. A professional-type spray gun is being used here, but aerosol spray primer is available inexpensively from auto parts stores

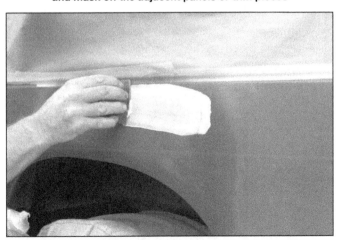

11 The primer will help reveal imperfections or scratches. Fill these with glazing compound. Follow the label instructions and sand it with 360 or 400-grit sandpaper until it's smooth. Repeat the glazing, sanding and respraying until the primer reveals a perfectly smooth surface

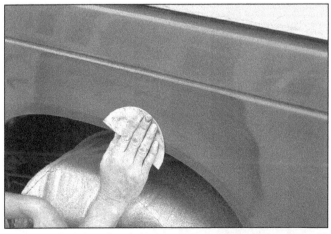

12 Finish sand the primer with very fine sandpaper (400 or 600-grit) to remove the primer overspray. Clean the area with water and allow it to dry. Use a tack rag to remove any dust, then apply the finish coat. Don't attempt to rub out or wax the repair area until the paint has dried completely (at least two weeks)

9.11a On some earlier vehicles, the hood bumpers are located on the hood itself; adjust the hood closing height by turning the hood bumpers in or out

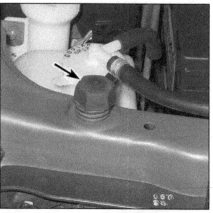

9.11b On other earlier vehicles, and on all newer vehicles, the bumpers are located on the radiator crossmember; to adjust the hood closing height, turn the bumpers in or out

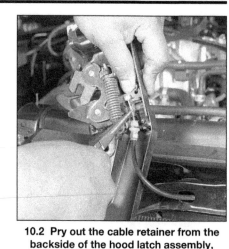

10.2 Pry out the cable retainer from the backside of the hood latch assembly, then disengage the cable

11   Finally, adjust the hood bumpers **(see illustrations)** on the hood, or on the radiator crossmember, so that the hood, when closed, is flush with the fenders.

12   The hood latch assembly, as well as the hinges, should be periodically lubricated with white, lithium-base grease to prevent binding and wear.

10   Hood release latch and cable - removal and installation

Latch

Refer to illustration 10.2

1   Scribe a line around the latch to aid alignment when installing, then detach the latch retaining bolts from the radiator support **(see illustration 9.10)** and remove the latch.

2   Disconnect the hood release cable by disengaging the cable from the latch assembly **(see illustration)**.

3   Installation is the reverse of the removal procedure. Make sure that the latch is adjusted so that the hood engages securely

when closed and the hood bumpers are slightly compressed.

Cable

Refer to illustrations 10.6a, 10.6b, 10.6c and 10.6d

4   Disconnect the hood release cable from the latch assembly as described above.

5   Attach a piece of stiff wire to the end of the cable, then follow the cable back to the firewall and detach all the cable retaining clips.

6   Working in the passenger compartment, lift the hood release lever upward to access the two release lever mounting screws. Detach the hood release lever, then disengage the cable from the back of the release lever **(see illustrations)**. 1998 and later Amigo models are equipped with a locking release lever. On these models, remove the trim piece that covers the lock cylinder and remove the release lever retaining screws **(see illustrations)**. If you want to replace the key lock cylinder, remove the two lock cylinder mounting screws.

7   Pull the cable and grommet rearward into the passenger compartment until you

can see the wire. Ensure that the new cable has a grommet attached, then remove the old cable from the wire and replace it with the new cable.

8   Working from engine compartment pull the cable back through the firewall.

9   When installing the cable, push on the grommet with your fingers from the passenger compartment to seat the grommet in the firewall correctly.

10   Installation is otherwise the reverse of removal.

11   Radiator grille - removal and installation

Refer to illustrations 11.1a and 11.1b

Warning: *Some models covered by this manual are equipped with airbags. Always disable the airbag system (see Chapter 12) before working in the vicinity of the impact sensors, steering column or instrument panel. Failure to follow these procedures may cause accidental deployment of the airbag, which could cause personal injury.*

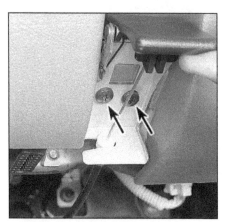

10.6a Lift the release lever and remove the lever retaining screws

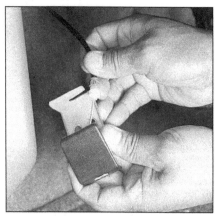

10.6b Disengage the cable from the lever, then pull the cable rearward into the passenger compartment

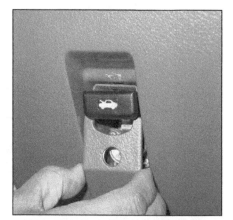

10.6c On 1998 and later Amigo models, remove the trim piece that covers the release lever lock cylinder . . .

10.6d ... and then remove the lever retaining screws; on these models, you can also replace the key lock cylinder, if necessary, by removing the two lock cylinder mounting screws

11.1a Remove the screw from the bottom of the grille

11.1b Pull out on the grille while using a screwdriver to release the grille retaining clips - lift upward on the tab to release the clip

1 The radiator grille is held in place by clips and screws. Remove the retaining screws and disengage the grille retaining clips with a small screwdriver **(see illustrations)**.
2 Once all the retaining clips are disengaged, pull the grille out and remove it.
3 Remove the grille retaining clips from the radiator support and reinstall the clips back into the grille.
4 To install, position the grille in place and press on the grille until the clips snap into place on the radiator support.
5 Reinstall the retaining screws.

12 Bumpers - removal and installation

Refer to illustrations 12.3a, 12.3b, 12.3c and 12.3d
1 Disconnect any electrical connections that would interfere with bumper removal.
2 Support the bumper with a jack or jack-

12.3a To detach the front bumper on 1997 and earlier models, remove the front bumper bracket-to-frame rail bolts (right side bolts not shown)

stands. Alternatively, have an assistant support the bumper as the bolts are removed.
3 Working from the backside of the bumper, remove the retaining bolts securing the bumper bracket to the outside of each frame rail **(see illustrations)**. Remove the

12.3b To detach the rear bumper on 1997 and earlier models, remove the frame rail-to-bumper bracket bolts (right side bolts not shown)

bumper from the vehicle.
4 If replacing the bumper fascia remove the nuts or bolts securing the bumper fascia to the bumper.
5 Installation is the reverse of removal.

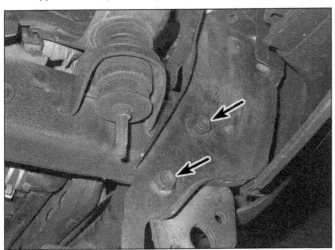

12.3c To detach the front bumper on 1998 and later models, remove the front bumper bracket-to-frame rail bolts (left side bolts not shown)

12.3d To detach the rear bumper on 1998 and later models, remove the frame rail-to-bumper bracket bolts (left side bolts not shown)

13.7a Working in the wheel opening remove the screws securing the inner fenderwell splash shield . . .

13.7b . . . then remove the fender to rocker panel bolt

13.7c Detach the retaining bolt located at the front of the fender below the side marker light

13.7d Open the door and remove the upper fender-to-door pillar bolt

13.7e The lower fender-to-door pillar bolt is accessible from the passenger compartment after removing the lower dashboard trim panel and the lower kick panel

13   Front fender - removal and installation

Refer to illustrations 13.7a, 13.7b, 13.7c, 13.7d, 13.7e and 13.7f

Note: *The (1997 and earlier) front fender assembly depicted in the accompanying photos is typical. The exact number and location*

of the front fender fasteners on 1998 and later models are similar but not identical to those shown here.

1   Loosen the front wheel lug nuts, raise the vehicle and support it securely on jackstands. Remove the wheel.

2   Remove the radiator grille (see Section 11).

3   Remove the front bumper (see Sec-

tion 12).

4   Remove the side marker and turn signal light assembly (see Chapter 12).

5   If removing the passenger side fender, remove the radio antenna (see Chapter 12).

6   Remove the lower dashboard trim panels (see Section 26).

7   Remove the fender mounting bolts and nuts **(see illustrations)**.

8   Detach the fender. It's a good idea to have an assistant support the fender while it's being moved away from the vehicle to prevent damage to the surrounding body panels.

9   Installation is the reverse of removal.

13.7f Detach the remaining bolts located in the hood opening

14   Door trim panel - removal and installation

Refer to illustrations 14.2, 14.3, 14.4a, 14.4b, 14.4c, 14.4d, 14.5a, 14.5b, 14.5c, 14.5d, 14.6 and 14.7

1   Disconnect the negative cable from the battery.

2   On manual window equipped models,

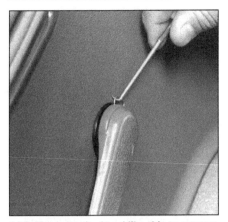

14.2 Use a hooked tool like this to remove the window crank retaining clip

14.3 Using a small screwdriver, pry out the armrest switch control plate

14.4a Remove the armrest/pull handle retaining screws

14.4b Remove the retaining screw and detach the inside door handle trim bezel

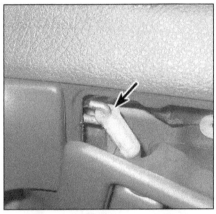

14.4c On 1998 and later models, the inside door handle and trim bezel are a one-piece assembly; after prying out the inside door handle/trim bezel assembly, rotate the assembly counterclockwise and disconnect the handle lever from the actuating rod

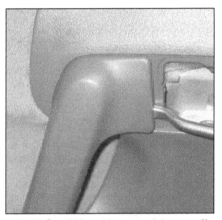

14.4d On 1998 and later models, pry off the pull handle cover

remove the window crank, using a hooked tool to remove the retainer clip **(see illustration)**. A special tool is available for this purpose, but it's not essential. With the clip removed, pull off the handle.

3   On power window equipped models, pry out the armrest switch control plate **(see illustration)** and disconnect the electrical connections.

4   On 1997 and earlier models, detach the armrest pull handle retaining screws **(see illustration)**, then remove the inside door

handle trim bezel **(see illustration)**. On 1998 and later models, the inside door handle and trim bezel are a one-piece assembly. On these models, pry out the bezel and then disconnect the door handle lever arm from the actuating rod **(see illustration)**. On 1998 and later models, pry off the pull handle cover

(see illustration).

5   Remove the remaining door panel retaining screws. On 1997 and earlier models, remove the screws from the edge of the door panel **(see illustrations)**. On 1998 and later models, remove the two screws from the pull handle **(see illustration)**. On all models, insert a wide putty knife, a thin screwdriver or a special trim panel removal tool between the trim panel and the head of the retaining clip to disengage the door panel retaining clips **(see illustration)**.

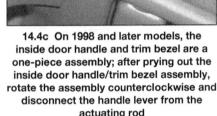

14.5a On 1997 and earlier models, pry out the plastic trim covers . . .

14.5b . . . and remove the retaining screws along the edge of the trim panel

14.5c On 1998 and later models, remove these two screws from the pull handle

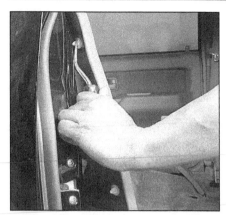

14.5d Using a trim removal tool or wide putty knife, detach the plastic clips securing the door trim panel to the door

14.6 Once all the retaining clips and screws have been removed, detach the trim panel and disconnect the electrical connector from the backside

14.7 Remove the door panel support bracket screws and carefully peel back the watershield to access the inner door cavity

6   Once all of the clips and screws are disengaged, detach the trim panel, disconnect the electrical connectors from the door-mounted speaker (if equipped) and from any power accessories (power windows, power door locks, etc.). Once everything has been detached or disconnected, remove the trim panel from the vehicle by gently pulling it up and out **(see illustration)**.

7   For access to the inner door, remove the door panel support bracket **(see illustration)**. Then peel back the watershield, taking care not to tear it. To install the trim panel, first press the watershield back into place. If necessary, add more sealant to hold it in place.

8   Installation is the reverse of removal.

---

15 Door latch, lock cylinder and handles - removal and installation

---

Door latch

Refer to illustrations 15.2a, 15.2b and 15.4

1   Raise the window then remove the door trim panel and watershield as described in (see Section 14).

15.2a Detach the plastic clips on the actuating rods leading to the latch (typical 1997 and earlier model)

2   Working through the large access hole, disengage the outside door handle-to-latch rod, outside door lock-to-latch rod, the inside lock-to-latch rod, the inside handle-to-latch rod and the lock solenoid-to-latch rod (if equipped) **(see illustrations)**.

3   All door locking rods are attached by

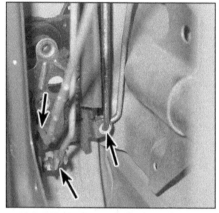

15.2b Typical latch assembly on 1998 and later models

plastic clips. The plastic clips can be removed by unsnapping the portion engaging the connecting rod and then pulling the rod out of its locating hole.

4   Remove the screws securing the latch to the door **(see illustration)**. Remove the latch assembly from the door.

5   Installation is the reverse of removal.

15.4 Remove the latch screws from the end of the door and pull the latch assembly through the access hole

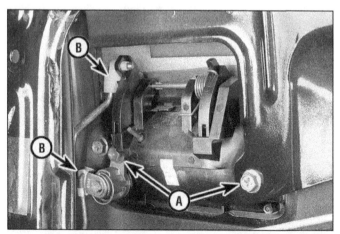

15.7 Detach the actuating rod clips (B) and the outside handle retaining bolts (A); both items can be reached through the access hole in the door frame (typical 1997 and earlier model)

15.12 Remove the inside handle retaining screws, then rotate the handle outward and detach the actuating rods from the backside (typical 1997 and earlier model); on 1998 and later models, the inside handle/trim bezel assembly must be removed before removing the trim panel (see illustration 14.4c)

16.3 Lower the window completely and remove the inner weatherstrip from the door window opening

Outside handle and door lock cylinder

Refer to illustration 15.7

6   To remove the outside handle and lock cylinder assembly, raise the window and remove the door trim panel and watershield (see Section 14).

7   Working through the access hole, disengage the plastic clips that secure the outside handle-to-latch rod and the outside door lock-to-latch rod **(see illustration)**.

8   Remove the outside handle retaining bolts **(see illustration 15.7)**.

9   Remove the handle and lock cylinder assembly from the vehicle.

10   Installation is the reverse of removal.

Inside handle

Refer to illustration 15.12

11   Remove the door trim panel (see Section 14) and peel away the watershield.

12   Unclip the door actuating rod guide, then remove the door handle retaining screws **(see illustration)**.

13   Pull the handle free from the door, then disconnect the actuating rod from the backside of the handle control and remove the handle from the door.

14   Installation is the reverse of removal.

---

16   Door window glass - removal and installation

Refer to illustrations 16.3, 16.4a and 16.4b

1   Remove the door trim panel and the plastic watershield (see Section 14).

2   Lower the window glass all the way down into the door.

3   Carefully pry the inner weatherstrip out of the door window opening **(see illustration)**.

4   Raise the window just enough to access the window retaining bolts through the hole in the door frame **(see illustrations)**.

5   Place a rag over the glass to help prevent scratching the glass and remove the two glass mounting bolts.

6   Remove the glass by pulling it up and out.

7   Installation is the reverse of removal.

---

17   Door window glass regulator - removal and installation

Refer to illustrations 17.4a, 17.4b and 17.4c

1   Remove the door trim panel and the plastic watershield (see Section 14).

2   Remove the window glass assembly (see Section 16).

3   On power operated windows, disconnect the electrical connector from the window regulator motor.

16.4a Raise the window just enough to access the glass retaining bolts through the hole in the door frame (1997 and earlier models)

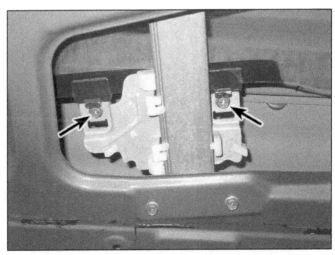

16.4b Door window glass retaining bolts (typical 1998 and later models)

17.4a Remove the window equalizer mounting bolts . . .

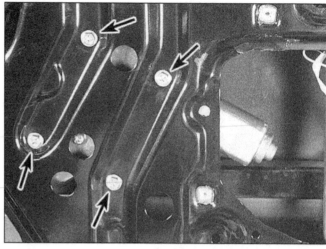

17.4b . . . then remove the window regulator bolts (typical 1997 and earlier model shown; later models similar)

4   On power operated windows, remove the equalizer arm bracket and the regulator mounting bolts **(see illustrations)**. On manually operated windows remove the regulator mounting bolts **(see illustration)**, then slide the regulator assembly out of the lower channel guide.

5   Pull the equalizer arm and regulator assemblies through the service hole in the door frame to remove it.

6   Installation is the reverse of removal.

18 Outside mirrors - removal and installation

Refer to illustrations 18.2a, 18.2b, 18.3a and 18.3b

1   Remove the door trim panel and the plastic watershield (see Section 14).

2   Pry off the mirror trim cover **(see illustrations)**.

3   Remove the mirror retaining bolts or screws and detach the mirror from the vehicle **(see illustrations)**.

4   Disconnect the electrical connector from the mirror (if equipped).

5   Installation is the reverse of removal.

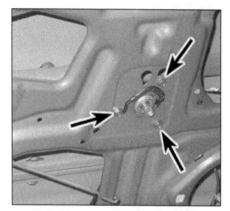

17.4c To remove a manual window regulator, remove these bolts (typical 1998 and later model shown; earlier models similar)

19 Door - removal, installation and adjustment

Removal and installation

Refer to illustrations 19.6, 19.8a and 19.8b

Caution: *The door is heavy and somewhat awkward to remove and install - at least two*

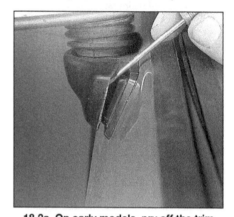

18.2a On early models, pry off the trim cover on the outside of the mirror

people should perform this procedure.

1   Raise the window completely in the door and then disconnect the negative cable from the battery.

2   Open the door all the way and support it on jacks or blocks covered with rags to prevent damaging the paint.

3   Remove the door trim panel and water deflector as described in Section 14.

4   Clearly label, then disconnect, all electri-

18.2b On later models, pry off the trim cover on the inside of the door

18.3a Outside mirror screw locations (typical early model shown)

18.3b Outside mirror bolt locations (typical later model shown)

19.6 Gently tap the pin for the door stop strut in the direction shown

19.8a Before loosening the door retaining bolts, draw a line around the hinge plate for a reinstallation reference

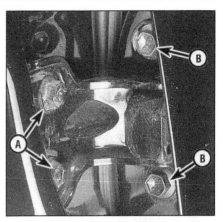

19.8b Open the front door to access the rear door hinge-to-body (A) and hinge-to-door (B) bolts

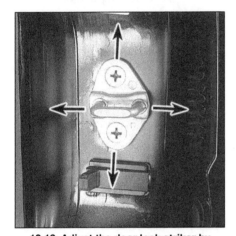

19.13 Adjust the door lock striker by loosening the mounting screws and gently tapping the striker in the desired direction

cal connections, ground wires and harness retaining clips attached to the door.

5   From the door side, detach the rubber conduit between the body and the door. Then pull the wiring harness through the conduit hole and remove it from the door.

6   Remove the door stop strut center pin **(see illustration)**.

7   Mark around the door hinges with a pen

or a scribe to facilitate realignment during reassembly.

8   With an assistant holding the door, remove the hinge to door bolts **(see illustrations)** and lift the door off.

9   Installation is the reverse of removal.

Adjustment

Refer to illustration 19.13

10   Having proper door to body alignment is a critical part of a well functioning door assembly. First check the door hinge pins for excessive play. Fully open the door and lift up and down on the door without lifting the body. If a door has 1/16-inch or more excessive play, the hinges should be replaced.

11   Make any door-to-body alignment adjustments by loosening the hinge-to-body bolts or hinge-to-door bolts and moving the door **(see illustration 19.8b)**. Proper body alignment is achieved when the top of the doors are parallel with the roof section, the front door is flush with the fender, the rear door is flush with the rear quarter panel and the bottom of the doors are aligned with the lower rocker panel. If these goals can't be reached by adjusting the hinge-to-body or hinge-to-door bolts, body alignment shims may have to be purchased and inserted

behind the hinges to achieve correct alignment.

12   To adjust the door closed position, scribe a line or mark around the striker plate to provide a reference point, then check that the door latch is contacting the center of the latch striker. If not adjust the up and down position first.

13   Finally adjust the latch striker sideways position, so that the door panel is flush with the center pillar or rear quarter panel and provides positive engagement with the latch mechanism **(see illustration)**.

20   Tailgate latch, handle and lock cylinder - removal and installation

Tailgate trim panel

Refer to illustrations 20.1a, 20.1b, 20.1c, 20.2a and 20.2b

1   On Rodeo and Passport models, remove the tailgate trim panel retaining screws and clips **(see illustrations)**. Then remove the tailgate access cover **(see illustration)** and peel back the plastic watershield.

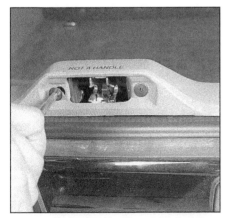

20.1a On Rodeo and Passport models, remove the screws located at the top of the tailgate trim panel . . .

20.1b . . . pry up to detach the remaining clips securing the tailgate trim panel . . .

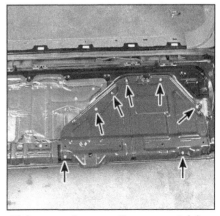

20.1c . . . and remove the cover retaining screws to access the inside of the tailgate

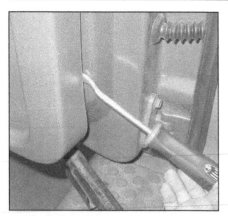

20.2a On Amigo models, pry off the tailgate trim panel . . .

20.2b . . . and peel off the plastic watershield

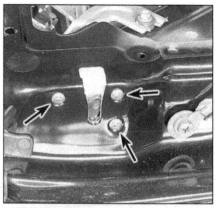

20.3 To detach the latch from the tailgate on Rodeo and Passport models, remove these retaining screws

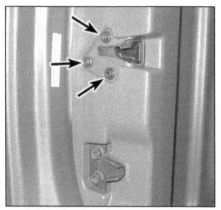

20.4a To detach the latch from the tailgate on Amigo models, remove these retaining screws

20.4b Disconnect the actuating rods from the latch mechanism and then remove the latch assembly (Amigo models)

20.6 Rear window latch screw locations (Rodeo and Passport models)

2   On Amigo models, pry off the tailgate trim panel and then remove the plastic watershield **(see illustrations)**.

Tailgate latch

Refer to illustrations 20.3, 20.4a and 20.4b
3   On Rodeo and Passport models, remove the latch mounting screws **(see illus-**

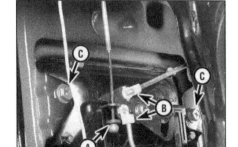

20.9 To remove the outside handle on Rodeo and Passport models, disconnect the release cable (A) and the actuating rods (B), then remove the retaining bolts (C)

tration) from each end of the tailgate. It may be necessary to use an impact-driver to loosen them. Remove the tailgate trim panel (see Step 1 or 2), disconnect the actuating rods from the latch and remove the latch from the tailgate.
4   On Amigo models, remove the latch mounting screws **(see illustration)** from the right end of the tailgate (Amigo models have only one latch). Remove the tailgate trim panel (see Step 1 or 2), disconnect the actuating rods from the latch **(see illustration)** and remove the latch from the tailgate.
5   Installation is the reverse of removal.

Window latch (Rodeo and Passport models only)

Refer to illustration 20.6
6   Remove the latch retaining bolts **(see illustration)**.
7   Disconnect the actuating rods from the latch and remove the latch from the door.
8   Installation is the reverse of removal.

Outside handle

Refer to illustrations 20.9 and 20.10
9   To remove the outside handle from the tailgate on Rodeo and Passport models, remove the tailgate trim panel (see Step 1),

disconnect the release cable and the actuating rods from the back of the handle and remove the handle mounting bolts **(see illustration)**.
10   To remove the outside handle from the tailgate on Amigo models, remove the tailgate trim panel (see Step 2), disconnect the actuating rod and remove the mounting bolts **(see illustration)**.
11   Installation is the reverse of removal.

20.10 To remove the outside handle on Amigo models, disconnect the actuating rod and remove the mounting bolts

20.12a To remove the lock cylinder from the tailgate on Rodeo and Passport models, disconnect the actuating rod (A), then remove the lock retaining clip (B) by sliding it off to the left

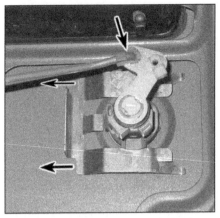

20.12b To remove the lock cylinder from the tailgate on Amigo models, disconnect the actuating rod and then remove the lock cylinder retaining clip by sliding it off to the left

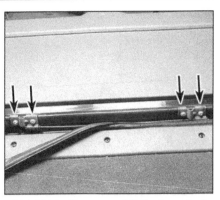

21.1 Typical torque rod (tailgate spring) mounting hardware locations (Rodeo and Passport models)

Lock cylinder

Refer to illustrations 20.12a and 20.12b

12   Detach the plastic clip securing the lock actuating rod **(see illustrations)**.
13   Remove the lock retaining clip securing the lock to the door.
14   Pull the lock cylinder out of the door.
15   Installation is the reverse of removal.

---

21   Tailgate - removal, installation and adjustment

Removal and installation

Refer to illustrations 21.1, 21.2a, 21.2b, 21.7, 21.8a and 21.8b
Caution: *The tailgate is heavy and somewhat awkward to remove and install - at least two people should perform this procedure.*
1   On Rodeo and Passport models, open the tailgate to identify the torque rod (tailgate spring) mounting hardware locations **(see illustration)**. Close the tailgate and then, working underneath the vehicle, remove the mounting hardware and the torque rod.

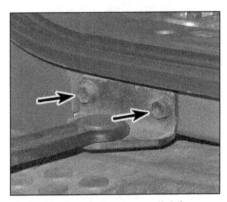

21.2a To detach the stopper link from an Amigo model, remove these bolts

2   On Amigo models, unbolt the tailgate stopper link from the body **(see illustration)**. If you want to remove or replace the stopper link assembly, remove the mounting bolts that secure the stopper link channel to the underside of the tailgate **(see illustration)**.
3   Remove the tailgate trim panel (see Step 1 or 2).
4   Disconnect the electrical connectors for all wiring leading to the tailgate.

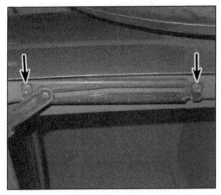

21.2b To detach the channel for the stopper link from the underside of the tailgate on Amigo models, remove these bolts

5   Cover the lower bumper area around the opening with pads or cloths to protect the painted surfaces when the tailgate is removed.
6   Open the tailgate.
7   On Rodeo and Passport models, remove the tailgate support cables from the door opening **(see illustration)**.
8   While an assistant supports the tailgate, detach the hinge-to-tailgate bolts **(see illustrations)** and remove the tailgate from the vehicle.
9   Installation is the reverse of removal.

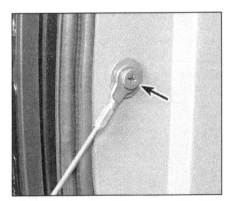

21.7 Remove the tailgate support cable retaining bolts from the door opening (Rodeo and Passport models only)

21.8a Remove the hinge-to-tailgate bolts (Rodeo and Passport models shown)

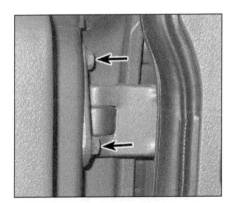

21.8b To detach the tailgate from the hinges on Amigo models, remove these bolts (upper hinge shown, lower hinge identical)

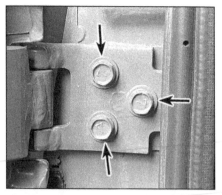

21.10 Loosen the hinge bolts (arrows) to adjust the tailgate (Amigo models shown)

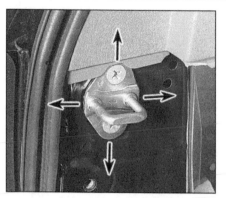

21.12 Remove the plastic trim panel, if necessary, to access the striker, loosen the screws, adjust the striker in the desired direction and then retighten the screws

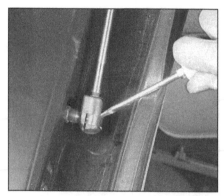

22.2 Use a small screwdriver to pry the clip out of its locking groove, then detach the end of the strut from the mounting stud

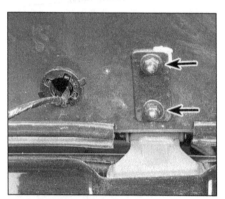

22.3 With the help of an assistant to hold the glass, remove the retaining nuts (arrows) from each hinge plate, then lift off the glass

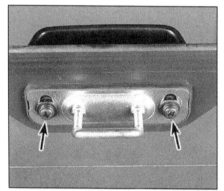

22.6 If the window does not close properly, it will be necessary to loosen the screws (arrows) to adjust the striker plate

23.3a Typical steering column cover screws on an earlier model (1993 model shown)

Adjustment

Refer to illustrations 21.10 and 21.12

10 Tailgate adjustments are made by loosening the hinge-to-body or hinge-to-tailgate bolts and moving the tailgate **(see illustration)**. Proper alignment is achieved when the top of the tailgate is aligned with the top of the rear quarter panel. If these goals can't be reached by adjusting the hinge to body or hinge to tailgate bolts, body alignment shims may have to be purchased and inserted behind the hinges to achieve correct alignment.

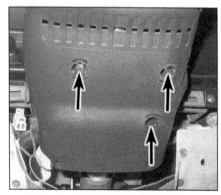

23.3b Typical steering column cover screws on 1998 and later models (Amigo model shown, Rodeo and Passport similar)

11 To adjust the tailgate closed position, first check that the latch is contacting the center of the latch striker assembly. If not, remove striker assembly and add or subtract shims to achieve correct alignment.
12 Finally, adjust the latch striker assembly as necessary (up and down or sideways) to provide positive engagement with the latch mechanism **(see illustration)**. Check that the outside of the tailgate is flush with rear quarter panel and readjust, if necessary.

22 Tailgate window (Rodeo and Passport models) - removal, installation and adjustment

Removal and installation

Refer to illustrations 22.2 and 22.3
Caution: *The tailgate window is heavy and somewhat awkward to hold - at least two people should perform this procedure.*
1 Open the tailgate window and support it securely.
2 Using a small screwdriver, detach the retaining clips at both ends of the support struts. Then pry or pull sharply to detach it from the vehicle **(see illustration)**.
3 While an assistant supports the window, detach the hinge to glass bolts **(see illustra-**

tion). Remove the tailgate window from the vehicle.
4 Installation is the reverse of removal.

Adjustment

Refer to illustration 22.6
5 Adjustments are made by loosening the hinge-to glass bolts and moving the window. Proper alignment is achieved when the edges of the glass are parallel with the rear quarter panel and the top of the tailgate.
6 Finally, adjust the latch striker assembly as necessary (up and down) to provide positive engagement with the latch mechanism **(see illustration)**.

23 Steering column cover - removal and installation

Refer to illustrations 23.3a and 23.3b
Warning: *Some models covered by this manual are equipped with airbags. Always disable the airbag system (see Chapter 12) before working in the vicinity of the impact sensors, steering column or instrument panel. Failure to follow these procedures may cause accidental deployment of the airbag, which could cause personal injury.*
1 Remove the steering wheel (see Chapter 10).

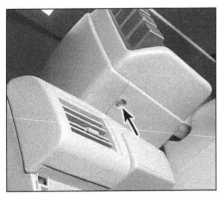

24.2a On 1995 and earlier models, detach the screw at each end of the instrument cluster bezel

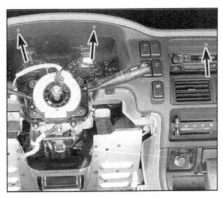

24.2b On 1996 and 1997 models, remove the screws securing the upper edge of the bezel (knee bolster removed)

24.2c To remove the instrument cluster bezel on 1998 and later models, remove the two upper cluster bezel screws . . .

24.2d . . . and remove the two lower bezel screws (knee bolster removed)

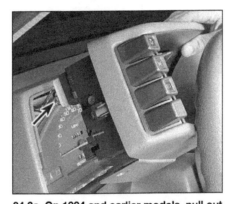

24.3a On 1994 and earlier models, pull out the instrument cluster and unplug the electrical connectors on the backside

24.3b On 1998 and later models, pull out the instrument cluster trim bezel and unplug this connector . . .

24.3c . . . and these two connectors

25.1 Unscrew the shift knob from the manual transmission and/or transfer case shift levers

2   Remove the driver side knee bolster (see Section 26).
3   Remove the steering column cover screws **(see illustrations)**.
4   Separate the cover halves and detach them from the steering column.
5   Installation is the reverse of removal.

24   Instrument cluster bezel - removal and installation

Refer to illustrations 24.2a, 24.2b, 24.2c, 24.2d, 24.3a, 24.3b and 24.3c
Warning: *Some models covered by this manual are equipped with airbags. Always disable the airbag system (see Chapter 12) before working in the vicinity of the impact sensors, steering column or instrument panel. Failure to follow these procedures may cause accidental deployment of the airbag, which could cause personal injury.*
1   On 1996 and later models, remove the driver-side knee bolster (see Section 26).
2   Remove the bezel retaining screws **(see illustrations)**.
3   Tilt the steering wheel down and pull the instrument cluster bezel outward to access the electrical connections on the backside **(see illustrations)**.
4   Disconnect all electrical connections

from the back of the cluster bezel and remove the bezel from the vehicle.
5   Installation is the reverse of removal.

25   Center console - removal and installation

Warning: *Some models covered by this manual are equipped with airbags. Always disable the airbag system (see Chapter 12) before working in the vicinity of the impact sensors, steering column or instrument panel. Failure*

to follow these procedures may cause accidental deployment of the airbag, which could cause personal injury.

1997 and earlier models

Shift lever console
Refer to illustrations 25.1 and 25.2
1   Unscrew the shift knob from the manual transmission and/or transfer case shift levers **(see illustration)**.
2   Pry out the plastic trim caps (if equipped) and remove the retaining screws

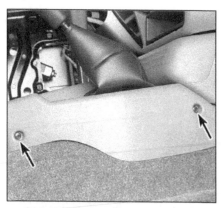

25.2 Detach the plastic trim caps (if equipped), then remove the retaining screws from each side of the shift lever console

25.7a On 1995 and earlier models, remove the screws and the plastic trim bezel from the front console

25.7b On 1996 and later models, remove the screws from each side of the console

located on each side of the shift console **(see illustration)**.

3   Lift the console up and over the shift lever.

4   Disconnect any electrical connections and remove the console from the vehicle.

5   Installation is the reverse of removal.

Front console

Refer to illustrations 25.7a, 25.7b and 25.8

6   Remove the shift lever console (see Steps 1 through 4).

7   Remove the retaining screws located on each side of the front console, then detach the plastic trim bezel **(see illustrations)**.

8   On 1995 and earlier vehicles remove the radio (see Chapter 12) and the remainder of the front console retaining screws **(see illustration)**.

9   Remove the console from the vehicle.

10   Installation is the reverse of removal.

Rear console (later Rodeo and Passport models)

Refer to illustrations 25.11, 25.12a, 25.12b and 25.13

11   Apply the parking brake and remove the

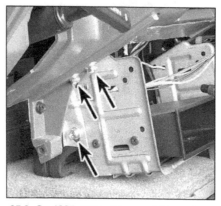

25.8 On 1995 and earlier models, remove the screws from each side of the console

trim bezel **(see illustration)**.

12   Pry out the plastic trim cap from the front of the console and remove the retaining screws **(see illustrations)**.

13   Remove the retaining screws located inside the console glove compartment **(see illustration)**.

14   Remove the rear half of the floor console by lifting it up and over the parking brake lever.

15   Installation is the reverse of removal.

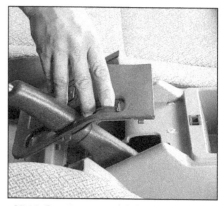

25.11 Pry out the trim bezel surrounding the parking brake lever, then disconnect the electrical connector from the switch assembly (1996 and 1997 Rodeo and Passport models)

1998 and later models

Refer to illustrations 25.16, 25.17, 25.19, 25.21 and 25.23

16   Open the rear console storage compartment lid and remove the two screws in the bottom of the storage compartment **(see illustration)**.

25.12a Pry out the trim cap on the front of the rear console . . .

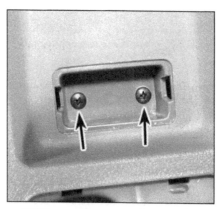

25.12b . . . then remove the retaining screws (1996 and 1997 Rodeo and Passport models)

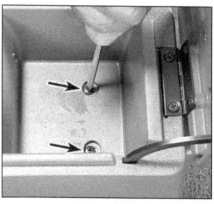

25.13 Open the rear console glove compartment and remove the retaining screws (1996 and 1997 Rodeo and Passport models)

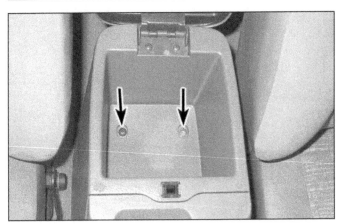

25.16 To remove the rear console on 1998 and later models, open the storage compartment lid and remove these two screws

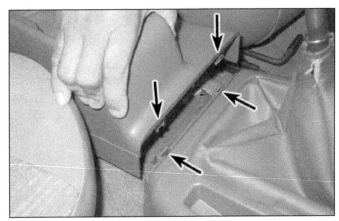

25.17 Pull up the forward edge of the center console to disengage the console locator clips from their corresponding slots on the rear edge of the front half of the console

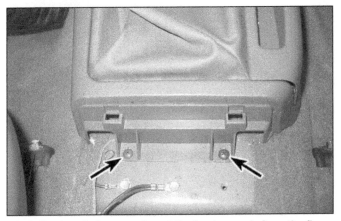

25.19 To detach the rear edge of the front console from the floor, remove these two screws

25.21 On 1998 and later models with a manual transmission, remove this small trim piece from the right side of the center console

17 Lift up the forward edge of the rear console to disengage the two console locator clips from their corresponding slots on the rear edge of the front console (see illustration).
18 Remove the rear console.
19 Remove the two mounting screws that secure the rear edge of the front console to the floor (see illustration).
20 Remove the shift lever knob and, if applicable, the transfer lever knob (see Chapter 7).

21 On models with a manual transmission, remove the trim piece from the right side of the center console (see illustration).
22 Remove the center trim bezel (see Section 26).
23 Remove the remaining four console mounting screws (see illustration).
24 Remove the front console.
25 Installation is the reverse of removal.

26 Dashboard trim panels - removal and installation

Warning: *Some models covered by this manual are equipped with airbags. Always disable the airbag system (see Chapter 12) before working in the vicinity of the impact sensors, steering column or instrument panel. Failure to follow these procedures may cause accidental deployment of the airbag, which could cause personal injury.*

Knee bolster

Refer to illustration 26.3a, 26.3b, 26.3c, 26.4 and 26.5

1 The knee bolster is located on the lower half of the instrument panel on the driver's side of the vehicle. Removal of the covers will allow access to a variety of electrical, heating and air conditioning components.
2 On 1996 and later vehicles, detach the hood release handle (see Section 10).
3 On 1997 and earlier models, detach the retaining screws along the perimeter of the

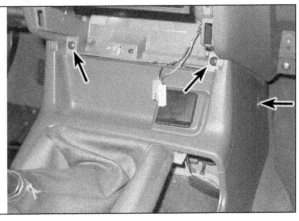

25.23 To detach the front console from the dash, remove the two console mounting screws from the forward upper edge of the console and remove the two screws from the left and right sides of the console (left console mounting screw not visible in this photo)

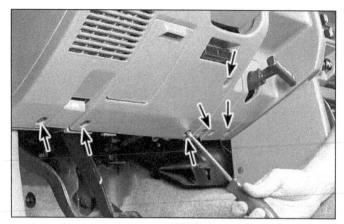

26.3a Remove the screws from the lower edge of the knee bolster (1995 and earlier model shown)

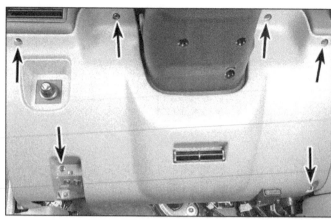

26.3b Remove the screws (arrows) from the outer edge of the knee bolster (1996 and later model shown)

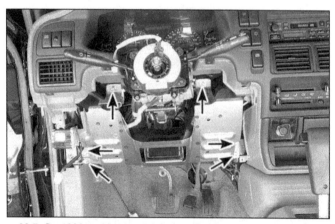

26.3c On 1998 and later models, remove this screw from the lower right corner of the knee bolster

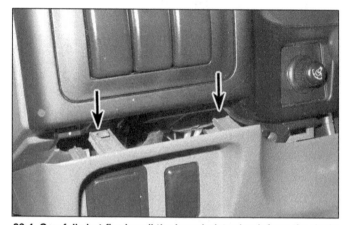

26.4 Carefully but firmly pull the knee bolster back from the dash; be careful not to break any of the locator tabs

knee bolster **(see illustrations)**. On 1998 and later models, remove the screw at the lower right corner of the bolster **(see illustration)**.

4   Pull outward on the lower edge of the knee bolster **(see illustration)**, disconnect any electrical connectors that would interfere with removal and detach it from the vehicle.

5   On 1996 and later models it will be necessary to remove a steel reinforcement panel to access any components underneath the instrument panel **(see illustration)**.

6   Installation is the reverse of removal.

Glove box

Refer to illustrations 26.7 and 26.8

7   On 1995 and earlier models, open the glove box and remove the glove box retaining

pins **(see illustration)**.

8   On 1996 and later models, remove the hinge screws from the lower edge of the glove box door **(see illustration)**.

9   Detach the glove box door from the instrument panel and remove it from the vehicle.

10   Installation is the reverse of removal.

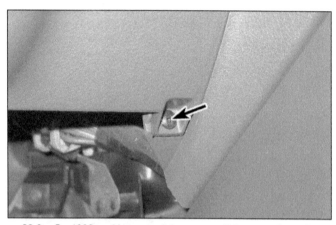

26.5 On 1996 and later models, you'll also have to remove the screws securing the knee bolster reinforcement panel to access any switches, flashers, etc. above it

26.7 On 1995 and earlier models, use a pair of needle nose pliers to compress the glove box retaining pin

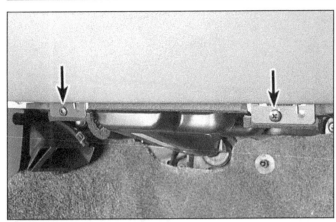

26.8 On 1996 and later models, remove the screws from the glove box door hinge

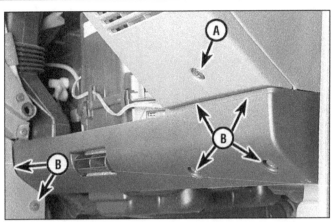

26.12 On 1995 and earlier models, remove the retaining screw from the speaker grille (A) then detach the screws from the lower reinforcement panel (B)

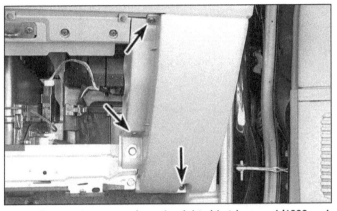

26.13 Detach the screws from the right side trim panel (1996 and later models)

26.17 Pry off the bezel from the front of the center trim panel

Passenger side trim panels

Refer to illustrations 26.12 and 26.13

11   Remove the glove box door from the instrument panel as described above.

12   On 1995 and earlier models, remove the screws securing the right speaker grille and the lower reinforcement panel **(see illustration)**. Pull the reinforcement panel outward and detach the air conditioning duct from the rear if equipped.

13   On 1996 and later models, remove the retaining screws from right trim panel **(see illustration)**.

14   Installation is the reverse of removal.

Center trim panel

1996 and 1997 models

Refer to illustration 26.17, 26.18 and 26.19

15   Remove the glove box door and the driver's side knee bolster as described above.

16   Remove the instrument cluster bezel (see Section 24).

17   Pry off the bezel on the front of the center trim panel **(see illustration)**.

18   Detach the retaining screws securing the package tray and remove it from the center trim panel **(see illustration)**.

19   Detach the remainder of the retaining screws and remove the center trim panel from the vehicle **(see illustration)**.

20   Installation is the reverse of removal.

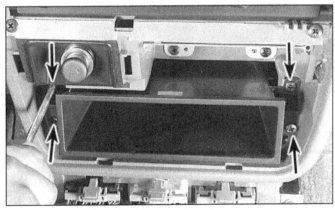

26.18 Remove the package tray screws

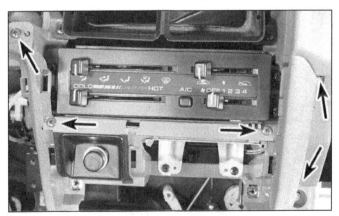

26.19 Detach the remaining center trim panel screws, then remove it from the vehicle

26.21 To detach the center trim panel on 1998 and later models, remove this screw from the cavity behind the ashtray

26.22 Pull out the center trim panel and disconnect any electrical connectors (1998 and later models)

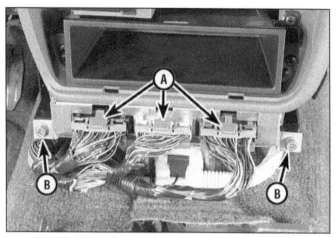

27.8 On 1996 and later models, disconnect the electrical connectors (A) and the retaining bolts (B) to remove the PCM

27.10 Typical instrument panel mounting bolt locations (1996 and 1997 model shown)

1998 and later models

Refer to illustrations 26.21 and 26.22

21   Open the ashtray and remove the screw behind it **(see illustration)**.

22   Pull out the center trim panel and disconnect any electrical connectors from the back **(see illustration)**.

23   Remove the center trim panel.

24   Installation is the reverse of removal.

27   Instrument panel - removal and installation

Refer to illustration 27.8

Warning: *Some models covered by this manual are equipped with airbags. Always disable the airbag system (see Chapter 12) before working in the vicinity of the impact sensors, steering column or instrument panel. Failure to follow these procedures may cause accidental deployment of the airbag, which could cause personal injury.*

1   Disconnect the negative battery cable.

2   Remove the driver's side airbag (if equipped) and the steering wheel (see Chapter 10).

3   Detach the hood release handle (see Section 10).

4   Remove the instrument cluster bezel (see Section 24).

5   Remove the center floor console (see Section 25).

6   Remove the dashboard trim panels and the glove box (see Section 26).

7   Remove the instrument cluster (see Chapter 12).

8   On 1996 and later models, remove the ECM **(see illustration)**.

1997 and earlier models

Refer to illustration 27.10

9   Remove the driver and passenger side kick panels.

10   Remove the bolts securing the lower half of the instrument panel **(see illustration)**.

11   On 1996 and later models, remove the bolts that attach the center of the instrument

panel to the cowl support tube.

12   Pry out the defroster grilles and remove the screws securing the upper edge of the instrument panel **(see illustration 27.10)**.

13   Pull the instrument panel towards the rear of the vehicle and detach any electrical connectors interfering with removal.

14   Lift the instrument panel over the steering column and remove it from the vehicle.

15   Installation is the reverse of removal.

1998 and later models

Refer to illustrations 27.16a, 27.16b, 27.17, 27.18, 27.19, 27.20, 27.21 and 27.22

16   Using a small screwdriver, pry off the three covers from the top of the dash and remove the three upper dash mounting nuts **(see illustrations)**.

17   Remove the left and right kick panels **(see illustration)**.

18   Detach the fuse panel from the left end of the dash and remove the lower left dash mounting bolt **(see illustration)**.

19   Remove the lower right dash mounting

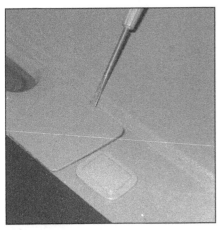

27.16a Pry out the three small access covers from the top of the dash (left cover shown; center and right covers identical) . . .

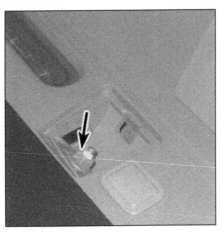

27.16b . . . and remove the three upper dash mounting nuts (left nut shown)

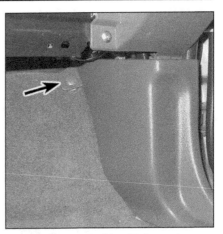

27.17 To detach either kick panel on a 1998 or later model, simply pull out this pop fastener and then carefully separate the kick panel from the body (right kick panel shown, left kick panel identical)

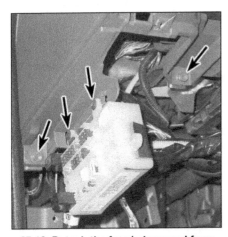

27.18 Detach the fuse/relay panel from the left end of the dash and remove the two lower left dash mounting bolts

27.19 Remove the lower right dash mounting bolt

27.20 Remove the two nuts that secure the steering column bracket to the underside of the dash

27.21 Remove these four screws from the center part of the dash

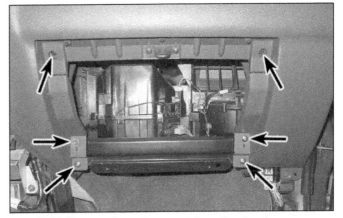

27.22 Remove these six screws from the glovebox area

bolt (see illustration).
20   Remove the two nuts that secure the steering column bracket to the underside of the dash (see illustration) and allow the steering column to hang down.

21   Remove the four screws from the center of the dash (see illustration).
22   Remove the six screws from the glove box cavity (see illustration).
23   Make a last-minute inspection to verify

that all electrical connectors are disconnected, all wiring harnesses are detached, and all fasteners are removed.
24   Remove the instrument panel assembly.
25   Installation is the reverse of removal.

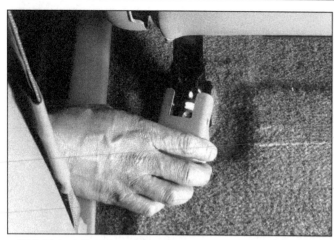

28.2 Detach the trim covers to access the seat retaining bolts

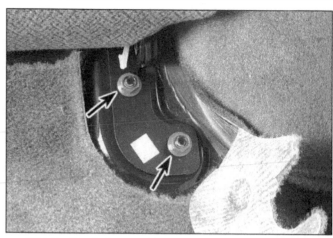

28.7 Peel back the carpeting to access the rear seat back retaining nuts

28   Seats - removal and installation

Front seat

Refer to illustration 28.2

1   Position the seat all the way forward or all the way to the rear to access the front seat retaining bolts.

2   Detach any bolt trim covers and remove the retaining bolts **(see illustration)**.

3   Tilt the seat upward to access the underneath, then disconnect any electrical connectors and lift the seat from the vehicle.

4   Installation is the reverse of removal.

Rear seat

Refer to illustrations 28.7

5   Detach the bolt trim covers and remove the seat cushion retaining bolts. Then lift up on the front edge and remove the cushion from the vehicle.

6   Lower the seat back and detach the carpeting after prying out the plastic retaining clips.

7   Detach the retaining bolts at the lower edge of the seat back **(see illustration)**.

8   Lift up on the lower edge of the seat back and remove it from the vehicle.

9   Installation is the reverse of removal.

29   Seat belts - check

1   Check the seat belts, buckles, latch plates and guide loops for any obvious damage or signs of wear.

2   Make sure the seat belt reminder light comes on when the key is turned on.

3   The seat belts are designed to lock up during a sudden stop or impact, yet allow free movement during normal driving. The retractors should hold the belt against your chest while driving and rewind the belt when the buckle is unlatched.

4   If any of the above checks reveal problems with the seat-belt system, replace parts as necessary. **Note:** *Before replacing any seat belt components, check with your local dealership service department; the seat belt system might be covered under the vehicle warranty.*

Chapter 12
Chassis electrical system

Contents

1   General information

The electrical system is a 12-volt, negative ground type. Power for the lights and all electrical accessories is supplied by a lead/acid-type battery which is charged by the alternator.

This Chapter covers repair and service procedures for the various electrical components not associated with the engine. Information on the battery, alternator, distributor and starter motor can be found in Chapter 5.

It should be noted that when portions of the electrical system are serviced, the cable should be disconnected from the negative battery terminal to prevent electrical shorts and/or fires.

2   Electrical troubleshooting - general information

A typical electrical circuit consists of an electrical component, any switches, relays, motors, fuses, fusible links or circuit breakers related to that component and the wiring and electrical connectors that link the component to both the battery and the chassis. To help you pinpoint an electrical circuit problem, wiring diagrams are included at the end of this Chapter.

Before tackling any troublesome electrical circuit, first study the appropriate wiring diagrams to get a complete understanding of what makes up that individual circuit. Trouble spots, for instance, can often be narrowed down by noting if other components related to the circuit are operating properly. If several components or circuits fail at one time, chances are the problem is in a fuse or ground connection, because several circuits are often routed through the same fuse and ground connections.

Electrical problems usually stem from simple causes, such as loose or corroded connections, a blown fuse, a melted fusible link or a bad relay. Visually inspect the condition of all fuses, wires and connections in a problem circuit before troubleshooting it.

If testing instruments are going to be utilized, use the diagrams to plan ahead of time where you will make the necessary connections in order to accurately pinpoint the trouble spot.

The basic tools needed for electrical troubleshooting include a circuit tester or voltmeter (a 12-volt bulb with a set of test leads can also be used), a continuity tester, which includes a bulb, battery and set of test leads, and a jumper wire, preferably with a circuit breaker incorporated, which can be used to bypass electrical components. Before attempting to locate a problem with test instruments, use the wiring diagram(s) to decide where to make the connections.

Voltage checks

Voltage checks should be performed if a circuit is not functioning properly. Connect one lead of a circuit tester to either the negative battery terminal or a known good ground. Connect the other lead to a electrical connector in the circuit being tested, preferably nearest to the battery or fuse. If the bulb of the tester lights, voltage is present, which means that the part of the circuit between the electrical connector and the battery is problem free. Continue checking the rest of the circuit in the same fashion. When you reach a point at which no voltage is present, the problem lies between that point and the last test point with voltage. Most of the time the problem can be traced to a loose connection. Keep in mind that some circuits receive voltage only when the ignition key is in the Accessory or Run position.

Finding a short

One method of finding shorts in a circuit is to remove the fuse and connect a test light or voltmeter in its place. There should be no voltage present in the circuit. Move the wiring harness from side to side while watching the test light. If the bulb goes on, there is a short to ground somewhere in that area, probably

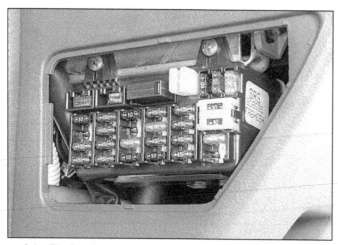

3.1a The interior fuse box is located on the lower half of the instrument panel, behind the fuse panel cover

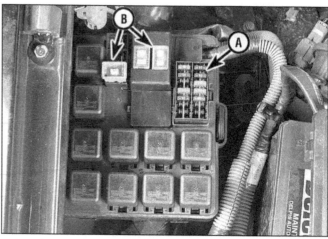

3.1b The engine compartment fuse box is located next to the battery; it contains fuses (A) as well as fusible links (B)

where the insulation has rubbed through. The same test can be performed on each component in the circuit, even a switch.

Ground check

Perform a ground test to check whether a component is properly grounded. Disconnect the battery and connect one lead of a self-powered test light, known as a continuity tester, to a known good ground. Connect the other lead to the wire or ground connection being tested. If the bulb goes on, the ground is good. If the bulb does not go on, the ground is not good.

Continuity check

A continuity check is done to determine if there are any breaks in a circuit - if it is passing electricity properly. With the circuit off (no power in the circuit), a self-powered continuity tester can be used to check the circuit. Connect the test leads to both ends of the circuit (or to the "power" end and a good ground), and if the test light comes on the circuit is passing current properly. If the light doesn't come on, there is a break somewhere in the circuit. The same procedure can be used to test a switch, by connecting the continuity tester to the power in and power out

sides of the switch. With the switch turned On, the test light should come on.

Finding an open circuit

When diagnosing for possible open circuits, it is often difficult to locate them by sight because oxidation or terminal misalignment are hidden by the electrical connectors. Merely wiggling an electrical connector on a sensor or in the wiring harness may correct the open circuit condition. Remember this when an open circuit is indicated when troubleshooting a circuit. Intermittent problems may also be caused by oxidized or loose connections.

Electrical troubleshooting is simple if you keep in mind that all electrical circuits are basically electricity running from the battery, through the wires, switches, relays, fuses and fusible links to each electrical component (light bulb, motor, etc.) and to ground, from which it is passed back to the battery. Any electrical problem is an interruption in the flow of electricity to and from the battery.

3   Fuses - general information

Refer to illustrations 3.1a, 3.1b and 3.3

The electrical circuits of the vehicle are

protected by a combination of fuses, circuit breakers and fusible links. The fuse blocks are located under the instrument panel on the left side of the dashboard, and in the engine compartment next to the battery **(see illustrations)**.

Each of the fuses is designed to protect a specific circuit, and the various circuits are identified on the fuse panel cover.

Miniaturized fuses are employed in the fuse blocks. These compact fuses, with blade terminal design, allow fingertip removal and replacement. If an electrical component fails, always check the fuse first. The best way to check the fuses is with a test light. Check for power at the exposed terminal tips of each fuse. If power is present at one side of the fuse but not the other, the fuse is blown. A blown fuse can also be identified through the clear plastic body. Visually inspect the element for evidence of damage **(see illustration)**.

Be sure to replace blown fuses with the correct type. Fuses of different ratings are physically interchangeable, but only fuses of the proper rating should be used. Replacing a fuse with one of a higher or lower value than specified is not recommended. Each electrical circuit needs a specific amount of protection. The amperage value of each fuse is molded into the fuse body.

If the replacement fuse immediately fails, don't replace it again until the cause of the problem is isolated and corrected. In most cases, this will be a short circuit in the wiring caused by a broken or deteriorated wire.

4   Fusible links - general information

Some circuits are protected by fusible links. The links are used in circuits which are not ordinarily fused, such as the ignition circuit.

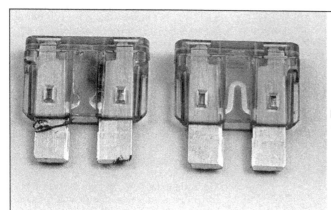

3.3 When a fuse blows, the element between the terminals melts (the fuse on the left is blown)

6.4 Most relays are marked on the outside to easily identify the control circuit and power circuits

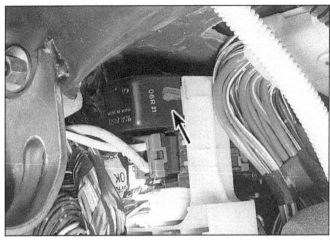

7.1 On earlier models, the turn signal and hazard flasher is located under the dash to the left of the steering column

The fusible links are located in the engine compartment fuse block **(see illustration 3.1b)** and are similar to fuses.

To replace a fusible link, first disconnect the negative cable from the battery. Unplug the burned-out link from the fuse block and replace it with a new one (available from your dealer or auto parts store). Always determine the cause for the overload which melted the fusible link before installing a new one.

5   Circuit breakers - general information

Circuit breakers protect components such as, power windows, power door locks and headlights.

On some models the circuit breaker resets itself automatically, so an electrical overload in a circuit breaker protected system will cause the circuit to fail momentarily, then come back on. If the circuit doesn't come back on, check it immediately. Once the condition is corrected, the circuit breaker will resume its normal function. Some circuit breakers must be reset manually.

6   Relays - general information and testing

General information

1   Several electrical accessories in the vehicle, such as the fuel injection system, horns, starter, and fog lamps use relays to transmit the electrical signal to the component. Relays use a low-current circuit (the control circuit) to open and close a high-current circuit (the power circuit). If the relay is defective, that component will not operate properly. The various relays are mounted in the engine compartment **(see illustration 3.1b)** and several locations throughout the vehicle. If a faulty relay is suspected, it can be removed and tested using the procedure

below or by a dealer service department or a repair shop. Defective relays must be replaced as a unit.

Testing

Refer to illustration 6.4

2   It's best to refer to the wiring diagram for the circuit to determine the proper connections for the relay you're testing. However, if you're not able to determine the correct test connections from the wiring diagrams, you may be able to determine the test connections from the information that follows.

3   On most relays, two of the terminals are the relay's control circuit (they connect to the relay coil which, when energized, closes the large contacts to complete the circuit). The other terminals are the power circuit (they are connected together within the relay when the control-circuit coil is energized).

4   The relays are marked as an aid to help you determine which terminals are the control circuit and which are the power circuit **(see illustration)**.

5   Connect a fused jumper wire between one of the two control circuit terminals and the positive battery terminal. Connect another jumper wire between the other control circuit terminal and ground. When the connections are made, the relay should click. On some relays, polarity may be critical, so, if the relay doesn't click, try swapping the jumper wires on the control circuit terminals.

6   With the jumper wires connected, check for continuity between the power circuit terminals as indicated by the markings on the relay.

7   If the relay fails any of the above tests, replace it.

7   Turn signal and hazard flashers - check and replacement

Refer to illustration 7.1

1   The turn signal flasher and hazard flash-

ers are contained in a single small canister-shaped unit located under the dash. On earlier models, the flasher unit **(see illustration)** is located near the steering column. On later models, the flasher unit is located farther to the left, near the fuse and relay panel. The turn signal flashes the turn signals and the hazard flasher flashes all four turn signals simultaneously when activated.

2   When the flasher unit is functioning properly, an audible click can be heard during its operation. If the turn signals fail on one side or the other and the flasher unit does not make its characteristic clicking sound, or if a bulb on one side of the vehicle flashes much faster than normal but the bulb at the other end of the vehicle (on the same side) doesn't light at all, a faulty turn signal bulb may be indicated.

3   If both turn signals fail to blink, the problem may be due to a blown fuse, a faulty flasher unit, a broken switch or a loose or open connection. If a quick check of the fuse box indicates that the turn signal fuse has blown, check the wiring for a short before installing a new fuse.

4   To replace the flasher, simply pull it out of the wiring harness.

5   Make sure that the replacement unit is identical to the original. Compare the old one to the new one before installing it.

6   Installation is the reverse of removal.

8   Steering column switches - check and replacement

Warning: *Some models covered by this manual are equipped with airbags. Always disable the airbag system (see Section 26) before working in the vicinity of the impact sensors, steering column or instrument panel. Failure to follow these procedures may cause accidental deployment of the airbag, which could cause personal injury.*

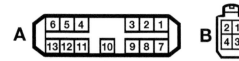

8.2a Combination switch connector identification, terminal guide and continuity chart (1995 and earlier models)

A   *Turn signal, hazard and dimmer switch*
B   *Cruise control switch*

Switch	Switch positions	Continuity between
Turn signal	Neutral	1 and 2
	Left	1 and 2; 4 and 5
	Right	1 and 2; 3 and 5
Hazard	On	2 and 11; 3, 4 and 5
Dimmer	Low beam	8 and 10
	High beam	9 and 10
	Flash to pass	7, 9 and 10
Cruise control	On	2 and 4
	Set coast*	1 and 3
	Resume accel*	2, 3 and 4

*Test with main switch in the On position

Check

Refer to illustrations 8.2a, 8.2b and 8.2c

1   Trace the wire from the combination switch to the main wiring harness connector and disconnect the connectors.

2   Using an ohmmeter or self-powered test light and the accompanying diagrams, check for continuity between the indicated switch terminals with the switch in each of the indicated positions **(see illustrations)**. If the continuity isn't as specified, replace the switch.

Replacement

Refer to illustration 8.7

Caution: *On models equipped with an airbag, the airbag coil is part of the combination switch - DO NOT attempt to separate the airbag coil from the combination switch*

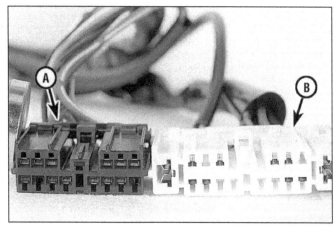

8.2b Combination switch connector identification, terminal guide and continuity chart (1996 and 1997 models)

A   *Headlight, turn signal, and dimmer switch*
B   *Cruise control and windshield wiper switch*

Switch	Switch positions	Continuity between
Headlight	Lo	7 and 8
	Hi	1, 7 and 8
Turn signal	Neutral	0
	Left	3 and 4
	Right	3 and 5
Dimmer	Low beam	13 and 16
	High beam	15 and 16
	Flash to pass	14, 15 and 16
Cruise control	Cruise	12 and 13
	Set coast	9 and 10; 12 and 13
	Resume accel	9 and 11; 12 and 13
Windshield wiper	Mist	3 and 5
	Off	5 and 6
with Intermitent	Int	2 and 3; 5 and 6
	Lo	3 and 5
	Hi	3 and 4
	Wash	1 and 7
Windshield wiper	Mist	1 and 7; 3 and 5
	Off	1 and 7; 5 and 6
without Intermitent	Lo	1 and 7; 3 and 5
	Hi	1 and 7; 3 and 4
	Wash	1 and 7

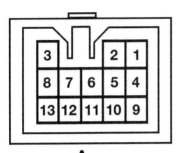
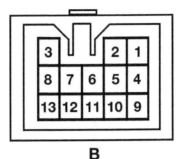

A　　　　　**B**

8.2c Combination switch connector identification, terminal guide and continuity chart (1998 and later models)

A *Headlight, dimmer, passing and turn signal switch*
B *Cruise control and windshield wiper switch*

Switch	Switch positions	Continuity between terminals
Headlights	OFF	No continuity
	ON	2, 3 and 9
Dimming switch	LO	4 and 5
	HI	1 and 4
Passing switch	OFF	No continuity
	ON	4 and 10
Turn signal	OFF	No continuity
	LEFT	6 and 7
	RIGHT	6 and 8
Cruise control	CANCEL - OFF	No continuity
	CANCEL - ON	9 and 12
	RES/ACC - OFF	No continuity
	RES/ACC - ON	9 and 11
	SET COAST - OFF	No continuity
	SET COAST ON	9 and 10
Windshield wipers	OFF	No continuity
	INT*	2 and 3
	LO	3 and 5
	HI	3 and 4
	MIST - OFF	No continuity
	MIST - ON	3 and 5
	WASH - OFF	No continuity
	WASH - ON	1 and 7

* If equipped with intermittent wipers.

47017-12-8.2c HAYNES

assembly.

3　Disconnect the cable from the negative terminal of the battery.

4　Remove the steering wheel (see Chapter 10).

5　Remove the steering column covers (see Chapter 11).

6　Unplug the electrical connectors from the combination switch.

7　Remove the retaining screws and pull the switch from the shaft **(see illustration)**.

8　Installation is the reverse the removal. If

equipped with an airbag, center the airbag coil as follows:

a) *Rotate the coil hub clockwise until it stops. Stop turning when resistance is felt. Do not force the coil against its stop or damage to the coil may result.*

b) *Rotate the coil hub counterclockwise 3 full turns.*

c) *Align the alignment mark on the coil hub with the mark on the coil body.*

9　Ignition switch and lock cylinder - check and replacement

Warning: *Some models covered by this manual are equipped with airbags. Always disable*

the airbag system (see Section 26) before working in the vicinity of the impact sensors, steering column or instrument panel. Failure to follow these procedures may cause accidental deployment of the airbag, which could cause personal injury.

Check

Refer to illustrations 9.2a and 9.2b

1　Trace the wiring from the ignition switch to the main wiring harness connector and disconnect the connector.

2　Using an ohmmeter or self-powered test light and the accompanying diagrams, check for continuity between the indicated switch terminals with the switch in each of the indicated positions **(see illustrations)**. If the continuity isn't as specified, replace the switch.

8.7 Combination switch retaining screws (1996 and later model shown, 1995 and earlier models similar)

Switch positions (key inserted)	Continuity between
Lock	7 and 8
Off	7 and 8
Acc	1 and 2; 7 and 8
On	1 and 2; 1 and 3 7 and 8
Start	1 and 3; 1 and 6 1 and 5; 7 and 8

9.2a Ignition switch terminal guide and continuity chart (1995 and earlier models)

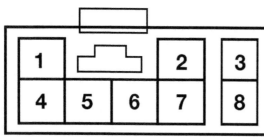

Switch positions (key inserted)	Continuity between
Lock	5 and 7
Off	5 and 7
Acc	1 and 4; 5 and 7
On	1 and 4; 1 and 8
	3 and 6; 5 and 7
Start	1 and 2; 1 and 8
	3 and 6; 5 and 7

9.2b   Ignition switch terminal guide and continuity chart (1996 and 1997 models)

9.2c   Ignition switch terminal guide and continuity chart (1998 and later models)

47017-12-9.2c HAYNES

Switch positions (key inserted)	Continuity between terminals
LOCK	No continuity
OFF	No continuity
ACC	2 and 3
ON	1, 2 and 3
START	3 and 4; 5 and 7

Replacement

Refer to illustrations 9.9 and 9.12

3   Disconnect the cable from the negative terminal of the battery.

4   Place the ignition key in the OFF position.

5   Remove the steering wheel (see Chapter 10).

6   Remove the steering column cover (see Chapter 11).

7   Remove the combination switch from the steering column (see Section 8).

Ignition switch

8   Unplug the ignition switch wiring harness connector.

9   Detach the switch retaining screw **(see illustration)** remove the switch assembly from the steering column.

10   Installation is reverse of the removal.

Lock cylinder

11   Use a small screwdriver or punch to depress the lock cylinder retaining pin.

12   Pull straight out on the lock cylinder assembly to remove it from the vehicle **(see illustration)**.

13   To install the lock cylinder, depress the retaining pin and guide the lock cylinder into the steering column housing until the retaining pin to extends itself back into the locating

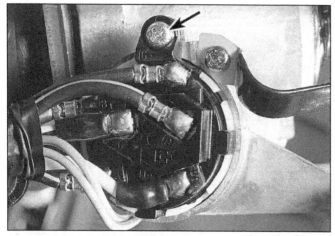

9.9   Detach the retaining screw to remove the ignition switch from the steering column housing

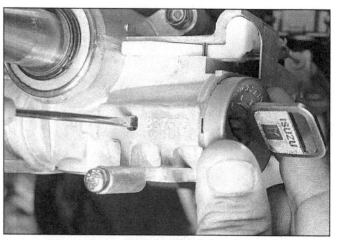

9.12   With the lock cylinder in the OFF position, use a small screwdriver to depress the retaining pin, then pull straight out to remove the lock cylinder

10.3a On 1995 and earlier models, disconnect the electrical connector(s) (A) and remove the retaining screw(s) (B) securing the switches to each side of the of the instrument cluster bezel

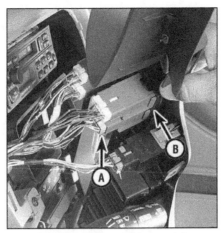

10.3b On 1996 and later models, disconnect the electrical connector(s) (A) and depress the retaining clip(s) (B) to remove the switches

hole in the steering column housing.

14   The remainder of the installation is the reverse of removal.

10   Instrument panel switches - removal and installation

Refer to illustrations 10.3a and 10.3b
Warning: *Some models covered by this manual are equipped with airbags. Always disable the airbag system (see Section 26) before working in the vicinity of the impact sensors, steering column or instrument panel. Failure to follow these procedures may cause accidental deployment of the airbag, which could cause personal injury.*

1   Remove the instrument cluster trim bezel (see Chapter 11).
2   Working on the reverse side of the bezel detach the electrical connector(s) from the backside of the switch.
3   Depress the retaining clips or remove the screws securing the switch, then pull the

switch out to remove it **(see illustrations)**.
4   Installation is the reverse of removal.

11   Instrument panel gauges - check

Warning: *Some models covered by this manual are equipped with airbags. Always disable the airbag system (see Section 26) before working in the vicinity of the impact sensors, steering column or instrument panel. Failure to follow these procedures may cause accidental deployment of the airbag, which could cause personal injury.*

Fuel, oil and temperature gauges

1   All tests below require the ignition switch to be turned to ON position when testing.
2   If the gauge pointer does not move from the empty, low or cold positions, check the fuse. If the fuse is OK, locate the particular sending unit for the circuit you're working on (see Chapter 4 for fuel sending unit location,

Chapter 2 for the oil sending unit location or Chapter 3 for the temperature sending unit location). Connect the sending unit connector to ground. If the pointer goes to the full, high or hot position replace the sending unit. If the pointer stays in same position, use a jumper wire to ground the sending unit terminal on the back of the gauge, if necessary, refer to the wiring diagrams at the end of this Chapter. If the pointer moves, the problem lies in the wire between the gauge and the sending unit. If the pointer does not move with the sending unit terminal on the back of the gauge grounded, check for voltage at the other terminal of the gauge. If voltage is present, replace the gauge.

12   Instrument cluster - removal and installation

Refer to illustrations 12.4a, 12.4b and 12.4c and 12.4d
Warning: *Some models covered by this manual are equipped with airbags. Always disable the airbag system (see Section 26) before working in the vicinity of the impact sensors, steering column or instrument panel. Failure to follow these procedures may cause accidental deployment of the airbag, which could cause personal injury.*

1   Disconnect the negative battery cable.
2   Remove the instrument cluster bezel (see Chapter 11).
3   On 1995 and earlier models, remove the driver's side knee bolster.
4   Remove the cluster mounting screws **(see illustrations)** and pull the instrument cluster towards the steering wheel.
5   On 1995 and earlier models, unscrew the speedometer cable from the backside of the instrument cluster.
6   Disconnect any electrical connectors that would interfere with removal.
7   Cover the steering column with a cloth to protect the trim covers then remove the instrument cluster from the vehicle.
8   Installation is the reverse of removal.

12.4a On 1995 and earlier models, remove the screws securing the front of the instrument cluster . . .

12.4b . . .then remove the screws securing the bottom of the instrument cluster located on each side of the steering column

12.4c 1996 and later instrument cluster mounting screw locations (steering wheel removed for clarity)

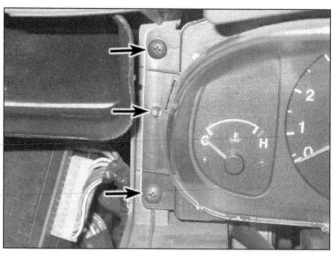

12.4d On 1998 and later models, the instrument cluster is positioned by locator pins; when installing the cluster, make sure that the locator pin at each end of the cluster fits through the hole in the cluster mounting flange (left end of cluster on a 1998 Amigo model shown)

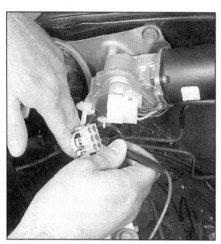

13.2 Use a voltmeter or test light to check for battery power at the wiper motor

13.8 Front wiper motor retaining bolt locations (arrows)

13 Wiper motor - removal and installation

Wiper motor circuit check

Refer to illustration 13.2

Note: *Refer to the wiring diagrams for wire colors and locations in the following checks. Keep in mind that power wires are generally larger in diameter and more brightly colored, whereas ground wires are usually smaller in diameter and darker in color. When checking for voltage, probe a grounded 12-volt test light to each terminal at a connector until it lights; this verifies voltage (power) at the terminal.*

1   If the wipers work slowly, make sure the battery is in good condition and is fully charged (see Chapter 1). If the battery is in good condition, remove the wiper motor (see below) and operate the wiper arms by hand.

Check for binding linkage and pivots. Lubricate or repair the linkage or pivots as necessary. Reinstall the wiper motor. If the wipers still operate slowly, check for loose or corroded connections, especially the ground connection. If all connections are good, replace the motor.

2   If the wipers fail to operate when activated, check the fuse. If the fuse is OK, connect a jumper wire between the wiper motor and ground, then retest. If the motor works now, repair the ground connection. If the motor still doesn't work, turn on the wipers and check for voltage at the motor **(see illustration)**. If there's no voltage at the motor, remove the motor and check it off the vehicle with fused jumper wires from the battery. If the motor now works, check for binding linkage (see Step 1 above). If the motor still doesn't work, replace it. If there's no voltage at the motor, check for voltage at the switch. If there's no voltage at the switch, check the

wiring between the switch and fuse panel for continuity. If the wiring is OK, the switch is probably bad.

3   If the wipers only work on one speed, check the continuity of the wires between the switch and motor. If the wires are OK, replace the switch.

4   If the interval (delay) function is inoperative, check the continuity of all the wiring between the switch and motor. If the wiring is OK, replace the interval module.

5   If the wipers stop at the position they're in when the switch is turned off (fail to park), check for voltage at the wiper motor when the wiper switch is OFF but the ignition is ON. If voltage is present, the limit switch in the motor is malfunctioning. Replace the wiper motor. If no voltage is present, trace and repair the limit switch wiring between the fuse panel and wiper motor.

6   If the wipers won't shut off unless the ignition is OFF, disconnect the wiring from the wiper control switch. If the wipers stop, replace the switch. If the wipers keep running, there's a defective limit switch in the motor; replace the motor.

7   If the wipers won't retract below the hoodline, check for mechanical obstructions in the wiper linkage or on the vehicle's body which would prevent the wipers from parking. If there are no obstructions, check the wiring between the switch and motor for continuity. If the wiring is OK, replace the wiper motor.

Wiper motor replacement

Front

Refer to illustrations 13.8 and 13.9

8   Remove the wiper motor retaining bolts **(see illustration)**, then pull the motor out from the firewall.

9   Remove the motor spindle nut from the backside of the motor **(see illustration)**.

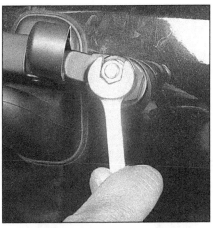

13.9 Pull the front wiper motor outward and unscrew the spindle nut (arrow)

13.12 Pull back the rear wiper arm cover, then remove the nut and pull the arm straight off its splined shaft

13.13 Remove the drive spindle retaining nut from the outside of the tailgate

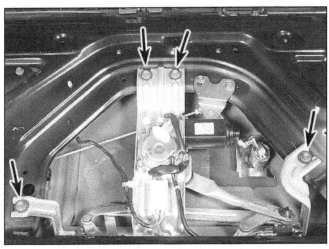

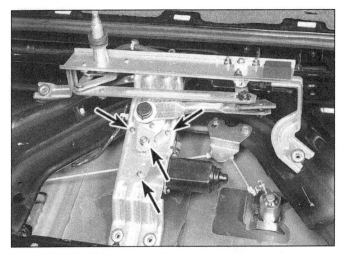

13.16 Remove the rear wiper linkage bracket retaining nuts (arrows)

13.18 Turn the linkage assembly over to access the wiper motor retaining screws and spindle nut (arrows)

10 Disconnect the electrical connector and remove the motor from the vehicle.

11 Installation is the reverse of removal.

Rear

Refer to illustrations 13.12, 13.13, 13.16 and 13.18

12 Pull the rear wiper arm cover back to access the wiper arm retaining nut. Detach the nut and pull the wiper arm straight off the shaft to remove it **(see illustration)**.

13 Remove the drive spindle retaining nut **(see illustration)**.

14 Lower the tailgate and remove the tailgate trim panel and the tailgate access cover (see Chapter 11).

15 Disconnect the electrical connector from wiper motor.

16 Remove the wiper linkage bracket mounting bolts **(see illustration)**.

17 Pull the linkage assembly out from the tailgate, then turn it over to access the wiper motor retaining screws.

18 Remove the retaining screws and the wiper motor spindle nut **(see illustration)**.

19 Installation is the reverse of removal.

14 Radio and speakers - removal and installation

Warning: *Some models covered by this manual are equipped with airbags. Always disable the airbag system (see Section 26) before working in the vicinity of the impact sensors, steering column or instrument panel. Failure to follow these procedures may cause accidental deployment of the airbag, which could cause personal injury.*

1 Disconnect the negative battery cable.

Radio

Refer to illustrations 14.3a, 14.3b and 14.3c

2 On 1995 and earlier models, remove the front console trim bezel. On 1996 and later models, remove the instrument cluster bezel (see Chapter 11).

3 Remove the retaining screws and pull the radio outward to access the backside, then disconnect the electrical connectors and the antenna lead and lift the radio out of the

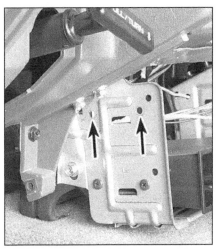

14.3a On 1995 and earlier models, the radio mounting screws are located on each side of the front console brace

vehicle **(see illustrations)**.

4 Installation is the reverse of removal.

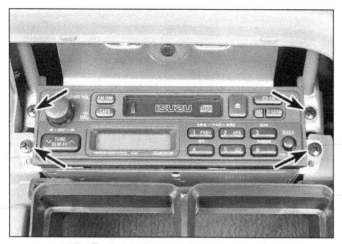

14.3b Typical radio mounting screw locations (1996 and 1997 models)

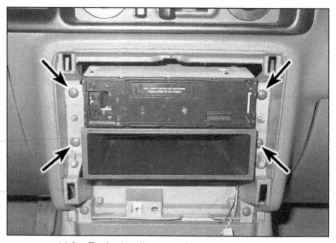

14.3c Typical radio mounting screw locations (1998 and later models)

14.6 On 1996 and later models, carefully pry off the speaker grille

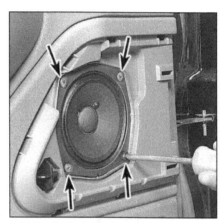

14.7a Typical speaker screw locations (1996 and 1997 model shown)

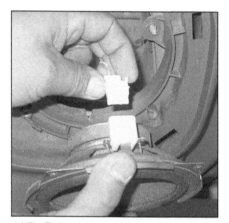

14.7b Pull out the speaker and unplug the electrical connector (1998 model shown, other models similar)

Speakers

Refer to illustrations 14.6, 14.7a and 14.7b

5   On 1995 and earlier models the speakers are mounted in the instrument panel. Remove the lower dash board panels to access the speakers (see Chapter 11).

6   On 1996 and later models, pry off the speaker grille from the door trim panel **(see illustration)**.

7   Remove the speaker retaining screws **(see illustration)**. Disconnect the electrical connector and remove the speaker from the vehicle **(see illustration)**.

8   Installation is the reverse of removal.

---

15   Antenna - removal and installation

Refer to illustration 15.2

1   Detach the radio and disconnect the antenna lead from the backside of the radio (see Section 14). Detach any retaining clips under the instrument panel securing the antenna lead.

2   Working from the outside of the vehicle, use a small wrench and remove the mast **(see illustration)**.

3   Using a pair of snap ring pliers or similar tool, remove the antenna base retaining nut.

4   Pull the antenna up and out to remove it.

5   Installation is the reverse of removal.

---

16   Rear window defogger - check and repair

1   The rear window defogger consists of a number of horizontal elements baked onto the glass surface.

2   Small breaks in the element can be repaired without removing the rear window.

Check

Refer to illustrations 16.4, 16.5 and 16.7

3   Turn the ignition switch and defogger system switches to the ON position.

4   When measuring voltage during the next two tests, wrap a piece of aluminum foil around the tip of the voltmeter negative probe and press the foil against the heating element with your finger **(see illustration)**.

5   Check the voltage at the center of each heating element **(see illustration)**. If the voltage is 6-volts, the element is okay (there is no break). If the voltage is 12-volts, the element

is broken between the center of the element and the positive end. If the voltage is 0-volts the element is broken between the center of the element and ground.

6   Connect the negative lead to a good body ground. The reading should stay the same.

7   To find the break, place the voltmeter positive lead against the defogger positive

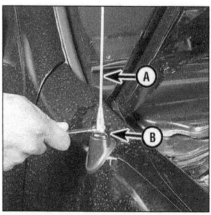

15.2 Detach the antenna mast (A) then the antenna base retaining nut (B)

16.4 When measuring the voltage at the rear window defogger grid, wrap a piece of aluminum foil around the negative probe of the voltmeter and press the foil against the wire with your finger

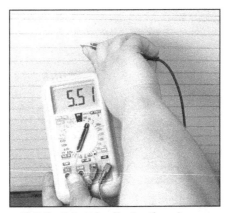

16.5 To determine if a heating element has broken, check the voltage at the center of each element - if the voltage is 6-volts, the element is unbroken - if the voltage is 12-volts, the element is broken between the center and the positive end - if there is no voltage, the element is broken between the center and ground

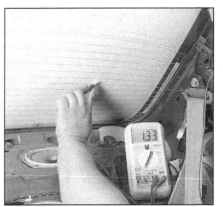

16.7 To find the break, place the voltmeter positive lead against the defogger positive terminal, place the voltmeter negative lead with the foil strip against the heating element at the positive terminal end and slide it toward the negative terminal end - the point at which the voltmeter reading changes abruptly is the point at which the element is broken

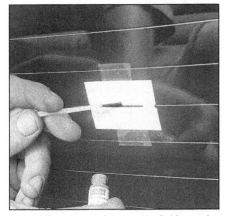

16.13 To use a defogger repair kit, apply masking tape to the inside of the window at the damaged area, then brush on the special conductive coating

terminal. Place the voltmeter negative lead with the foil strip against the heating element at the positive terminal end and slide it toward the negative terminal end. The point at which the voltmeter deflects from zero to several volts is the point at which the heating element is broken **(see illustration)**.

Repair

Refer to illustration 16.13

8   Repair the break in the element using a repair kit specifically recommended for this purpose, such as Dupont paste No. 4817 (or equivalent). Included in this kit is plastic conductive epoxy.

9   Prior to repairing a break, turn off the system and allow it to cool off for a few minutes.

10   Lightly buff the element area with fine steel wool, then clean it thoroughly with rubbing alcohol.

11   Use masking tape to mask off the area being repaired.

12   Thoroughly mix the epoxy, following the instructions provided with the repair kit.

13   Apply the epoxy material to the slit in the masking tape, overlapping the undamaged area about 3/4-inch on either end **(see illustration)**.

14   Allow the repair to cure for 24 hours before removing the tape and using the system.

17 Headlights - removal and installation

1   Disconnect the negative cable from the battery.

Sealed beam type

2   Remove the radiator grille (Chapter 11).

3   Remove the headlight retainer screws, taking care not to disturb the adjustment screws.

4   Remove the retainer and pull the headlight out far enough to unplug the connector.

5   Remove the headlight.

6   To install the headlight, connect the wiring harness connector, place the headlight in position and install the retainer and screws. Tighten the screws securely.

7   Install the radiator grille.

Halogen bulb-type

Warning: *Halogen gas-filled bulbs are under pressure and may shatter if the surface is scratched or the bulb is dropped. Wear eye protection and handle the bulbs carefully, grasping only the base whenever possible. Do not touch the surface of the bulb with your fingers because the oil from your skin could cause it to overheat and fail prematurely. If you do touch the bulb surface, clean it with rubbing alcohol.*

8   Reach behind the headlight assembly and unplug the electrical connector.

1997 and earlier models

Refer to illustrations 17.9 and 17.10

9   Disengage the bulb holder retaining clip **(see illustration)**.

10   Grasp the bulb holder securely and pull it out of the housing **(see illustration)**.

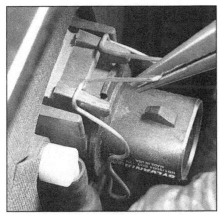

17.9 Remove the bulb holder retaining clip (1997 and earlier models)

17.10 Pull the bulb assembly straight out of the housing (1997 and earlier models)

17.11 To disconnect the electrical connector from the bulb holder, push down on this locking tang and pull off the connector (1998 and later models)

17.12 To release the socket retaining ring, turn it counterclockwise 1/4-turn (1998 and later models)

17.13 Pull the bulb holder out of the headlight housing (1998 and later models)

1998 and later models

Refer to illustrations 17.11, 17.12 and 17.13

11   Disconnect the electrical connector from the bulb holder **(see illustration)**.
12   Release the socket retaining ring by turning it counterclockwise 1/4-turn **(see illustration)**.
13   Pull the bulb holder out of the headlight housing **(see illustration)**.

All models

14   Pull the old bulb out of the socket.
15   Without touching the glass with your bare fingers (wear a pair of lightweight cotton gloves), insert the new bulb into the socket.
16   The remainder of installation is the reverse of removal.
17   Test headlight operation to make sure that reassembly was correct.

---

18   Headlights - adjustment

Refer to illustrations 18.1 and 18.3
Note: *The headlights must be aimed correctly. If adjusted incorrectly they could blind the driver of an oncoming vehicle and cause a serious accident or seriously reduce your ability to see the road. The headlights should be checked for proper aim every 12 months and any time a new headlight is installed or front end body work is performed. It should be emphasized that the following procedure is only an interim step that will provide temporary adjustment until the headlights can be adjusted by a properly equipped shop.*
1   Headlights have two spring loaded adjusting screws, one on the top or bottom controlling up-and-down movement and one on the side controlling left-and-right movement **(see illustration)**.
2   There are several methods of adjusting the headlights. The simplest method requires masking tape, a blank wall and a level floor.
3   Position masking tape vertically on the wall in reference to the vehicle centerline and the centerlines of both headlights **(see illustration)**.

18.1 The headlight vertical adjustment screw is located at the bottom of the headlight and the horizontal screw is on the side of the headlight (1994 and later model shown, 1993 and earlier similar)

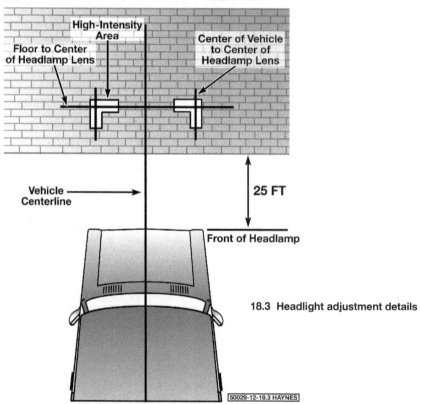

18.3 Headlight adjustment details

19.3a Headlight housing retaining bolts (1994 through 1997 models)

19.3b Headlight housing retaining bolts (one bolt not visible in this photo) (1998 and later models)

4   Position a horizontal tape line in reference to the centerline of all the headlights. **Note:** *It may be easier to position the tape on the wall with the vehicle parked only a few inches away.*

5   Adjustment should be made with the vehicle parked 25 feet from the wall, sitting level, the gas tank half-full and no unusually heavy load in the vehicle.

6   Starting with the low beam adjustment, position the high intensity zone so it is two inches below the horizontal line and two inches to the side of the headlight vertical line, away from oncoming traffic. Adjustment is made by turning the top or bottom adjusting screw to raise or lower the beam. The adjusting screw on the side should be used in the same manner to move the beam left or right.

7   With the high beams on, the high intensity zone should be vertically centered with the exact center just below the horizontal line. **Note:** *It may not be possible to position the headlight aim exactly for both high and low beams. If a compromise must be made, keep in mind that the low beams are the most used and have the greatest effect on driver safety.*

8   Have the headlights adjusted by a dealer service department or service station at the earliest opportunity.

---

19   Headlight housing (1994 and later models) - removal and installation

---

Refer to illustrations 19.3a and 19.3b
Warning: *These vehicles are equipped with halogen gas-filled headlight bulbs which are under pressure and may shatter if the surface is damaged or the bulb is dropped. Wear eye protection and handle the bulbs carefully, grasping only the base whenever possible. Do not touch the surface of the bulb with your fingers because the oil from your skin could cause it to overheat and fail prematurely. If you do touch the bulb surface, clean it with*

20.3 Check for power at the horn terminal with the horn button depressed

rubbing alcohol.
1   Remove the headlight bulb (see Section 17).
2   Remove the radiator grille (see Chapter 11) and the side marker light.
3   Remove the retaining bolts **(see illustrations)**, detach the housing and withdraw it from the vehicle.
4   Installation is the reverse of removal.

---

20   Horn - check and replacement

Check

Refer to illustrations 20.3
Note: *Check the fuses before beginning electrical diagnosis.*
1   Disconnect the electrical connector from the horn.
2   To test the horn, connect battery voltage to the two terminals with a pair of jumper wires. If the horn doesn't sound, replace it.
3   If the horn does sound, check for voltage at the terminal when the horn button is depressed **(see illustration)**. If there's voltage at the terminal, check for a bad ground at the horn.

20.9a Disconnect the electrical connector, remove the bolt and detach the horn (1997 and earlier models)

4   If there's no voltage at the horn, check the relay (see Section 6). Note that most horn relays are either the four-terminal or externally grounded three-terminal type.
5   If the relay is OK, check for voltage to the relay power and control circuits. If either of the circuits is not receiving voltage, inspect the wiring between the relay and the fuse panel.
6   If both relay circuits are receiving voltage, depress the horn button and check the circuit from the relay to the horn button for continuity to ground. If there's no continuity, check the circuit for an open. If there's no open circuit, replace the horn button.
7   If there's continuity to ground through the horn button, check for an open or short in the circuit from the relay to the horn.

Replacement

Refer to illustrations 20.9a and 20.9b
8   Remove the radiator grille (see Chapter 11).
9   To replace the horn(s), disconnect the electrical connector and remove the bracket bolt **(see illustrations)**.
10   Installation is the reverse of removal.

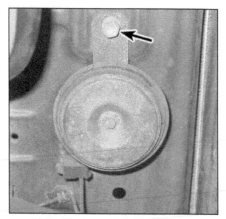

20.9b   Typical 1998 and later horn installation

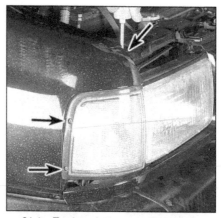

21.1a   To detach the turn signal/side marker light assembly, remove these three retaining screws (1997 and earlier model shown, 1998 and later models similar)

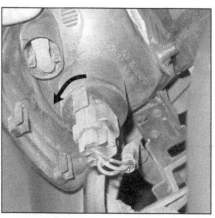

21.1b   To detach the bulb holder from the side marker light assembly, rotate the bulb holder counterclockwise . . .

21.1c   . . . and pull the bulb holder out of the side marker light assembly; to remove the bulb from the holder, push it in, turn it counterclockwise and pull it out (1998 and later model shown; 1997 and earlier models similar)

21.3   Remove the turn signal lens retaining screws to access the bulb

21.6a   To remove the tail light assembly on 1997 and earlier models, remove these screws

21   Bulb replacement

Front side marker lights

Refer to illustrations 21.1a, 21.1b and 21.1c
1   Remove the screws that secure the side

marker light housing **(see illustration)** and pull out the side marker assembly. Rotate the bulb holder counterclockwise and pull out the bulb holder **(see illustrations)**. To remove the bulb from the holder, push it in, turn it counterclockwise and pull it out.
2   Installation is the reverse of removal.

Front turn signal lights

Refer to illustration 21.3
3   Remove the screws securing the turn signal lens **(see illustration)**.
4   Rotate the bulb counterclockwise to remove it.
5   Installation is the reverse of removal.

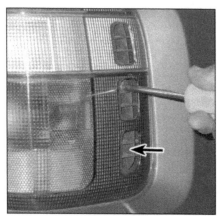

21.6b   To remove the tail light assembly on 1998 and later models, remove these two screws . . .

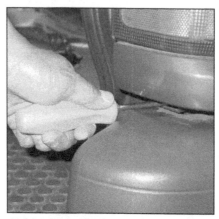

21.6c   . . . remove this screw, which can be accessed through the gap between the tail light assembly and the bumper cover . . .

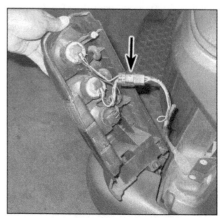

21.6d   . . . and then remove the tail light assembly and unplug the electrical connector

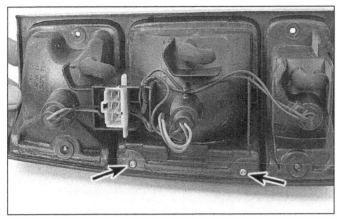

21.8 On 1997 and earlier models, detach the lens retaining screws from the back of the tail light housing, then remove the lens to access the bulbs

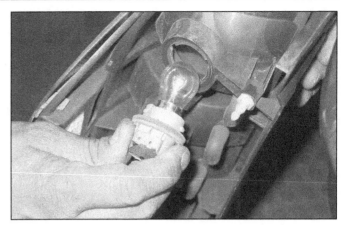

21.9a On 1998 and later models, push in the bulb socket, turn it counterclockwise and then pull it out . . .

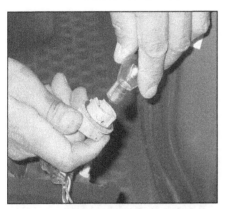

21.9b . . . and then push the bulb into the socket, turn it counterclockwise and pull it out of the socket

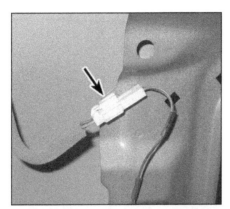

21.12 After removing the trim panel from the tailgate, disconnect the high-mount brake light electrical connector

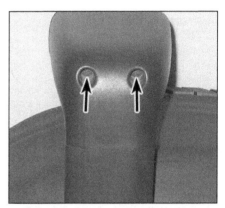

21.13 To detach the high-mount brake light assembly, remove these two screws

Rear tail light/brake light/turn signal lights

Refer to illustrations 21.6a, 21.6b, 21.6c, 21.6d, 21.8, 21.9a and 21.9b

6   Remove the tail light assembly mounting screws and remove the tail light assembly

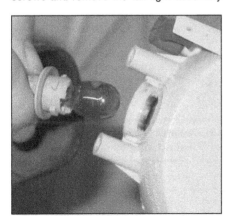

21.14 To detach the bulb socket from the high-mount brake light assembly, turn it counterclockwise and pull it out; to remove the bulb from the socket, turn the bulb counterclockwise and pull it out of the socket

(see illustrations).
7   Unplug the electrical connector(s) from the tail light housing.
8   On 1997 and earlier models, remove the screws located on reverse side of the housing and detach the lens **(see illustration)**. Push in and rotate the turn signal, brake light and tail light bulbs counterclockwise to remove them from the holder.
9   On 1998 and later models, push in the

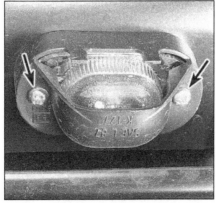

21.17 To remove the license plate light, detach the screws, then pull it down and out to access the bulb

bulb socket, turn it counterclockwise and pull it out of the tail light assembly and then push the bulb into the socket, turn it counterclockwise and pull it out of the socket **(see illustrations)**.
10   Installation is the reverse of removal.

High-mount brake light

Refer to illustrations 21.12, 21.13 and 21.14

11   Remove the trim panel from the tailgate (see Section 20 in Chapter 11).
12   Disconnect the high-mount brake light electrical connector **(see illustration)**.
13   Remove the high-mount brake light retaining screws **(see illustration)**.
14   Pull out the high-mount brake light assembly, turn the bulb socket counterclockwise and pull it out of the high-mount brake light assembly **(see illustration)**.
15   To replace the bulb, push it into the socket, turn it counterclockwise and pull it out of the socket.
16   Installation is the reverse of removal.

License plate light

Refer to illustration 21.17

17   Remove the screws and detach the bulb and lens assembly **(see illustration)**, then remove the bulb from the socket.
18   Installation is the reverse of removal.

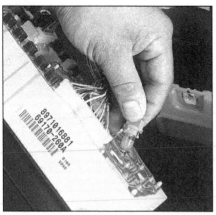

21.20 Remove the instrument cluster bulbs by rotating them 1/4-turn counterclockwise and pulling straight out

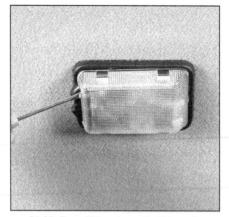

21.22 Pry off the dome light lens to access the bulb

21.23 If you have to pry the old interior light bulb loose, do so only on the metal ends of the bulb, NOT on the glass part

Instrument cluster lights

Refer to illustration 21.20

19   Remove the instrument cluster (see Section 12). (The bulbs are installed on the backside of the cluster.)

20   To remove an instrument cluster light bulb, rotate it counterclockwise and pull it out **(see illustration)**.

21   Installation is the reverse of removal.

Interior light

Refer to illustrations 21.22 and 21.23

22   Pry off the interior light lens **(see illustration)**.

23   Remove the interior light bulb **(see illustration)** from the terminals. If the bulb is difficult to remove, pry it out with a small screwdriver, but pry only on the metal ends of the bulb, not on the glass part of the bulb. Prying against the glass might shatter it.

24   Installation is the reverse of removal.

Spotlight bulb

Refer to illustrations 21.25 and 21.26

25   Pry off the spotlight bulb lens **(see illustration)**.

26   Remove the spotlight bulb **(see illustration)** from the terminals. If the bulb is difficult to remove, pry it out with a small screwdriver, but pry only on the metal ends of the bulb, not on the glass part of the bulb. Prying against the glass might shatter it.

27   Installation is the reverse of removal.

---

22   Electric side view mirrors - description and check

1   Most electric rear view mirrors use two motors to move the glass; one for up and down adjustments and one for left-right adjustments.

2   The control switch has a selector portion which sends voltage to the left or right side mirror. With the ignition ON but the engine OFF, roll down the windows and operate the mirror control switch through all functions (left-right and up-down) for both the left

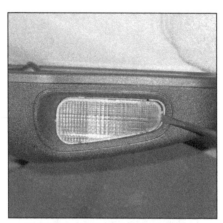

21.25 Pry off the spotlight bulb lens

and right side mirrors.

3   Listen carefully for the sound of the electric motors running in the mirrors.

4   If the motors can be heard but the mirror glass doesn't move, there's probably a problem with the drive mechanism inside the mirror. Remove and disassemble the mirror to locate the problem.

5   If the mirrors don't operate and no sound comes from the mirrors, check the fuse (see Chapter 1).

6   If the fuse is OK, remove the mirror control switch from its mounting without disconnecting the wires attached to it. Turn the ignition ON and check for voltage at the switch. There should be voltage at one terminal. If there's no voltage at the switch, check for an open or short in the circuit between the fuse panel and the switch.

7   If there's voltage at the switch, disconnect it. Check the switch for continuity in all its operating positions. If the switch does not have continuity, replace it.

8   Re-connect the switch. Locate the wire going from the switch to ground. Leaving the switch connected, connect a jumper wire between this wire and ground. If the mirror works normally with this wire in place, repair the faulty ground connection.

9   If the mirror still doesn't work, remove

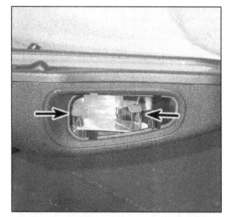

21.26 If you have to pry the old spotlight bulb loose, do so only at the metal ends of the bulb, NOT on the glass part

the mirror and check the wires at the mirror for voltage. Check with ignition ON and the mirror selector switch on the appropriate side. Operate the mirror switch in all its positions. There should be voltage at one of the switch-to-mirror wires in each switch position (except the neutral "off" position).

10   If there's not voltage in each switch position, check the circuit between the mirror and control switch for opens and shorts.

11   If there's voltage, remove the mirror and test it off the vehicle with jumper wires. Replace the mirror if it fails this test.

---

23   Cruise control system - description and check

1   The cruise control system maintains vehicle speed with a vacuum actuated servo located in the engine compartment, which is connected to the accelerator pedal by a cable. The system consists of the cruise control unit, brake switch, control switches, vacuum hose and vehicle speed sensor. Some features of the system require special testers and diagnostic procedures which are beyond the scope of this manual. Listed below are

some general procedures that may be used to locate common problems.

2 Locate and check the fuse (see Section 3).

3 Have an assistant operate the brake lights while you check their operation (voltage from the brake light switch deactivates the cruise control).

4 If the brake lights don't come on or don't shut off, correct the problem and retest the cruise control.

5 Visually inspect the control cable between actuator assembly and accelerator pedal for free movement, replace if necessary.

6 The cruise control system uses a speed sensing device. The speed sensor is located in the speedometer. (To test the speed sensor, see Chapter 6).

7 Test drive the vehicle to determine if the cruise control is now working. If it isn't, take it to a dealer service department or an automotive electrical specialist for further diagnosis.

24 Power window system - description and check

1 The power window system operates electric motors, mounted in the doors, which lower and raise the windows. The system consists of the control switches, the motors, regulators, glass mechanisms and associated wiring.

2 The power windows can be lowered and raised from the master control switch by the driver or by remote switches located at the individual windows. Each window has a separate motor which is reversible. The position of the control switch determines the polarity and therefore the direction of operation.

3 The circuit is protected by a fuse and a circuit breaker. Each motor is also equipped with an internal circuit breaker, this prevents one stuck window from disabling the whole system.

4 The power window system will only operate when the ignition switch is ON. In addition, many models have a window lockout switch at the master control switch which, when activated, disables the switches at the rear windows and, sometimes, the switch at the passenger's window also. Always check these items before troubleshooting a window problem.

5 These procedures are general in nature, so if you can't find the problem using them, take the vehicle to a dealer service department or other properly equipped repair facility.

6 If the power windows won't operate, always check the fuse and circuit breaker first.

7 If only the rear windows are inoperative, or if the windows only operate from the master control switch, check the rear window lockout switch for continuity in the unlocked position. Replace it if it doesn't have continuity.

8 Check the wiring between the switches

and fuse panel for continuity. Repair the wiring, if necessary.

9 If only one window is inoperative from the master control switch, try the other control switch at the window. **Note:** *This doesn't apply to the drivers door window.*

10 If the same window works from one switch, but not the other, check the switch for continuity.

11 If the switch tests OK, check for a short or open in the circuit between the affected switch and the window motor.

12 If one window is inoperative from both switches, remove the trim panel from the affected door and check for voltage at the switch and at the motor while the switch is operated.

13 If voltage is reaching the motor, disconnect the glass from the regulator (see Chapter 11). Move the window up and down by hand while checking for binding and damage. Also check for binding and damage to the regulator. If the regulator is not damaged and the window moves up and down smoothly, replace the motor. If there's binding or damage, lubricate, repair or replace parts, as necessary.

14 If voltage isn't reaching the motor, check the wiring in the circuit for continuity between the switches and motors. You'll need to consult the wiring diagram for the vehicle. If the circuit is equipped with a relay, check that the relay is grounded properly and receiving voltage.

15 Test the windows after you are done to confirm proper repairs.

25 Power door lock system - description and check

1 The power door lock system operates the door lock actuators mounted in each door. The system consists of the switches, actuators, a control unit and associated wiring. Diagnosis can usually be limited to simple checks of the wiring connections and actuators for minor faults which can be easily repaired. Since this system uses an electronic control unit in-depth diagnosis should be left to a dealership service department. The door lock control unit is located behind the instrument panel, to the right of the fuse box.

2 Power door lock systems are operated by bi-directional solenoids located in the doors. The lock switches have two operating positions: Lock and Unlock. When activated, the switch sends a ground signal to the door lock control unit to lock or unlock the doors. Depending on which way the switch is activated, the control unit reverses polarity to the solenoids, allowing the two sides of the circuit to be used alternately as the feed (positive) and ground side.

3 Some vehicles may have an anti-theft system incorporated into the power locks. If you are unable to locate the trouble using the following general Steps, consult your a dealer

service department.

4 Always check the circuit protection first. Some vehicles use a combination of circuit breakers and fuses.

5 Operate the door lock switches in both directions (Lock and Unlock) with the engine off. Listen for the click of the actuators operating.

6 Test the switches for continuity. Replace the switch if there's not continuity in both switch positions.

7 Check the wiring between the switches, control unit and actuators for continuity. Repair the wiring if there's no continuity.

8 Check for a bad ground at the switches or the control unit.

9 If all but one lock solenoids operate, remove the trim panel from the affected door (see Chapter 11) and check for voltage at the actuator while the lock switch is operated One of the wires should have voltage in the Lock position; the other should have voltage in the Unlock position.

10 If the inoperative actuator is receiving voltage, replace the actuator.

11 If the inoperative actuator isn't receiving voltage, check for an open or short in the wire between the lock actuator and the control unit. **Note:** *It's common for wires to break in the portion of the harness between the body and door (opening and closing the door fatigues and eventually breaks the wires).*

26 Airbag - general information

Refer to illustration 26.1

Some models are equipped with a Supplemental Restraint System (SRS), more commonly known as an airbag. This system is designed to protect the driver, and the front seat passenger, from serious injury in the event of a head-on or frontal collision. It consists of an airbag module in the center of the steering wheel and the right side of the instrument panel and a sensing/diagnostic module mounted in the center of the vehicle **(see illustration)**.

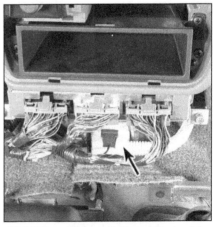

26.1 The sensing and diagnostic module (arrow) is located behind the front console

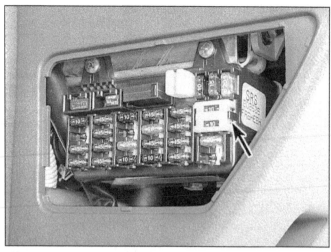

26.7a Remove the SRS (airbag) fuses which are identified by the yellow cover in the passenger compartment fuse block

26.7b Disconnect the driver's side airbag connector at the base of the steering column

Airbag module

Steering wheel-mounted

The airbag inflator module contains a housing incorporating the cushion (airbag) and inflator unit, mounted in the center of the steering wheel The inflator assembly is mounted on the back of the housing over a hole through which gas is expelled, inflating the airbag almost instantaneously when an electrical signal is sent from the system. A coil assembly on the steering column under the module carries this signal to the module.

This coil assembly can transmit an electrical signal regardless of steering wheel position. The igniter in the airbag converts the electrical signal to heat and ignites a small explosive charge, which inflates the airbag.

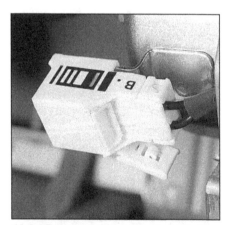

26.7c Remove the instrument panel glove compartment to access and disconnect the passenger side airbag connector

Instrument panel-mounted

The airbag is mounted above the glove compartment and designated by the letters SRS (Supplemental Restraint System). It consists of an inflator containing an igniter, an airbag assembly, a reaction housing and a trim cover.

The airbag is considerably larger that the steering wheel-mounted unit and is supported by the steel reaction housing. The trim cover is textured and painted to match the instrument panel and has a molded seam which splits when the airbag inflates.

Sensing and diagnostic module

The sensing and diagnostic module supplies the current to the airbag system in the event of the collision, even if battery power is cut off. It checks this system every time the vehicle is started, causing the "AIR BAG" light to go on then off, if the system is operating properly. If there is a fault in the system, the light will go on and stay on, flash, or the dash will make a beeping sound. If this happens, the vehicle should be taken to your dealer immediately for service.

Disabling the system

Refer to illustrations 26.7a, 26.7b and 26.7c

Whenever working in the vicinity of the steering wheel, steering column or near other components of the airbag system, the system should be disarmed **(see illustrations)**. To do this, perform the following steps:

a) *Remove the airbag fuses from the passenger compartment fuse block or disconnect the battery.*

b) *Disconnect the yellow airbag connector at the base of the steering column and behind the instrument panel glove compartment.*

Enabling the system

a) *Turn the ignition switch to the LOCK position.*

b) *Connect the yellow airbag connector at the base of the steering column and behind the instrument panel glove compartment*

c) *Install the airbag fuses in the passenger compartment fuse block or reconnect the battery.*

---

27 Wiring diagrams - general information

Since it isn't possible to include all wiring diagrams for every year covered by this manual, the following diagrams are those that are typical and most commonly needed.

Prior to troubleshooting any circuits, check the fuse and circuit breakers (if equipped) to make sure they're in good condition. Make sure the battery is properly charged and check the cable connections (see Chapter 1).

When checking a circuit, make sure that all connectors are clean, with no broken or loose terminals. When unplugging a connector, do not pull on the wires. Pull only on the connector housings themselves.

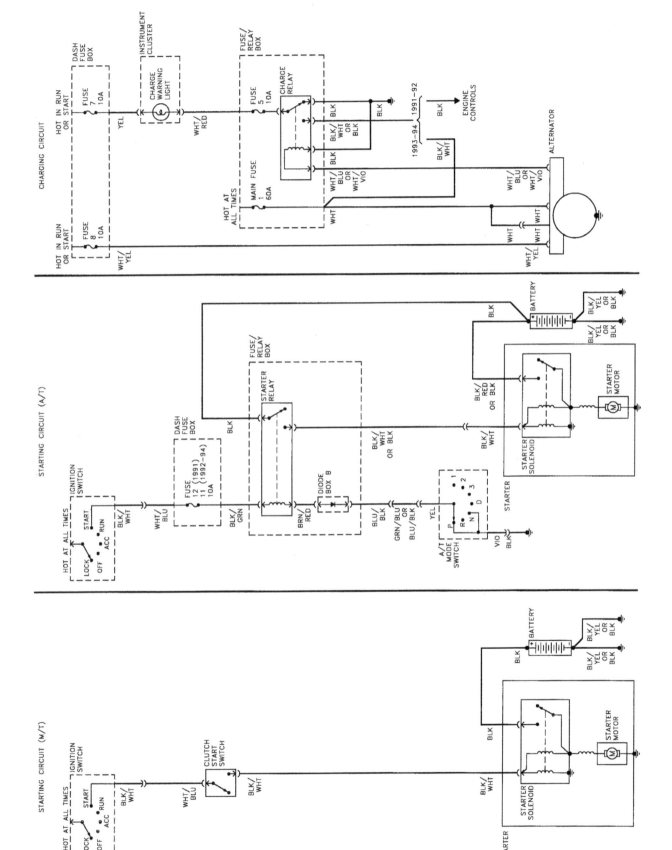

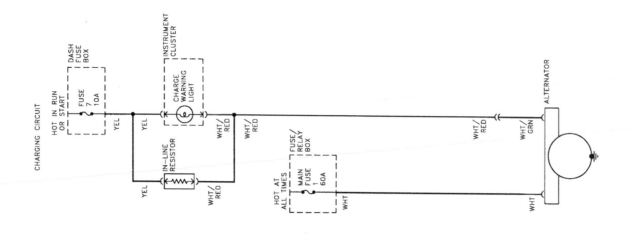

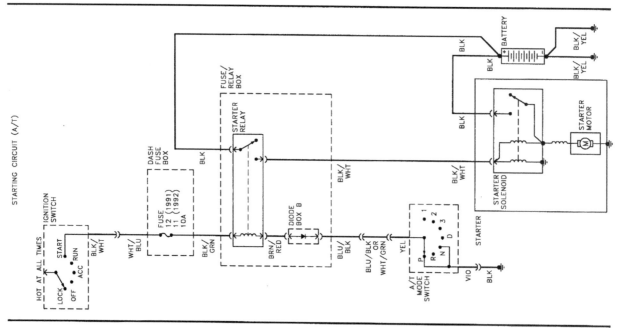

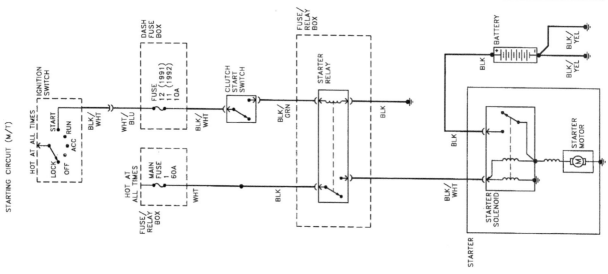

Typical 1991 and 1992 3.1L V6 engine starting and charging system

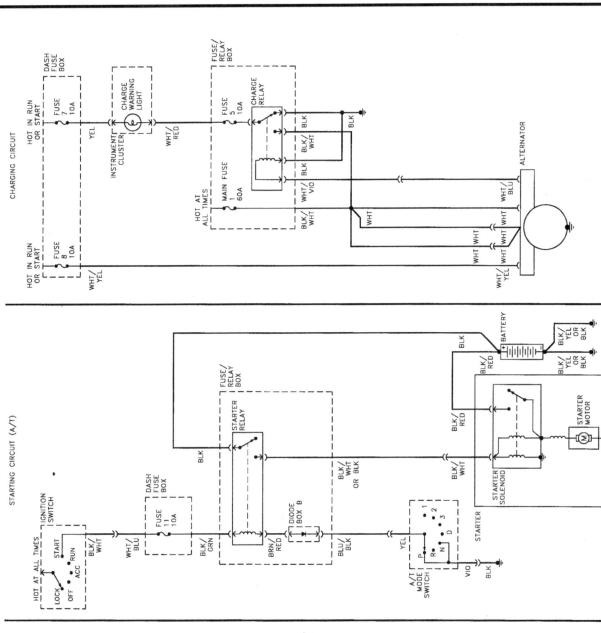

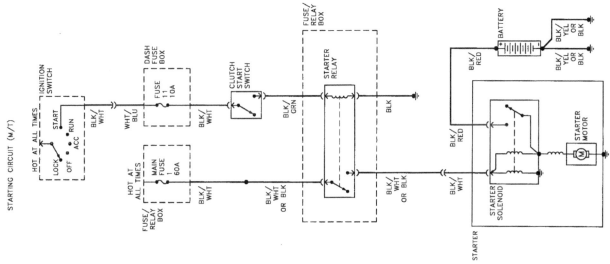

Typical 1993 and 1994 3.2L V6 engine starting and charging system

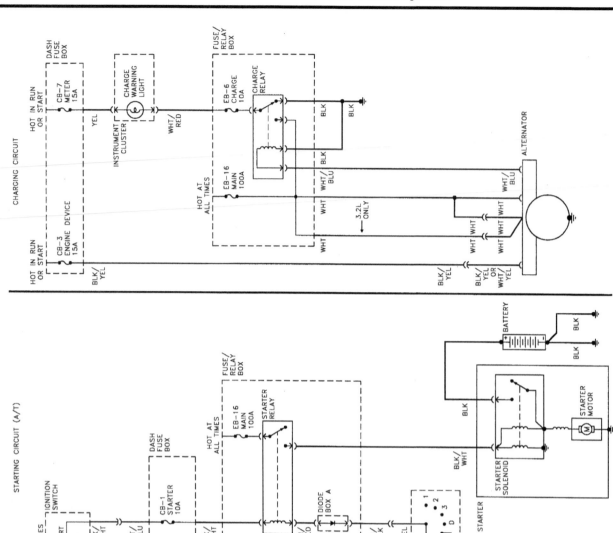

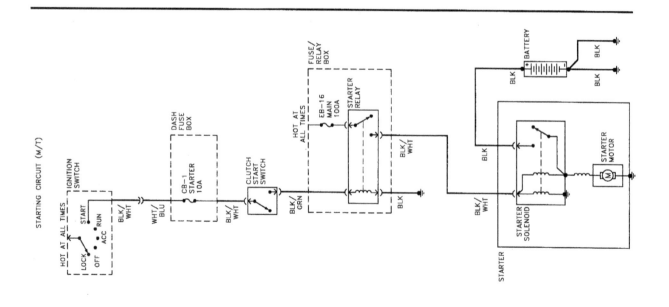

Typical 1995 starting and charging system

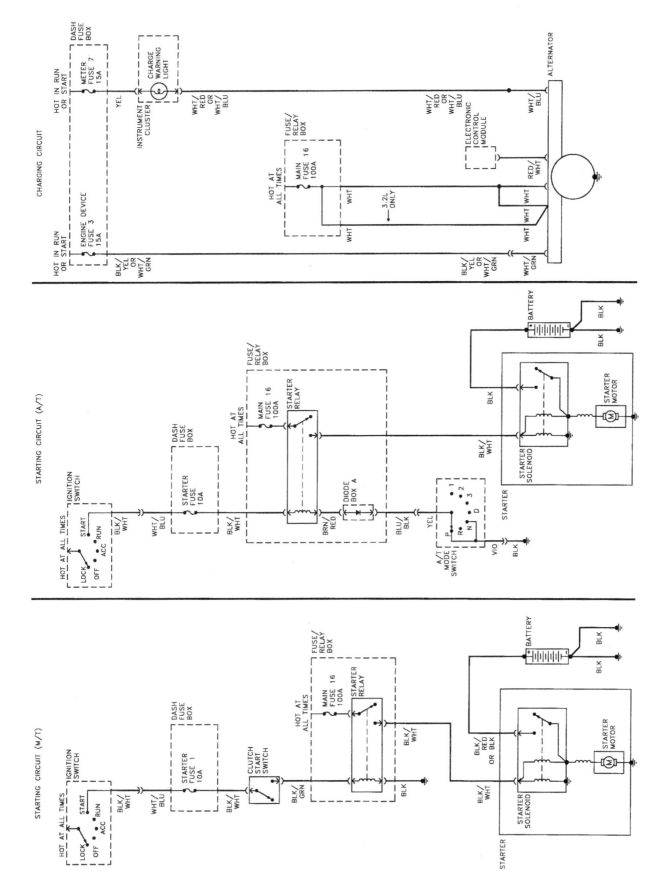

Typical 1996 and later starting and charging system

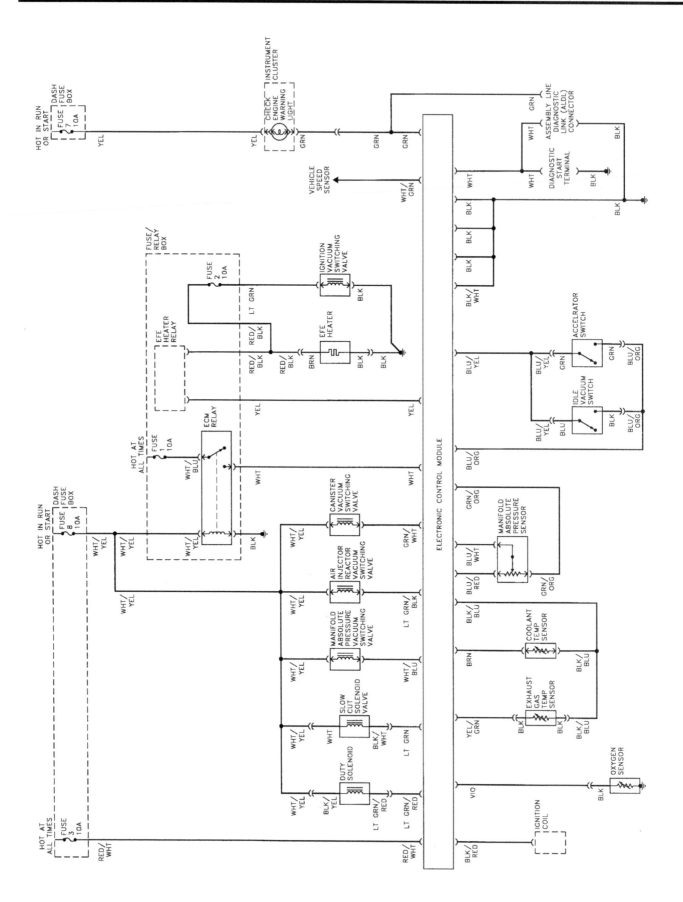

Typical 1989 and 1990 feedback carburetor engine control system

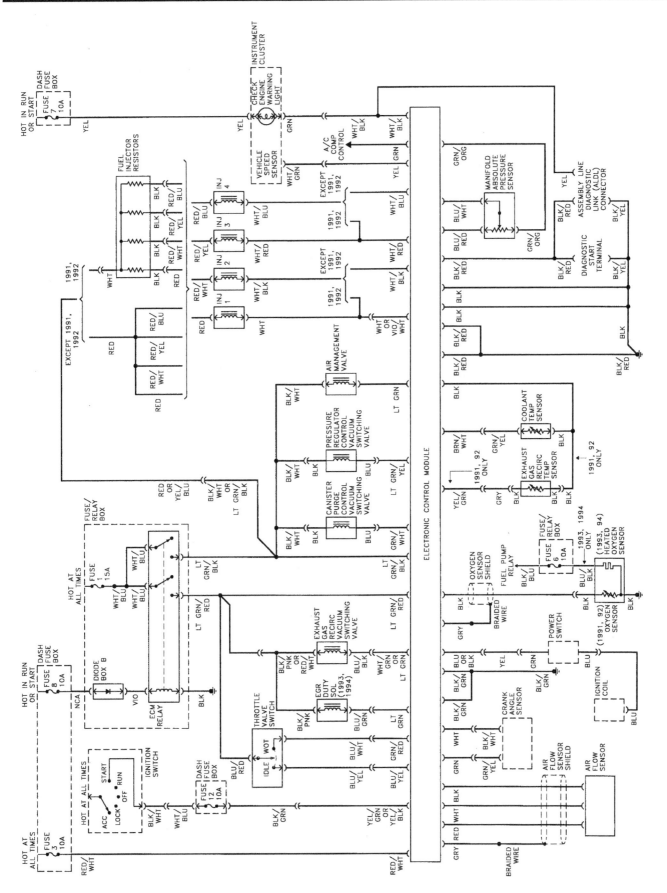

Typical 1991 through 1994 four-cylinder engine control system

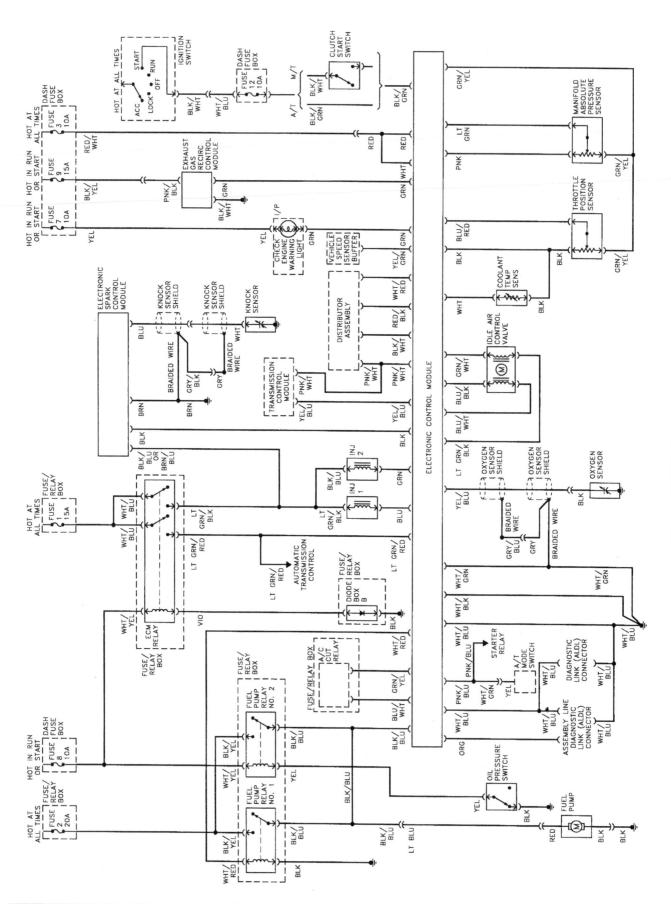

Typical 1991 and 1992 3.1L V6 engine control system

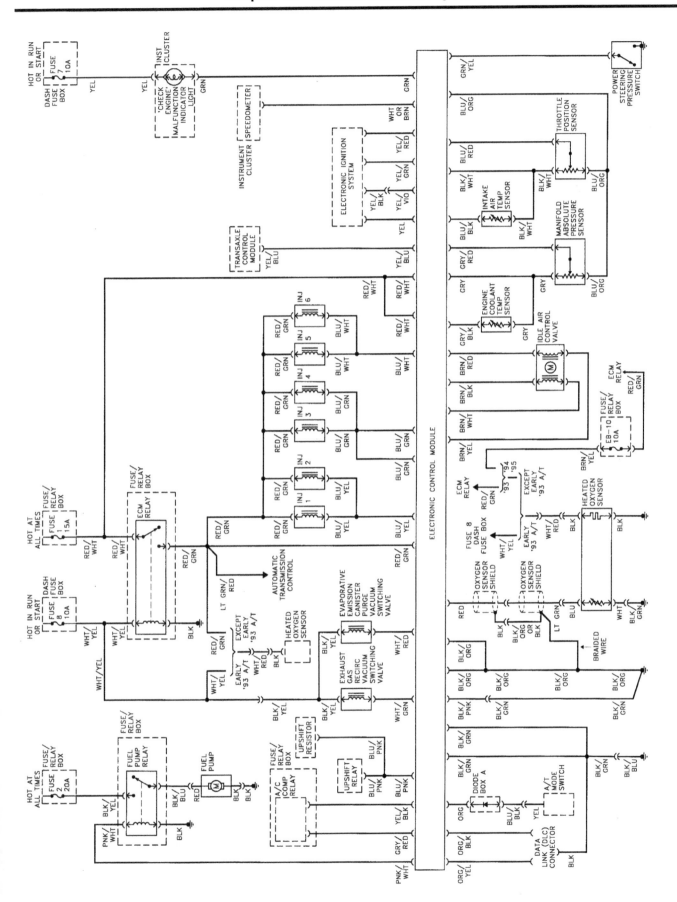

Typical 1993 through 1995 3.2L V6 engine control system

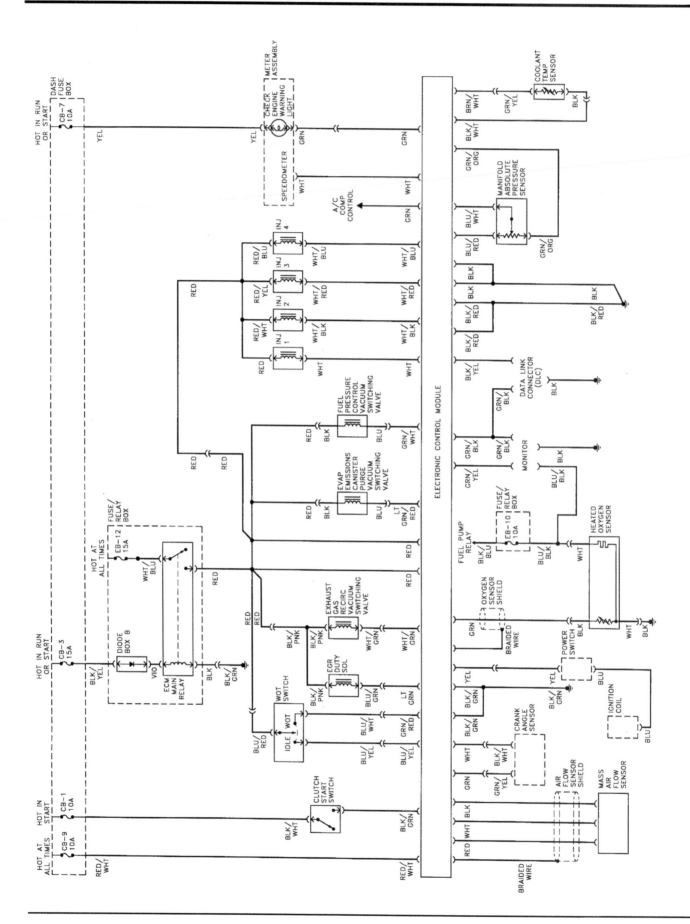

Typical 1995 four-cylinder engine control system

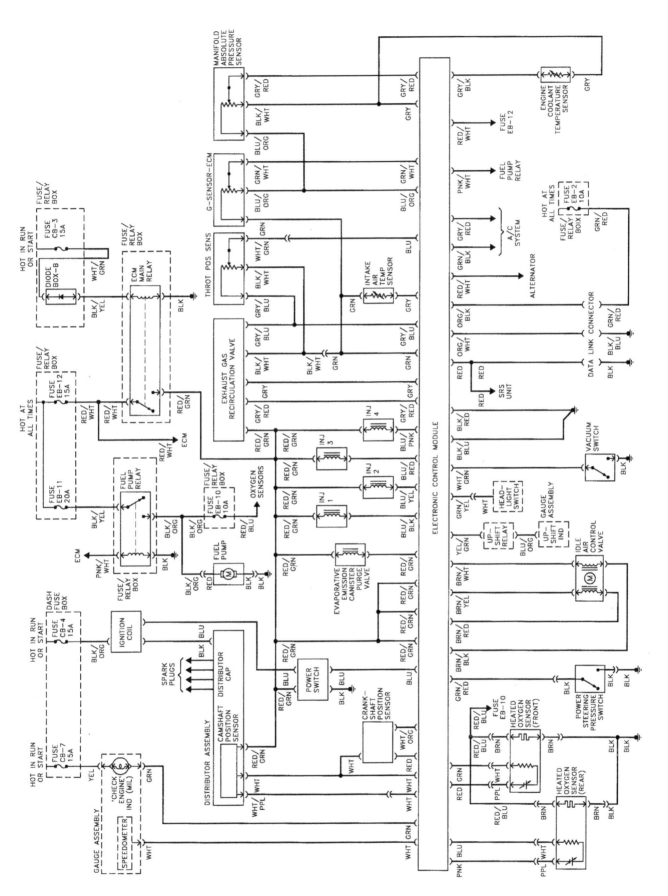

Typical 1996 and 1997 four-cylinder engine control system

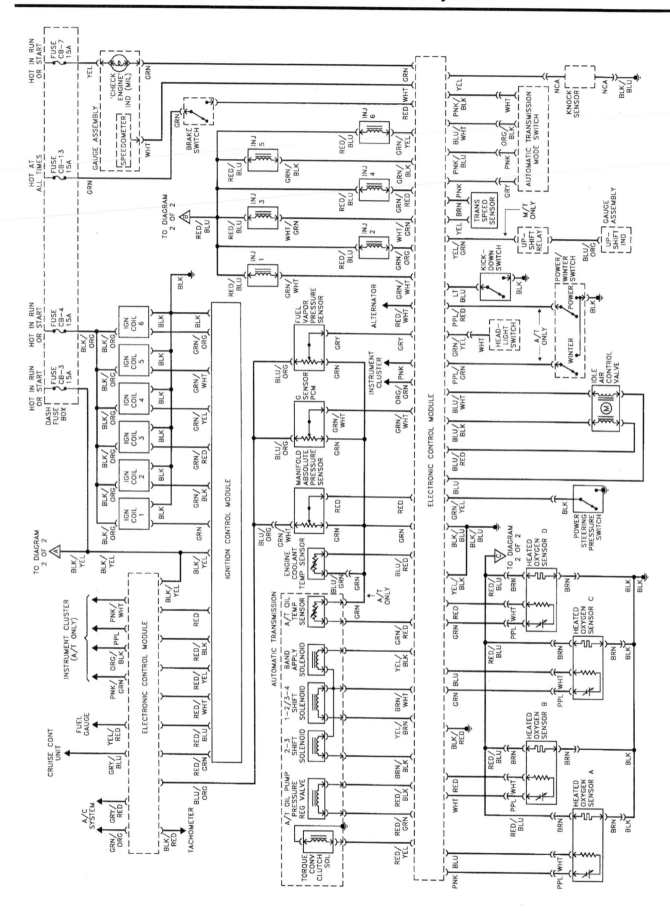

Typical 1996 and 1997 3.2L V6 engine control system (part 1 of 2)

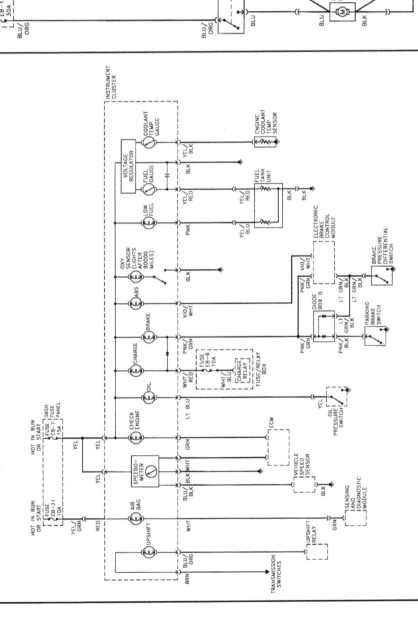

Typical 1993 through 1997 3.2L V6 engine cooling system

Typical 1996 and 1997 3.2L V6 engine control system (part 2 of 2)

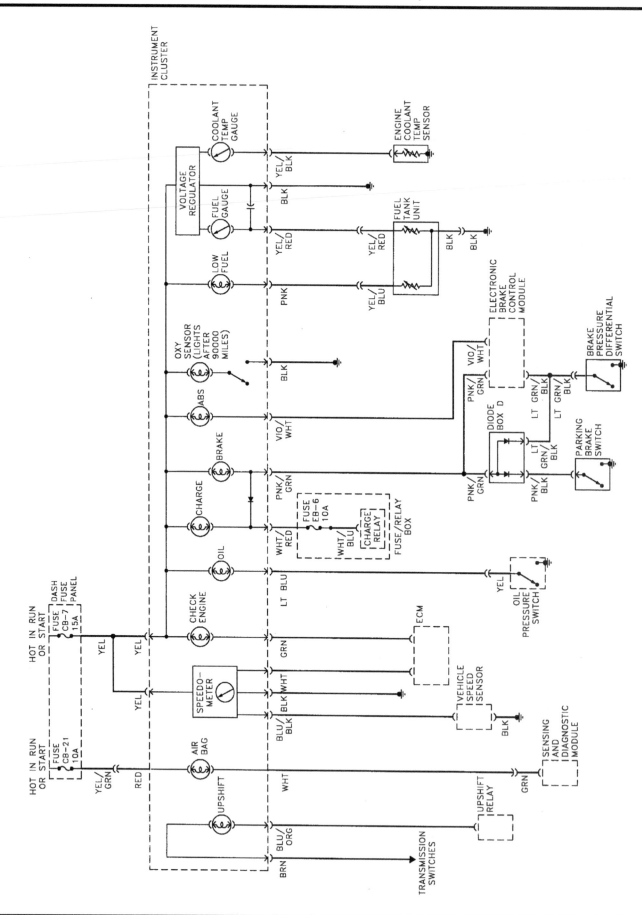

Typical four-cylinder engine warning system

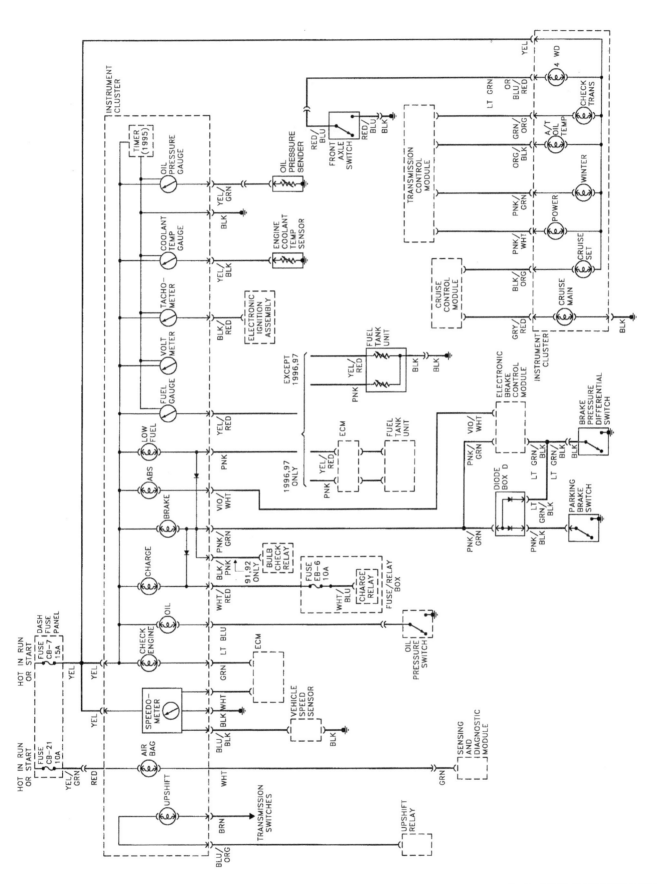

Typical V6 engine warning system

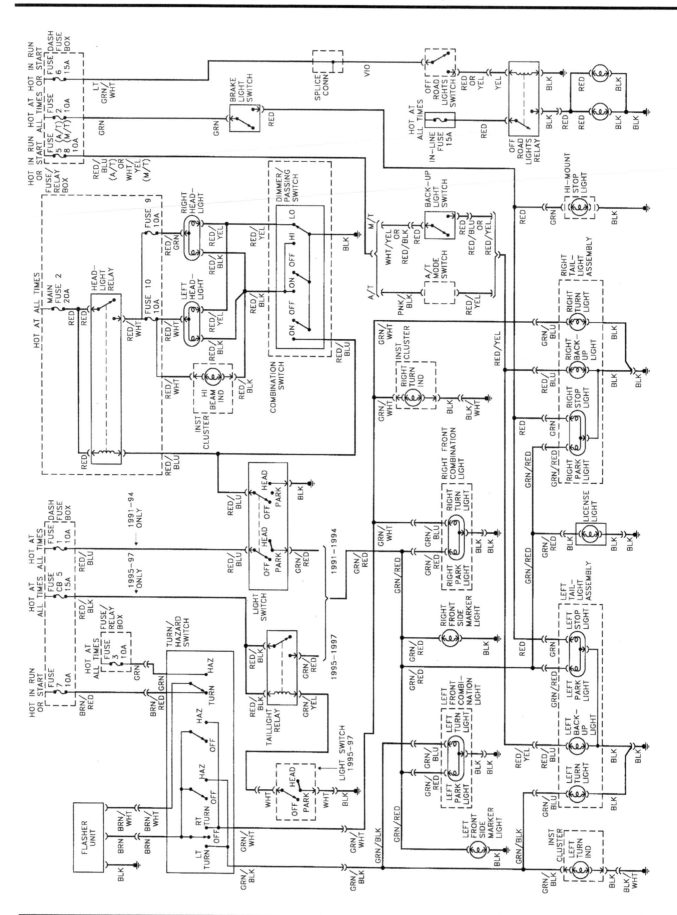

Typical exterior lighting system

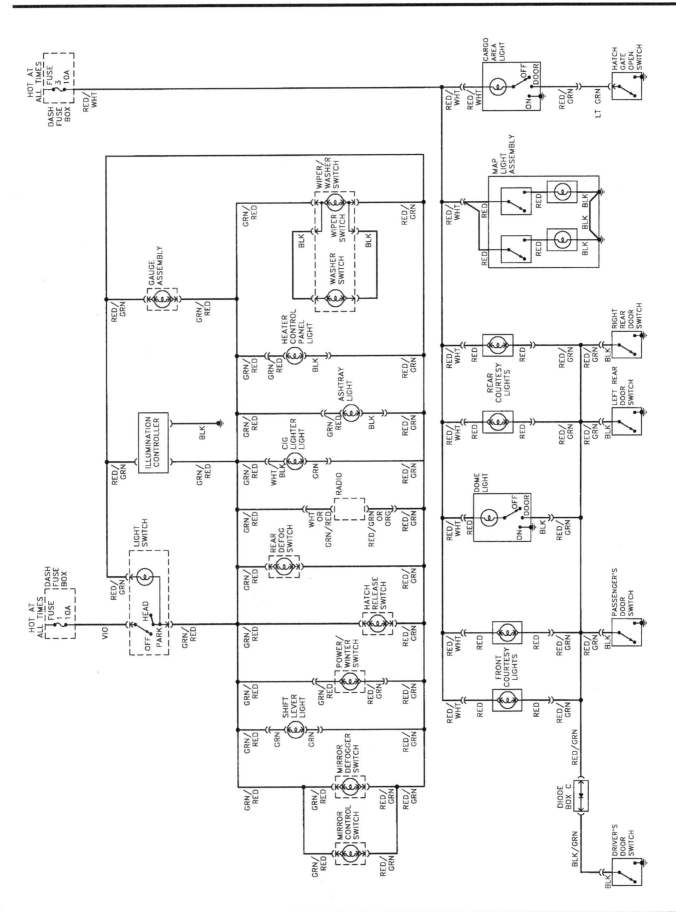

Typical 1994 and earlier interior lighting system

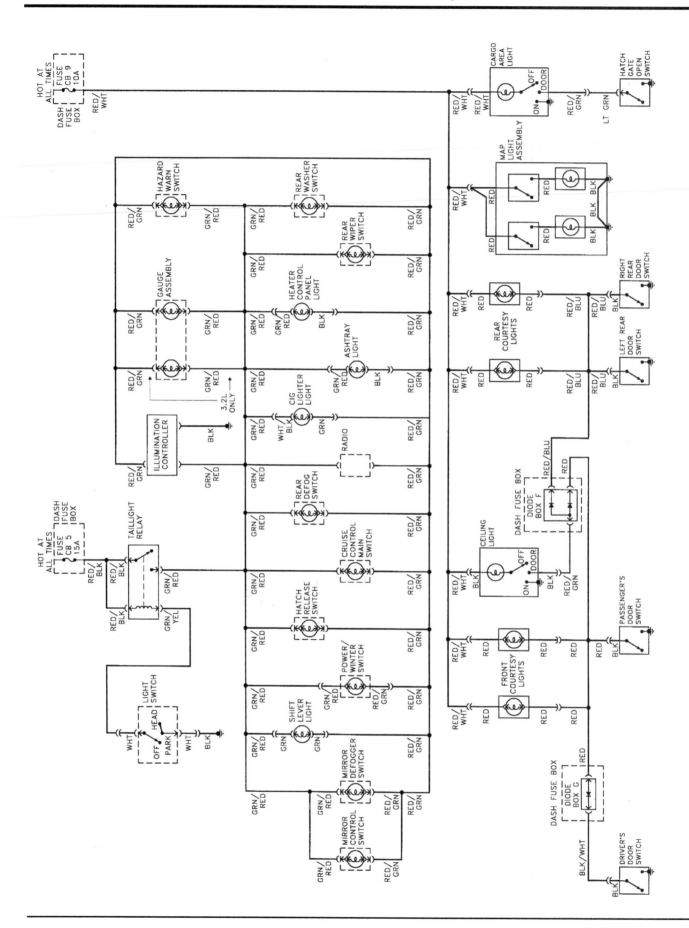

Typical 1995 and later interior lighting system

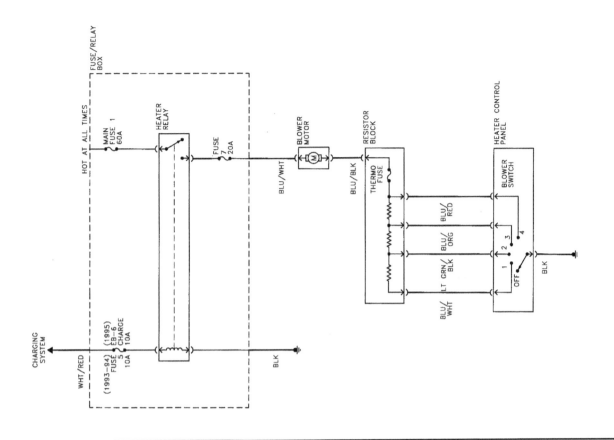

Typical 1993 through 1995 heating system

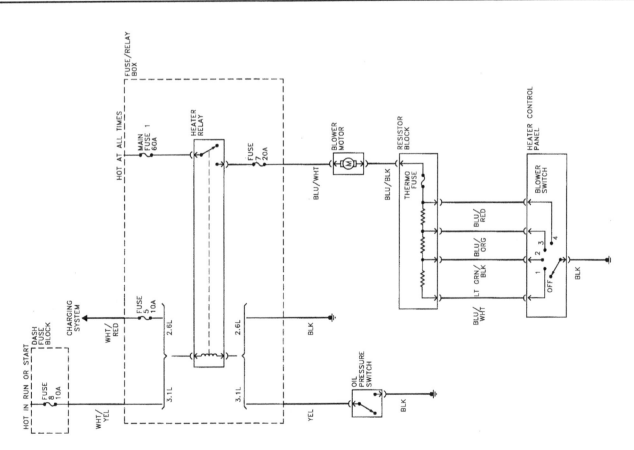

Typical 1991 and 1992 heating system

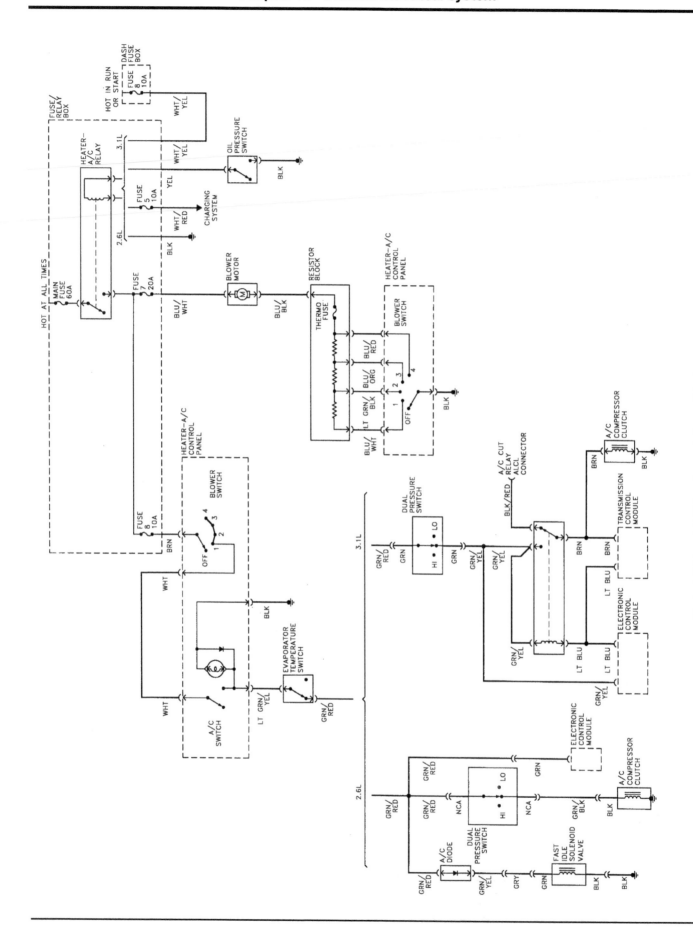

Typical 1991 and 1992 heating and air conditioning system

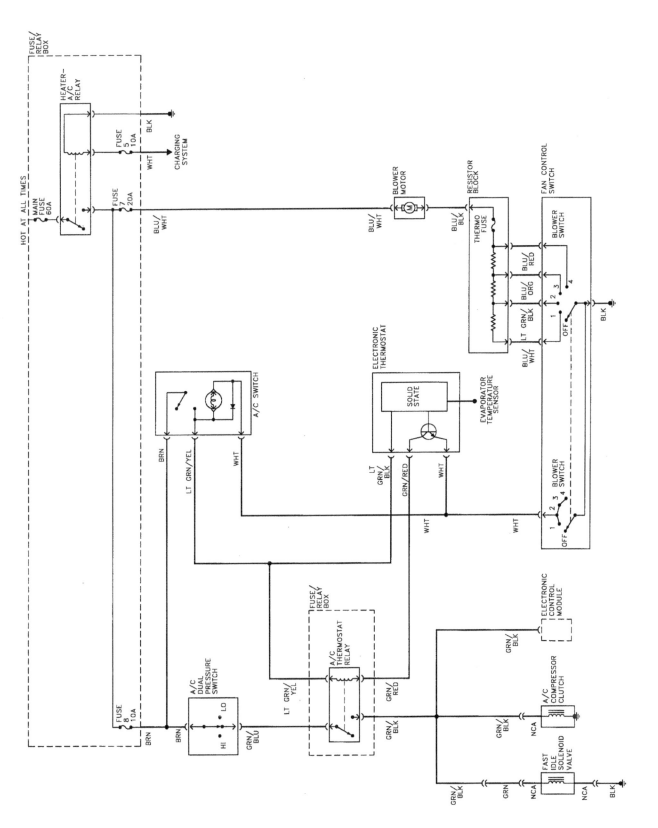

Typical 1993 and 1994 four-cylinder engine heating and air conditioning system

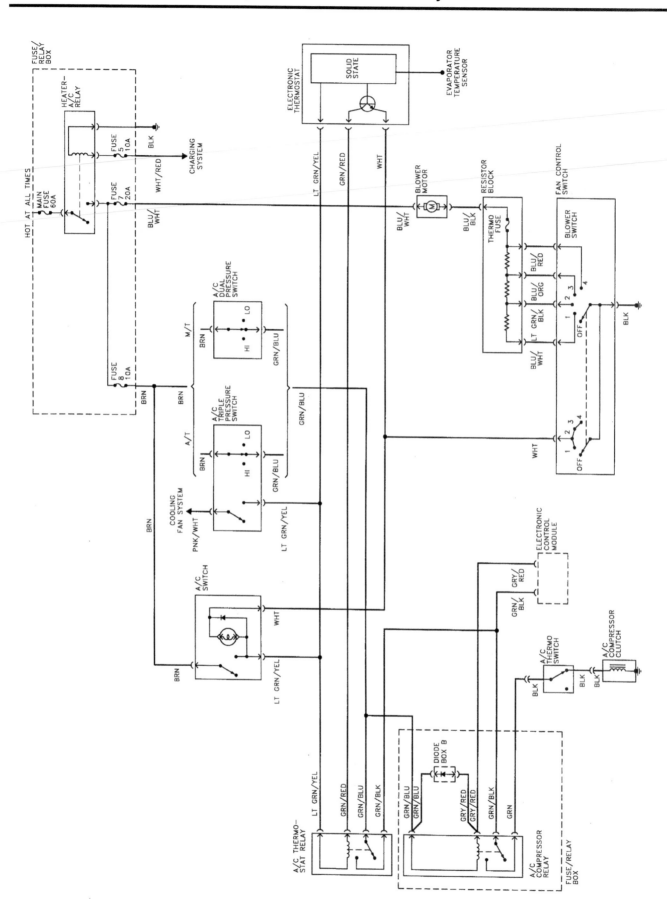

Typical 1993 and 1994 V6 engine heating and air conditioning system

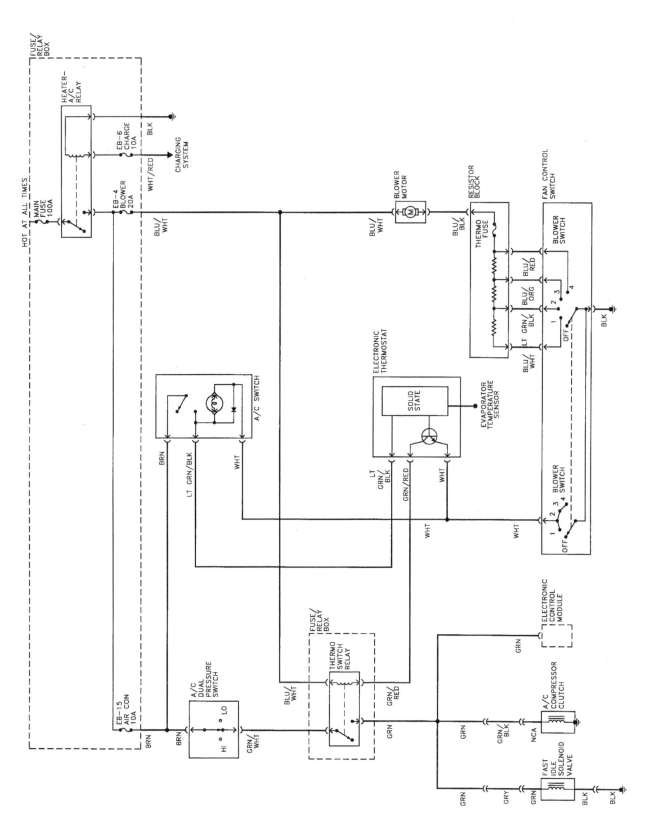

Typical 1995 four-cylinder engine heating and air conditioning system

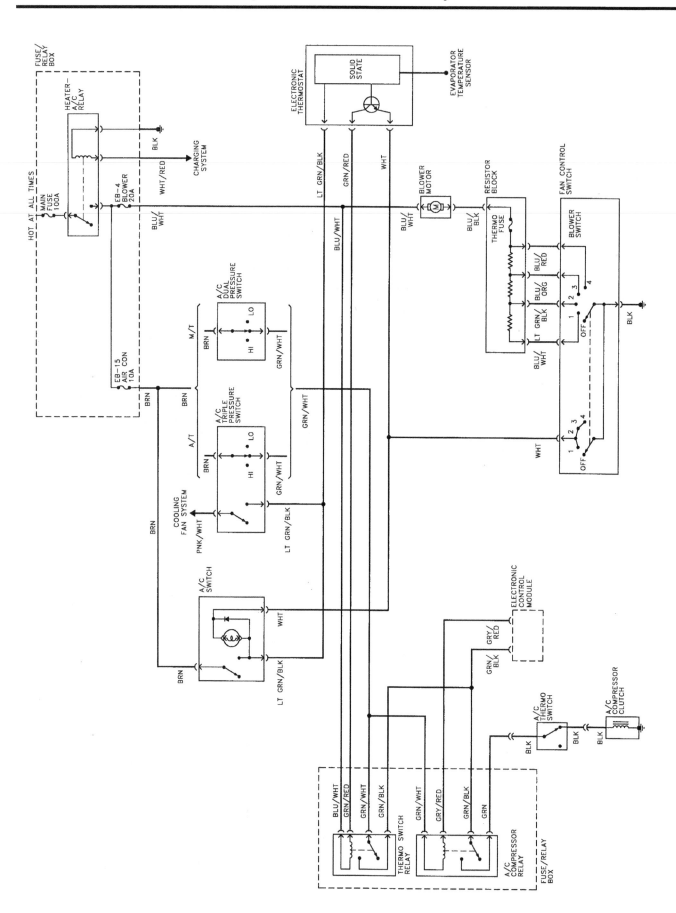

Typical 1995 V6 engine heating and air conditioning system

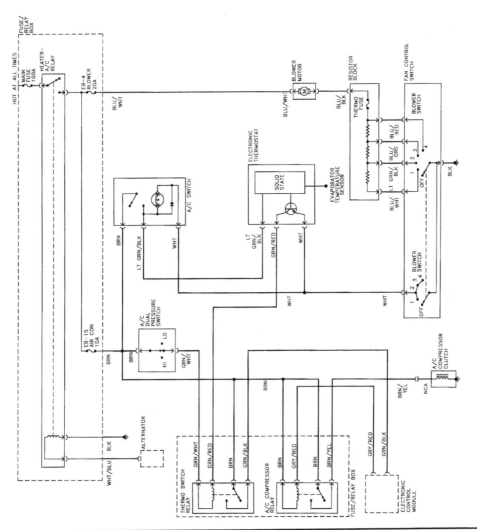

Typical 1996 and later four-cylinder engine heating and air conditioning system

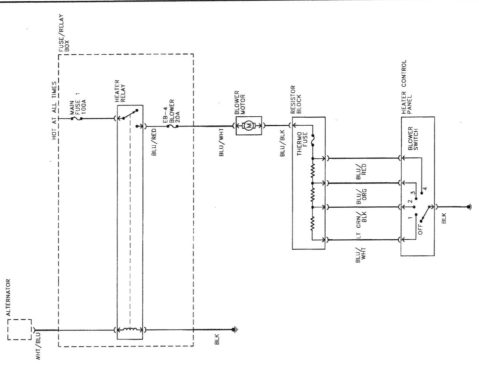

Typical 1996 and 1997 heating system

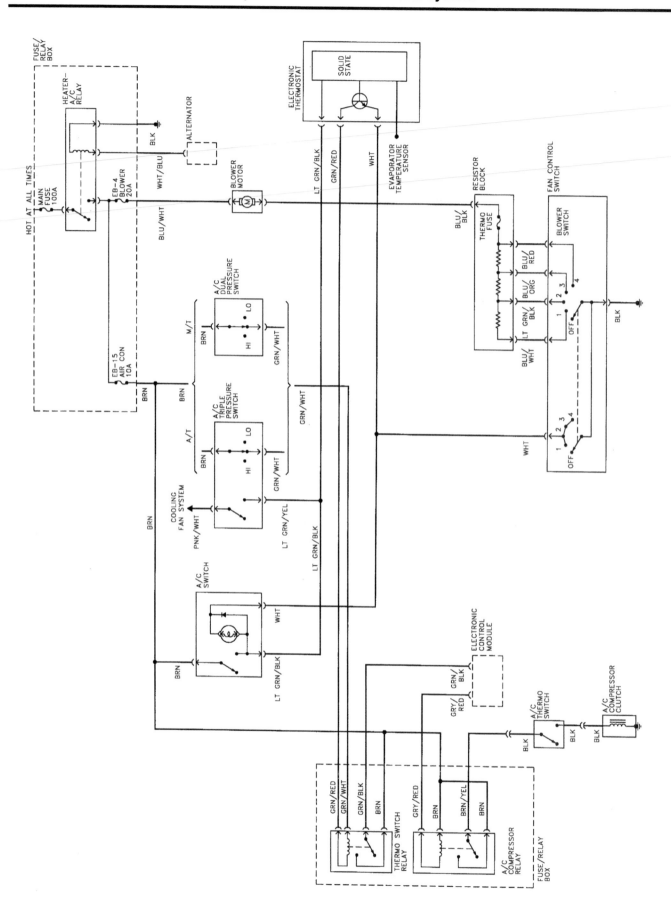

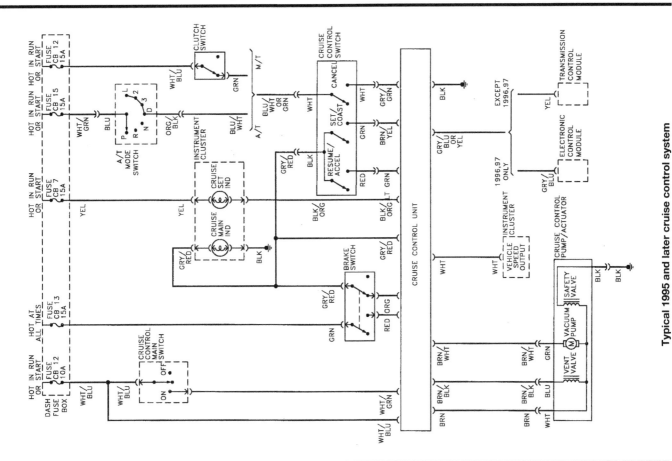

Typical 1995 and later cruise control system

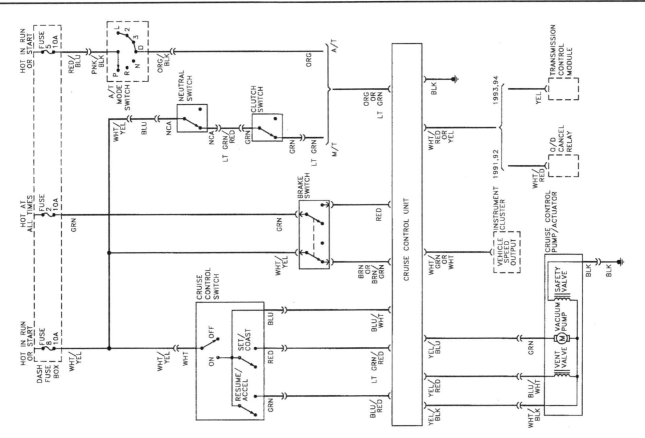

Typical 1994 and earlier cruise control system

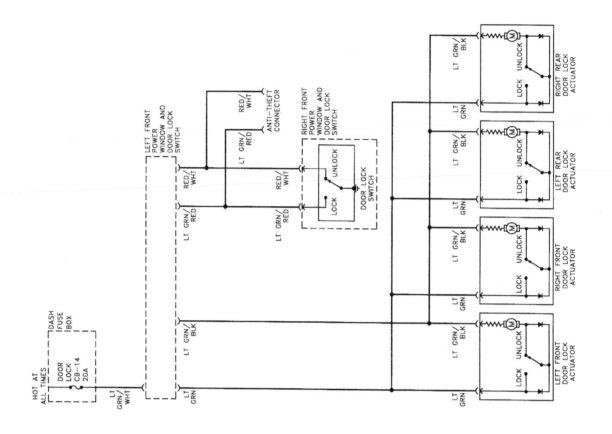

Typical 1995 and later door lock system

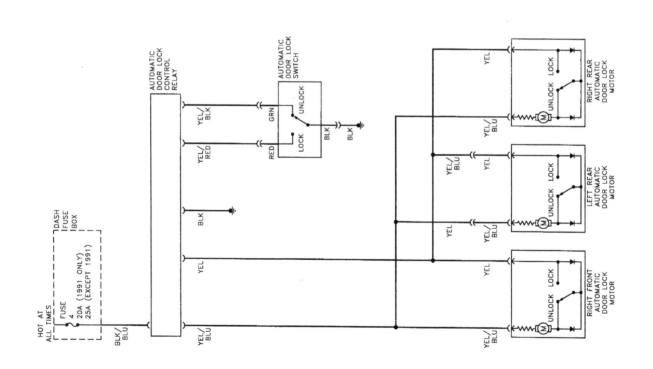

Typical 1994 and earlier door lock system

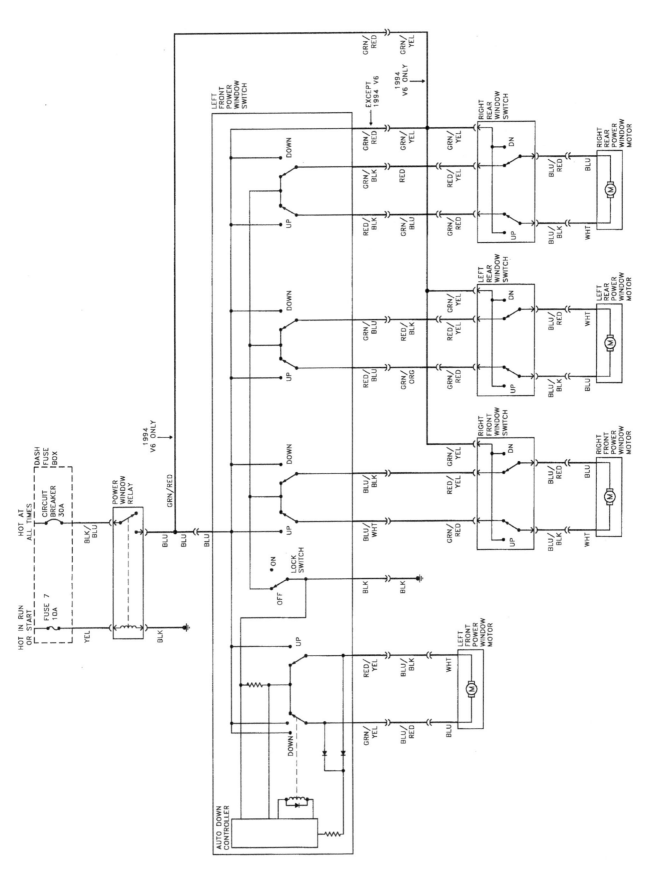

Typical 1994 and earlier power window system

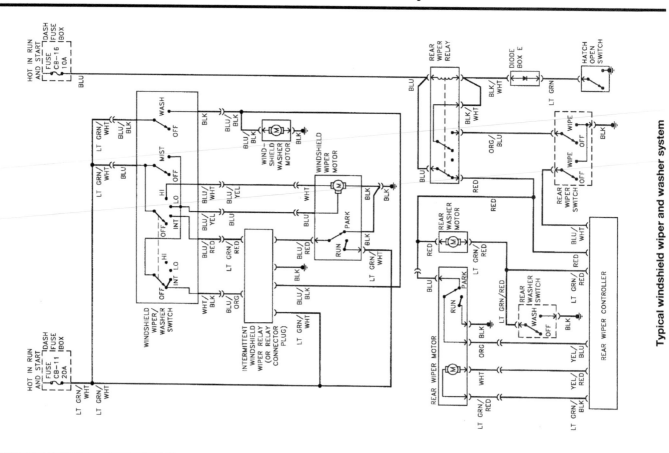

Typical windshield wiper and washer system

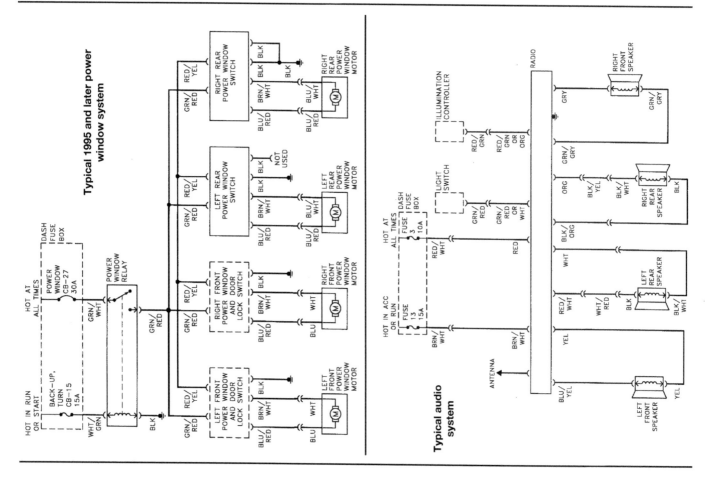

Typical 1995 and later power window system

Typical audio system

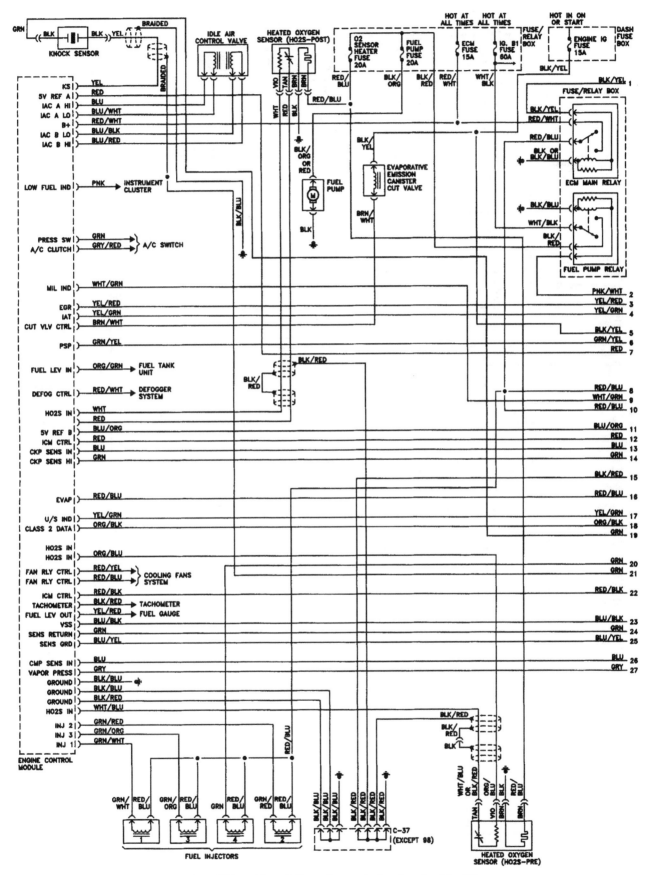

Typical 1998 and later 2.2L engine control system (part 1 of 3)

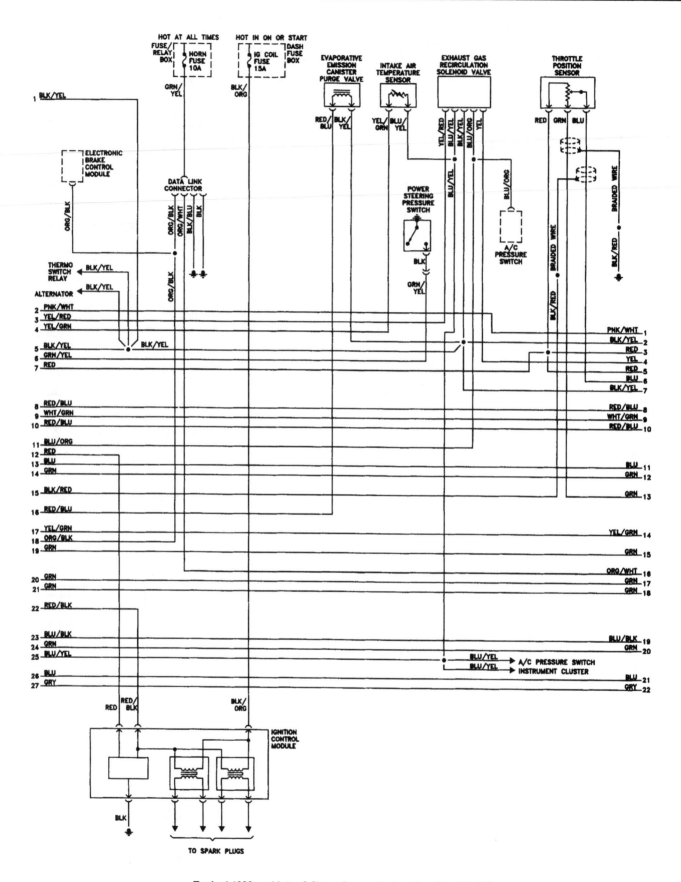

Typical 1998 and later 2.2L engine control system (part 2 of 3)

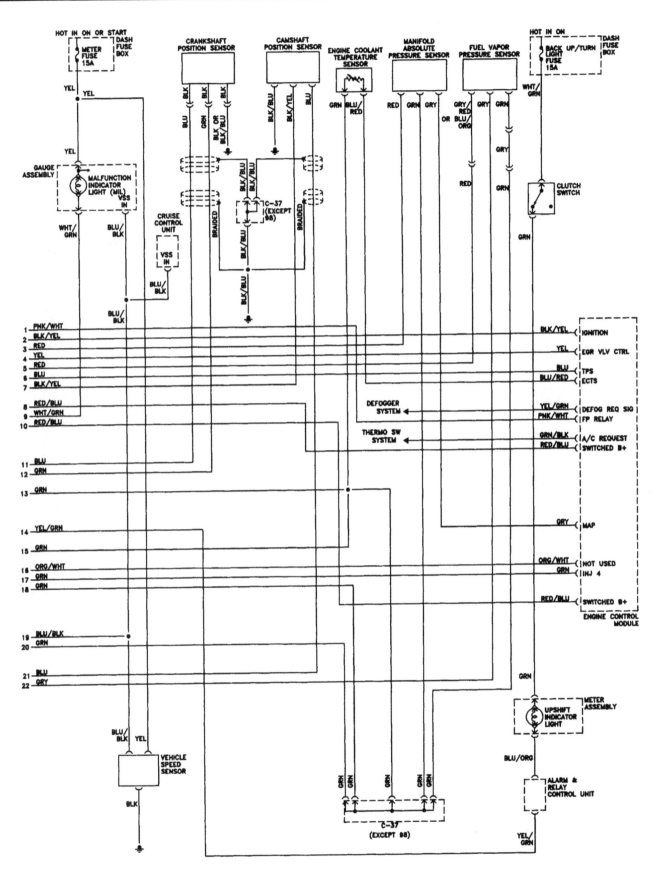

Typical 1998 and later 2.2L engine control system (part 3 of 3)

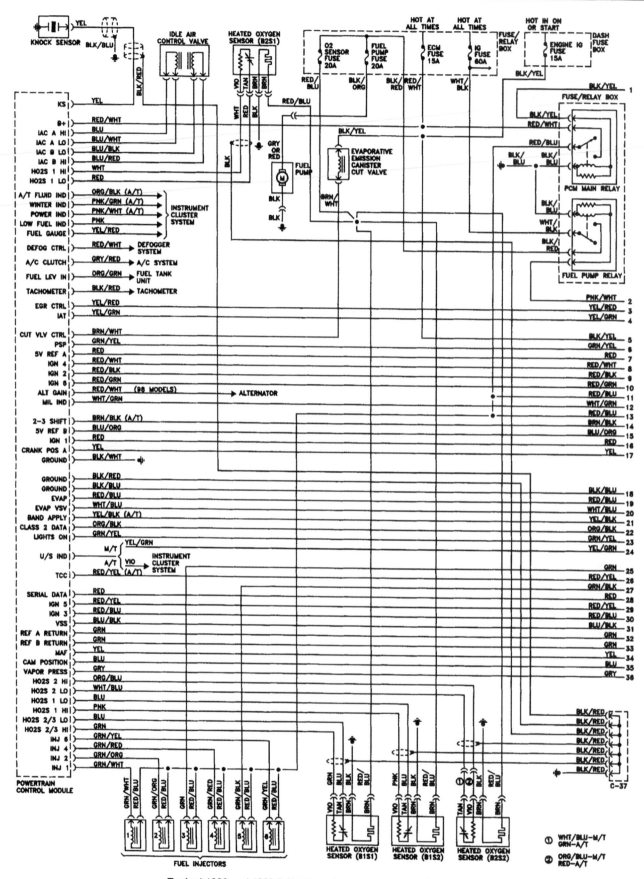

Typical 1998 and 1999 3.2L V6 engine control system (part 1 of 4)

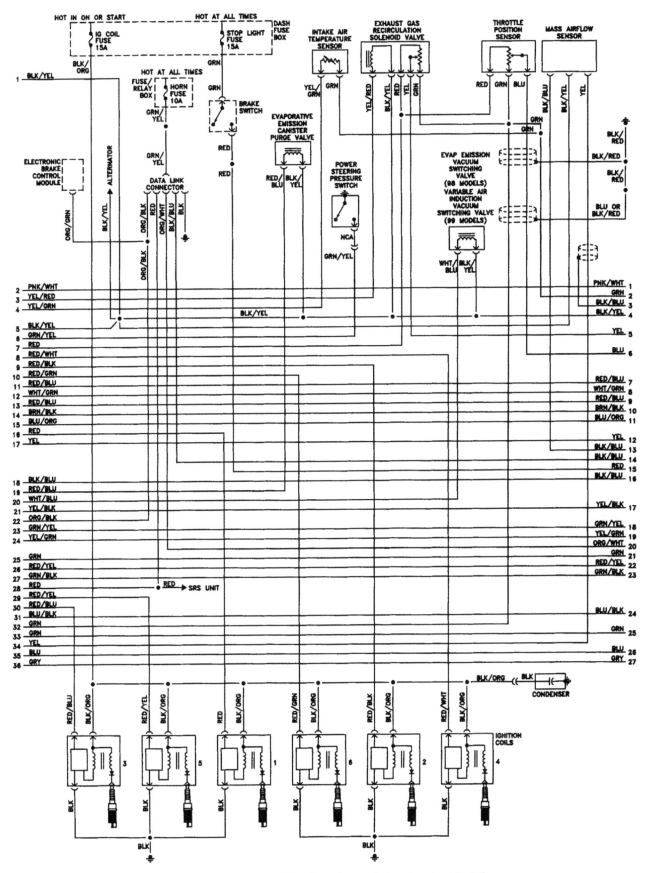

Typical 1998 and 1999 3.2L V6 engine control system (part 2 of 4)

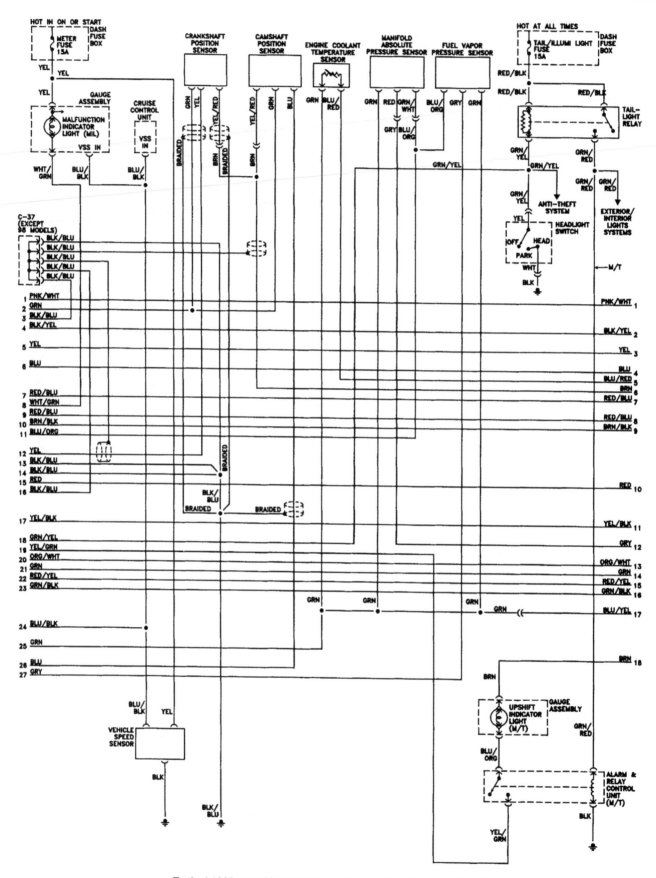

Typical 1998 and 1999 3.2L V6 engine control system (part 3 of 4)

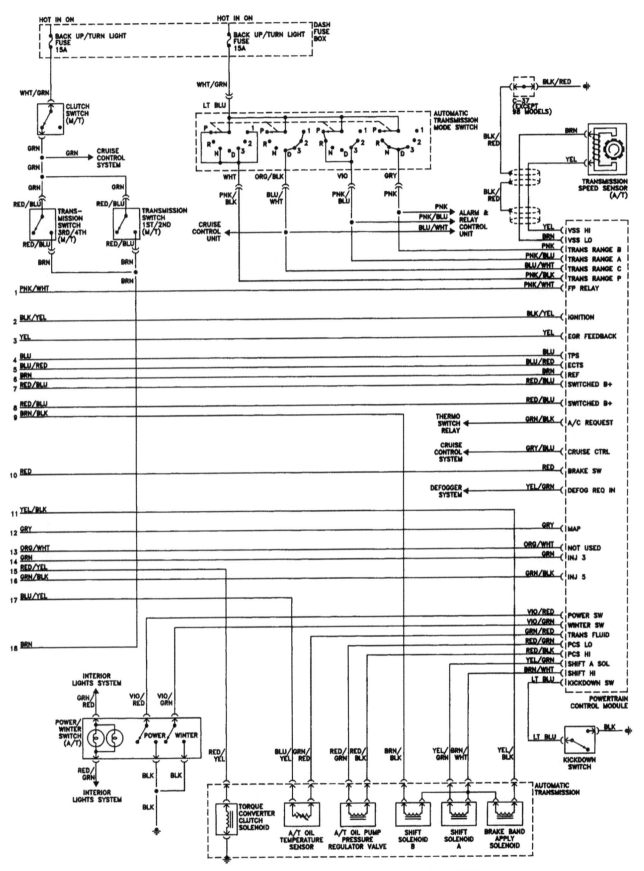

Typical 1998 and 1999 3.2L V6 engine control system (part 4 of 4)

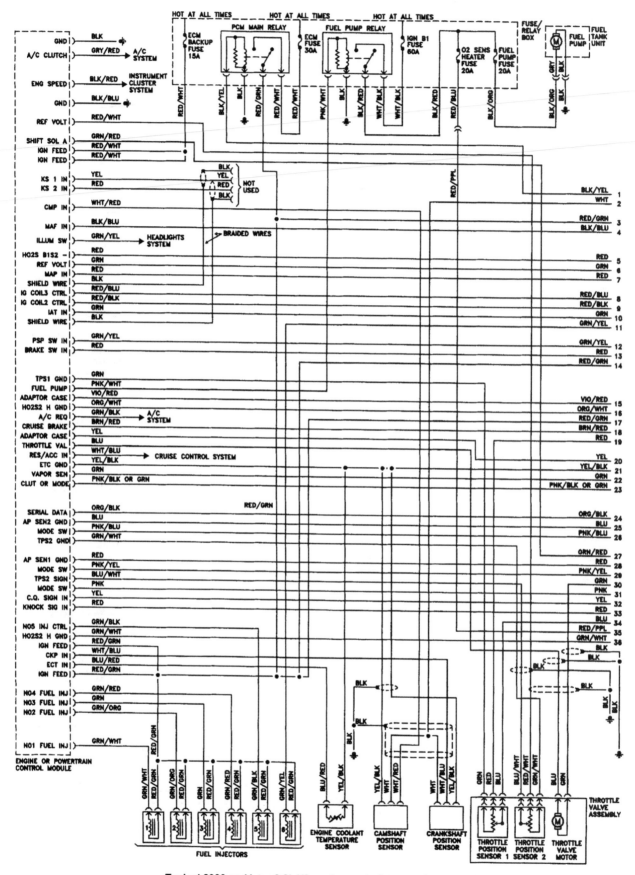

Typical 2000 and later 3.2L V6 engine control system (part 1 of 4)

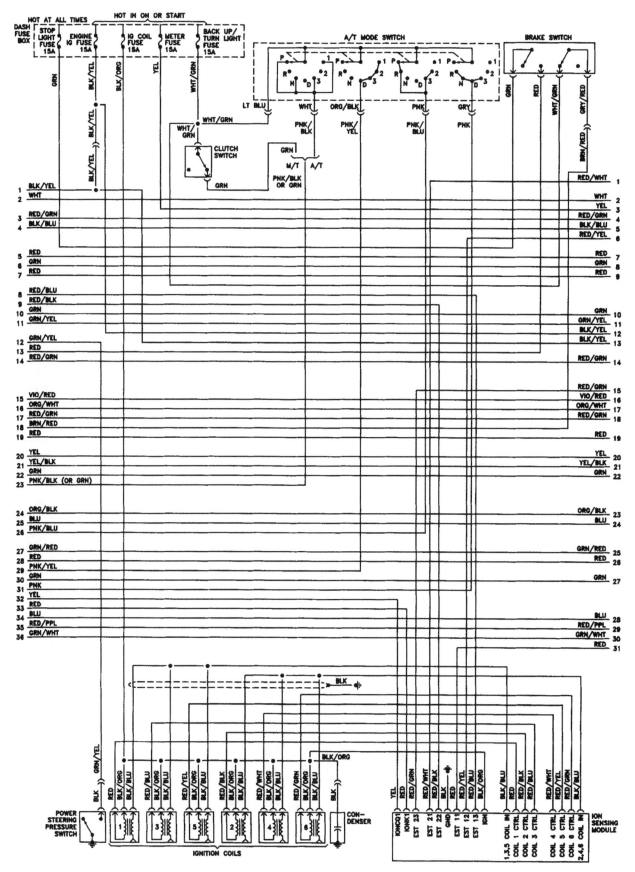

Typical 2000 and later 3.2L V6 engine control system (part 2 of 4)

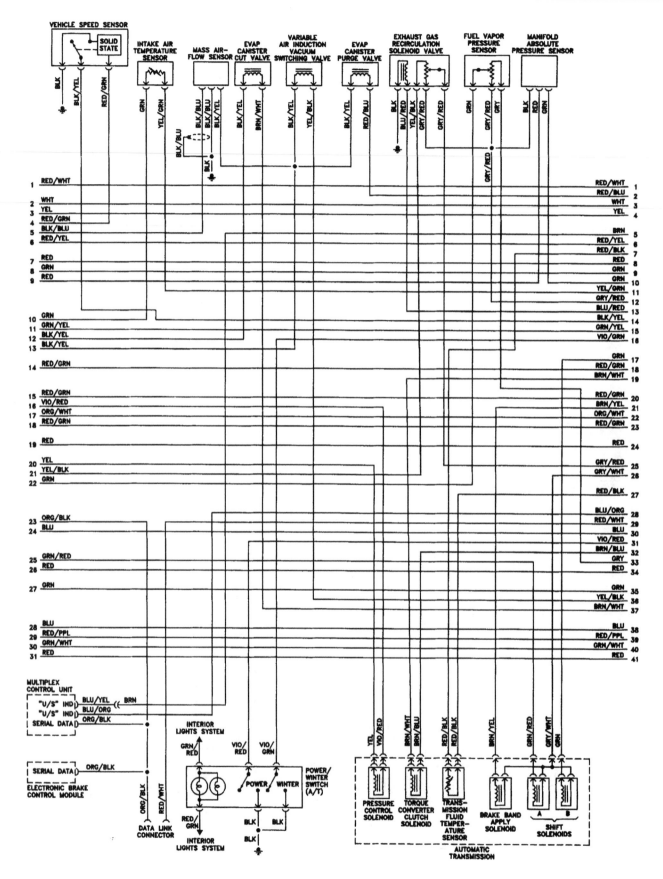

Typical 2000 and later 3.2L V6 engine control system (part 3 of 4)

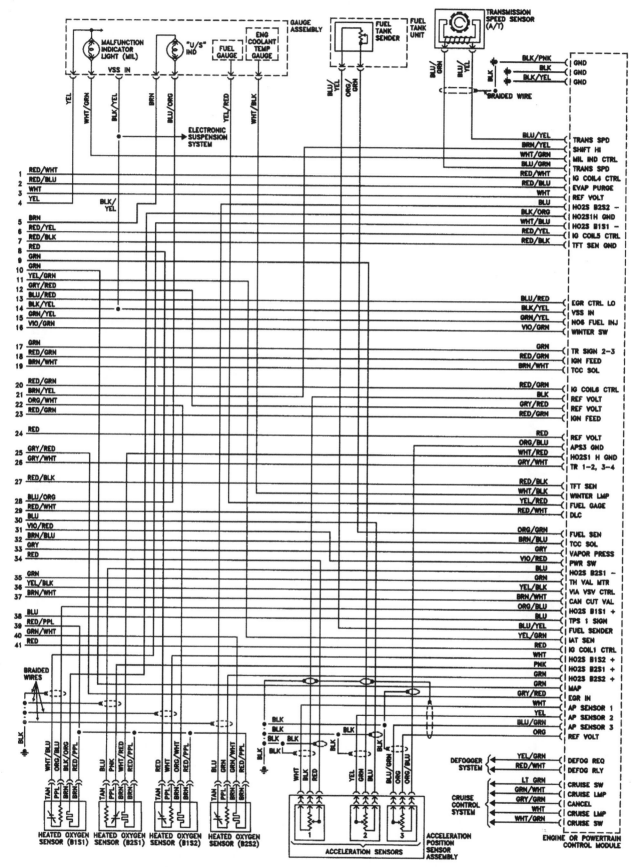

Typical 2000 and later 3.2L V6 engine control system (part 4 of 4)

Notes

Index

A

B

C

E

Haynes Automotive Manuals

ACURA
- 12020 Integra '86 thru '89 & Legend '86 thru '90
- 12021 Integra '90 thru '93 & Legend '91 thru '95
  Integra '94 thru '00 - see HONDA Civic (42025)
  MDX '01 thru '07 - see Honda Pilot (42037)
- 12050 Acura TL all models '99 thru '08

AMC
- 14020 Concord/Hornet/Gremlin/Spirit '70 thru '83
- 14025 (Renault) Alliance & Encore '83 thru '87

AUDI
- 15020 4000 all models '80 thru '87
- 15025 5000 all models '77 thru '83
- 15026 5000 all models '84 thru '88
  Audi A4 '96 thru '01 - see VW Passat (96023)
- 15030 Audi A4 '02 thru '08

AUSTIN-HEALEY
  Sprite - see MG Midget (66015)

BMW
- 18020 3/5 Series '82 thru '92
- 18021 3-Series including Z3 models '92 thru '98
- 18022 3-Series including Z4 models '99 thru '05
- 18023 3-Series '06 thru '14
- 18025 320i all 4-cylinder models '75 thru '83
- 18050 1500 thru 2002 except Turbo '59 thru '77

BUICK
- 19010 Buick Century '97 thru '05
  Century (front-wheel drive) - see GM (38005)
- 19020 Buick, Oldsmobile & Pontiac Full-size (Front wheel drive) '85 thru '05
- 19025 Buick, Oldsmobile & Pontiac Full-size (Rear wheel drive) '70 thru '90
- 19027 Buick LaCrosse '05 thru '13
  Regal - see GENERAL MOTORS (38010)
  Skyhawk - see GM (38020, 38025)
  Skylark - see GM (38020, 38025)
  Somerset - see GENERAL MOTORS (38025)

CADILLAC
- 21015 CTS & CTS-V '03 thru '14
- 21030 Cadillac Rear Wheel Drive '70 thru '93
  Cimarron, Eldorado & Seville - see GM (38015, 38030, 38031)

CHEVROLET
- 10305 Chevrolet Engine Overhaul Manual
- 24010 Astro & GMC Safari Mini-vans '85 thru '05
- 24015 Camaro V8 all models '70 thru '81
- 24016 Camaro all models '82 thru '92
  Cavalier - see GM (38016)
  Celebrity - see GM (38005)
- 24017 Camaro & Firebird '93 thru '02
- 24018 Camaro '10 thru '15
- 24020 Chevelle, Malibu, El Camino '69 thru '87
- 24024 Chevette & Pontiac T1000 '76 thru '87
  Citation - see GENERAL MOTORS (38020)
- 24027 Colorado & GMC Canyon '04 thru '12
- 24032 Corsica & Beretta all models '87 thru '96
- 24040 Corvette all V8 models '68 thru '82
- 24041 Corvette all models '84 thru '96
- 24042 Corvette '97 thru '13
- 24044 Cruze '11 thru '19
- 24045 Full-size Sedans Caprice, Impala, Biscayne, Bel Air & Wagons '69 thru '90
- 24046 Impala SS & Caprice and Buick Roadmaster '91 thru '96
  Impala '00 thru '05 - see LUMINA (24048)
- 24047 Impala & Monte Carlo all models '06 thru '11
  Lumina '90 thru '94 - see GM (38010)
- 24048 Lumina & Monte Carlo '95 thru '05
  Lumina APV - see GM (38035)
- 24050 Luv Pick-up all 2WD & 4WD '72 thru '82
- 24051 Malibu '13 thru '19
- 24055 Monte Carlo all models '70 thru '88
  Monte Carlo '95 thru '01 - see LUMINA (24048)
- 24059 Nova all V8 models '69 thru '79
- 24060 Nova/Geo Prizm '85 thru '92
- 24064 Pick-ups '67 thru '87 - Chevrolet & GMC
- 24065 Pick-ups '88 thru '98 - Chevrolet & GMC
- 24066 Pick-ups '06 thru '08 - Chevrolet & GMC
- 24067 Chevy Silverado & GMC Sierra '07 thru '14
- 24068 Chevy Silverado & GMC Sierra '14 thru '19
- 24070 S-10 & S-15 Pick-ups '82 thru '93
- 24071 S-10 & Sonoma Pick-ups '94 thru '04
- 24072 Chevrolet TrailBlazer, GMC Envoy & Oldsmobile Bravada '02 thru '09
- 24075 Sprint '85 thru '88, Geo Metro '89 thru '01
- 24080 Vans - Chevrolet & GMC '68 thru '96
- 24081 Chevrolet Express & GMC Savana Full-size Vans '96 thru '19

CHRYSLER
- 10310 Chrysler Engine Overhaul Manual
- 25015 Chrysler Cirrus, Dodge Stratus, Plymouth Breeze, '95 thru '00
- 25020 Full-size Front-Wheel Drive '88 thru '93
  K-Cars - see DODGE Aries (30008)
- 25025 Chrysler LHS, Concorde & New Yorker, Dodge Intrepid, Eagle Vision, '93 thru '97
- 25026 Chrysler LHS, Concorde, 300M, Dodge Intrepid '98 thru '04
- 25027 Chrysler 300 '05 thru '18, Dodge Charger '06 thru '18, Magnum '05 thru '08 & Challenger '08 thru '18
- 25030 Chrysler & Plymouth Mid-size '82 thru '95
  Rear-wheel Drive - see DODGE (30050)
- 25035 PT Cruiser all models '01 thru '10
- 25040 Chrysler Sebring '07 thru '06, Dodge Stratus '01 thru '06 & Dodge Avenger '95 thru '00
- 25041 Chrysler Sebring '07 thru '10, 200 '11 thru '17 Dodge Avenger '08 thru '14

DATSUN
- 28005 200SX all models '80 thru '83
- 28012 240Z, 260Z & 280Z Coupe '70 thru '78
- 28014 280ZX Coupe & 2+2 '79 thru '83
  300ZX - see NISSAN (72010)
- 28018 510 & PL521 Pick-up '68 thru '73
- 28020 510 all models '78 thru '81
- 28022 620 Series Pick-up all models '73 thru '79
  720 Series Pick-up - see NISSAN (72030)

DODGE
- 30008 Aries & Plymouth Reliant '81 thru '89
- 30010 Caravan & Plymouth Voyager '84 thru '95
- 30011 Caravan & Plymouth Voyager '96 thru '02
- 30012 Challenger & Plymouth Sapporro '78 thru '83
- 30013 Caravan, Chrysler Voyager & Town & Country '03 thru '07
- 30014 Grand Caravan & Chrysler Town & Country '08 thru '18

- 30020 Dakota Pick-ups all models '87 thru '96
- 30021 Durango '98 & '99 & Dakota '97 thru '99
- 30022 Durango '00 thru '03 & Dakota '00 thru '04
- 30023 Durango '04 thru '09 & Dakota '05 thru '11
- 30025 Dart, Challenger/Plymouth Barracuda & Valiant 6-cylinder models '67 thru '76
- 30030 Daytona & Chrysler Laser '84 thru '89
  Intrepid - see Chrysler (25025, 25026)
- 30034 Neon all models '95 thru '99
- 30035 Omni & Plymouth Horizon '78 thru '90
- 30036 Dodge & Plymouth Neon '00 thru '05
- 30040 Pick-ups full-size models '74 thru '93
- 30041 Pick-ups full-size models '94 thru '01
- 30042 Pick-ups full-size models '02 thru '08
- 30043 Pick-ups full-size models '09 thru '18
- 30045 Ram 50/D50 Pick-ups & Raider and Plymouth Arrow Pick-ups '79 thru '93
- 30050 Dodge/Plymouth/Chrysler RWD '71 thru '89
- 30055 Shadow & Plymouth Sundance '87 thru '94
- 30060 Spirit & Plymouth Acclaim '89 thru '95
- 30065 Vans - Dodge & Plymouth '71 thru '03

EAGLE
  Talon - see MITSUBISHI (68030, 68031)
  Vision - see CHRYSLER (25025)

FIAT
- 34010 124 Sport Coupe & Spider '68 thru '78
- 34025 X1/9 all models '74 thru '80

FORD
- 10320 Ford Engine Overhaul Manual
- 10355 Ford Automatic Transmission Overhaul
- 11500 Mustang '64-1/2 thru '70 Restoration Guide
- 36004 Aerostar Mini-vans '86 thru '97
- 36006 Contour & Mercury Mystique '95 thru '00
- 36008 Courier Pick-up all models '72 thru '82
- 36012 Crown Victoria & Mercury Grand Marquis '88 thru '11
- 36016 Escort & Mercury Lynx '81 thru '90
- 36020 Escort & Mercury Tracer '91 thru '02
- 36022 Escape '01 thru '17, Mazda Tribute '01 thru '11 & Mercury Mariner '05 thru '11
- 36024 Explorer & Mazda Navajo '91 thru '01
- 36025 Explorer & Mercury Mountaineer '02 thru '10
- 36026 Explorer '11 thru '17
- 36028 Fairmont & Mercury Zephyr '78 thru '83
- 36030 Festiva & Aspire '88 thru '97
- 36032 Fiesta all models '77 thru '80
- 36034 Focus all models '00 thru '11
- 36035 Focus '12 thru '14
- 36045 Ford Fusion '06 thru '14 & Mercury Milan '06 thru '10
- 36048 Mustang V8 all models '64-1/2 thru '73
- 36049 Mustang II 4-cylinder, V6 & V8 '74 thru '78
- 36050 Mustang & Mercury Capri '79 thru '93
- 36051 Mustang all models '94 thru '04
- 36052 Mustang '05 thru '14
- 36054 Pick-ups and Bronco '73 thru '79
- 36058 Pick-ups and Bronco '80 thru '96
- 36059 F-150 '97 thru '03, Expedition '97 thru '17, F-250 '97 thru '99, F-150 Heritage '04 & Lincoln Navigator '98 thru '17
- 36060 Super Duty Pick-up & Excursion '99 thru '10
- 36061 F-150 full-size '04 thru '14
- 36062 Pinto & Mercury Bobcat '75 thru '80
- 36063 F-150 full-size '15 thru '17
- 36064 Super Duty Pick-ups '11 thru '16
- 36066 Probe all models '89 thru '92
- 36070 Ranger & Bronco II gas models '83 thru '92
- 36071 Ranger '93 thru '11 & Mazda Pick-ups '94 thru '09
- 36074 Taurus & Mercury Sable '86 thru '95
- 36075 Taurus & Mercury Sable '96 thru '07
- 36076 Taurus '08 thru '14, Five Hundred '05 thru '07, Mercury Montego '05 thru '07 & Sable '08 thru '09
- 36078 Tempo & Mercury Topaz '84 thru '94
- 36082 Thunderbird & Mercury Cougar '83 thru '88
- 36086 Thunderbird & Mercury Cougar '89 thru '97
- 36090 Vans all V8 Econoline models '69 thru '91
- 36094 Vans full size '92 thru '14
- 36097 Windstar '95 thru '03, Freestar & Mercury Monterey Mini-van '04 thru '07

GENERAL MOTORS
- 10360 GM Automatic Transmission Overhaul
- 38005 Buick Century, Chevrolet Celebrity, Olds Cutlass Ciera & Pontiac 6000 '82 thru '96
- 38010 Buick Regal, Chevrolet Lumina, Oldsmobile Cutlass Supreme & Pontiac Grand Prix front wheel drive '88 thru '07
- 38015 Buick Skyhawk, Cadillac Cimarron, Chevrolet Cavalier, Oldsmobile Firenza Pontiac J-2000 & Sunbird '82 thru '94
- 38016 Chevrolet Cavalier/Pontiac Sunfire '95 thru '05
- 38017 Chevrolet Cobalt & Pontiac G5 '05 thru '11
- 38020 Buick Skylark, Chevrolet Citation, Olds Omega, Pontiac Phoenix '80 thru '85
- 38025 Buick Skylark & Somerset, Olds Achieva, Calais & Pontiac Grand Am '85 thru '98
- 38026 Chevrolet Malibu, Olds Alero & Cutlass, Pontiac Grand Am '97 thru '03
- 38027 Chevrolet Malibu '04 thru '12
- 38030 Cadillac Eldorado, Seville, Oldsmobile Toronado & Buick Riviera '71 thru '85
- 38031 Cadillac Eldorado, Seville, DeVille, Fleetwood, Oldsmobile Toronado & Buick Riviera '86 thru '93
- 38032 Cadillac DeVille '94 thru '05, Seville '92 thru '04 & Cadillac DTS '06 thru '10
- 38035 Chevrolet Lumina APV, Olds Silhouette & Pontiac Trans Sport all models '90 thru '96
- 38036 Chevrolet Venture, Olds Silhouette, Pontiac Trans Sport & Montana '97 thru '05
- 38040 Chevrolet Equinox '05 thru '17, GMC Terrain '10 thru '17 & Pontiac Torrent '06 thru '09

GEO
  Metro - see CHEVROLET Sprint (24075)
  Prizm - '85 thru '92 see CHEVY (24060), '93 thru '02 see TOYOTA Corolla (92036)
- 40030 Storm all models '90 thru '93
  Tracker - see SUZUKI Samurai (90010)

GMC
  Vans & Pick-ups - see CHEVROLET

HONDA
- 42010 Accord CVCC all models '76 thru '83
- 42011 Accord all models '84 thru '89
- 42012 Accord all models '90 thru '93
- 42013 Accord all models '94 thru '97
- 42014 Accord all models '98 thru '02
- 42015 Accord '03 thru '12 & Crosstour '10 thru '14
- 42016 Accord '13 thru '17

- 42020 Civic 1200 all models '73 thru '79
- 42021 Civic 1300 & 1500 CVCC '80 thru '83
- 42022 Civic 1500 CVCC all models '75 thru '79
- 42023 Civic all models '84 thru '91
- 42024 Civic & del Sol '92 thru '95
- 42025 Civic '96 thru '00, & CR-V '97 thru '01 & Acura Integra '94 thru '00
- 42026 Civic '01 thru '11 & CR-V '02 thru '11
- 42027 Civic '12 thru '15 & CR-V '12 thru '16
- 42030 Fit '07 thru '13
- 42035 Odyssey all models '99 thru '10
  Passport - see ISUZU Rodeo (47017)
- 42037 Honda Pilot '03 thru '08, Ridgeline '06 thru '14 & Acura MDX '01 thru '07
- 42040 Prelude CVCC all models '79 thru '89

HYUNDAI
- 43010 Elantra all models '96 thru '19
- 43015 Excel & Accent all models '86 thru '13
- 43050 Santa Fe all models '01 thru '12
- 43055 Sonata all models '99 thru '14

INFINITI
  G35 '03 thru '08 - see NISSAN 350Z (72011)

ISUZU
  Hombre - see CHEVROLET S-10 (24071)
- 47017 Rodeo '91 thru '02, Amigo '89 thru '94 & '98 thru '02 & Honda Passport '95 thru '02
- 47020 Trooper '84 thru '91 & Pick-up '81 thru '93

JAGUAR
- 49010 XJ6 all 6-cylinder models '68 thru '86
- 49011 XJ6 all models '88 thru '94
- 49015 XJ12 & XJS all 12-cylinder models '72 thru '85

JEEP
- 50010 Cherokee, Comanche & Wagoneer Limited all models '84 thru '01
- 50011 Cherokee '14 thru '19
- 50020 CJ all models '49 thru '86
- 50025 Grand Cherokee all models '93 thru '04
- 50026 Grand Cherokee '05 thru '19 & Dodge Durango '11 thru '19
- 50029 Grand Wagoneer & Pick-up '72 thru '91
- 50030 Wrangler all models '87 thru '17
- 50035 Liberty '02 thru '12 & Dodge Nitro '07 thru '11
- 50050 Patriot & Compass '07 thru '17

KIA
- 54050 Optima '01 thru '10
- 54060 Sedona '02 thru '14
- 54070 Sephia '94 thru '01, Spectra '00 thru '09, Sportage '05 thru '20
- 54077 Sorento '03 thru '13

LEXUS
  ES 300/330 - see TOYOTA Camry (92007, 92008)
  RX 300/330/350 - see TOYOTA Highlander (92095)

LINCOLN
  Navigator - see FORD Pick-up (36059)
- 59010 Rear-Wheel Drive Continental '70 thru '87, Mark Series '70 thru '92 & Town Car '81 thru '10

MAZDA
- 61010 GLC (rear wheel drive) '77 thru '83
- 61011 GLC (front wheel drive) '81 thru '85
- 61012 Mazda3 '04 thru '11
- 61015 323 & Protegé '90 thru '03
- 61016 MX-5 Miata '90 thru '14
- 61020 MPV all models '89 thru '98
- 61030 Pick-ups '72 thru '93
  Pick-ups '94 thru '09 - see Ford (36071)
- 61035 RX-7 all models '79 thru '85
- 61036 RX-7 all models '86 thru '91
- 61040 626 (rear-wheel drive) all models '79 thru '82
- 61041 626 & MX-6 (front-wheel drive) '83 thru '92
- 61042 626 '93 thru '01 & MX-6/Ford Probe '93 thru '02
- 61043 Mazda6 '03 thru '13

MERCEDES-BENZ
- 63012 123 Series Diesel '76 thru '85
- 63015 190 Series 4-cylinder gas models '84 thru '88
- 63020 230, 250 & 280 6-cylinder SOHC '68 thru '72
- 63025 280 123 Series gas models '77 thru '81
- 63030 350 & 450 all models '71 thru '80
- 63040 C-Class: C230/C240/C280/C320/C350 '01 thru '07

MERCURY
- 64200 Villager & Nissan Quest '93 thru '01
  All other titles, see FORD listing.

MG
- 66010 MGB Roadster & GT Coupe '62 thru '80
- 66015 MG Midget & Austin Healey Sprite Roadster '58 thru '80

MINI
- 67020 Mini '02 thru '13

MITSUBISHI
- 68020 Cordia, Tredia, Galant, Precis & Mirage '83 thru '93
- 68030 Eclipse, Eagle Talon & Plymouth Laser '90 thru '94
- 68031 Eclipse '95 thru '05 & Eagle Talon '95 thru '98
- 68035 Galant '94 thru '12
- 68040 Pick-up '83 thru '96 & Montero '83 thru '93

NISSAN
- 72010 300ZX all models incl. Turbo '84 thru '89
- 72011 350Z & Infiniti G35 all models '03 thru '08
- 72015 Altima all models '93 thru '06
- 72016 Altima '07 thru '12
- 72020 Maxima all models '85 thru '92
- 72021 Maxima all models '93 thru '08
- 72025 Murano '03 thru '14
- 72030 Pick-ups '80 thru '97 & Pathfinder '87 thru '95
- 72031 Frontier, Xterra & Pathfinder '96 thru '04
- 72032 Frontier & Xterra '05 thru '14
- 72037 Pathfinder '05 thru '14
- 72040 Pulsar all models '83 thru '86
- 72042 Rogue all models '08 thru '20
- 72050 Sentra all models '82 thru '94
- 72051 Sentra & 200SX all models '95 thru '06
- 72060 Stanza all models '82 thru '90
- 72070 Titan pick-ups '04 thru '10 & Armada '05 thru '14
- 72080 Versa all models '07 thru '19

OLDSMOBILE
- 73015 Cutlass V6 & V8 gas models '74 thru '88
  For other OLDSMOBILE titles, see BUICK, CHEVROLET or GM listings.

PLYMOUTH
  For PLYMOUTH titles, see DODGE.

PONTIAC
- 79008 Fiero all models '84 thru '88
- 79018 Firebird V8 models except Turbo '70 thru '81
- 79019 Firebird all models '82 thru '92
- 79025 G6 all models '05 thru '09
- 79040 Mid-size Rear-wheel Drive '70 thru '87
  Vibe '03 thru '10 - see TOYOTA Corolla (92037)
  For other PONTIAC titles, see BUICK, CHEVROLET or GM listings.

PORSCHE
- 80020 911 Coupe & Targa models '65 thru '89
- 80025 914 all 4-cylinder models '69 thru '76
- 80030 924 all models including Turbo '76 thru '82
- 80035 944 all models including Turbo '83 thru '89

RENAULT
  Alliance & Encore - see AMC (14025)

SAAB
- 84010 900 all models including Turbo '79 thru '88

SATURN
- 87010 Saturn all S-series models '91 thru '02
  Saturn Ion '03 thru '07 - see GM (38017)
- 87020 Saturn L-series all models '00 thru '04
- 87040 Saturn VUE '02 thru '09

SUBARU
- 89002 1100, 1300, 1400 & 1600 '71 thru '79
- 89003 1600 & 1800 2WD & 4WD '80 thru '94
- 89080 Impreza '02 thru '11, WRX '02 thru '14 & WRX STI '04 thru '14
- 89100 Legacy all models '90 thru '99
- 89101 Legacy & Forester '00 thru '09
- 89102 Legacy '10 thru '16 & Forester '12 thru '16

SUZUKI
- 90010 Samurai/Sidekick & Geo Tracker '86 thru '01

TOYOTA
- 92005 Camry all models '83 thru '91
- 92006 Camry '92 thru '96 & Avalon '95 thru '96
- 92007 Camry, Avalon, Solara, Lexus ES 300 '97 thru '01
- 92008 Camry, Avalon, Lexus ES 300/330 '02 thru '06 & Solara '02 thru '08
- 92009 Camry, Avalon & Lexus ES 350 '07 thru '17
- 92015 Celica Rear-wheel Drive '71 thru '85
- 92020 Celica Front-wheel Drive '86 thru '99
- 92025 Celica Supra all models '79 thru '92
- 92030 Corolla all models '75 thru '79
- 92032 Corolla rear-wheel drive models '80 thru '87
- 92035 Corolla front-wheel drive models '84 thru '92
- 92036 Corolla & Geo/Chevrolet Prizm '93 thru '02
- 92037 Corolla '03 thru '19, Matrix '03 thru '14, & Pontiac Vibe '03 thru '10
- 92040 Corolla Tercel all models '80 thru '82
- 92045 Corona all models '74 thru '82
- 92050 Cressida all models '78 thru '82
- 92055 Land Cruiser FJ40/43/45/55 '68 thru '82
- 92056 Land Cruiser FJ60/62/80/FZJ80 '80 thru '96
- 92060 Matrix '03 thru '11 & Pontiac Vibe '03 thru '10
- 92065 MR2 all models '85 thru '87
- 92070 Pick-up all models '69 thru '78
- 92075 Pick-up all models '79 thru '95
- 92076 Tacoma '95 thru '04, 4Runner '96 thru '02 & T100 '93 thru '08
- 92077 Tacoma all models '05 thru '18
- 92078 Tundra '00 thru '06, Sequoia '01 thru '07
- 92079 4Runner all models '03 thru '09
- 92080 Previa all models '91 thru '95
- 92081 Prius all models '01 thru '12
- 92082 RAV4 all models '96 thru '12
- 92085 Tercel all models '87 thru '94
- 92090 Sienna all models '98 thru '10
- 92095 Highlander '01 thru '19 & Lexus RX330/330/350 '99 thru '19
- 92179 Tundra '07 thru '19 & Sequoia '08 thru '19

TRIUMPH
- 94007 Spitfire all models '62 thru '81
- 94010 TR7 all models '75 thru '81

VW
- 96008 Beetle & Karmann Ghia '54 thru '79
- 96009 New Beetle '98 thru '10
- 96016 Rabbit, Jetta, Scirocco & Pick-up gas models '75 thru '92 & Convertible '80 thru '92
- 96017 Golf, GTI & Jetta '93 thru '98 & Cabrio '95 thru '02
- 96018 Golf, GTI & Jetta '99 thru '05
- 96019 Jetta, Rabbit, GLI, GTI & Golf '05 thru '11
- 96020 Rabbit, Jetta, Pick-up diesel '77 thru '84
- 96021 Jetta '11 thru '18 & Golf '15 thru '19
- 96023 Passat '98 thru '05, Audi A4 '96 thru '01
- 96030 Transporter 1600 all models '68 thru '79
- 96035 Transporter 1700, 1800, 2000 '72 thru '79
- 96040 Type 3 1500 & 1600 '63 thru '73
- 96045 Vanagon Air-Cooled all models '80 thru '83

VOLVO
- 97010 120, 130 Series & 1800 Sports '61 thru '73
- 97015 140 Series all models '66 thru '74
- 97020 240 Series all models '76 thru '93
- 97040 740 & 760 Series all models '82 thru '88
- 97050 850 Series all models '93 thru '97

TECHBOOK MANUALS
- 10205 Automotive Computer Codes
- 10206 OBD-II & Electronic Engine Management
- 10210 Automotive Emissions Control Manual
- 10215 Fuel Injection Manual '78 thru '85
- 10225 Holley Carburetor Manual
- 10230 Rochester Carburetor Manual
- 10305 Chevrolet Engine Overhaul Manual
- 10320 Ford Engine Overhaul Manual
- 10330 GM and Ford Diesel Engine Repair Manual
- 10331 Duramax Diesel Engines '01 thru '19
- 10332 Cummins Diesel Engine Performance
- 10333 GM, Ford & Chrysler Engine Performance
- 10334 GM Engine Performance
- 10340 Small Engine Repair Manual, 5 HP & Less
- 10341 Small Engine Repair Manual, 5.5 thru 20 HP
- 10345 Suspension, Steering & Driveline Manual
- 10355 Ford Automatic Transmission Overhaul
- 10360 GM Automatic Transmission Overhaul
- 10405 Automotive Body Repair & Painting
- 10410 Automotive Brake Manual
- 10411 Automotive Anti-lock Brake (ABS) Systems
- 10425 Automotive Heating & Air Conditioning
- 10435 Automotive Tools Manual
- 10445 Welding Manual

Over a 100 Haynes motorcycle manuals also available

10/22
